Berichte des German Chapter of the ACM

Band 2: **Niedereichholz, Datenbanktechnologie**
Tagung II/1979 am 21./22. 9. 1979 in Bad Nauheim. 240 Seiten, DM 38,–/ÖS 297,–/SFr. 38,–

Band 3: **Remmele/Schecher, Microcomputing**
Tagung III/1979 am 24./25. 10. 1979 in München. 280 Seiten, DM 44,–/ÖS 343,–/SFr. 44,–

Band 4: **Schneider, Portable Software**
Tagung I/1980 am 18. 1. 1980 in Erlangen. 176 Seiten, DM 36,–/ÖS 281,–/SFr. 36,–

Band 6: **Hauer/Seeger, Hardware für Software**
Tagung III/1980 am 10./11. 10.1980 in Konstanz. 303 Seiten, DM 54,–/ÖS 421,–/SFr. 54,–

Band 7: **Nehmer, Implementierungssprachen für nichtsequentielle Programmsysteme**
Tagung I/1981 am 20. 2. 1981 in Kaiserslautern. 208 Seiten, DM 38,–/ÖS 217,–/SFr. 38,–

Band 8: **Schlier, Personal Computing**
Tagung II/1981 am 12. 10. 1981 in Freiburg i. Br. 195 Seiten, DM 40,–/ÖS 312,–/SFr. 40,–

Band 10: **Kulisch/Ullrich, Wissenschaftliches Rechnen und Programmiersprachen**
Fachseminar am 2./3. 4. 1982 in Karlsruhe. 231 Seiten, DM 52,–/ÖS 406,–/SFr. 52,–

Band 11: **Langmaack/Schlender/Schmidt, Implementierung PASCAL-artiger Programmiersprachen**
Tagung II/1982 am 12. 7. 1982 in Kiel. 221 Seiten, DM 46,–/ÖS 359,–/SFr. 46,–

Band 13: **Schneider, Proceedings of the International Computing Symposium 1983 on Application Systems Development**
March 22 – 24, 1983 Nürnberg. 528 Seiten, DM 90,–/ÖS 702,–/SFr. 90,–

Band 14: **Balzert, Software-Ergonomie**
Tagung I/1983 am 28./29. 4. 1983 in Nürnberg. 422 Seiten, DM 72,–/ÖS 562,–/SFr. 72,–

Band 17: **Remmele/Schecher, Microcomputing II**
Tagung III/1983 vom 25. bis 27. 10. 1983 in München. 358 Seiten, DM 64,–/ÖS 461,–/SFr. 64,–

Band 18: **Morgenbrod/Sammer, Programmierumgebungen und Compiler**
Tagung I/1984 vom 2. bis 4. 4. 1984 in München. 293 Seiten, DM 56,–/ÖS 437,–/SFr. 56,–

Band 19: **Morgenbrod/Remmele, Entwurf großer Software-Systeme**
Workshop vom 8. bis 11. 5. 1984 in Grassau. 464 Seiten, DM 82,–/ÖS 640,–/SFr. 82,–

Band 20: **Gorny/Kilian, Computer-Software und Sachmängelhaftung**
Workshop am 29./30. 11. 1984 in Hannover. 208 Seiten, DM 48,–/ÖS 375,–/SFr. 48,–

Band 21: **Kölsch/Schmidt/Schweiggert, Wirtschaftsgut Software**
Tagung I/1985 am 26./27. 3. 1985 in Ulm. 318 Seiten, DM 58,–/ÖS 453,–/SFr. 58,–

Band 22: **Molzberger/Zemanek, Software-Entwicklung: Kreativer Prozeß oder formales Problem?**
Seminar am 20. 3. 1985 in Neubiberg. 176 Seiten, DM 42,–/ÖS 328,–/SFr. 42,–

Band 23: **Klopcic/Marty/Rothauser, Arbeitsplatzrechner in der Unternehmung**
Tagung II/1985 am 12./13. 9. 1985 in Zürich. 355 Seiten, DM 66,–/ÖS 515,–/SFr. 66,–

Band 24: **Bullinger, Software-Ergonomie '85 Mensch-Computer-Interaktion**
Tagung III/1985 am 24./25. 9. 1985 in Stuttgart. 482 Seiten, DM 78,–/ÖS 609,– SFr. 78,–

Band 25: **Wedekind/Kratzer, Büroautomation '85**
Tagung IV/1985 vom 2. bis 4. 10. 1985 in Erlangen. 280 Seiten, DM 56,–/ÖS 437,–/SFr. 56,–

Band 26: **Wippermann, Software-Architektur und modulare Programmierung**
Tagung I/1986 am 24./25. 2. 1986 in Kaiserslautern. 181 Seiten, DM 36,–/ÖS 281,–/SFr. 36,–

Fortsetzung 3. Umschlagseite

Berichte des German Chapter
of the ACM   45

H.-D. Böcker (Hrsg.)
Software-Ergonomie '95

Berichte des German Chapter of the ACM

Im Auftrag des German Chapter
of the ACM herausgegeben durch den Vorstand

Chairman
Dr. Sabine Jausel-Hüsken, c/o CORPHIS Informations-Management Consulting GmbH, Kirchgasse 47, 65183 Wiesbaden

Vice Chairman
Prof. Dr.-Ing. Peter Gorny, Universität Oldenburg, 26111 Oldenburg

Treasurer
Eckhard Jaus, Gemsenweg 12, 71229 Leonberg

Secretary
Prof. Dr. Günter Riedewald, Universität Rostock, Einsteinstraße 21, 18052 Rostock

Band 45

Die Reihe dient der schnellen und weiten Verbreitung neuer, für die Praxis relevanter Entwicklungen in der Informatik. Hierbei sollen alle Gebiete der Informatik sowie ihre Anwendungen angemessen berücksichtigt werden.

Bevorzugt werden in dieser Reihe die Tagungsberichte der vom German Chapter allein oder gemeinsam mit anderen Gesellschaften veranstalteten Tagungen veröffentlicht. Darüber hinaus sollen wichtige Forschungs- und Übersichtsberichte in dieser Reihe aufgenommen werden.

Aktualität und Qualität sind entscheidend für die Veröffentlichung. Die Herausgeber nehmen Manuskripte in deutscher und englischer Sprache entgegen.

Software-Ergonomie '95

**Mensch – Computer – Interaktion
Anwendungsbereiche lernen voneinander**

Herausgegeben von

Dr. Heinz-Dieter Böcker
Gesellschaft für Mathematik
und Datenverabeitung (GMD),
Institut für Integrierte
Publikations- und Informations-
syteme (IPSI), Darmstadt

B. G. Teubner Stuttgart 1995

Die Deutsche Bibliothek – CIP-Einheitsaufnahme

Software-Ergonomie '95 : Mensch – Computer – Interaktion –
Anwendungsbereiche lernen voneinander ; [gemeinsame Fachtagung
des German Chapter of the ACM ... vom 20. bis 23. Februar 1995
in Darmstadt] / hrsg. von Heinz-Dieter Böcker.
Stuttgart : Teubner, 1995
  (Berichte des German Chapter of the ACM ; Bd. 45)
   ISBN 978-3-519-02686-0   ISBN 978-3-322-94752-9 (eBook)
   DOI 10.1007/978-3-322-94752-9
   ISSN 0724-9764
NE: Böcker, Heinz-Dieter [Hrsg.] ; Association for Computing Machinery /
  German Chapter : Berichte des German ...

Gesamtherstellung: Präzis-Druck GmbH, Karlsruhe

Programmkomitee

Udo Bleimann, FH Darmstadt
Heinz-Dieter Böcker (Vorsitz), GMD, Darmstadt
Klaus Böhm, ZGDV, Darmstadt
Wolfgang Dzida, GMD, Sankt Augustin
Jürgen Friedrich, Universität Bremen
Peter Gorny, Universität Oldenburg
Volker Haarslev, Universität Hamburg
Andreas M. Heinecke, Universität Hamburg
Michael Herczeg, Alcatel-SEL AG, Stuttgart
Angelika Herzig, Schweizerische Kreditanstalt, Zürich
Martin Hofmann, SAP AG, Walldorf
Christoph Kasten, Projektträger 'Arbeit und Technik', Bonn
Reinhard Keil-Slawik, Universität Paderborn
Helmut Krcmar, Universität Hohenheim
Heidi Krömker, Siemens AG, Erlangen
Karl-Heinz Rödiger, Universität Bremen
Matthias Schneider-Hufschmidt, Siemens AG, München
Johannes Springer, RWTH Aachen
Norbert Streitz, GMD, Darmstadt
Gerd Szwillus, Universität Paderborn
Manfred Tscheligi, Universität Wien
Claus Unger, Fernuniversität Hagen
Kaisa Väänänen, ZGDV, Darmstadt
Hartmut Wandke, Humboldt-Universität, Berlin
Jens Wandmacher, TH Darmstadt
Hans-Peter Wiedling, ZGDV, Darmstadt
Jürgen Ziegler, FhG-IAO, Stuttgart

Vorwort des Herausgebers

In der Einleitung zum Tagungsband der Software-Ergonomie '93 stellt Karl-Heinz Rödiger zutreffend fest, daß der Anteil der Informatiker unter den Vortragenden der Tagung über die Jahre gesehen rapide abgenommen hat, und spürt möglichen Gründen für diese Entwicklung nach. Zwar hat die Tagung Software-Ergonomie ihren festen Platz im Bewußtsein der an Software-Ergonomie und Mensch-Computer Interaktion interessierten deutschsprachigen Wissenschaftler, Tatsache ist aber auch, daß es bisher in Deutschland nicht gelungen ist, sie als eine der nordamerikanischen CHI vergleichbare Tagungsserie zu etablieren, die breite Wissenschafts-, Hersteller- und Anwenderkreise zu mobilisieren imstande ist. Das wenig ausgeprägte Interesse der Informatiker an der Software-Ergonomie als Tagung steht in auffallendem Gegensatz zu der Beobachtung, daß Fragen der Benutzungsschnittstelle und Mensch-Computer Interaktion in stark zunehmendem Maße in sehr vielen Fachgruppen der Gesellschaft für Informatik thematisiert werden. Es lag daher nahe, bei der Software-Ergonomie '95 den Versuch zu machen, wenigstens einige dieser Gruppen unter dem Motto "Anwendungsbereiche lernen voneinander" zusammenzubringen. Die Zusammensetzung und die ungewöhnliche Größe des Programmkomitees spiegelt dieses Bemühen wider.

Daneben gibt es noch einen zweiten, eher formalen Grund, der auf die Tagungskonzeption nicht ohne Einfluß bleiben konnte. Die Tagung Software-Ergonomie wird, beginnend mit dem Jahre 1983, in zweijährigem Turnus vom German Chapter of the ACM veranstaltet, und seit der 4. Tagung im Jahre 1989 tritt die Gesellschaft für Informatik, vertreten duch den Fachausschuß "Ergonomie in der Informatik", als Mitveranstalter auf. Seit dem Sommer 1993 umfaßt dieser Fachausschuß zwei Fachgruppen, da zur Fachgruppe "Software-Ergonomie" die Fachgruppe "Entwicklungswerkzeuge für Benutzungsschnittstellen" hinzugekommen ist.

Ohne die gewachsenen Bindungen zur Arbeitswissenschaft und Psychologie zu vernachlässigen, wendet die Software-Ergonomie '95 sich daher wieder stärker den Informatikern zu. Dies kommt in der inhaltlichen Spannbreite der angenommenen Beiträge zum Ausdruck, insbesondere aber in der Auswahl der eingeladenen Vorträge, die sich zum Teil mit Fragen auseinandersetzen, die am Rande oder gelegentlich auch knapp außerhalb des traditionellen Gebietes der Software-Ergonomie liegen.

Auch in anderer Hinsicht dokumentiert die Software-Ergonomie '95 eine Entwicklungsdynamik, und es ist dabei vielleicht nicht nur ein glücklicher Zufall, daß das

Institut für Integrierte Publikations- und Informationssysteme (IPSI) der GMD (als Arbeitgeber des Herausgebers) einer der Mitveranstalter dieser Tagung ist: zum ersten Mal wurden elektronische Medien in großem Umfang dazu benutzt, die Tagung als einen sich entwickelnden Prozeß zu dokumentieren. Beginnend im Sommer 1994 wurden auf der Basis der Technologie des World-Wide Web (WWW) Informationen bereitgestellt, die, ständig aktualisiert, den jeweils augenblicklichen Stand der Tagungsvorbereitung darstellten. Zu jedem Zeitpunkt war es Interessierten möglich, sich aktuell unter der Adresse

http://www.darmstadt.gmd.de/gmdda.ann.html

über die Tagungsvorbereitungen zu informieren. Wir beabsichtigen, diese Informationsquelle über den Zeitpunkt der Tagung hinaus bereitzustellen.

Schon immer war die "Software-Ergonomie" eine Tagung, die vielfältige Veranstaltungsformen beinhaltete. Neben eingeladenen und eingereichten Vorträgen fanden wiederholt Tutorien, Poster-, Diskussions- und Demonstrationsveranstaltungen statt. Damit wurde dem weitverbreiteten Wunsch vieler Teilnehmer nach "aktiveren" Formen der Beteiligung entsprochen. In dem hier vorliegenden Tagungsband wird eine dieser Formen, die der themenorientierten Arbeitsgruppen, auch für Nicht-Tagungsteilnehmer festgehalten und dokumentiert.

Den Kolleginnen und Kollegen im Programmkomitee danke ich für die Hilfe bei der inhaltlichen Ausgestaltung der Tagung, insbesondere der Bewertung und Auswahl der Beiträge und der Umsetzung von neuartigen Veranstaltungsformen. Zur Erstellung der Druckvorlagen für diesen Tagungsband stellte uns Karl-Heinz Rödiger dankenswerterweise die bei der Software-Ergonomie '93 verwendeten Formatierungshilfen zur Verfügung, so daß dieser Tagungsband in der gleichen Form wie der damalige erscheinen kann.

Die engagierte Kooperationsbereitschaft der Mitarbeiter aller beteiligten Institutionen hat die Mühen der Tagungsorganisation erträglich bleiben lassen: Die Gesellschaft für Informatik und das German Chapter of the ACM bildeten das finanzielle Rückgrat, die Gesellschaft für Mathematik und Datenverarbeitung stellte die organisatorische Infrastruktur, und die Technische Hochschule Darmstadt ermöglichte die Durchführung der Tagung in ihren Räumlichkeiten. Stellvertretend auch für viele andere möchte ich Herrn Dietmar Effenberger und Frau Ute Kischel danken, die mir bei der Organisation der Tagung in allen Phasen eine unschätzbare Hilfe gewesen sind.

Darmstadt, im Dezember 1994   Heinz-Dieter Böcker

Inhaltsverzeichnis

Workshops (Kurzfassungen)

Is There a Wearable Computer In Your Future?
(Extended Abstract)

Len Bass
Carnegie Mellon University
Pittsburgh, Pa 15213 USA
ljb@cs.cmu.edu[1]

Abstract.

Wearable computers are beginning to appear in research environments and soon will be commercially available. We present several different philosophies behind wearable computers and then focus on a particular project at Carnegie Mellon University involving the development of a wearable designed for the maintenance of large vehicles. This computer is discussed in terms of its project use (both solo and collaborative), its development process (rapid)and its physical characteristics. The paper concludes by discussing some implications of wearable computers on the current user interface evaluation techniques, on user interface development environments and on models for cooperative work.

1   Introduction

With the decreasing size and increasing power of most of the components of a modern computer, it was only a matter of time until computers that can be worn on the body began appearing. Currently, there are several different approaches to body worn computer that have appeared in research laboratories.

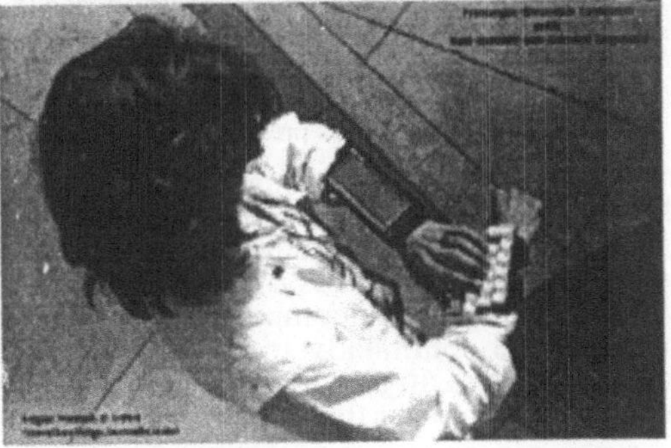

Figure 1 - General Purpose Wearable Computer

Figure 1 shows a picture of a wearable computer that was developed at the University of Toronto [3]. This configuration consists of a display/CPU combination that is worn on one arm and a one handed input device that is worn on the other. This device is an example of attempting to bring general purpose computing to the body. Any program that can be driven with keyboard input (which is most programs) and whose output is legible on the display (currently a

---

[1] This work is sponsored by the U.S. Department of Defense

limited set but sure to grow) can be executed. This would be especially true if one of the new light-weight mouse input devices were incorporated into the keyboard.

Figures 2 and 3 show a picture of a wearable computer that was developed at Carnegie Mellon University [5]. The configuration shown in Figure 2 consists of a Virtual Vision goggles (currently VGA resolution) together with an input device and an embedded processor. Figure 3 shows an expanded picture of the input device and the processor. The interesting thing about the input device is that it consists of a dial and a pressure switch. This is not a general purpose input device but is specialized together with the software for the particular task of assisting maintenance of large vehicles by providing a hypertext version of maintenance manuals.

Figure 2 - Maintenance task oriented wearable computer

Figure 3- Processor, input dial and pressure switch from Figure 2 in close up

This dichotomy between general purpose computers and task oriented computers is not a new one (think of your VCR) but the trends of miniaturization and decreasing cycle times (time from concept to implementation) will make task oriented computers much more frequent. The focus on task orientated computers is also different from the focus on ubiquitious computing[6]. In ubiquitious computing, each person is assumed to be continually interacting with hundreds of nearby wireless interconnected computers. In task oriented computing, the user is assumed to be interacting with a single computer that supports a single task.

In the remainder of this paper, we will explain in more detail the task oriented computers being constructed by CMU, we will describe the future envisioned for this family of computers and we will explore some of the implications of the construction of this class of computers on Human Computer Interaction.

2   CMU Task Oriented Computers

The pictured iteration has a weight of 1.5 pounds including batteries, has a 386 processor and is an embedded system without an operating system.

The use of a dial enables the worker to manipulate the system wearing gloves or through the cloth of a pocket. It also facilitates the protection of the device from the grease and corrosive chemicals that are pervasive in a maintenance environment.

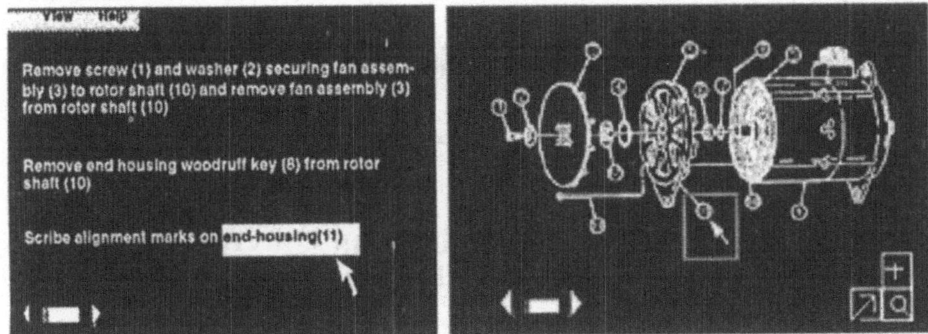

Figure 4 - Interface illustration of hypertext link to the corresponding schematic

The computer pictured is the sixth iteration in a process that heavily utilizes undergraduate students, takes nine months from conception through fabrication to final delivery and involves Industrial Designers, Human-Computer Interaction professionals, software, hardware and mechanical engineers. On some iterations, functionality is added to the system without extreme concern for size and performance. On the other iterations, the functionality achieved in the previous iteration is made smaller, lighter and more efficient.

3 Future for CMU Wearable Computers

The current version of the wearable computer is used only in solo situations. The goal is to extend the maintenance scenario to encompass collaborative problem solving where an expert is located in a remote location and views the maintenance scene through a video connection. The maintenance worker will have a lipstick video camera as a portion of the computer configuration and that camera will enable the two workers to collaborate to solve particular problems.

We also intend to add additional modalities to the wearable. In addition to the current dial, a speech recognition capability is currently under development and text to speech output is being investigated.

The goal, then, is to have a collection of hardware and software components so that a hands free application can be configured very quickly. The software for the different modalities will operate on a plug and play basis, the hardware components will plug into a common bus, the form factor will be chosen from a variety of available form factors and the system will operate in either solo or cooperative mode with a comfortable user interface.

4 Getting There From Here

The future we have sketched involves several different types of developments. Those of interest to the Human Computer Interaction community include the development of plug and play software modules for different modalities, understanding the user interface issues associated with mixed modalities, understanding the user interface issues associated with cooperative work where one participant is an expert consultant and the other is engaged in manual operations on some physical artifact, and understanding how the aesthetics of the form relate to the aesthetics of the user interface.

The next sections deal with some of these issues.

4.1 Plug and Play Software Modules for Different Modalities

Even given the single task of maintenance, there are differing input and output possibilities and differing scenarios of usage. For this reason, the software architecture should be designed to have a clear separation between the application dependent portions (independent of input/output considerations) and the input/output dependent portions (without specific application knowledge).

At the application level, then, we can focus on those aspects that are specific to each application area for the which a wearable might be used and at the user interface level, attention can be paid to making the various modalities work in a plug and play fashion.

This leads to a software architecture that has the form where each input or output modality has a library that manages the specifics of that modality and gives a common interface to the virtual device layer. The virtual device layer manages all of the decisions about what modalities are included in the current system. This yields a layered software architecture that has the following flavor..
The development of the complete set of libraries will be an evolutionary process. The libraries will be developed as needed by the various applications that we develop. Initial modalities will be the display device, dial and speech. Special attention will be focussed on those modalities that will be highly used., e.g., the display library.

4.2 Understand the User Interface

We are approaching understanding the user interface through a variety of field studies. Two have so far been completed. One involves the US Marines doing maintenance on amphibious vehicles. This application consists of performing a maintenance inspection to determine the necessity for parts replacement. A second field study involved a controlled experiment utilizing students at the Pittsburgh Institute of Aeronautics. In this experiment, a particular maintenance task was performed by students in four different conditions: alone using paper manuals, alone using computer based manuals, collaborating with a remote expert using audio communication and computer based manuals, and collaborating with a remote expert using both audio and video communication and computer based manuals.

The results of these field studies are informing both the design of the user interface for the next iteration and the longer term problem of deciding on the appropriate strategies for invoking expert assistance.

Two other field studies are in the planning stages. One involves actual lead mechanics from a variety of airlines who are being trained on the maintenance of a new Boeing aircraft and the other will result from a relationship with Daimler Benz that is just beginning.

4.3 Compatibility of Form and the User Interface

The size of the housing and the use of a dial for an input device are the results of requirements derived from the necessary electronics and the use of the device. The style chosen is the result of user investigations and application of industrial design techniques. The various portions of the physical components are in harmony with each other. The style of the user interface should also be in harmony with the physical style of the device.

5 Implications for Human Computer Interaction

In the past, most computers have been designed as general purpose devices. This is because, in large part, computers have been expensive to construct and the cycle time for a new computer has been long. Special purpose computers often have a small potential market and it did not pay to make the investment to produce them. With the decrease in cycle time and the increased availability of off the shelf miniaturized components, the construction of a specialized task oriented computer will much more frequently be cost effective.

5.1 Issues of Style

Currently, the only method we have to discuss style is through style manuals. That is, there are not tools that will test a description of a window boundary to determine whether it is Motif compatible. This, by itself, isn't particularly interesting but if the problem is framed "determine if two windows have the same style", it becomes more interesting. If the problem is framed "determine if the style of this user interface is the same as the style of this form, then we have a problem worthy of some effort. The solution to this problem assumes that we have some notational mechanism to describe style, that there is some concept of what it means to be of a certain style and the style concept extends between the look and feel of the user interface to the size shape and feel of the form.

5.2 Limited Resource Availability

Another problem, possibly another version of the style description problem, is how to build knowledge of the target system into the construction system. Wearable computers for many years, will have to deal with issues of limited display real estate, limited resolution, limited memory and limited power availability. Thus, prototyping and system construction tools should be able to understand the impact of particular choices on a target machine that has different capabilities. For example, when using a WYSIWYG builder on a computer with a high resolution display, I would like to be able to specify that the target system has VGA resolution and have font size, etc. be automatically constrained. An example is understanding the implications of selecting speech as an input modality on power, on the number of slots required during design and so on.

5.3 Multi-Modality

We sketched a layered software architecture that will support a plug and play collection of different modalities. Such an architecture will not support an intermixing of modalities. This is called the "put that there" problem [1]. For example, "Put that there" is spoken and the "that" and the "there" are gestures pointing to items on the display. In order for true multi-modality to be implementable, knowledge of the input to one modality has to be available to the others.

5.4 Rapid Development

The wearable computers are developed on a nine month cycle. This conforms to the school year and is, relatively, inelastic. Given that each cycle begins with a technology assessment and that the students all must be trained in the method used, the effective cycle time is about six months. This is from understanding the concepts to final delivery. Standard user interface design and evaluation methodologies are too time consuming to fit into this cycle. A normal evaluation takes at least several weeks for set up including detailed story boarding or prototype construction, selection of customers, and evaluation of results. Given construction times, the number of user evaluation iterations is strictly limited. Any design methodology that involves a through exploration of alternatives is also too expensive in terms of time.

What is needed are "lightweight" techniques such as those proposed by Jacob Nielsen[4]. These techniques are needed both during the user interface design process and during the evaluation process. In particular, one question to ask of a design process is at what point in the methodology does the selected technology for the computer become an element. If it is late in the method, then the method is likely too expensive to be useful in a situation where rapid deployment is important. A question to ask of an evaluation method is the cost (in terms of time) or using the method and the probability that the results are indicative of a particular class of users.

5.5 Computer Supported Cooperative Work

CSCW environments have traditionally been broken down by time and synchronicity[2]. This taxonomy is far too limited to be useful in the design of a cooperative wearable system with an expert helper. There is clearly a difference between cooperative authoring between peers using equivalent full featured work stations and a maintenance problem solving between an expert and a less skilled employee where one participant has a full featured work station and the other has a wearable computer. Although some systems have been constructed that are intended for one to one instruction, no general principles have as yet emerged to guide others in the construction of such systems.

6   Conclusion

Wearable computers constructed using a rapid development process are the logical extension of current trends both in development and in miniaturization. These trends will make task oriented computers realistic and, consequently, call into question some of our current assumptions about how one builds and evaluates a computer system. In particular, user interfaces must be constructed and evaluated in conjunction with the form and the electronics of the rest of the system. This paper gives some of the implications of these trends.

Acknowledgements

The wearable project at CMU was started by Dan Siewiorek. At this point, it is quite a large project and thanks are due to Dick Martin, Bethany Smith, David Miller, Asim Smailagic, John Stivoric, Jane Siegel, Bonnie John, Bob Kraut and Kathleen Carley both for making the computer a reality and in having their visions of what it could be.

References

[1] Bolt, R. Put-that-there: voice and gesture at the graphics interface. Computer Graphics 14,3 (July 1980), 262-270. Proceedings ACM SIGGRAPH, 1980.

[2] Greif, I. and Sarin, S. Data sharing in group work, Proceedings Conference on Computer Supported Work, pp 175-191, 1986.

[3] Matias, E., MacKenzie, I., and Buxton, W., Half-QWERTY: A one-handed keyboard facilitating skill transfer from QWERTY. Proceedings INTERCHI '93: Human Factors in Computing Systems, pp88-94. ACM, New York, 1993.

[4] Nielsen, J. Usability Engineering. Academic Press, San Diego, Calif., 1993.

[5] Smailagic, A. and Siewiorek, D. A case study in embedded systems design: VuMan2 wearable computers, IEEE Design and Test of Computers, 10, 4(Sept 1993).

[6] Weiser, M. Some computer science issues in ubiquitous computing, Communications of the ACM 36, 7(July 1993) 75-85.

New Perspectives on Working, Learning, and Collaborating and Computational Artifacts in Their Support

Gerhard Fischer

Center for Life-Long Learning and Design (L^{3}D)

Department of Computer Science and Institute of Cognitive Science

University of Colorado at Boulder

Abstract

Human-computer interaction has refocused many research efforts within computer science from a technology-centered view to a user-centered view. Work-centered design transcends user-centered design by acknowledging that not only are most people novice or generic users of computer systems, but skilled workers in specific domains, as well. These efforts need to be augmented by a learner- and collaboration-centered design perspective that emphasizes the dynamic and collaborative, rather than the static and individualistic, nature of human knowledge and work.

The current research efforts of reinventing and reengineering computational environments support the integration of working, learning, and collaboration. New conceptual frameworks *and* computational environments are also needed that will serve as "objects-to-think-with" to demonstrate, communicate, and open up for critiquing the emerging new understanding. In this paper, I explore how new perspectives transcend the dominant current understanding and discuss attempts to turn these perspectives into reality.

1   Introduction

> Wisdom is not a product of schooling but the
> life-long attempt to acquire it. (A. Einstein)

Learning is a new form of labor [34] and working is often (and needs to be) a collaborative effort among colleagues and peers. In the emerging knowledge society [4], an educated person will be someone who is willing to consider learning as a lifelong process. More and more knowledge, especially advanced knowledge, is acquired well past the age of formal schooling, and in many situations through educational processes that do not center on the traditional school. Seen in this context, working, learning, and collaboration become intimately intertwined rather than being three different and distinct activities.

In our research in human-computer interaction (HCI) over the last fifteen years, we have created conceptual frameworks and innovative systems and have conducted assessment studies to address problems of working, learning, and collaboration with computational artifacts. The content domains of our work are design activities in which design is understood very broadly as the process of determining how things ought to be. Design can be seen as a fundamental activity within all professions [29]. Civil engineers design bridges, lawyers design strategies for cases, politicians design policies, educators design curricula and lessons, and software engineers design software. Design is a collaborative, argumentative process without optimal solutions but with trade-offs. It is impossible for design processes to account for every aspect that might affect the designed artifact. Therefore design must be treated as an evolutionary process in which all stakeholders continue to learn new information and insights as the process unfolds.

The intertwining among learning, working, and collaboration becomes obvious in design, particularly in high-technology fields. Designers are constantly working, learning, and collaborating, which results from a growing recognition that in the information age, change is unavoidable and obsolescence is guaranteed. Learning can no longer be considered a process that occurs only in schools. We need to rethink, reinvent, and reengineer companies, universities, schools, and the relationship among them by exploring new relationships among learning, working, and collaboration.

This article argues for three major issues:

1.) Working, learning, and collaborating need to be integrated, which requires us to move beyond the current boundaries of HCI research perspectives and developments.

2.) Representations are needed that support mutual understanding and allow design stakeholders to work, learn, and collaborate. The design of modern computational environments is characterized by a "symmetry of ignorance" of these stakeholders. None of those involved (e.g., software designer(s), domain designer(s), client(s), customer(s)) as individuals or as one group will have the knowledge necessary for the development of a computational environment.

3.) HCI needs to be viewed as a design science. The HCI community should not be content with reflecting on and evaluating designs developed by other communities, but should itself invent and develop new conceptual

frameworks and computational artifacts. To demonstrate our own adherence
to this goal, short examples of our system-building efforts are included.

2 Beyond Human-Computer Interaction

In this section, I will briefly summarize some of the major shortcomings of current
HCI research and development efforts with respect to the goal of intertwining
working, learning, and collaborating (a detailed discussion of these arguments can be
found in [9]).

2.1 Human-Computer Interaction Is More than User Interfaces

> Applying the Macintosh style to poorly designed applications and
> machines is like trying to put Béarnaise sauce on a hotdog! (A. Kay)

Human-computer interaction is more than "screen-deep." The interface is
important—but if we change only interfaces and not the systems behind them we will
only be able to scratch the surface. We should strive for "interfaceless systems" in
which nothing stands between users and their tasks (and in which system objects
become "ready-at-hand" in a Heideggerian sense). Human-computer interaction
should be concerned with tasks, with shared understanding, with explanations,
justifications, and argumentation about actions, and not just with interfaces.
Although the usual concerns of interface designers (creating more legible types,
designing better scroll bars, developing models of keystroke use) are all important,
they are secondary considerations. The essential challenges lie in improving the ways
people can use computers to work, think, communicate, learn, critique, explain,
argue, debate, observe, decide, calculate, simulate, and design. In the future, the
emphasis has to be on humans and their tasks—not on computers, interfaces, and
tools.

2.2 Make Systems Useful and Usable

> If ease of use was the only valid criterion, people would
> stick to tricycles and never try bicycles. (Engelbart)

Useful computers that are not usable are of little help; but so are usable computers
that are not useful [7]. One of the major goals of human-computer interaction
research must be to achieve these two goals—usefulness and usability—
simultaneously. The "versus" relationship between useful and usable can be turned

into an "and" relationship (1) by using familiar representations (based on previous knowledge and analogous situations), (2) by exploiting the strengths of human information processing, (3) by integrating knowledge in the head with knowledge in the world, and (4) by designing "better" systems that take advantage of the unique possibilities of computational media (specifically by focusing on "on demand" notions, such as learning on demand, using on demand, and detail on demand).

High-functionality computer systems (such as Unix, Word, Canvas, and Mathematica) illustrate the tension between usefulness and usablity by creating a "tool-mastery" burden that can outweigh the advantage of the broad functionality offered. Many approaches that have represented major advances in human-computer interaction, such as direct manipulation [18] (bridging the interface gulf by representing the world of the computer as a collection of objects that are directly analogous to objects in the real world), lose some of their power in high-functionality systems in which the complex and abundant functionality can neither be represented explicitly on the screen nor be explored by browsing mechanisms.

2.3 A Broader View of Communication and Collaboration Processes

Intellectual teamwork [16] is an increasingly important part of knowledge workers' activities and innovative computational environments are needed that help communities of practice [19] to perform better. Communication and coordination processes provide a focus for a number of research efforts. A communication and coordination perspective illustrates the requirement to include support for communication with:

- ourselves (e.g., capturing our thoughts of the past, allowing us to create personalized information environments that extend the knowledge we can keep in our head [1,21]),
- our tools (e.g., knowing which tools exist, how they can be used, and how they can be tailored to our specific needs),
- domain knowledge (e.g., knowing the major concepts and strategies of a particular domain),
- our colleagues (e.g., supporting short-term as well as long-term, indirect collaboration [10]),
- other humans (e.g., supporting interdisciplinary computer-supported cooperative work), and
- our agents and critics (e.g., in the context of cooperative and distributed problem-solving systems).

2.4 Support Human Problem-Domain Interaction

> Interfaces get into the way. I don't want to focus my energies
> on an interface. I want to focus on the job. (D. Norman)

To bring tasks to the forefront, computers must become "invisible" [32,33]. If the most important role for computation in the future is to provide people with a powerful medium for expression, then the medium should support them in working on the task rather than require them to focus their intellectual resources on the medium itself. To achieve this goal, human-computer interaction needs to advance to *human problem-domain interaction* [11], requiring that the major abstractions of a given domain are modeled in the computer. This will enable users to describe things briefly because the systems allow them to interact with domain-oriented concepts. To achieve human problem-domain interaction, we have to sacrifice generality for the power of specialized interactions. The domain-oriented design of artifacts supports the grounding of interaction, creates representations for mutual understanding, and allows referential anchoring for working, learning, and collaborating.

Human problem-domain interaction puts owners in charge by allowing them to communicate with the systems at a level that is *situated* within their own world [30]. By supporting representations for mutual understanding [5], such as prototypes, mock-ups, scenarios, images, or visions of the future, human problem-domain interaction makes it easier for owners of problems to participate in the design process because the representations of the evolving artifacts are less abstract and less alienated from practical-use situations. By keeping owners in the loop, human problem-domain interaction supports the integration of problem framing and problem solving [25,26] and allows the ongoing evolution of computational environments. By making information relevant to the task at hand [14], systems are able to deliver relevant knowledge, in the context of a problem or a service, at appropriate moments for consideration by human professionals.

3 Working, Learning, and Collaborating

3.1 Integration of Working and Learning

Learning is part of living, a natural consequence of being alive and in touch with the world, and not a process separate from the rest of life. What learners need, therefore, is not only instruction but *access* to the world (in order to connect the knowledge in their head with the knowledge in the world [21]) and a chance to play a meaningful part in it. Education should be a distributed lifelong process by which one learns material as one needs it. School learning and workplace learning [27] need to be integrated. We refer to workplace learning not as it is currently practiced (i.e., companies imitating school learning by sending their employees to decontextualized classrooms), but as it could be or should be. Examples include apprenticeship-style relationships, such as internships for doctors and Ph.D. students. In such learning situations, problems are not given, but need to be framed. Collaboration is critical and learning is firmly integrated with working. Figure 1 compares school and workplace learning along a number of dimensions.

	Schools	**Workplace**
EMPHASIS ON:	"basic" skills	education embedded in ongoing work activities
POTENTIAL DRAWBACKS:	decontextualized, not situated	important concepts are not encountered
PROBLEMS ARE:	given	constructed
NEW TOPICS:	defined by curricula	arise accidentally from work situations
STRUCTURE:	pedagogic or "logical" structure	work activity
ROLES:	expert-novice model	reciprocal learning
TEACHERS/ TRAINERS:	expound subject matter	engage in work practice
MODE:	instructionism (knowledge absorption)	constructionism (knowledge construction)

Figure 1: A comparison of school and workplace learning

3.2 Motivation

One of the benefits of integrating working and learning is the potential increase in motivation. Motivation to learn new things is critically influenced by optimal flow, a continual feeling of challenge, direct engagement, the right tools for the job, and a focus on the task [3]. Users are willing and motivated to learn when the following conditions hold: (1) they actively desire and control learning, (2) they are successful in finding and using new information, (3) they can see the immediate benefit of learning something new to their current working situation, and (4) their environments are intrinsically motivating and allow them to achieve interesting results with a reasonably small effort.

3.3 Collaboration

Many collaboration technologies (e.g., most CSCW systems) employ the computer as a medium with few interpretable components. Future computational environments need to integrate humans and computational resources more creatively. Computational environments that can interpret objects, actions, and artifacts (not only from a tool perspective, but from a domain perspective) can make information and resources available at the bidding of the user, whereas persons become a skill resource only when they consent to do so and they can also restrict time, place, and methods as they choose. The big expectation of the National Information Infrastructure, namely that Nobel Prize winners will be accessible by every school child, ignores the fact that most Nobel Prize winners will have anything but time to respond to every question directed their way.

To increase the computational support of collaborative environments, a limited shared context must be established. General-purpose information spaces can have only a limited notion of users' tasks at hand. *Domain-oriented design environments* [8] exploit domain semantics and the design context to actively notify designers when there is information they should know. Many current design systems are limited because they function only as "keepers" of the artifact, in which one deposits representations of the artifact being designed. Our experience has shown that designers integrate designing and discussing in such a way as to make separate interpretation difficult [22]. Talking *about* an artifact means talking *within* the context of the artifact (and not in separated e-mail conversations or design rationale handbooks). Later interpretation of discussions requires reconstruction of the context in which they were originally elicited. The most important implication of this view is that design artifacts must not be artificially separated from communication about them.

A new type of collaboration ("grass-roots communities") will be possible through exploiting the resources of virtual communities [23]. Systems such as the world-wide web provide collective expertise (as well as nonsense). The growing megabytes of content are contributed by volunteers and the combination of free expression, lack of central control, many-to-many communication access, and volunteer effort has created a new kind of social organization.

4   Instrumental Versions of Integrated Environments

The integration of working and learning and new approaches toward learning [24] emphasize that learning (1) is a process of knowledge construction, not one of knowledge recording or absorption; (2) is knowledge dependent; and (3) is highly tuned to the situation in which it takes place. This requires computational environments that are *simultaneously* worker/learner-directed and supportive. These requirements are satisfied neither by intelligent tutoring systems nor by interactive learning environments or application programs.

In our research, *domain-oriented design environments* have emerged as systems serving the integration of working, learning, and collaborating by modeling problem domains. They (1) allow users to focus on their tasks (and not just on the interface), (2) increase the usefulness without sacrificing usability, (3) facilitate human problem-domain interaction, and (4) support short-term and indirect, long-term collaboration. In the context of these research efforts, we have explored topics such as design by composition, design by modification, the integration of problem framing and problem solving, the use of critics to increase the back-talk of situations, and the reconceptualization of breakdowns as sources for creativity. *Critiquing* and *proactivity* both support the integration of learning, working, and collaborating, but are founded on different role distributions between designers and computational environments.

4.1 Critiquing

Critics in design environments [10,12,15] are programs that "look over the shoulder" of users as they perform tasks in computational environments and signal breakdowns and offer critiques from time to time. Critics compute their advice by using domain knowledge to examine the actions users perform (e.g., information spaces visited) and the products they create (e.g., constructions and specifications). In critiquing, humans select (partial) goals and communicate some of them to the system, attempt to achieve these goals, and retain control of the interactions. Critics detect potential problems and provide information relevant to the identified problems. Users evaluate the critiques and decide how to respond.

4.2 Proactivity

Proactivity [31] uses a different role distribution between humans and computers and integrates learning with working in a different manner. Proactivity allows designers to delegate certain tasks to the system—and in performing these tasks the system uses knowledge that may be unknown, and yet be of interest to the designer. For designers, work-centered events are the triggers of learning episodes; the system provides them with contextualized information in the process of solving a problem today, which might be relevant for future problem solutions. The relevance of this information is still determined by the user.

ProNet, a proactive domain-oriented design environment [31], will be used as a concrete example. ProNet supports the delegation of specific subtasks and provides tightly coupled linkages between the evolving artifact and the knowledge and argumentation behind the artifact. It provides numerous opportunities to explore and learn task-relevant domain knowledge in the context of the construction of an actual artifact. It was created (1) to support learning on demand; (2) to explore whether a detailed, dynamic, contextualized, user-specified artifact is a more compelling vehicle for learning about new concepts than the "static," uncontextualized, theoretical examples presented in textbooks; (3) to investigate the question whether people will actually take advantage of the computational mechanisms provided; and (4) to transcend the boundaries of pencil and paper technologies and on-line tutorials in which the learner mostly reads or at best does exercises suggested by the computer rather than being engaged in self-directed, authentic activities.

4.3 A Scenario

The following abbreviated scenario (for details, see [31]) is provided to illustrate how proactivity was used by non-expert, inexperienced network designers to learn while working.

Problem Context and Goals. The computer network designers investigated a networking problem based a real networking situation at the University of Colorado Biology Department. In this problem, three existing networks were located in five rooms. The workstations shown were Macintoshes using a proprietary AppleTalk protocol on LocalTalk media, but the networks were considered to be too slow. Figure 2 shows the three networks that were part of the initial problem. All wires in this figure were actually red in color, indicating AppleTalk on LocalTalk cables. The designers tried to achieve the following goals: (1) all 18 computers and the printer located in the five rooms should be able to communicate via the network, (2) devices on all three networks should be able to use the server and printer located in the room in the top center of the design, and (3) the overall network should be faster than the current design, yet easy to install and maintain.

Observed solution sequence and activities. After studying the problem statement and looking over the initial problem state, the following major activities were observed:

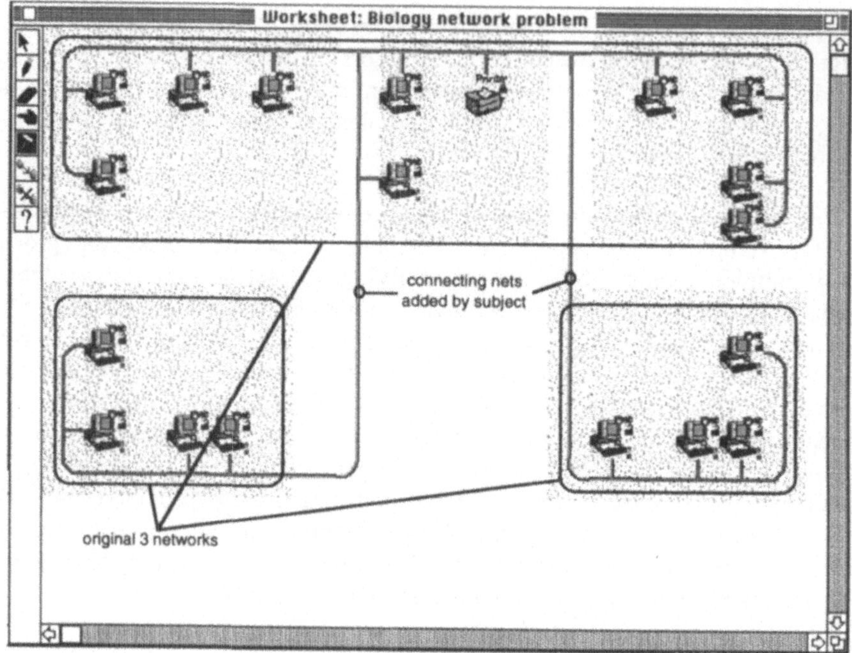

Figure 2: Problem state after construction tasks to connect the subnets.

1.) *Accessing domain knowledge*: A designer used the domain knowledge hypermedia to look up the word "topology" in the glossary, read about topological design issues for several minutes, remarked, "Good, this is what I needed to know...," and went on to construction tasks.

2.) *Construction activities*: Designers selected the wire item in the gallery and then used the worksheet drawing tool to connect the three subnets and form one large network, as shown in Figure 2.

3.) *Using proactivity to obtain design details*: Designers then enabled proactivity so the system would compute design details using the global priorities. Figure 3 shows the resulting design. Three new details appeared in the upper part of the design. From left to right these are "router," "gateway," and "routing gateway," respectively. All wires in the design changed from red to green, except for one segment of wire between the routing gateway and the gateway where the printer is connected to the network. This one segment remained red.

Learning from unfamiliar design details. Designers noticed this change in network color and requested a local explanation of what the green wire meant by using the

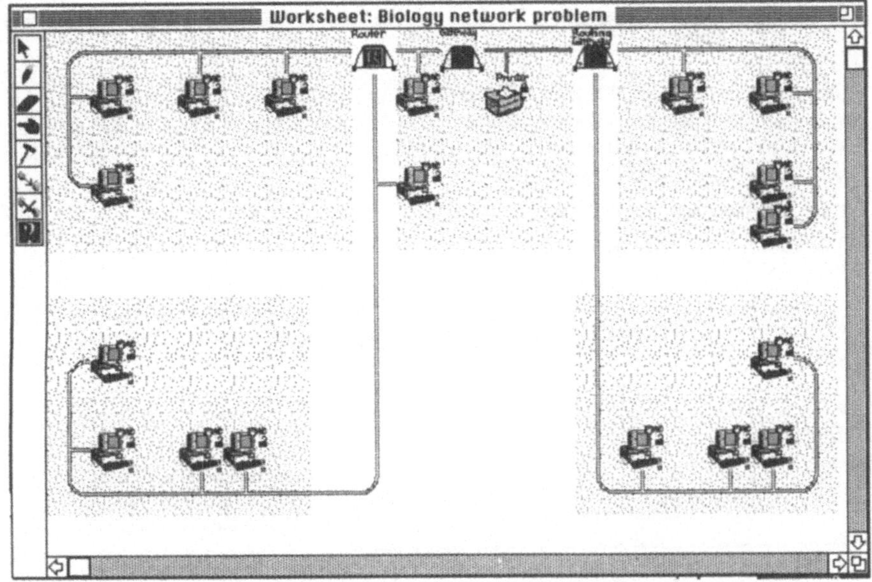

Figure 3: The problem state after enabling proactivity.

query tool labeled with "?". After finding that EtherTalk protocol on Thin-Ethernet media was used, they accessed domain knowledge about the pros and cons of using this protocol and media. They then requested and read a local explanation of the routing gateway that appeared in the top right of the design. Figure 4 shows the local explanation of the device, illustrating the contextualized explanation.

Changing artifact specifications. Designers requested a local explanation for the printer in the design because its icon indicated that it was "locked" on "user-specified priorities." They then selected (in a specification dialog box not shown here) "allow flexibility" for both "how should network requirements be computed next time?" and "current protocol requirements," indicating that the system should proactively compute the best protocol and media using the global design priorities.

Allowing proactivity to obtain new details. After printer specifications were "unlocked," the system proactively computed new network and media requirements for the printer and updated the rest of the design.

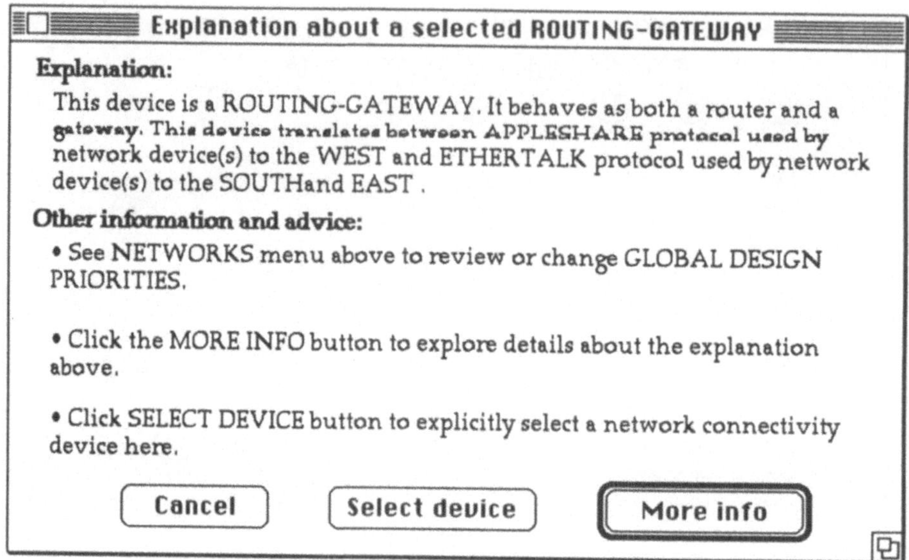

Figure 4: Local explanation of a new routing gateway design detail.

5   Design for Mutual Understanding and Change

If a lion could talk, we could not understand him. (L.Wittgenstein)

Our work on domain-oriented design environments is based on an understanding of design activities as intrinsically collaborative and ongoing. Complexity in design arises from the need to synthesize different perspectives on a problem, the management of large amounts of information potentially relevant to a design task, and understanding the design decisions that have determined the long-term evolution of a designed artifact. Design activities require collaboration among stakeholders, because they are characterized by a *symmetry of ignorance*, meaning that no individual stakeholder (or individual group of stakeholders) knows all the relevant knowledge.

A major challenge of system building is to identify and refine the system requirements of domain workers, and to implement appropriate functionality. Representations in the context of our work are created not primarily for computational processing but (1) to elicit knowledge from domain workers that is often tacit and therefore not easily expressible in abstract situations, and (2) to communicate the intentions and background between system builders and domain workers.

Communication among stakeholders (environment developers, domain designers and clients) is difficult because they use different languages. Explicit representations ground collaborative design by providing a context for communication. Representations help to detect communication breakdowns caused by unfamiliar terminology and tacit background assumptions, and turn the breakdowns into opportunities to create a shared understanding.

5.1 Shared Context in Design Environments

An important component of a shared context is the *intent* of the collaborators. A shared understanding of intent promotes mutual intelligibility by serving as a resource for assessing the relevance of information within the context of collaboration. In our design environments, design activities, including the communication of intent, are centered around artifacts. By capturing the intentions and priorities of designers and associating them with the artifacts, design environments can locate stored artifacts and information relevant to a designer's task at hand and can provide the designer with resources for assessing the relevance of delivered information.

Domain-oriented design environments address these problems in three ways. First, a domain-orientation allows a default intent to be assumed, namely, the creation of a "good" artifact in the given domain. Second, a construction situation (see Figure 2) can be "parsed" by the system, providing the system with information about the artifact under construction. Third, a specification component allows designers to explicitly communicate high-level design intentions to the system, thereby establishing a shared context between the designers and the design environment.

5.2 Seeding, Evolutionary Growth, and Reseeding

Design problems are intrinsically ill-defined, open-ended, and "wicked," making it impossible to predict, let alone collect, all the potentially relevant information in advance. Design environments must capture information continuously over the lifetime of the system [17] and make that information available to designers when it is relevant to their particular tasks. We have developed a process model for the evolution of domain-oriented design environments [13] consisting of three phases: seeding, evolutionary growth, and reseeding (see Figure 5).

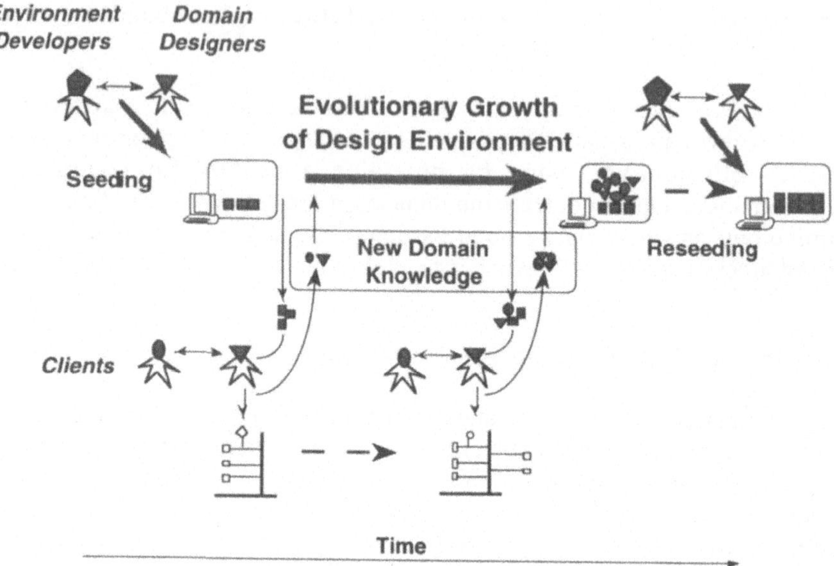

Figure 5: A process model for the development and evolution of domain-oriented design environments

A *seed* is a collection of knowledge and procedures created through a collaboration between environment developers and domain designers. To design new artifacts that are useful for skilled domain workers, either (1) environment developers have to understand the domain concepts and the use activities of the domain designers, (2) domain designers have to understand the possibilities and limitations of computational artifacts, or (3) domain designers must be able to give complete descriptions of their demands, which we know is not possible. Seeds should stimulate, focus, and mediate discussion between the stakeholders, and they should provide mechanisms to capture additional knowledge during the incremental growth phase. There is no absolute requirement for the completeness, correctness, or specificity of the information in the seed. In fact, it is often its shortcomings in these respects that provoke input from designers.

Evolutionary growth during system use is a process of adding information related directly or indirectly to the artifact being designed. Thus, the artifact is the foundation for evolutionary growth. During the growth phase the designers who use the system are primarily focused on their task at hand. Information input is highly situation specific—tied to a specific artifact and stated in particular rather than in general.

Information will grow over time, order will eventually break down, and the system will begin to degrade in its usefulness.

Reseeding is necessary when evolutionary growth stops proceeding smoothly. During reseeding, the system's information is restructured, generalized, and formalized to serve future design tasks. The reseeding process creates a forum to discuss what design information captured in the context of specific design projects should be incorporated into the extended seed to support the next cycle of evolutionary growth and reseeding. Tools contained in design environments support reseeding by making suggestions about how the information can be formalized [28].

5.3 End-User Modifiability

> Convivial tools are those which give each person who uses them
> the greatest opportunity to enrich the environment
> with the fruits of his or her vision. (I. Illich)

The message derived from the "seeding - evolutionary growth - reseeding" model is that no matter how much software designers try to anticipate and provide for what users will need, the effort always falls short because (1) it is impossible to know in advance what is needed, (2) knowledge is tacit, and (3) the world changes. It is an empirical fact that all successful software undergoes changes.

The approach described by our model documents well how large software systems, such as Symbolics' Genera, Unix, the X Window System, have evolved over time. In such systems, users develop new techniques and extend the functionality of the system to solve problems not anticipated by the system's original authors. New releases of the system often incorporate ideas and code produced by users. In the same way that these software systems are extensible by programmers who use them, design environments need to be extended by domain designers who are neither interested nor trained in the (low-level) details of computational environments [20].

End-users may wish to have functionality that fits their needs, but the creation of this functionality is a difficult task. Two major approaches, namely programmable design environments and collaborative work practices, make end-user programming a more realistic challenge.

Programmable design environments [6] are designed to cope with complexity from different angles by integrating a number of distinct elements: (a) an "application-enriched" programming environment, (b) a "critiquing component" that monitors the user's work and occasionally offers suggestions for changes or tutorial assistance, (c) a "catalog" of illustrative or exemplary work that the user can employ as a starting

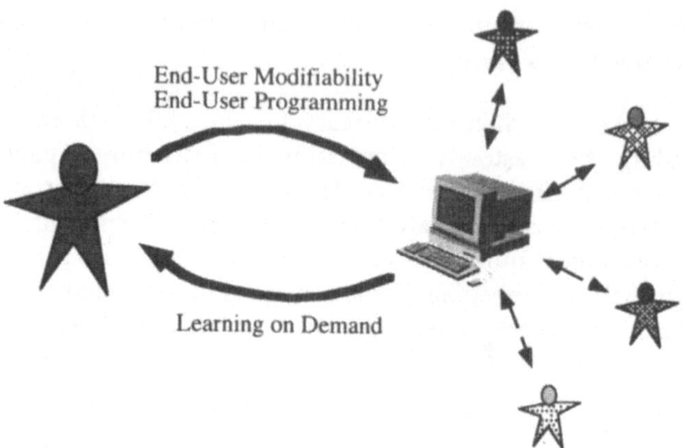

Figure 6: Learning on Demand and End-User Modifiability

point for his or her own work, and (d) embedded tutorial components that the user can access for learning about the application or domain. The first of these elements is primarily aimed at alleviating the problems of complexity faced by the experienced user, whereas the last three of these elements might be viewed collectively as alleviating complexity for the less experienced user.

Collaborative work practices [20] can support end-user computing. In any organization that deals with the same computational artifacts there will eventually be power-users and local developers who will acquire the knowledge to tailor environments to the specific needs of a group. Social resources provided by the community of practice will assist the individual to cope with changes and new demands.

6   Lessons Learned From Our System-Building Efforts

Domain-oriented design environments address the research issue articulated in section 2, "Beyond Human-Computer Interaction": (1) they deal not only with interfaces and tools, but with artifacts and design knowledge of domains; (2) they make an attempt to close the gap between useful and usable by hiding low-level computational details; (3) they support a variety of collaboration processes; and (4) they support human problem-domain interaction. They are instrumental versions of systems that are simultaneously user-directed and computationally supportive.

The critiquing paradigm offers learning opportunities (e.g., by supporting learning on demand [6]) and collaboration opportunities (e.g., by investigating the thinking and the perspectives of other designers as those are embedded in the critiquing itself as well as in argumentation and cases associated with the problem at hand [10]).

Proactivity addresses the production paradox (productive work cannot be done without learning, and learning is prohibited by a lack of time [2]) by helping users get real work done while providing opportunities to learn from the help that is provided. As new details appear in the design, the user can request an explanation of what they are and why they are needed and use these devices to access other related domain knowledge. A shared understanding between designer and system is achieved through the construction and the specification. The construction environment supports experiential cognition allowing the user to create or modify design artifacts and then observe the proactive responses of the system to these actions. Local explanations and argumentative hypermedia allow the designer to reflect on pros and cons, alternatives, and decisions by accessing a web of factual knowledge related to the problem being solved.

7   Human-Computer Interaction: A Design Science

Der Worte sind genug gewechselt,
lasst mich doch endlich Taten sehen. (Goethe, "Faust")

Many researchers and developers in the HCI community have been historically content with evaluating and assessing computational artifacts created by other groups. I strongly believe that HCI needs to become much more a design science, intertwining theory, system-building efforts, and assessment, as illustrated in Figure 7. The HCI community should not only reflect but also propose and develop new designs and new methods and new environments to be used in design. We should not only reflect but enable change. We cannot be content to leaving things as they are or assume that other people should change them.

A design science always has to be prescriptive; it cannot remain only descriptive. Whereas we have to build on as much descriptive knowledge as possible (e.g., strengths and weaknesses of human information processes, properties of technologies, work practices as they exist), we need to decide which kind of computational artifacts we want to have to serve the purposes of working, learning, and collaboration. Our prescriptions will not only be technological, but we have to be (or will implicitly be) social reformers, because information technologies will change how people learn, work, and collaborate. We have to find a careful balance between tradition and transcendence [5]. Designers can prompt and support change in

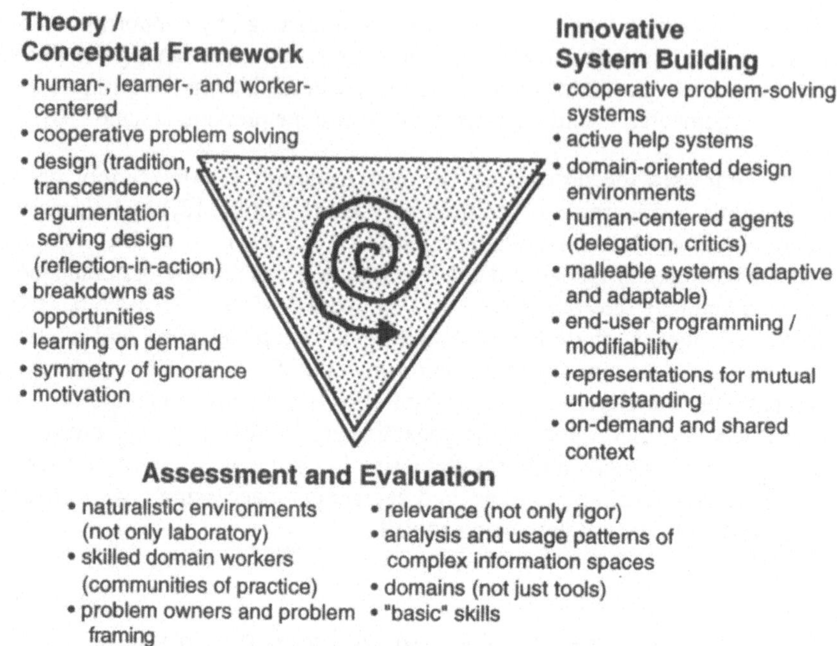

Figure 7: The intertwining between conceptual frameworks/theory, innovative systems, and assessment and evaluation

communities of practice but they cannot and should not predetermine it. Design and use mutually shape one another in iterative, social processes, as indicated in Figures 5 and 6. By seeding a design environment, we not only seed an artifact, but we seed a community of practice—and the community of practice changes in using this artifact.

8   Conclusions

> This is not the end. It is not even
> the beginning. But it is, perhaps,
> the end of the beginning. (W. Churchill)

The dominant HCI research issues of the last decade have been (1) WIMPs (windows, icons, menus, and pointers); (2) an emphasis on interfaces; and (3) a focus on beginners. Important contributions have been made, as can readily be observed by anyone who compares today's interfaces to an ASCII terminal connected to a mainframe computer of ten years ago. These major achievements should not be considered the end, but rather the beginning. As HCI researchers, we—in close collaboration with others—not only have to reinvent and reengineer the computational artifacts and media, but we have to change the underlying processes of working, learning, and collaborating. The major argument behind the current business reengineering debate is that investments in information technology have delivered disappointing results because companies tend to use technologies to mechanize old ways of doing business. The same argument holds for education and collaboration: we use technology as add-on to existing practices, rather than to fundamentally rethink what education and collaboration should be all about in the next century. The old frameworks such as instructionism, curriculum, memorization, and decontextualized learning are not changed by technology itself whether we deal with intelligent tutoring systems, multimedia, or world-wide networks. We have to actively contribute to new frameworks, such as lifelong learning, integration of working and learning, authentic problems, self-directed learning, (intrinsic) motivation, collaborative learning, organizational learning, new content, and new unique properties of computational media.

9   Acknowledgments

The author would like to thank the members of the Center for Lifelong Learning and Design (particularly Hal Eden and Jim Sullivan) at the University of Colorado who have made major contributions to the conceptual framework and systems described in this paper. The research was supported by (1) the National Science Foundation, Grant RED-9253425, (2) the ARPA HCI program, Grant N66001-94-C-6038, (3) Nynex, Science and Technology Center, (4) Software Research Associates (SRA), and (5) PFU. During the academic year 1994/95, the author is supported by the "SEL-Stiftungsprofessor" of the Technical University Darmstadt.

10   References

[1]   Bush, V.: As We May Think. In: Atlantic Monthly. 176 (1945) pp. 101-108.

[2]   Carroll, J. M.; Rosson, M. B.: Paradox of the Active User. In: Interfacing Thought: Cognitive Aspects of Human-Computer Interaction. J. M. Carroll, (Ed.): The MIT Press: Cambridge, MA. (1987) pp. 80-111.

[3]   Csikszentmihalyi, M.: Flow: The Psychology of Optimal Experience. HarperCollins Publishers: (1990).

[4]   Drucker, P. F.: The Age of Social Transformation. In: The Atlantic Monthly. (1994) pp. 53-80.

[5]   Ehn, P.: Work-Oriented Design of Computer Artifacts. Almquist & Wiksell International: Stockholm, Sweden, (1988).

[6]   Eisenberg, M.; Fischer, G.: Programmable Design Environments: Integrating End-User Programming with Domain-Oriented Assistance. In: Human Factors in Computing Systems, CHI'94 Conference Proceedings. Boston, MA. (1994) pp. 431-437.

[7]   Fischer, G.: Making Computers more Useful and more Usable. In: Proceedings of the 2nd International Conference on Human-Computer Interaction (Honolulu, Hawaii). Elsevier Science Publishers: New York. (1987) pp. 97-104.

[8]   Fischer, G.: Domain-Oriented Design Environments. In: Automated Software Engineering, Vol. 9, No. 2 (1994) pp 177-203

[9]   Fischer, G.: Beyond Human-Computer Interaction. In: Mensch Computer Kommunikation. H.-D. Boecker, W. Glatthaar, T. Strothotte (Eds.), Springer Verlag, Berlin-Heidelberg-New York (1993) pp. 274-287

[10]   Fischer, G. et al: Supporting Indirect, Collaborative Design with Integrated Knowledge-Based Design Environments. In: Human Computer Interaction, Special Issue on Computer Supported Cooperative Work. 7 (1992) pp. 281-314.

[11]   Fischer, G.; Lemke, A. C.: Construction Kits and Design Environments: Steps Toward Human Problem-Domain Communication. In: Human-Computer Interaction. 3 (1988) pp. 179-222.

[12]   Fischer, G. et al: The Role of Critiquing in Cooperative Problem Solving. In: ACM Transactions on Information Systems. 9 (1991) pp. 123-151.

[13]   Fischer, G. et al: Seeding, Evolutionary Growth and Reseeding: Supporting Incremental Development of Design Environments. In: Human Factors in Computing Systems, CHI'94 Conference Proceedings (Boston, MA). (1994) pp. 292-298.

[14]   Fischer, G.; Nakakoji, K.: Beyond the Macho Approach of Artificial Intelligence: Empower Human Designers - Do Not Replace Them. In: Knowledge-Based Systems Journal. 5 (1992) pp. 15-30.

[15]   Fischer, G. et al: Embedding Critics in Design Environments. In: The Knowledge Engineering Review Journal. Cambridge University Press: 8 (1993) pp. 285-307.

[16]   Galegher, P. et al, (Ed.): Intellectual Teamwork. Lawrence Erlbaum Associates: Hillsdale, NJ. (1990).

[17]   Henderson, A.; Kyng, M.: There's No Place Like Home: Continuing Design in Use. In: Design at Work: Cooperative Design of Computer Systems. J. Greenbaum and M. Kyng, (Ed.): Lawrence Erlbaum Associates: Hillsdale, NJ. (1991) pp. 219-240.

[18]   Hutchins, E. L. et al: Direct Manipulation Interfaces. In: User Centered System Design, New Perspectives on Human-Computer Interaction. D. A. Norman and S. W. Draper, (Eds): Lawrence Erlbaum Associates: Hillsdale, NJ. (1986) pp. 87-124.

[19]   Lave, J.; Wenger, E.: Situated Learning. Cambridge University Press: Cambridge, UK, (1991).

[20]   Nardi, B. A.: A Small Matter of Programming. The MIT Press: Cambridge, MA, (1993).

[21]   Norman, D. A.: Things That Make Us Smart. Addison-Wesley Publishing Company: Reading, MA, (1993).

[22]   Reeves, B. N.: The Role of Embedded Communication and Artifact History in Collaborative Design. Ph.D. Thesis: University of Colorado, 1993.

[23] Rheingold, H.: The Virtual Community: Homesteading on the Electronic Frontier. Harper Perennial: (1994).

[24] Resnick, L. B., (Ed.): Knowing, Learning, and Instruction: Essays in Honor of Robert Glaser. Lawrence Erlbaum Associates: Hillsdale, NJ. (1989) .

[25] Rittel, H.: Second-Generation Design Methods. In: Developments in Design Methodology. N. Cross, (Ed.): John Wiley & Sons: New York. (1984) pp. 317-327.

[26] Schoen, D. A.: The Reflective Practitioner: How Professionals Think in Action. Basic Books: New York, (1983).

[27] Scribner, S.; Sachs, P.: On The Job Training: A Case Study. In: National Center on Education and Employment. (1990) pp. 1-4.

[28] Shipman, F.: Supporting Knowledge-Base Evolution with Incremental Formalization. Ph.D. Thesis: University of Colorado, 1993.

[29] Simon, H. A.: Cognitive Science: The Newest Science of the Artificial. In: Perspectives on Cognitive Science. D. A. Norman, (Ed.): Ablex Publishing Corporation, Lawrence Erlbaum Associates: Norwood, NJ - Hillsdale, NJ. (1981).

[30] Suchman, L. A.: Plans and Situated Actions. Cambridge University Press: Cambridge, UK, (1987).

[31] Sullivan, J.: A Proactive Computational Approach for Learning While Working. Ph.D. Thesis: University of Colorado at Boulder, 1994.

[32] Uhlich, E.: Von der Benutzungsoberflaeche zur Arbeitsgestaltung. In: Software-Ergonomie'93. K. H. Roediger, (Ed.): B.G. Teubner: Stuttgart, Germany. (1993) pp. 19-29.

[33] Volpert, W.: Von der Software-Ergonomie zur Informatik. In: Software-Ergonomie'93. K. H. Roediger, (Ed.): B.G. Teubner: Stuttgart, Germany. (1993) pp. 51-65.

[34] Zuboff, S.: In The Age Of The Smart Machine. Basic Books, Inc: New York, (1988).

Gerhard Fischer
Department of Computer Science, Campus Box 430
University of Colorado
Boulder, CO 80309-0430 USA
e-mail: gerhard@cs.colorado.edu

Anwendungsbereiche lernen voneinander
... und woraus lernen wir?

Karl-Heinz Rödiger
Universität Bremen

Design is fun; work is not
(Harold Thimbleby 1991, p. 141)

Zusammenfassung

Ausgehend vom Motto dieser Tagung, das Lernen nahelegt, wird an vier Beispielen deutlich gemacht, daß die Bereitschaft zum wissenschaftlichen Diskurs in der Software-Ergonomie nicht gerade ausgeprägt ist. Bei den Beispielen handelt es sich um Publikationen aus den letzten beiden Jahren, in denen teilweise fundamentale Kritik am Zustand der Disziplin geübt wird. Sie werden gemeinsam mit einer richtungweisenden Arbeit, die ebenfalls die bisherige Arbeit für die Zukunft infragestellt, zusammenfassend präsentiert und in ihrer Bedeutung für das Fachgebiet diskutiert. Die Stille um diese Beiträge legt die Frage nahe, woraus wir lernen.

Im zweiten Teil des Beitrags wird über zwei Untersuchungen berichtet, die darüber Aufschluß geben, wo die Software-Ergonomie mit ihren Gestaltungsanstrengungen zwischen styleguide- und partizipationsorientierter Entwicklung steht. Hieran knüpfen sich einige Überlegungen zum Gestaltungsbegriff in der Informatik an.

1 Woraus lernen wir?

»Anwendungsbereiche lernen voneinander« heißt das von mir mitzuverantwortende, nichtsdestotrotz matte Motto unserer diesjährigen Tagung. Matt nenne ich das Motto, weil es ein wenig halbherzig zum Lernen auffordert, die lernenden Personen außen vorläßt und nicht angibt, wie dieser Lernprozeß vonstatten gehen soll. Dabei ist, wie im Beitrag gezeigt wird, wahrlich schon Gelegenheit genug zum Lernen. Insgesamt hebt das Motto auf eine Anstrengung ab, die aus meiner Sicht dem Wissenschaftsbetrieb schon seit längerem abhanden gekommen ist. Schreiben statt Lesen, Dozieren statt Lernen, die Lösung bereithalten statt nach dem Weg und den Zielen zu fragen, scheinen mir die Grundhaltungen der heutigen Umtriebigkeit zu sein.

Einmal in die Welt entlassen, läßt sich ein solches Motto nur noch schwer zurückholen. Die Anwendungsbereiche sollen voneinander lernen, aber woraus lernen wir? Schon die Einladung zur Tagung offenbart, daß wir nicht viel gelernt haben. Da wird immer noch das Residuum der Informatik beschworen, das sich ausschließlich dem Rechner, seiner Hard- und Software-Entwicklung widmen darf. Wer eine Dichotomie zwischen den »menschenorientierten Disziplinen« und den »computerorientierten Disziplinen der Informatik« konstruiert, glaubt offensichtlich immer noch, daß es diese von jeglicher Anwendung und Benutzung freien Räume informatischen Tuns gäbe. Diese Haltung erinnert mich fatal an Peter Dennings task force report »Computing as a Discipline« (Denning, P.J. et al. 1989), in dem festgestellt wird, daß »the fundamental question underlying all of computing is, "What can be (efficiently) automated?"«. Die Auswertung berechenbarer Funktionen ist die Aufgabe von Rechnern; die von Informatikern beschränkt sich nach dieser Auffassung darauf, solche Funktionen zu konstruieren.

Es mag an den Bedingungen und der Verfaßtheit des heutigen Wissenschaftsbetriebs liegen, daß Beiträge zur Eröffnung einer Tagung nicht zu einer kritischen Bestandsaufnahme genutzt werden. Dabei gäben die Befunde über den Zustand und den Erfolg der Disziplin Software-Ergonomie reichlichen Anlaß für tiefe Nachdenklichkeit. Nicht wenige Kritiker führen an, daß es in unserem Gegenstandsbereich nur zwei wirkliche Innovationen gegeben hätte, und beide seien ohne Zutun der Software-Ergonomie-Forscher zustandegekommen: 1981 mit dem Xerox Star und 1984 mit dem Macintosh von Apple. Auch die Style Guides als Marksteine ergonomischen Gestaltungswillens sind Produkte von Anbietern, die sie eher mit Blick auf Marktanteile geschaffen haben, denn mit der Absicht, wissenschaftliche Erkenntnisse umzusetzen. Alles andere im Fach, so die Kritiker, sei Nachlaufforschung, die nun, ex post, begründen könne, warum z.B. Direkte Manipulation so erfolgreich ist.

Ich will mich hier nicht mit dieser Rundum-Kritik an unseren Forschungsbemühungen beschäftigen. Eine Auseinandersetzung damit gerät - führt man sie nicht präzise und detailliert, wozu an dieser Stelle der Raum fehlt - sehr schnell in den Bereich der Behauptungen und Rechthabereien. Ich gestehe jedoch, daß ich diese Position verstehe. Mehr als solcherart grundsätzlichen Infragestellens der Software-Ergonomie berühren mich vier differenziertere Kritiken an unserem Lehr- und Forschungsgebiet aus jüngerer Zeit; bei dreien irritiert mich zusätzlich die Resonanz, die diese Kritiken in unseren Reihen ausgelöst hat: nahezu keine. Bei der vierten gab es bisher keine Chance zur Stellungnahme: sie ist gerade erst erschienen, war jedoch für jeden, der in diesem Gebiet forscht, abzusehen.

2   Fehlende Wirkung

1992 schreibt Peter Gorny in einer »Zwischenbilanz Menschengerechte Gestaltung von Software«: "Software-ergonomische Methoden und Werkzeuge - z.B. softwareergonomische Analysen (Aufgaben- und Benutzeranalysen), Gestaltungsansätze (Dialoggestaltung und Informationspräsentation) und Evaluationskomponenten (Evaluation der Benutzbarkeit) - finden nur als wissenschaftliche Fragestellungen Einzug in AuT-Projekte (Arbeit und Technik, der Verf.). In der normalen Softwareentwicklungspraxis wird auf ihre Anwendung verzichtet" (Gorny 1992, S. 13). Walter Volpert hat diese Kritik in seinem Beitrag zur Software-Ergonomie-Tagung 1993 aufgegriffen und einen »merkwürdigen Widerspruch « konstatiert: »Die Bildschirm-Oberflächen werden hübscher, aber die Arbeitsplätze werden nicht besser« (Volpert 1993, S. 53).

Die deutschsprachige Software-Ergonomie hat ihr eigenes Profil in einem ganzheitlichen Gestaltungsansatz gefunden. In jedem grundlegenden Beitrag, in dem das Gebiet ausdifferenziert wird, wird dieser Ansatz entweder mit dem "Zwiebelmodell" (Rödiger 1985, S. 460) oder der "modifizierten Leavitt-Raute" (Oberquelle 1991, S. 11) veranschaulicht. Dennoch scheint, wie die Zitate belegen, von diesem ganzheitlichen Ansatz in der Praxis nicht viel übrig zu bleiben. Ausgenommen natürlich in den Arbeit und Technik-geförderten Forschungsvorhaben, sie sind von der Projektbeantragung her schon einem solchermaßen ganzheitlichen Ansatz verpflichtet. Leider wird auch in diesen Projekten des öfteren diagnostiziert, daß nach Wegfall der Fördermittel auch die ganzheitlichen Gestaltungsmaßnahmen entfallen.

Man muß nicht sehr tief nachgrübeln, um hierfür Erklärungen zu finden. Wenn sich schon mit wenig Aufwand Akzeptanz bei neuen Systemen erreichen läßt, warum dann viel treiben? Für mich erstaunlicher, ja erschreckender ist die Feststellung, daß auch auf dem Gebiet der Arbeitsmittelgestaltung, der technischen Auslegung von Benutzungsschnittstellen, in der Praxis eine hohe Unsicherheit herrscht und viele Fehler gemacht werden, die wir schon ausgerottet gesehen hatten. Ich beziehe mich hierbei auf meine vielfältigen Praxiskontakte, die mir einen Einblick in die Qualität software-ergonomischer Lösungen erlauben. Es werden nach wie vor Masken entworfen, die wir schon 1983 kritisiert haben; Menüs sind unstrukturiert und verstoßen gegen Grundsätze der Objektorientierung; Fenster werden kreiert, die weder aufgabenangemessen noch Styleguide-konform sind (mit einem besonders schlechten Beispiel geht hier die Fa. Microsoft voran); und in sog. Fehlermeldungen taucht immer noch der erhellende Hinweis »Fehler« auf.

Daß die Praxis einmal wieder »nichts gelernt« hat, können wir ihr sicher vorwerfen, hilft der Reputation unserer Disziplin aber auch nicht auf die Beine. Eine zurückhaltendere Grundhaltung sollte uns dazu treiben, die Fehler bei uns zu suchen. Davon sehe ich einige:

- Transfer software-ergonomischen Wissens haben wir bisher nach der hydraulischen Theorie des Lernens (P.M. Davies 1969) betrieben: Wissen ist eine Flüssigkeit, die sich vornehmlich in unseren Köpfen und in Büchern befindet; und diese Flüssigkeit verspritzen wir am liebsten vor einer großen Hörerschaft. Daß sich in einem so praxisorientierten Feld, wie der Softwareentwicklung, damit nur geringe Lernerfolge erzielen lassen, sollte sich herumgesprochen haben.

- Auch wenn ich gerade Bücher der hydraulischen Theorie zugeordnet habe: Das software-ergonomische Lehrbuch, das das Feld in toto darstellt und dazu noch Handlungsanleitung bietet, ist noch immer nicht geschrieben. Das praxisorientierte deutschsprachige Pendant zu Tog on Interface (Tognazzini 1992) fehlt ebenfalls.

- Transfer läßt sich auch nicht allein über Normen und Standards, die wir zur Zeit reichlich - so innerhalb der ISO 9241 allein 17 Teile - produzieren, erzielen. Sie sind wegen ihrer zeitlosen Allgemeinheit eher zur Abschreckung der Praxis geeignet.

- Wenn wir über software-ergonomische Gestaltung reden, zeigen wir schlechte Beispiele und hoffen, daß unsere Hörer aus der Kritik an ihnen die gute Gestalt herausfinden. Wir befinden uns hier in einem grundsätzlichen Dilemma, das Christopher Alexander auch in anderen Gestaltungsdisziplinen konstatiert hat: »... the procedure suggests no direct practical way of identifying good fit. We recognize bad fit whenever we see a high spot marked by ink. But in practice we see good fit only from a negative point of view, ... (Alexander 1964, p. 22).

Wie aber macht man's dann? Die Patentlösung muß auch ich schuldig bleiben; die besten Erfahrungen habe ich gemacht, wenn wir uns auf Entwicklung eingelassen haben. Wir haben für die Praxis selbst entwickelt und versucht, gute Beispiele zu geben. Außerdem haben wir uns beratend an Entwicklungsprozessen beteiligt, gemeinsam gute Schnittstellen entwickelt und diese dann verbessert.

Peter Gorny hat mit seiner Bilanz an den Grundfesten unseres Tuns, dem Streben nach auch praktischer Wirksamkeit gerüttelt. Wo bleibt der Aufschrei, der Protest? Was mir in unserem Arbeitsgebiet fehlt, sind nicht die schnellen Lösungen zu diesem Problem, die Transfer-Projekte, die von diesem Dilemma leben, sondern der Diskurs zu diesem Vorwurf.

3   Ungesicherte Forschungsergebnisse

1993 schreiben Nachreiner & Mesenholl in einer »Bilanzierung vorliegender Forschungsergebnisse zur Arbeit an Bildschirmgeräten«: Eine methodische Überprüfung der Ergebnisse zeige als erstes Defizit den »insgesamt eher unbefriedigenden Stand der Forschungen zur Software-Ergonomie - von der Gegenstandsbestimmung über die Definition der Konzepte, die berücksichtigten Kriterien, die angewandten Methoden und die untersuchten Fragestellungen« (Nachreiner & Mesenholl 1993, S. 604). Und sie kommen zu dem Schluß, daß die Software-Ergonomie »das Bild einer wenig strukturierten Disziplin« bietet (a.a.O., S. 602). Vornehm hält sich Friedhelm Nachreiner auch in einem weiteren Beitrag zurück, »einzelne Arbeiten, Personen oder Forschungsgruppen zu kritisieren oder an den Pranger zu stellen« (Nachreiner 1994, S. 53). Vielmehr geht es ihm darum, »eine sich etablierende Disziplin auf ihre nachlässige und z.T. nicht verantwortbare Behandlung ihrer Methodenprobleme hinzuweisen« (l.c.).

Es mag wie "Nachklappen" klingen, aber Nachreiners Original-Bilanz an den Projektträger Arbeit und Technik liegt mir nicht vor. 1990 habe ich mehrere Vorträge über Objektorientierung gehalten und mir dazu auch einschlägige Arbeiten zu objektorientierten Benutzungsoberflächen angeschaut, um herauszufinden, ob es Indizien für eine mögliche Überlegenheit dieses Prinzips gegenüber Funktionsorientierung im Bereich der Schnittstellengestaltung gibt. Was ich fand, war, daß in der mir vorliegenden Literatur Objektorientierung mit Direkter Manipulation oder mit graphischen Oberflächen gleichgesetzt wurde, allesamt aber das Hohelied auf die sog. neuen Interaktionstechniken sangen. Ich habe mich damals nicht getraut, von Methodenproblemen der Software-Ergonomie zu sprechen. Daß es diesen Arbeiten sehr wohl an interner Validität fehlte, war mir bewußt. Vorgetragen habe ich damals, daß ich wegen dieser Probleme aus allen mir vorliegenden Arbeiten keinen Vorteil einer objektorientierten Auslegung von Benutzungsschnittstellen gegenüber funktionsorientierten feststellen könne, die Arbeiten letztlich nur einen Schluß zulassen: Wer sich in seiner Oberfläche auskennt - sei sie nun objekt- oder funktionsorientiert -, ist darin gut und allen Novizen überlegen (sic!). Die damalige Euphorie um graphische versus zeichenorientierte Oberflächen, die mit den anderen Sachverhalten immer wieder vermengt wurde, habe ich nicht verstanden.

Unter einer Nachwirkung dieser nie wissenschaftlich überprüften Euphorie leiden wir heute noch: Es ist der sinnlose Drang nach Symbolen dort, wo Zeichen angemessener wären. Wenn jemand glaubt, ein Bild sage immer mehr als tausend Worte, so möge er sich die Beispiele von Barbara Lauter (Lauter 1987) oder die Vorschläge der DIN E 40 107 Teil 1 anschauen. Um diese Symbole in der täglichen Arbeit handhaben zu können, müßte ich mir deren Bedeutung mittels Zeichen auf einem Post-it notieren und an den Bildschirm kleben.

Doch zurück zu Nachreiners Bilanzierung: Wiederum irritiert mich deren Rezeption unter den Software-Ergonomie-Forschern. Kein Aufschrei, kein Protest, vor allem aber auch kein Diskurs zum Zustand unseres Forschungsgebiets.

4   Unzureichende Lehre

In seinem Beitrag »Informatik auf der Mauer« setzt sich Jörg Pflüger mit den in jüngerer Zeit erschienenen Artikeln zur Gegenstandsbestimmung der Informatik und zur Curriculardebatte auseinander. Er folgert aus einer Analyse des Gegenstandsbereichs, »daß wir zu den ingenieurwissenschaftlichen und mathematischen Säulen als drittes Standbein eine geistes- und sozialwissenschaftliche Fundierung in die Informatikausbildung aufnehmen müssen« (Pflüger 1994, S. 255). Man könnte ob dieser Feststellung eines theoretischen Informatikers frohlocken, würde er nicht folgendermaßen fortfahren: »Es kann ... nicht genügen, den anderen Fächern ein weiteres hinzuzugesellen, sondern es muß darum gehen, die Spanne zwischen Verstehen, Formalismus und Technik als solche zu lehren« (l.c.). Und mit einem Seitenhieb auf die Angewandte Informatik in Bremen: »Allzuoft wird dabei nur der kleinste gemeinsame Nenner einer bastelnden Unverbindlichkeit ausgebildet« (a.a.O, S. 256). Nicht nur die Bremer Lehrenden haben bisher angenommen, daß sie den zuvor formulierten Ansprüchen halbwegs gerecht würden.

Bis hierhin kann man Pflügers Kritik noch nachvollziehen und mit ihr in eine neue Curriculardebatte insbesondere auch um die Ausbildung in Software-Ergonomie eintreten. Zu einer solchen Debatte fordert auch Horst Oberquelle auf, der in einem nachdenklichen Beitrag nach neuen Inhalten und Formen der Ausbildung in Software-Ergonomie sucht (Oberquelle 1994). Mit dem, was Oberquelle unter Orientierungswissen behandelt und vermitteln will, kommte er Pflügers Ansprüchen an eine dritte Säule nahe: »Das Orientierungswissen umfaßt alle die Teile, die mit Menschenbild, Wertmaßstäben und Bewertungen zu tun haben« (a.a.O., S. 25).

Mit der Feststellung »Im Hinblick auf die heutige technikorientierte Ausbildung müßte also vordringlich eine andere "Sprache" vermittelt werden, in der Differenzen und Konflikte verstehbar sind und Irrtümer ausgedrückt werden können« (Pflüger a.a.O., S. 256, Hervorhebung im Original) kommt der Autor zu dem überraschenden Schluß: »Es erschiene mir nicht abwegig, wenn Informatiker lernen würden, Gedichte zu interpretieren und historische Prozesse zu beurteilen« (l.c.).

Mit dieser Forderung nach der Einübung des »hermeneutischen Blicks« geraten wir an die Grenzen dessen, was im Rahmen der üblichen Hochschullehre zu leisten ist. Ich interpretiere Pflüger nicht so eng, daß ich Friedrich Schillers Balladen oder Ernst Jandls Lautgedichte zum Gegenstand von Übungen und Leistungsnachweisen machen müßte. Sein Anliegen des Sicheinlassens auf das Differente, des Verstehenwollens anderer Sprachwelten, des Umgehens mit Mehrdeutigkeiten läßt sich meines Erachtens auch in einem praktischen Softwareentwicklungsprojekt für einen den Studierenden fremden Gegenstandsbereich einlösen. Allerdings bedarf es hierzu geraumer Zeit, die dieses Sicheinlassen ermöglicht.

In dieser Beziehung habe ich mit dem Bremer Projektstudium über eine Laufzeit von vier Semestern gute Erfahrungen gemacht. In meinem gerade laufenden Projekt "CAD im Schneiderhandwerk" lernen die Studierenden die Probleme und die Sprache des Handwerks; sie müssen damit umgehen, daß in den Betrieben Schnitte nach unterschiedlichen Methoden konstruiert werden; sie lernen, daß nicht alles, was technisch machbar ist, auch gewollt wird. Sie haben die dramatischen Veränderungen in der Textilbranche kennengelernt. Sie lernen zur Zeit, daß "easy-to-use"-Oberflächen in Handwerksbetrieben eine andere Bedeutung haben als an der Hochschule. Und sie haben vor allem eines schon gelernt: Daß sich die Arbeitsweisen von (zukünftigen) Benutzern nicht mit den Moden der Softwareentwicklung verändern. Unzweifelhaft ist Objektorientierung, wenn sie von den entsprechenden Sprachmitteln gestützt wird, ein gutes Programmierprinzip, das auch dem Stand der Kunst entspricht. Unzweifelhaft ist objektorientierte Abstraktion ein gutes Modellierungsprinzip im Entwurfsprozeß; sicherlich kann ich auch mit den entsprechenden Werkzeugen eine gute Systemspezifikation entwickeln. Daraus aber nun den Schluß zu ziehen, Benutzer, gleich welcher Anwendungsdomäne, hätten immer schon objektorientiert gearbeitet, die ganze Welt bestünde nur aus Objekten oder Objektorientierung sei ein "natürliches Prinzip", ist mehr als keck. Wer sich mit dem unverstellten Blick auf die Praxis einläßt, wird hierfür keinen Beleg finden.

Nicht überall in den Informatik-Studiengängen hat man für dieses Sicheinlassen auf andere Erfahrungswelten vier Semester Zeit; und auch in Bremen ist nicht jedes Projekt in diesem Sinne ein Erfolg. Sprache und Probleme von Anwendungsbereichen lernt man nicht über Vorlesungen verstehen. Es macht einen großen Unterschied, ob ich von den Problemen des Handwerks erzähle oder ob die Studierenden in die Betriebe gehen können und ihre Erfahrungen durch teilnehmende Beobachtung erwerben. Hierzu aber benötigt man ein "Feld", das diese zusätzlichen Belastungen erträgt, besser noch unterstützt.

Vor der anderen Forderung Pflügers, »daß die Vermittlung der drei Sichtweisen (mathematisch, ingenieur- und sozialwissenschaftlich, der Verf.) wesentlich in einer Konfrontation ihrer Methoden besteht« (Pflüger 1994, S. 256) muß ich als jemand, der ebenso wie Oberquelle Theorien und Methoden der Sozialwissenschaften »autodidaktisch« erworben hat und vermutlich »mehr schlecht als recht« vermittelt, passen. Da ich den »selektiven Umgang mit den Nachbardisziplinen« (Kubicek 1984) ablehne, der mich möglicherweise dazu treibt, nur das zu lehren, was ich verstanden habe oder was in meine Denkschemata paßt, darf ich eine Veranstaltung dieses Inhalts nur zusammen mit den Kollegen der entsprechenden Disziplinen (Arbeitswissenschaft, Psychologie) anbieten.

5   Unzeitgemäße Forschungsgebiete

In einer kleinen Nebenbemerkung seines schon zitierten Aufsatzes bemerkt Horst Oberquelle, ein besonderes Problem ergäbe sich aus der Tatsache, »daß die heute ausgebildeten Software-Ergonomie-Experten in ihrer beruflichen Praxis kaum noch solche Systeme bauen werden, die wir heute einigermaßen verstanden haben (z.B. GUIs), die genormt werden und die wir in unseren Lehrveranstaltungen behandeln« (Oberquelle 1994, S. 24). Er hebt damit auf einen Beitrag Marc Weisers zur zukünftigen Arbeit mit und an Informatiksystemen ab.

Aus einer Kritik am heutigen PC-Einsatz formuliert Weiser seine Perspektive "allgegenwärtiger Computernutzung": »... rather than being a tool through which we work, and thus disappearing from our awareness, the computer too often remains the focus of attention« (Weiser 1993, p. 76). Und weiter: »Getting the computer out of the way is not easy. This is not a graphical user interface (GUI) problem, but is a property of the whole context of usage of the machine and the attributes of its physical properties ... The problem is not one of "interface"« (l.c.). Damit wird nicht nur die Adäquatheit unserer Lehre in Software-Ergonomie infrage gestellt; damit stehen unsere Forschungsanstrengungen der vergangenen Jahre, die in der Hauptsache um den PC, um graphische Benutzungsschnittstellen oder die Arbeit mit PCs im allgemeinen kreisten, zur Disposition.

Es wäre nicht das erste Mal, daß aus dem Xerox Palo Alto Research Center (PARC) nicht nur Ideen für die zukünftige Arbeit, sondern ein hard- und softwaremäßig realisiertes System käme, daß unsere Arbeits- und Lebenswelt für die kommenden Jahre bestimmen sollte. Man erinnere sich in diesem Zusammenhang: Im April 1981 kam das Xerox 8010 Star Information System auf den Markt, das in Architektur und Handhabung viele völlig neue Eigenschaften aufwies, die uns heute so vertraut sind: Client-Server-Architektur, Lokales Netz, graphikfähiger Bildschirm, Maus, Objektorientierung, Symbole statt Zeichen zur Kennzeichnung von Objekten, Direkte Manipulation etc. Die von D.C. Smith et al., Mitarbeitern des Star-Projekts bei Xerox, vorgelegten Entwurfsprinzipien, stehen so oder sehr ähnlich heute in jedem Styleguide für graphische Benutzungsoberflächen: »Familiar user's conceptual model, seeing and pointing versus remembering and typing, what you see is what you get, universal commands, consistency, simplicity, modeless interaction, user tailorability« (Smith et al. 1982, p. 248).

Den Arbeiten aus dem PARC mag man noch entgegenhalten, daß sie zur Zeit noch mehr von Visionen getragen ist, deren "Beforschen" wegen der uns größtenteils fehlenden, bei Xerox maßgeschneidert entwickelten Hardware schwer fällt. Auch will ich nicht verkennen, daß mit den "aktiven Kennzeichen" (active badges) und dem Beispiel der Verortung von Personen tiefe Eingriffe in die Persönlichkeitsrechte verbunden sind, die ich mit meiner Auffassung von verantwortlichem Handeln in der Informatik nicht vereinen kann (Capurro et al. 1993). Dennoch sollte uns schon der "Abgesang" auf unsere zentralen Forschungsgegenstände dazu bringen, uns mit diesen Perspektiven verstärkt zu beschäftigen.

Grundsätzlicher sollte uns eine Kritik treffen, die Herbert Kubicek und Wolfgang Taube im Dezember 1994 unter dem Titel »Die gelegentlichen Nutzer als Herausforderung für die Systementwicklung« im Informatik-Spektrum publiziert haben. Ausgehend von der Annahme, daß arbeitsorientierte Informatiksysteme »vermutlich sogar einen relativ abnehmenden Anteil am Gesamteinsatz der Informationstechnik, zumindest wenn man den gängigen Arbeitsbegriff unterstellt und Computerspiele oder Bürgerinformationssysteme nicht ohne weiteres darunter subsumiert« (Kubicek & Taube 1994) kritisieren sie die grundsätzliche Ausrichtung der Software-Ergonomie: Sie habe als zentrales Paradigma »Softwaregestaltung ist Arbeitsgestaltung«, von daher einen Begriff von Benutzern, der mit denen alltagsorientierter Systeme wenig gemein hat, und würde deshalb für diese Systeme keine Konzepte bereithalten.

An zwei Beispielen (Electronic Cash und elektronischer Bibliothekskatalog) demonstrieren die Autoren zwei typische Verhaltensweisen, mit denen Software-Entwickler auf die mangelnde Kenntnis über den Gegenstandsbereich und dessen

Nutzern reagieren: "Der Schluß von sich auf andere" und "Der Schluß von einer Gruppe auf eine andere". Beide Projekte entwickelten sich als Fehlschläge. Aus einer Analyse von Arbeiten zur Benutzerklassifikation kommen die Autoren zu dem Schluß, daß eine solche Typisierung nicht weiterhilft, da sie einerseits statisch ist, sich damit andererseits Benutzer nur schwer definieren lassen, die neben ihrem professionellen Umgang mit einem arbeitsorientierten Informatiksystem an alltagsorientierten Systemen beispielsweise als gelegentliche Benutzer auftreten.

Kubicek & Taube plädieren für eine situative Sicht auf Nutzung, in der uns die aus der Leavitt-Raute (Oberquelle 1991) bekannten Einflußgrößen auf Technikentwicklung Mensch, Arbeit und Organisation in verallgemeinerter Form als Persönlichkeit, Anwendungsinhalt und Milieu entgegentreten. In einer solchen Sicht treten Arbeit und Arbeitende als eine Spezialisierung auf. Die Autoren haben keine neuen Gestaltungsdimensionen erfunden; sie haben uns darauf hingewiesen, daß sich unser Forschungsfeld öffnen muß, wenn wir den gewandelten Formen der Rechnernutzung gerecht werden wollen. Außerdem haben sie mit dem Vorschlag, die Nutzungssituation, beschrieben durch das Tripel Persönlichkeit, Inhalt und Milieu, in den Fokus zu nehmen, einen Weg zu benutzungsangemessener Forschung und Entwicklung gewiesen.

In der Tat haben wir uns in der Vergangenheit im wesentlichen auf Arbeit und arbeitsorientierte Informatiksysteme, oft sogar mit der weiteren Einschränkung Büro, beschränkt. Wir haben in der Software-Ergonomie-Forschung den Wandel in der Nutzung des Computers vom Werkzeug im Arbeitsleben zum Medium in allen gesellschaftlichen Bereichen zu wenig zur Kenntnis genommen (Coy 1994). Einen Großteil zukünftiger Computernutzung wird nicht das Rechnen, das computing, nicht die abhängige Erwerbsarbeit in Organisationen einnehmen, sondern die mediale Nutzung von Dienstleistungsangeboten in den sog. Infobahnen. Schon heute haben sich Standards in der Verwendung des Internets oder von World Wide Web (WWW) anarchisch und neben den bekannten software-ergonomischen Gestaltungsgrundsätzen herausgebildet.

Unser Nutzungsverhalten hat sich durch Rechnernetze schon verändert. Es wird sich durch den Ausbau dieser Netze weitergehend verändern; z.B. bezogen auf unsere Freizeitgestaltung nachhaltig dadurch, daß uns immer mehr Dienstleistungsunternehmen ihre Arbeit machen lassen, seien es nun Banken, Versicherer, Reiseveranstalter oder Transportunternehmen. Für diese Anwendungen, mit denen wir uns nun selbst herumschlagen dürfen, haben wir in der Vergangenheit - siehe Kubicek und Taube - zu wenig getan.

6　Wo bleibt das Positive?

Ich höre sie schon, die Rufe nicht nur der Veranstalter nach der positiven Wende im Vortrag, die doch noch Anlaß zu Ermutigung und Aufbruchstimmung bietet. Ich will sie nicht schuldig bleiben, allein schon, um zu demonstrieren, daß wir nicht nur kritisieren, sondern gelegentlich auch entwickeln. Es gilt, von zwei Untersuchungen zu berichten, die unter einem bestimmten Blickwinkel Antwort auf die Frage nach einer Vorgehensweise bei der software-ergonomischen Gestaltung geben sollten.

Ziel der Untersuchungen war herauszufinden, wie brauchbar die herstellereigenen Styleguides einschließlich der sie unterstützenden Werkzeuge bei der Entwicklung interaktiver Software sind. Als Derivat dieser Fragestellung sollte eine Vorgehensweise zwischen den Polen styleguide-, gleichbedeutend mit ausschließlich entwicklerbezogener, und partizipationsorientierter Entwicklung erarbeitet werden. Diesen Fragen wurde in zwei Diplomarbeiten nachgegangen.

Hierzu wurde in der ersten Arbeit (Rosebrock & Siegler 1993) zunächst in enger Kooperation mit Angestellten einer Bank ein Prototyp für den Bereich Kontokorrent entwickelt und evaluiert. Die Studenten brachten in diese Kooperation ihre software-ergonomischen Kenntnisse ein. Die Angestellten verfügten neben ihren Fachkenntnissen über langjährige Erfahrungen mit zeichenorientierten Bankanwendungen. Diese Fähigkeiten wurden in der Entwicklung eines graphikorientierten Prototypen zusammengeführt. Rosebrock und Siegler haben diesen Prototypen dann so genau, wie mit Hilfe des Common User Access (CUA) in der Version von 1991 (IBM 1991) und den sie unterstützenden Werkzeugen (IBM Toolkit) möglich, umgesetzt und das Ergebnis wie den Werkzeugeinsatz bewertet.
In einer zweiten Diplomarbeit hat Szczepanek (Szczepanek 1994) den gleichen, von den Bankangestellten mitentwickelten Prototypen mit Hilfe des Windows Application Design Guide, den weiteren Microsoft-Dokumenten und mit den sie unterstützenden Werkzeugen Visual C++ und Application Studio unter Windows umgesetzt. Auch er hat sowohl das Ergebnis seiner Implementierung wie den Werkzeugeinsatz bewertet.

Aus beiden Arbeiten lassen sich folgende Ergebnisse zusammenfassen:

- Der in der engen Kooperation mit den bankfachlichen Experten zustandegekommene Prototyp ist sowohl der CUA- wie der Windows-Implementierung bezüglich Aufgabenangemessenheit und Ästhetik überlegen. Allerdings hat auch die Kooperation einen grundsätzlichen Design-Fehler nicht verhindern können.

- Die CUA-Implementierung erfüllt die Anforderungen der Bankangestellten besser als die unter Windows.

- Der CUA ist als Styleguide für Entwickler hilfreicher als die entsprechenden Microsoft-Dokumente.

- Die Werkzeugunterstützung zur Entwicklung graphischer Benutzungsoberflächen von Microsoft ist besser als die von IBM. Komfortable Werkzeuge zur Entwicklung von CUA-Oberflächen werden nur von Fremdherstellern angeboten.

- Zwischen den Windows-Dokumenten Application Design Guide, Interactive Design Guide und Visual Design Guide gibt es zahlreiche Widersprüche. Immerhin aber bietet Microsoft im Gegensatz zu IBM Online-Unterstützung bei der Entwicklung an.

Zwar ist es als Nebenprodukt dieser Untersuchungen durchaus von Interesse, die Anbieter bezüglich ihrer Styleguides und der sie unterstützenden Werkzeuge miteinander zu vergleichen; im Mittelpunkt der Untersuchungen stand allerdings die Frage nach der Gebrauchstauglichkeit der Styleguides und nach einer generellen Vorgehensweise im Entwicklungsprozeß. Hierauf bezogen liegen deutliche Belege für die Vorzüge kooperationsorientierter Entwicklung vor, zeigt sich doch, daß die zwischen Bank- und Informatikexperten ausgehandelte Dialogschnittstelle den styleguide-konformen Produkten in mehreren Punkten deutlich überlegen ist. Der Begriff der Kooperationsorientierung wird hier im Sinne von Weltz & Ortmann (1992) als gemeinsamer Lern-, Entscheidungs- und Entwicklungsprozeß und in deutlicher Abgrenzung zu einem Partizipationsbegriff verstanden, der auf Manipulation und Wissensenteignung (Volpert 1992, S. 177) ausgerichtet ist.

Wenn auch Fallstudien, wie diejenigen, über die hier berichtet wird, durchaus Fragen nach der Übertragbarkeit und Generalisierbarkeit der Ergebnisse aufwerfen, so sind sie doch geeignet, zwei sehr einfach anmutende, schon länger diskutierte Thesen zu erhärten.

1 Mit Normen und Styleguides allein lassen sich allenfalls aufgabenunabhängige (Teile von) Benutzungsschnittstellen angemessen entwickeln.

2 Benutzer-Kooperation verhindert keine Design-Fehler; zur aufgabenangemessenen Gestaltung ist sie jedoch unabdingbar.

Man kann diese Thesen auch zu einer zusammenfassen: Das Problem der angemessenen Gestaltung von Benutzungsoberflächen ist trotz fünfzehn Jahren Forschung und Entwicklung zur Software-Ergonomie, trotz aller Fortschritte und trotz der bislang vorliegenden Standards ungelöst und wird vermutlich auch weiterhin ungelöst bleiben. Warum soll in der Informatik etwas zu leisten sein, was andere mit Gestaltung befaßte Disziplinen trotz wesentlich längerer Tradition nicht geschafft haben?

Stadtplaner bescheren uns Einkaufsstraßen, in den sich nach Geschäftsschluß niemand mehr aufhalten möchte; im neuen Bahnhof Kassel Wilhelmshöhe waren keine Toiletten vorgesehen; den neuerbauten Düsseldorfer Hauptbahnhof mußte man nachträglich wieder aufreißen, um Fahrstühle für Behinderte einzubauen; Bauingenieure dienten uns Spannbeton an, der nun mit hohem Aufwand saniert werden muß; alle drei großen Flugzeug-Unternehmen mußten schon Konstruktionsfehler einräumen. Man kann Gebrauchsgegenstände kaufen, die hervorragend designed aber dysfunktional sind (vice versa): Mein vorheriger Anrufbeantworter pflegte von Zeit zu Zeit Anrufe mit dem Text »Ich danke für Ihren Anruf« einzuleiten, um die Verbindung dann zu unterbrechen, ohne daß der Anrufer irgendeine Chance der Rückäußerung gehabt hätte.

Was aber begeistert uns an den Reisterassen von Banaue (Philippinen)? Es ist die grandiose Verbindung von Funktionalität und Ästhetik: Das Wasser fließt nach einem ausgeklügelten System, Reis kann auch an den steilsten Hängen angebaut werden, und die Felder in ihrer scheinbar willkürlichen Verteilung und ihren unterschiedlichen Bearbeitungs- und Reifezuständen sind ein Augenschmaus.

»Irren ist menschlich« überschreibt Donald Norman sein Kapitel über Design-Fehler und begründet, aus welchen Situationen heraus Menschen Fehler machen (Norman 1989). Da solche Situationen immer wieder eintreten können, werden auch immer wieder Fehler gemacht. Insofern relativiert sich zumindest ein Teil der immerwährenden Diskussion um die Softwarekrise (Spillner 1994), den um die mangelnde Produktqualität. Auch andere Disziplinen mit wesentlich längerer Lehr- und Forschungserfahrung kennen diese Probleme und haben sie nicht endgültig lösen können. Wir sollten uns jedoch die Erfahrungen der anderen mit Gestaltung befaßten Disziplinen zu eigen machen, daraus lernen und unsere eigenen Methoden und Techniken ständig zu verbessern.

7   Zur Gestaltungsdiskussion in der Informatik

In der Informatik wird entlang einem wie auch immer beschaffenen Vorgehensmodell ermittelt, analysiert, definiert, spezifiziert, entworfen, implementiert, getestet,
eingesetzt und gewartet. All diese Tätigkeiten im Softwareentwicklungsprozeß kennen auch ein Ergebnis: Anforderungsdefinition, Systemspezifikation, Entwurf, im-
plementiertes System, getestetes System, eingesetztes System, gewartetes System.
Und irgendwo dazwischen wird offensichtlich auch gestaltet, glaubt man der Infor-
matik-Literatur und meinem Beitrag, in dem dieser Begriff vielfältig vorkommt. Wo
aber passiert das zwischen Analyse, Spezifikation, Entwurf und Implementierung?
Und was ist das Ergebnis? Gestaltung offensichtlich nicht, der Begriff bezeichnet
einen Prozeß, allenfalls einen Zustand. Ist es die Gestalt, die gute Gestalt, das gut
gestaltete System? Was aber ist das? Wie ist es definiert? Etwa durch »Die zehn
Tugenden guter Software« der Gütegemeinschaft Software (1987)?

In der Informatik gibt es seit geraumer Zeit eine Diskussion um den Gestaltungsbegriff. Schon früh hieß es »Softwaregestaltung ist Arbeitsgestaltung« (Hacker
1987), womit allerdings mehr die Wirkung von Gestaltung in der Softwareentwicklung, denn ihr Wesen beschrieben wurde. Wenn Wolfgang Coy in der Informatik
eine universelle Gestaltungswissenschaft sieht, so ist diese Sicht gegen technikzen-
trierte Auffassungen von Informatik gerichtet, gibt jedoch ebenfalls keine Auskunft
darüber, was Gestaltung sein könnte, und wo im Entwicklungsprozeß sie Platz
greift (Coy 1989). Auch Arno Rolf verwendet einen breiten Gestaltungsbegriff, der
Technikfolgenabschätzung mit einschließt, ohne zu explizieren, wie der in der
Informatik gefüllt werden könnte und wo er in der Softwareentwicklung ansetzt
(Rolf 1992). Walter Volpert tritt gegen den »gewissermaßen imperialistischen«
Anspruch der Informatik an, alle mit Gestaltung befaßten Disziplinen vereinnahmen
zu wollen: »Wenn sie ... alle diese Disziplinen in sich aufsaugen möchte, so muß
sie dies auch substantiell, also hinsichtlich der Gesamtheit der Erkenntnisse tun.
Wenn sie sich aber als einen Teil der universellen Gestaltungswissenschaft sieht,
dann muß sie ihren eigenen Beitrag und Schwerpunkt definieren« (Volpert 1992,
S. 173). Frieder Nake definiert Gestalten als »das Herstellen mit Ungewißheit, das
unsichere Herstellen. Das Herstellen mit einer Gewißheit will ich im Gegensatz
hierzu die *Konstruktion* nennen« (Nake 1992, S. 200, Hervorhebung im Original).
Dirk Siefkes versteht unter Gestalten »die unruhige Einheit aus Verstehen und
Herstellen« und »Gestalten ist Auf-menschliche-Weise-Machen« (Siefkes 1992,
S. 112).

Diese Übersicht über die Gestaltungsdiskussion in der Informatik beansprucht
keine Vollständigkeit. Sie zeigt jedoch schon, daß hier mit sehr unterschiedlichen
Gestaltungsbegriffen argumentiert wird, die verschiedene Bedeutungshöfe und

Konnotationen haben. Sie sind nur schwer gegeneinander zu diskutieren und wegen ihrer unterschiedlichen Ansätze kaum für die Softwareentwicklungspraxis tauglich. Neben emphatischen Begriffen des Gestaltens (Siefkes), stehen solche die auf beabsichtigte Ziele (Rolf), (unbeabsichtigte) Wirkungen (Hacker), Werte (Nake) oder Konsequenzen (Volpert) abheben. Die Vielfalt der Bedeutungsgehalte von Technikgestaltung hat gerade Hans Dieter Hellige (Hellige 1994) in einer schönen Arbeit erhellt.

Wo aber im Software-Lebenszyklus findet Gestaltung statt? Hier liegt meines Erachtens die eigentliche Krux des Gestaltungsbegriffs in der Informatik. Gestaltung läßt sich nicht irgendwo zwischen Entwurf und Implementierung verorten und dort gegebenenfalls gezielt beeinflussen. Entgegen der reinen Lehre, nach der auch die Systemspezifikation noch frei von allen Gedanken an das WIE der technischen Realisierung sein soll, weiß jeder Entwickler, daß man sich zu keinem Zeitpunkt völlig frei von Gedanken an die (auch falsche) Realisierung machen kann. Und so entsteht dann schon bei der Analyse der frevlerische Gedanke an das Icon, das genau den Sachverhalt symbolisieren soll, den die Bankkaufleute gerade beim Interview vortragen. Schnell 'materialisiert' sich der Gedanke dann im ersten Prototypen, die Betroffenen sind ob der Symbolkraft des technischen Geräts beeindruckt, und so pflanzt sich der Gestaltungsfehler durch alle Systemversionen fort: Das Icon differenzierte nicht und muß durch Text ergänzt werden.

Gestaltung ist orthogonal zu allen Aktivitäten im Softwareentwicklungsprozeß und 'passiert' oder findet bewußt statt zu jedem Zeitpunkt, an dem wir mit Spielräumen, mit Unsicherheit, wie es Nake nennt, umgehen dürfen. Dieses Geschenk des Umgehen-Dürfens mit Spielräumen nimmt uns glücklicherweise keine Norm und kein Styleguide. Unglücklicherweise ist damit auch die Freiheit verbunden, Fehler machen zu dürfen. Und ein zweites: Unsere stärksten 'Waffen' bei der Bewältigung von Komplexität sind Abstraktion und Dekomposition. Im Prozeß der Problembewältigung lösen wir uns von vermeintlich unwichtigen, Benutzern aber bedeutungsvollen Details, wir dekomponieren das problembeladene Ganze in seine bewältigbaren Teile. Das 'Look' und manchmal auch das 'Feel' von Fenstern, Radioknöpfen und Schiebern vermitteln uns Styleguides, die gute Gestalt des Ganzen nicht. Als Informatiker haben wir gelernt die Töne zu setzen, nicht aber, wie daraus die (nicht nur funktionale) Melodie wird. Wir sind der berechenbaren Funktion verpflichtet und verlieren darüber das Gefühl für die Ästhetik. Vielleicht sollten wir doch einmal Christian von Ehrenfelsens Arbeit »Über 'Gestaltqualitäten'« lesen.

Von anderen Gestaltungsdisziplinen können wir lernen, daß jede Gestaltungssituation neu ist, neue Anforderungen an den Gestalter stellt und von neuen, oft widersprüchlichen Anforderungen seitens der Anwender und Benutzer begleitet wird. Rahmenbedingungen und Kontexte, innerhalb derer gestaltet werden soll, ändern sich ebenso wie die Anwendungsbereiche. Die Situation, in der gestaltet werden muß, mithin Gestaltung, ist immer konkret; die Gestaltungsbegriffe und -kriterien, die dem Gestalter Leitlinie sein sollen, hingegen sind abstrakt. Begriffe und Gestaltungsgrundsätze können Richtschnur sein, verhindern aber schlechte Gestalt nicht.

Gestaltungsaufgaben sind komplexe Aufgaben, weil heterogene und zum Teil divergierende Anforderungen im Prozeß zusammengeführt werden müssen, und weil sich Gestaltung in einem Bereich immer auch auf andere Bereiche auswirkt. Man kann (noch) kein Informatiksystem spurlos in eine Arbeits- oder Alltagssituation integrieren, Marc Weisers Vorstellung vom »getting the computer out of the way« ist noch Vision; immer sind mit der Benutzung eines Computers auch Veränderungen an Inhalten, Abläufen, an Qualifikationsanforderungen etc. verbunden. Walter Volpert hat unterstrichen, daß man "nicht nicht gestalten" kann, daß mit der Einführung eines Informatiksystems immer Eingriffe in Arbeits- und Lebensverhältnisse verbunden sind (Volpert 1993, S. 59).

An dieser Stelle spanne ich den Bogen von der Gestaltungsdiskussion zurück zu Lehre und Forschung. Denn den Hinweisen und Warnungen der auf Analyse spezialisierten Sozialwissenschaftler können wir, der Synthese verpflichteten Gestalter stolz das Zwiebelmodell, die Leavitt-Raute und das magische Dreieck von Kubicek und Taube entgegenhalten: Wir haben diese Interdependenzen bedacht und in unseren Schaubildern vorgesehen. In unseren Vorgehensmodellen haben wir alle notwendigen Aktivitäten, deren Ergebnistypen und die Interventionspunkte vorgesehen. Wir könnten zufrieden sein, wenn damit nicht die Lösung des Problems nur vorgetäuscht würde: »What does make design a problem in real world cases is that we are trying to make a diagram for forces whose field we do not understand« (Alexander 1964, p. 21). Um sie zu verstehen, benötigen wir nicht die Gestalter aus den Anwendungsbereichen sondern die Analytiker aus den anderen Disziplinen.

8   Literatur

Alexander, Ch.: Notes on the Synthesis of Form, Cambridge, MA 1964

Capurro, R. et al.: Ethische Leitlinien der Gesellschaft für Informatik, Informatik-Spektrum 16 (1993), S. 238-240

Coy, W.: Brauchen wir eine Theorie der Informatik?, Informatik-Spektrum 12 (1989), S. 256-266

Coy, W.: Computer als Medium - Drei Aufsätze, Universität Bremen, Fachbereich Mathematik/ Informatik, Bericht 3/94, Bremen 1994

Denning, P.J. et al.: Computing as a Discipline, Communications of the ACM 32 (1989), pp. 9—23

Gorny, P.: Zwischenbilanz Menschengerechte Gestaltung von Software, Universität Oldenburg, Fachbereich Informatik, Oldenburg 1992

Gütegemeinschaft Software e.V.: Das Anwenderbrevier - Die zehn Tugenden guter Software, Frankfurt 1987

Hacker, W.: Software-Ergonomie: Gestalten rechnergestützter Arbeit?!, in: Schönpflug, W. und M. Wittstock (Hrsg.), Software-Ergonomie '87, Stuttgart 1987, S. 31-45

Hellige, H.D.: Technikgestaltung: Ein Begriff als Programm. Geschichte, Systematik und Probleme, Universität Bremen, Forschungszentrum Arbeit und Technik, Bremen 1994

IBM: Systems Application Architecture: Common User Access - Guide to User Interface Design, Cary, NC 1991

IBM: Systems Application Architecture: Common User Access - Advanced Interface Design Refernce, Cary, NC 1991

Kubicek, H.: Defizite der Informatik hinsichtlich ihrer Folgenbewältigung und die politische Bewältigung dieser Defizite, Universität Trier, Trier 1984

Kubicek, H. und W. Taube: Die gelegentlichen Nutzer als Herausforderung für die Systementwicklung, Informatik-Spektrum 17 (1994) Nr.6 (im Druck)

Lauter, B.: Software-Ergonomie in der Praxis, München 1987

Nachreiner, F.: Methodenprobleme der Software-Ergonomie, in: Hartmann, A. et al. (Hrsg.), Menschengerechte Groupware - Software-ergonomische Gestaltung und partizipative Umsetzung, Stuttgart 1994, S. 51-63

Nachreiner, F. und E. Mesenholl: Defizite der Software-Ergonomie. Ergebnisse einer Bilanzierung vorliegender Forschungsergebnisse zur Arbeit an Bildschirmgeräten, in: Coy, W. et al. (Hrsg.), Menschengerechte Software als Wettbewerbsfaktor, Stuttgart 1993, S. 602-615

Nake, F.: Informatik und die Maschinisierung von Kopfarbeit, in: Coy, W. et al. (Hrsg.), Sichtweisen der Informatik, Braunschweig 1992, S. 181-201

Norman, D.A.: Dinge des Alltags - Gutes Design und Psychologie für Gebrauchsgegenstände, Frankfurt 1989

Oberquelle, H.: MCI - Quo vadis ? Perspektiven für die Gestaltung und Entwicklung der Mensch-Computer-Interaktion, in: Ackermann, D. und E. Ulich (Hrsg.), Software-Ergonomie '91, Stuttgart 1991, S. 9-24

Oberquelle, H.: Software-Ergonomie lehren - Software-Ergonomie lernen, Ergonomie & Informatik (1994) Nr. 22, S. 24-26

Rödiger, K.-H.: Beiträge der Softwareergonomie zu den frühen Phasen der Softwareentwicklung, in: Bullinger, H.-J. (Hrsg.), Software-Ergonomie '85, Stuttgart 1985, S. 455-464

Rödiger, K.-H.: Einschränkung des Handlungsspielraums durch den Einsatz von Dialogsystemen - Ergebnisse einer Felduntersuchung, in: Nullmeier, E. und K.-H. Rödiger (Hrsg.), Dialogsysteme in der Arbeitswelt, Mannheim 1987, S. 229-244

Rolf, A.: Informatik als Gestaltungswissenschaft - Bausteine für einen Sichtwechsel, in: Langenheder, W., G. Müller und B. Schinzel (Hrsg.), Informatik cui bono?, Berlin 1992, S. 40-48

Rosebrock, J. und M. Siegler: Bewertung des Common User Access (CUA), Diplomarbeit, Universität Bremen, Fachbereich Mathematik/Informatik, Bremen 1993

Siefkes, D.: Sinn im Formalen? Wie wir mit Maschinen und Formalismen umgehen, in: Coy, W. et al. (Hrsg.), Sichtweisen der Informatik, Braunschweig 1992, S. 97-114

Smith, D.C., C. Irby, R. Kimball, B. Verplank, and E. Harslem: Designing the Star User Interface, Byte (1982) No. 4, pp. 242-282

Spillner, A.: Kann eine Krise 25 Jahre dauern?, Informatik-Spektrum 17 (1994), S. 48-52

Szczepanek, U.: Bewertung des Windows Application Design Guide, Diplomarbeit, Universität Bremen, Fachbereich Mathematik/Informatik, Bremen 1994

Thimbleby, H.: User Interface Design, Wokingham 1991

Tognazzini, B.: Tog on Interface, Reading, MA 1992

Volpert, W.: Erhalten und Gestalten - Von der notwendigen Zähmung des Gestaltungsdrangs, in: Coy, W. et al. (Hrsg.), Sichtweisen der Informatik, Braunschweig 1992, S. 171-180

Volpert, W.: Von der Software-Ergonomie zur Arbeitsinformatik, in: Rödiger, K.-H. (Hrsg.), Software-Ergonomie '93, Stuttgart 1993, S. 51-65

Weiser, M.: Some Computer Science Issues in Ubiquitous Computing, Communications of the ACM 36 (1993) No. 7, pp. 75-85

Weltz, F. und R.G. Ortmann: Das Softwareprojekt - Projektmanagement in der Praxis, Frankfurt 1992

Unterstützungswerkzeuge für Entwickler von graphischen Benutzungsoberflächen

Frank Bachmann, Michael Porschen, Manfred Ramm,
Harald Reiterer, Stefan Schäfer, Helmut Simm
Gesellschaft für Mathematik und Datenverarbeitung, Universität Wien

Zusammenfassung

Durch den Einsatz von rechnerbasierten Unterstützungswerkzeugen, die in ein Entwicklungs-werkzeug für graphische Benutzungsoberflächen integriert sind, werden Entwickler in Frage-stellungen der software-ergonomischen Gestaltung unterstützt. Damit wird den Entwicklern die Möglichkeit eines bedarfsorientierten Erlernens und Anwendens von Ergonomie wissen geboten. Die verschiedenen Unterstützungswerkzeuge machen von objekt-orientierten, multimedialen und wissensbasierten Techniken Gebrauch, um das Ergonomiewissen zu vermitteln.

1   Einleitung

Im folgenden werden Ergebnisse des Gemeinschaftsprojektes "User Interface Design Assistance (IDA)" der GMD und der FHD Darmstadt vorgestellt [25]. Ziel des Vorhabens ist es, Entwicklern von graphischen Benutzungsoberflächen rechnerbasierte Unterstützungswerkzeuge zur Verfügung zu stellen, die sie bei der Umsetzung von software-ergonomischem Gestaltungswissen unterstützen. Anknüpfungspunkte für die Durchführung des Projektes waren einerseits die zunehmende Bedeutung von software-ergonomischen Forderungen aufgrund der von der EU-Kommission verabschiedeten "Richtlinie über die Mindestvorschriften bezüglich der Sicherheit und des Gesundheitsschutzes bei der Arbeit an Bild-schirmgeräten [8]" und andererseits empirische Untersuchungen, welche die unzureichende Verbreitung software-ergonomischer Erkenntnisse in der Praxis festgestellt haben [4, 19]. Diese Untersuchungen zeigten, daß ein hoher Prozentsatz der Softwareentwickler über keinerlei oder nur geringfügige software-ergo-nomische Kenntnisse verfügen. Zur Überwindung dieses Defizits wurde von der überwiegenden Mehrheit der befragten Entwickler eine rechnerbasierte Vermittlung software-ergonomischer Gestaltungsanforderungen, während des Entwicklungsprozesses, präferiert. Dazu benötigen die Entwickler leistungsfähige Entwicklungswerkzeuge (z.B. User Interface Management Systems, UIMS), die sie im Entwicklungsprozeß unterstützen und sie gleichzeitig bei der Einhaltung der software-ergonomischen Anforderungen anleiten.

Durch die im Rahmen des hier vorgestellten Projektes realisierten und in ein UIMS integrierten Unterstützungswerkzeuge soll erreicht werden, daß Entwickler

- software-ergonomische   Kenntnisse   während   des Implementierungsprozesses, gleichzeitig mit der Handhabung des Entwicklungswerkzeuges, vermittelt bekommen (learning on demand),

- vorgefertigte, ergonomischen Anforderungen entsprechende Dialogbausteine einer Benutzungsoberfläche verwenden können (reusability),

- bei der Anwendung von ergonomischen Standards und Richtlinien angeleitet werden (usability) und

- die ergonomische Qualität der entworfenen Benutzungsoberflächen schon im Entwicklungsprozeß überprüfen lassen können (quality assurance).

2 Vorgehensweise bei der Entwicklung der Unterstützungswerkzeuge

Es war nicht Ziel des Vorhabens ein neues Entwicklungswerkzeug für graphische Benutzungsoberflächen zu entwickeln, sondern marktgängige mit zusätzlichen Unterstützungsleistungen auszustatten. Dazu konnte ein Hersteller eines UIMS (ISA GmbH, Stuttgart) als Kooperationspartner gewonnen werden. Von diesem wurden bestimmte Erweiterungen an dem UIMS vorgenommen sowie definierte Schnittstellen offengelegt. Dies waren unabdingbare Voraussetzungen für eine Integration der verschiedenen Unterstützungswerkzeuge.

Damit konkrete Anforderungen an die Unterstützungswerkzeuge erfaßt und sie im praktischen Einsatz evaluiert werden konnten, wurden sie für ausgewählte Anwendungsdomänen entwickelt. Dazu wurden eine Reihe von Industriekontakten geknüpft und durch Workshops, Firmenbesuchen, Analysen vor Ort etc. vertieft. Diese Industriekontakte führten zu Kooperationsprojekte mit einigen Industriefirmen (Software AG, Darmstadt; SAP AG, Walldorf/Baden; Hoechst AG, Frankfurt/Main). Im Rahmen dieser Kooperationsprojekte wurden die verschiedenen Unterstützungswerkzeuge auf PCs (DOS/Windows) und Workstations (UNIX/Motif) prototypisch realisiert. Die umfassende Evaluierung dieser Prototypen im praktischen Einsatz wird derzeit durchgeführt. Die dabei gewonnen Erkenntnisse bilden die Grundlage für eine evolutionäre Weiterentwicklung der Unterstützungswerkzeuge.

Als Quelle zur Ableitung des software-ergonomischen Wissens wurden Gestaltungsanforderungen aus Normen (z.B. [12]), Style Guides (z.B. [11, 18, 21]) und herstellerunabhängigen Gestaltungsrichtlinien (z.B. [14, 17]) entnommen.

3 Typischer Einsatz der Unterstützungswerkzeuge

Abb. 1 zeigt eine typische Einsatzsituation für die verschiedenen Unterstützungswerkzeuge. Die getrennte Präsentation der IDA Unterstützungswerkzeuge und des UIMS ist der prototypischen Realisierung geschuldet. Die enge Kooperation mit dem Herstellern des UIMS soll es später ermöglichen, die verschiedenen Leistungen der Unterstützungswerkzeuge in das UIMS zu integrieren.

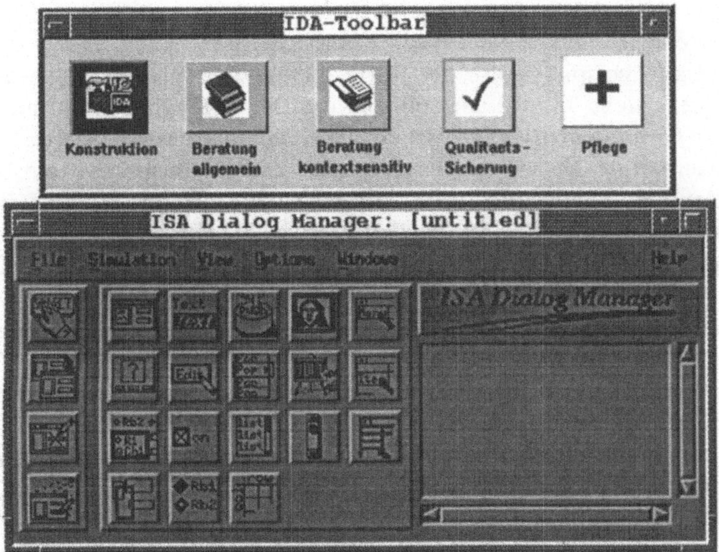

Abb. 1: Typische Einsatzsituation der Unterstützungswerkzeuge

Ein Entwickler implementiert - unter Verwendung eines UIMS - eine graphische Benutzungsoberfläche einer Softwareapplikation. Dazu nutzt er eine Klassenbibliothek mit vorgefertigten Dialogbausteinen. Zugang zu dieser Bibliothek bietet ihm das Unterstützungswerkzeug *Konstruktion*, das über ein Icon in der "IDA-Toolbar" aktivierbar ist [23]. Die Bibliothek hat für den Entwickler eine Art "Baukastenfunktion", indem ihm vorgefertigte Klassen von Dialogbausteine zur Verfügung gestellt werden (reusability). Diese Dialogbausteine entsprechen software-ergonomischen Gestaltungsforderungen (siehe Abschnitt 2) und enthalten einfache und komplexe Dialogobjekte (z.B. Push Button, Tabelle, Dialogfenster), sowie typische Dialogabläufe (z.B. Hilfe, Fehler, Auswahldialoge). Dialogbausteine können auch komplette, anwendungsspezifische Fensterarchitekturen repräsentieren (z.B. Hauptfenster mit Tochterfenstern und Dialogfenstern), die auf einem der gängigen Architekturkonzepten basieren (z.B. Single Document Interface). Durch die Anwendung von ergonomischen Dialogbausteine im Entwicklungsprozeß wird sowohl eine konsistente als auch eine aufgabenangemessene und benutzerfreundliche Realisierung der Benutzungsoberfläche sichergestellt (usability). Eine wichtiges Kriterium für den Erfolg der Bibliothek stellt die Zugänglichkeit zu den vordefinierten Dialogbausteinen dar. Die Zugänglichkeit beeinflußt maßgeblich die Nützlichkeit für den Entwickler. Durch verschiedene Navigations- und Präsentationsinstrumente (z.B. Browser, Suchfunktionen, Animationen) wird der Entwickler beim Finden geeigneter Dialogbausteine unterstützt. Kommt der Entwickler zur Erkenntnis, daß der gefundene Dialogbausteine seinen Wünschen entspricht, kann er eine Instanz der Objektklasse erzeugen.

Ist dem Entwickler beispielsweise der Verwendungszweck eines ausgewählten Dialogbausteines nicht klar, kann er die online verfügbare multimediale *kontextsensitive Beratung* aktivieren, die ebenfalls mittels eines Icons in der "IDA-Toolbar" aktivierbar ist. Die online-Beratung setzt zur Wissenspräsentation Texte, Graphiken, Animationen, Sprache und interaktive Beispiele ein und bietet vielfältige Navigationstechniken. Neben dem Verwendungszweck von vorgefertigten Dialogbausteinen kann sich der Entwickler auch in anschaulicher Weise allgemeines ergonomisches Wissen zur Implementierung von graphischen Benutzungsoberflächen vermittelt lassen. Der Einstieg in die *allgemeine Beratung* erfolgt in diesem Fall durch Aktivieren des Icons "Beratung allgemein" in der "IDA-Toolbar". Im Kapitel 4 erfolgt eine ausführliche Darstellung der verwendeten Konzepte und Techniken der online-Beratung.

Möchte der Entwickler seine bisherigen Implementierungen einer automatischen software-ergonomischen Qualitätssicherung unterziehen, kann er die, auf wissensbasierten Techniken beruhende, *Qualitätssicherung* über die "IDA-Toolbar" aktivieren [26]. Diese wertet Entwicklungsergebnisse aufgrund einer expliziten Anforderung durch den Entwickler aus (software quality assurance). Das Ergonomiewissen der verschiedenen Quellen (siehe Abschnitt 2) wurde mittels Regeln in die Wissensbasis aufgenommen. Nach Aktivierung der Qualitätssicherung werden die Daten der Benutzungsoberfläche abgespeichert und in die Qualitätssicherung eingelesen. Anschließend werden automatisch die Inferenzmechanismen aktiviert, die das Abarbeiten der Regeln über die Objekte der Benutzungsoberfläche steuern. Nach der Regelabarbeitung erhält der Entwickler Informationen zu den gefundenen ergonomischen Abweichungen. Zur weiteren Bearbeitung hat der Entwickler mehrere Aktionen zur Auswahl. Er kann sowohl in die Beratung als auch in die Konstruktion verzweigen, um Hinweise zur Fehlerkorrektur zu erhalten. Bei bestimmten Objekten kann auch in der Qualitätssicherung eine automatische Korrektur vorgenommen werden.

Ergonomiewissen unterliegt einem stetigen Wandel aufgrund neuer wissenschaftlicher Erkenntnisse und des technischen Fortschritts. Damit ist die Notwendigkeit offensichtlich, Möglichkeiten zur *Pflege* des in den IDA-Unterstützungswerkzeugen vorhandenen Ergonomiewissens bereitzustellen. Dazu wird ein weiteres Unterstützungswerkzeug angeboten, das über die "IDA-Toolbar" mittels des Pflegeicons aktiviert werden kann. Zum einen ist das in jedem einzelnen Unterstützungswerkzeug vorhandene Wissen zu pflegen, zum anderen ist die Konsistenz zwischen den Wissensbeständen der einzelnen Unterstützungswerkzeuge zu gewährleisten.

4 Vermittlung ergonomischen Wissens mittels hypermedialer Techniken

Die IDA-Beratung wendet sich an Entwickler von Benutzungsoberflächen, die bislang wenig Erfahrung im ergonomischen Design von Benutzungsoberflächen besitzen. Es soll aber auch Experten ein Nachschlagewerk zur Verfügung gestellt werden. Die Beratung verfolgt vor allem tutorielle Zwecke. Kurzfristig soll fehlendes ergonomisches Wissen, während des Entwurfprozesses der Benutzungsoberfläche, bedarfsorientiert vermittelt werden (learning on demand). Längerfristig wird ein permanenter Wissenstransfer angestrebt, so daß die Konsultation der Beratung immer seltener erfolgen muß. Mit diesen Zielsetzungen muß sich die IDA-Beratung der Konkurrenz konventioneller Methoden der Wissensgewinnung (z. B. Lesen von Style Guides, Befragen von Experten) stellen. Das Beratungssystem erweitert durch den Einsatz von multimedialen Techniken die Möglichkeiten verfügbarer online-Hilfesysteme (z.B. Animationen, Videos, Ton, Sprache). Die multimediale IDA-Beratung muß die Akzeptanz durch die Entwickler mit Hilfe einer konsequenten Ausnutzung der Vorteile der rechnerbasierten Realisierung erreichen. Dieser Anspruch diente als Leitlinie für die Implementierung und schlägt sich in drei grundlegenden Gestaltungsempfehlungen nieder [22]:

- Verwendung von möglichst wenig Text,
- Darstellung von Abläufen und Vorgängen durch dynamische Medien,
- interaktive Einbindung des Entwicklers.

4.1 Strukturierung der Inhalte

Zur Strukturierung der Inhalte der IDA-Beratung, deren Quellen bereits in Abschnitt 2 genannt wurden, boten sich zwei Alternativen an: eine hierarchische oder eine netzwerkartige Struktur. Da das Ergonomiewissen aus hierarchisch strukturierten Büchern entnommen wurde, fiel die Entscheidung zugunsten einer hierarchischen Struktur als primäre Strukturierung. Damit sollen den Entwicklern ein Anknüpfen an vertraute Medien der Wissensvermittlung erleichtert werden. Abb. 2 zeigt einen Ausschnitt dieser Inhaltsstruktur.

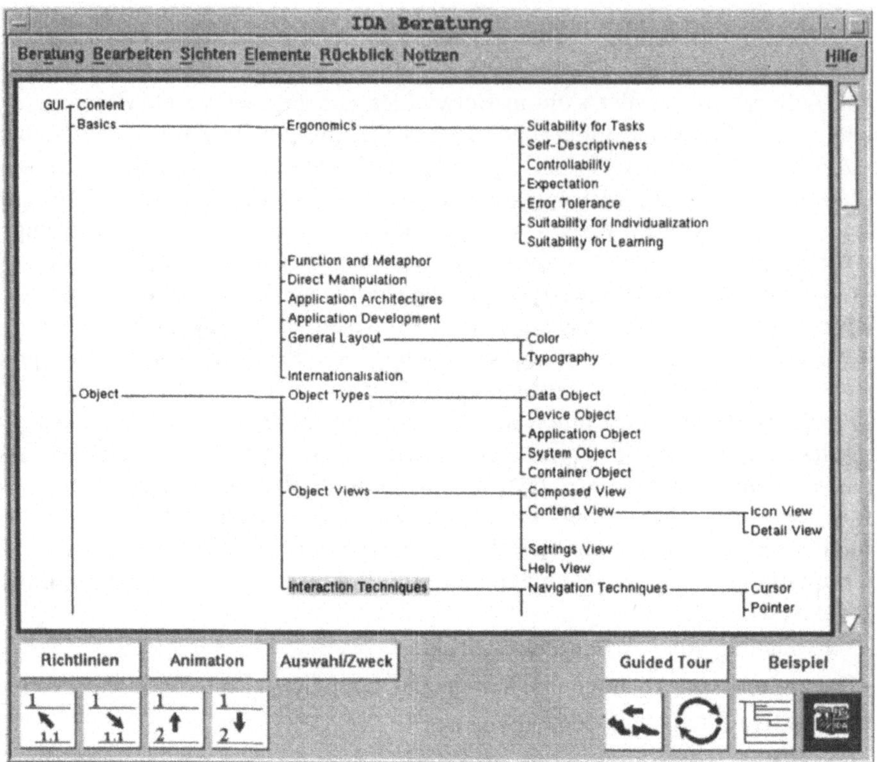

Abb. 2: Ausschnitt aus dem Inhaltsverzeichnis der IDA-Beratung

Die Inhaltsstruktur wird in allen Komponenten der IDA-Entwicklungsumgebung in gleicher Weise repräsentiert (z.B. Strukturierung der Inhalte der Klassenbibliothek, Strukturierung der Wissensbasis der Qualitätssicherung), um eine globale Orientierung des Entwicklers zu unterstützen. Gleichzeitig kann die Struktur der Querverweise in konventionellen Büchern auf die Hypertext-Links zwischen den einzelnen Beratungsthemen abgebildet werden. Dies liefert eine zusätzliche, netzwerkartige Struktur, die die hierarchische Struktur überlagert. Als Problem bei dieser Art der Strukturierung der Inhalte hat sich die Partitionierung in einzelne Beratungsthemen herausgestellt. Bei dieser vom Hypertext-Konzept geforderten Unterteilung stellte sich häufig das Problem zu entscheiden, wann eine weitere Differenzierung von einzelnen Beratungsthemen notwendig ist. Außerdem hat sich eine eindeutige Zuordnung von Beratungsthemen zu bestimmten Positionen in der hierarchischen Struktur als sehr schwierig erwiesen. Diese Probleme resultieren auch aus fortwährenden Weiterentwicklung des Ergonomiewissens und verlangen eine permanente Anpassung der Struktur.

4.2   Präsentation der Inhalte

Bei der Präsentation der Inhalte der IDA-Beratung an der Benutzungsoberfläche hat sich die Verwendung von Metaphern als nützlich erwiesen [16]. Aufgrund der Konkurrenz zu schriftlichen Richtliniensammlungen bot sich die Wahl der Buch- bzw. Ringbuch-Metapher an. Die Implementierung der IDA-Beratung unter OSF/Motif, mit der Bezeichnung Multimedia Interface Design Documents for Advice (MIDA), verwendet die Buch-Metapher [2]. Für die Realisierung der IDA-Beratung unter Microsoft Windows, mit der Bezeichnung Interface Design Advice Tool (IDAT), fiel die Entscheidung auf die eng verwandte Ringbuch-Metapher [16]. Diese unterstützt eine speziellen Explorationstechnik. Danach werden beim Verfolgen von Querverweisen die Seiten, die Ausgangspunkt von Querverweisen sind, aus dem Ringbuch entnommen und auf einen separaten Stapel abgelegt. Mit Hilfe der Seiten auf diesem Stapel kann der Entwickler dann stets die Eckpunkte seines aktuellen Explorationspfades rekonstruieren. Ziel dieser Technik ist eine bessere Orientierungsunterstützung bei der Navigation durch den Themenraum. Im Rahmen der in Durchführung befindlichen empirischen Evaluation soll die Eigung der verschiedenen Metaphern für Zwecke der Wissensvermittlung untersucht werden.

Abb. 3 zeigt eine typische Benutzungssituation der IDAT-Beratung. Der Stapel der durch das Verfolgen von Querverweisen herausgelegten Seiten wird dort durch die zurückliegenden und bis auf die Titelleiste verdeckten Fenster repräsentiert (Overlapped-Windows-Technik). Innerhalb eines Fensters erfolgt die Präsentation einer Seite auf fünf Karteikarten, die durch Klicken auf die zugehörigen Karteikartenreiter in den Vordergrund gebracht werden können. Bei der IDAT-Beratung wurde eine strikte Komplexitätsreduzierung verfolgt [5]. Ziel war es, im Sinne einer kognitiv tieferen Informationsverarbeitung die Aufmerksamkeit des Entwicklers auf den Inhalt und nicht die Interaktionselemente und -techniken zu richten. Daher befinden sich in den Themenfenstern der IDAT-Beratung keine Pull-Down-Menüs oder Push-Buttons. Die gesamte Funktionalität verbirgt sich in einem Pop-Up-Menü, das durch Drücken der rechten Maustaste aktiviert wird (siehe Abb. 3 Mitte). Zusätzlich werden die wichtigsten Navigationsfunktionen auch in einer Toolbar angeboten (siehe Abb. 3 links oben). Eine ausschließliche Bedienung per Tastatur ist ebenfalls möglich. Das Funktionalitätsangebot basiert auf Ergebnissen einer Befragung von Entwicklern der Software AG [6].

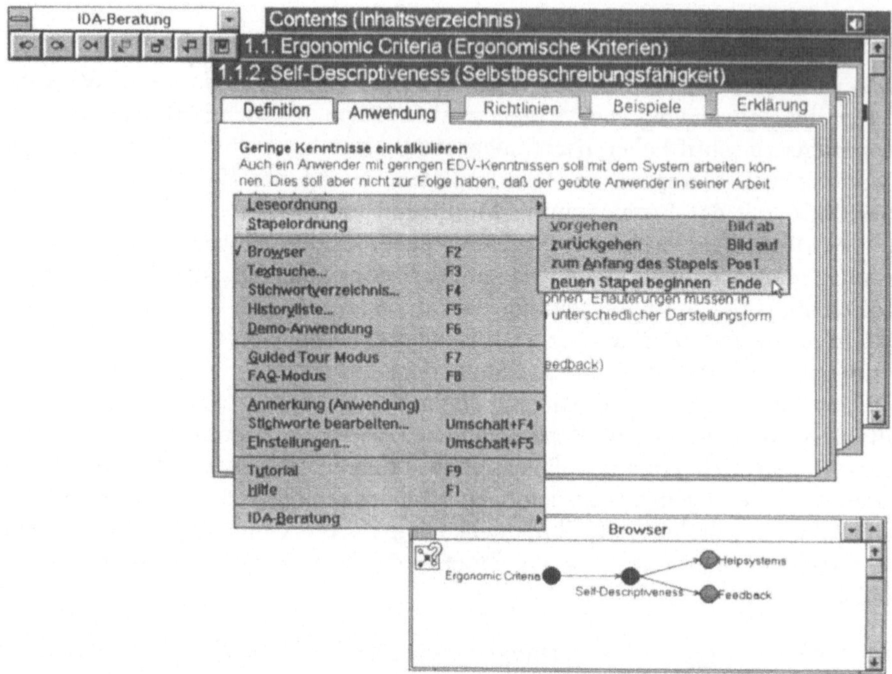

Abb. 3: Typische Benutzungssituation der IDAT-Beratung unter Windows

Im Vergleich dazu liegt dem MIDA-Beratungssystem die Buch-Metapher
zugrunde. Abb. 4 zeigt das MIDA-Hauptfenster, in dem auch ein graphischer
Browser anzeigt werden kann. Damit der Entwickler nicht mit zuviel Inhalten
konfrontiert wird, wird der Inhalt in Teile gegliedert. Es wird unterschieden
zwischen allgemeinen Erklärungen, sowie Richtlinien. Die allgemeinen
Erklärungen werden permanent im Hauptfenster angezeigt und geben eine kurze
Information zum Thema an. Verwandte Themen und Abweichungen in den
verschiedenen Style Guides werden ebenfalls im Hauptfenster angezeigt. Die
Richtlinien werden in einem eigenen Fenster präsentiert (Abb. 4 rechts oben). Der
Entwickler kann die angezeigten Richtlinien interaktiv explorieren, indem er mit
der Maus in die Beispielgraphik klickt. Zur Unterscheidung der Quellen der verschiedenen Richtlinien werden diese farbig angezeigt. So wird auch die besondere
Wichtigkeit einer Richtlinie visualisiert (Muß/Kann). Bei Beratungsthemen, bei
denen eine textuelle und graphische Präsentation nicht ausreicht, beispielsweise bei
Dialogabläufen, wird das Beratungsthema animiert bzw. als Videosequenz
aufgezeichnet und vertont. Diese Animationen können dann vom Entwickler, wie
von einem Videorecorder bekannt, abgespielt, angehalten und fortgesetzt werden.

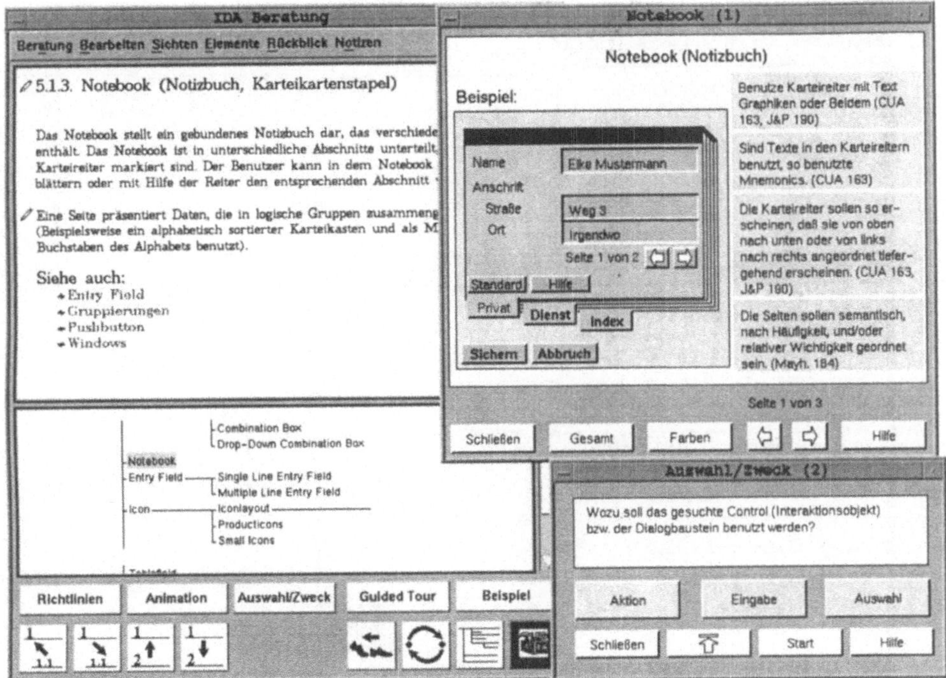

Abb. 4: Komponenten der MIDA-Beratung unter OSF/Motif

4.3   Navigationstechniken

Neben bekannten Navigationsfunktionen hypermedialer Systeme, wie z. B. Hotlinks, Historylisten, Inhalts- und Stichwortverzeichnisse können *Auswahldialoge* als zusätzliche Navigationsmöglichkeit genutzt werden. Abb. 4 zeigt ein Beispiel des Auswahldialogs (Fenster Auswahl/Zweck), der dem Entwickler bei der richtigen Auswahl von Dialogobjekten unterstützt. Der Entwickler wird zunächst gefragt, ob das Dialogobjekt zur Eingabe, Auswahl oder Aktion dienen soll. Es werden anschließend weitere Fragen gestellt, bis zum Schluß ein Beratungsthema zum vorgeschlagenen Dialogobjekt im Hauptfenster des Beratungssystems angezeigt wird.

Eine mehr themenübergreifende Sicht auf die Inhalte liefert die *Guided Tour*. Sie beschreibt einen vordefinierten Weg durch das Beratungssystem und erlaubt damit eine themenübergreifende Präsentation dessen Inhalte.

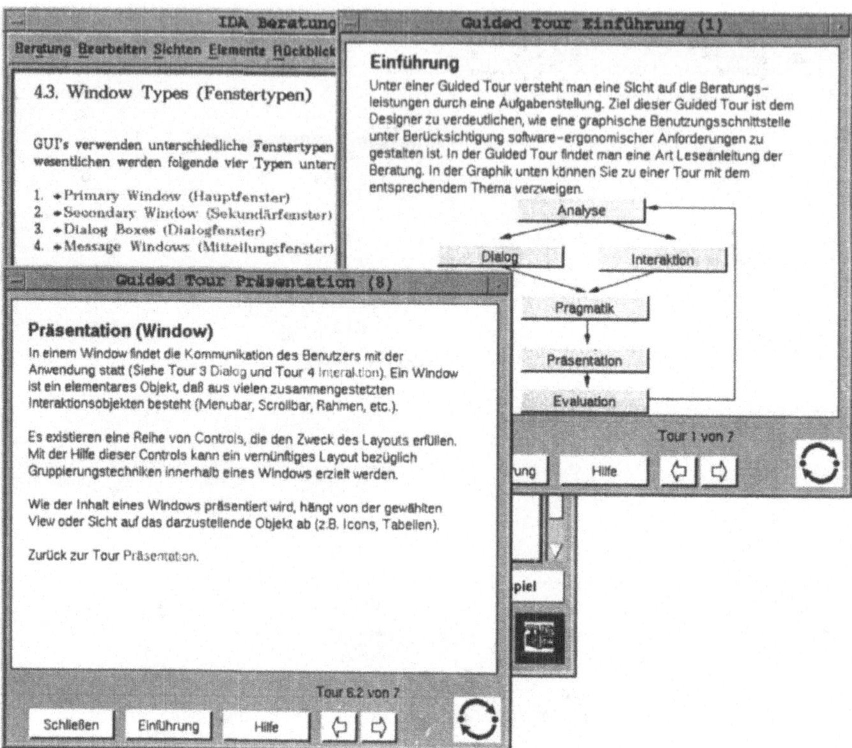

Abb. 5: Guided Tour der MIDA-Beratung unter OSF/Motif

Anhand der Guided Tour wird ein kompletter Oberflächendesignzyklus erklärt (Abb. 5 zeigt das Startfenster rechts oben). Deshalb wendet sich die Guided Tour vor allem an Anfänger im ergonomischen Oberflächendesign. Das Durcharbeiten eines bestimmten Abschnittes der Guided Tour erfolgt durch Aktivierung der gewünschten Phase des Designzyklus. Wird beispielsweise die Phase Präsentation im Startfenster ausgewählt, so werden in einem neuen Fenster (Abb. 5 vorne), die durchzuführenden Aktivitäten beschrieben. Anhand von verschiedenfarbigen Hotlinks kann die Navigation beeinflußt werden. Wird auf einen grünen Hotlink geklickt, verzweigt der Entwickler innerhalb der Guided Tour. Wird hingegen ein blauer Hotlink ausgewählt, so wird das entsprechende Beratungsthema im Hauptfenster der Beratung (Abb. 5 im Hintergrund), angezeigt. Die gesamte Funktionalität des Beratungssystems wird weiterhin zur Verfügung gestellt, so daß eine parallele Nutzung der Beratung und Guided Tour möglich ist.

5. Verwandte Forschungsansätze

Eine Auswertung einschlägiger Forschungsvorhaben hat gezeigt, daß die Entwicklung von Unterstützungsleistungen für Entwickler von Benutzungsoberflächen bereits Gegenstand von zahlreichen Forschungsvorhaben war und ist.

Im *Department of Computer Science der University of Colorado in Boulder* werden seit einigen Jahren sogenannte Kritiksysteme entwickelt. Vor allem im Bereich Designunterstützung wurden eine Reihe von prototypischen Systeme entwickelt, die unter Verwendung von wissensbasierten Techniken, Hypertextsystemen und Bibliotheken wiederverwendbarer Designergebnisse dem Designer Unterstützung bieten [9].

Am *Department of Computer and Information Science der University of Linköping* wurde ein UIMS mit einem Expertensystem verbunden, das mittels einer Wissensbasis die entworfenen graphischen Benutzungsoberflächen nach Style Guide spezifischen Anforderungen analysiert. Ergebnis des Qualitätssicherungsprozesses sind Kommentare, die der Entwickler anschließend berücksichtigen kann. [15]

Am *College of Computing des Georgia Institute of Technology in Atlanta* wird an der automatischen Generierung von Benutzungsoberflächen gearbeitet. Basierend auf einem objekt-orientierten Datenmodell werden automatisch - unter Beachtung von Style Guide spezifischen Anforderungen - Benutzungsoberflächen generiert und mittels eines UIMS visualisiert [1].

Am *Fraunhofer-Institut für Arbeitswirtschaft und Organisation in Stuttgart* wurde ebenfalls an der automatischen Generierung von Benutzungsoberflächen gearbeitet. Basierend auf einem Entity-Relationsship-Datenmodell, ergänzt um sogenannte Aufgabensichten und Dialognetzen, wurde mittels eines regelbasierten Systems automatisch die erste Version der Benutzungsoberfläche - unter Verwendung von software-ergonomischen Designregeln - generiert und mittels eines UIMS visualisiert [13].

Im *Fachbereich Informatik der Universität Oldenburg* wird dem Entwickler von Benutzungsoberflächen ergonomisches Wissen zur Verfügung gestellt. Anhand der Methode MUSE und eines wissensbasierten Beratungswerkzeuges soll sowohl Bewertungs- als auch Beratungsleistungen zur Verfügung gestellt werden [10].

Am *Lehrstuhl für Software-Technik der Ruhr-Universität Bochum* wird aus einem objekt-orientierten Datenmodell mittels verschiedener wissensbasierter Generatoren, die auch Ergonomiewissen enthalten, automatisch eine Beschreibung einer Benutzungsoberfläche generiert. Diese wird dann mit Hilfe eines UIMS visualisiert und steht anschießend dem Entwickler zur weiteren Bearbeitung zur Verfügung [3].

Am *Institut für Angewandte Psychologie der Humbolt-Universität Berlin* wird im Rahmen eines Projektes ein Interface-Ratgeber entwickelt. Dieser soll sowohl als

Textbuch [27] als auch als rechnerbasiertes Unterstützungswerkzeug verfügbar sein.

Im *Fachbereich Informatik der Technischen Hochschule Darmstadt* wird im Rahmen des Vorhabens Diades-II ein objekt-orientiertes Entwurfswerkzeug, das auch Ergonomiewissen beinhaltet, für interaktive Benutzungsschnittstellen auf der Grundlage koagierender Agenten entwickelt [7].

6 Resümee und Ausblick

Die bisherigen Projektergebnisse zeigen, daß rechnerbasierte Unterstützungswerkzeuge einen vielversprechenden Ansatz zur Vermittlung und Umsetzung von software-ergonomischen Gestaltungsforderungen darstellen. Sie weisen somit einen möglichen Weg zur konsequenten Umsetzung von software-ergonomischen Gestaltungsforderungen, direkt im Implementierungsprozeß von graphischen Benutzungsoberflächen. Diese prospektive Gestaltungsorientierung ist ein entscheidender Vorteil dieses Ansatzes gegenüber der korrektiven Gestaltungsorientierung traditioneller software-ergonomischer Evaluationsverfahren (z.B. EVADIS, [24]). Entscheidend für die Akzeptanz derartiger Unterstützungswerkzeuge ist einerseits das Gelingen der Integration in das primäre Entwicklungswerkzeug des Entwicklers und andererseits das unmittelbare Auffinden und Anwenden von Wissensinhalten. Dies berührt Fragen von geeigneten Retrieval Strategien und Mechanismen (z.B. Browser, Suchfunktionen, Guided Tour) sowie von geeigneten Präsentationsformen von Wissensinhalten (z.B. objekt-orientiert, multimedial, wissensbasiert). Durch die Anwendung einer Vielzahl von Retrieval Strategien und Wissenspräsentationsformen im beschriebenen Vorhaben wurde vorerst versucht, einen möglichst maximalen Rahmen aufzuspannen.

Der nächste Schritt besteht in der Evaluation der verschiedenen Konzepte hinsichtlich Eignung und Akzeptanz. Die in Durchführung befindlichen empirischen Evaluationen der verschiedenen Unterstützungswerkzeuge bei den Industriepartner sollen unter anderem Antworten auf folgende Fragestellungen liefern: Eignung der einzelnen Unterstützungswerkzeuge für Zwecke der Konstruktionsunterstützung, der multimedialen-Beratung und der automatisierten Qualitätssicherung; Vorteile und Grenzen der Wissenspräsentation in den jeweiligen Unterstützungswerkzeugen; erzielbare synergetische Effekte durch das Ineinandergreifen der verschiedenen Unterstützungswerkzeuge; Akzeptanz der zusätzlich zum eigentlichen Entwicklungswerkzeug zu nutzenden Unterstützungswerkzeuge durch die Entwickler.

Basierend auf den Evaluationsergebnissen ist eine Reimplementierung der Unterstützungswerkzeuge geplant, die geeignete und akzeptierte Konzepte und Mechanismen beibehält bzw. vertieft und ungeeignete bzw. nicht akzeptierte verwirft.

7   Literatur

[1]   Baar, D., Foley, J., Mullet, K.: Coupling Application Design and User Interface Design. In: CHI'92 Porceddings, Reading: Addison-Wesley 1992, pp.259-266

[2]   Bachmann, F.: Entwicklung des multimedialen Beratungssystem MIDA zur Gestaltung von Benutzungsoberflächen unter software-ergonomischen Gesichtspunkten, Diplomarbeit, Universität Bonn, Bonn, 1994.

[3]   Balzert, H.: Der JANUS-Dialogexperte: Vom Fachkonzept zur Dialogstruktur. Software-technik Trends, Band 13, Heft 3, 1993, S.62-72

[4]   Beimel, J., Hüttner, J. und Wandke, H.: Kenntnisse von Programmierern auf dem Gebiet der Software-Ergonomie: Stand und Möglichkeiten zur Verbesserung, unpublished paper of a lecture on the Fachtagung der Sektion Arbeits-, Betriebs und Organisationspsychologie des Berufsverbandes Deutscher Psychologen "Arbeits- Betriebs- und Organisationspsychologie vor Ort" 25.-27.5.1992, Bad Lauterbach.

[5]   Büchner, S.: Komplexitätsreduzierung der Benutzeroberfläche der IDA-Beratungs- und Erklärungsebene, Ergonomie-Studienarbeit, Universität/Gesamthochschule Essen, 1994.

[6]   Diehl, U.: Entwicklung einer Online-Beratung zur ergonomischen Gestaltung von Windows Anwendungen — demonstriert am Beispiel der Entire Office Workstation der Software AG, Diplomarbeit, Fachhochschule Darmstadt, Darmstadt, 1994.

[7]   Dilli, I., Hoffmann, H.-J., Koschorek, D.: Experience with an object-oriented design approach for a programming environment under a blackboard architecture. In: Lasker G.E. (ed.) Advances in Human Systems and Information Technology, The International Institute for Advanced Studies in Systems Research and Cybernetics 1992, pp.268-276

[8]   Europäische Gemeinschaft: Richtlinie über die Mindestvorschriften bezüglich der Sicherheit und des Gesundheitsschutzes bei der Arbeit an Bildschirmgeräten (EWG 90/270).

[9]   Fischer, G., Nakakoji, K., Ostwald, J., Stahl, G., Sumner, T.: Embedding Computer-Based Critics in the Contexts of Design. In: INTERCHI '93 Proceedings, Reading: Addison-Wesley 1993, pp.157-164

[10]   Gorny, P., u.a.: Projekt EXPOSE, Expertensystem zur phasenorientierten Software-Ergonomie-Beratung bei der Benutzungsschnittstellen-Entwicklung, 2. Zwischenbericht, Universität Oldenburg und Universität Rostock, 1994

[11]   IBM: Object-Oriented Interface Design, IBM Common User Access Guidelines. Carmel: Que Corporation. 1992.

[12]   ISO 9241: Ergonomic Requirements for Office Work with Visual Display Terminals.

[13]   Janssen, C. , Weisbecker, A., Ziegler, J.: Generation User Interfaces form Data Models and Dialogue Net Specifications. In: INTERCHI'93 Proceedings, Reading: Addison-Wesley 1993, pp.418-423

[14]   Jenz & Partner: Grafische Bediener-Oberflächen, Erlensee, Jenz & Partner, 1992.

[15]   Löwgren, J., Nordquist, T.: Knowledge-Based Evaluation as Design Support for Graphical User Interfaces. In: CHI'92 Porceddings, Reading: Addison-Wesley 1992, pp.181-188

[16]   Maaß, S., Oberquelle H.: Perspectives and Metaphors for Human-Computer Interaction. In Floyd, C., Züllighoven, H., Budde, R., Keil-Slawik, R. (Hrsg.): Software-Development and Reality Construction, Berlin: Springer, 1992, pp.233-251

[17]   Mayhew, D.: Principles and Guidelines in Software User Interface Design, Englewood Cliffs: Prentice Hall, 1992.

[18]   Mircorsoft: The Windows Interface, An Application Design Guide, Microsoft Press, 1992.

[19]   Molich, R., Nielsen, J.: Improving a human-computer dialogue, Communications of the ACM, vol. 33, no. 3, 1990, pp.338-348.

[20]   Müller, M.: Erweiterung des Hypertextsystems ToolBook um Hilfsmittel zur Reduzierung der Gefahr des Orientierungsverlustes, Diplomarbeit, Universität / Gesamthochschule Paderborn, Fachbereich Mathematik / Informatik und Wirtschaftswissenschaften, Paderborn, 1993.

[21]   Open Software Foundation: OSF/MOTIF Style Guide, Revision 1.2, London: Prentice-Hall, 1993.

[22]   Porschen, M.: Hypermediale Beratungssysteme — Herleitung und Evaluation eines werkzeugunabhängigen und plattformübergreifenden Entwicklungskonzeptes, Diplomarbeit, Universität Bonn, Bonn, 1994.

[23]   Ramm, M.: Generierung und Verwaltung einer Bibliothek von Modellvorlagen zur Gestaltung software-ergonomischer Benutzungsoberfllächen, Diplomarbeit, Universität Bonn, Bonn, 1994.

[24]   Reiterer, H., Oppermann, R.: Evaluation of user interfaces, EVADIS II - a comprehensive evaluation approach, Behaviour & Information Technology, Vol. 12, No. 3, pp.137-148.

[25]   Reiterer, H.: A User Interface Design Assistant Approach, in: Brunnstein K., Raubold E. (eds.): Applications and Impacts, Information Processing '94, Proceedings of the IFIP 13th World Computer Congress, Hamburg, Germany, 28 August - 2 September, 1994, IFIP Transactions A-52, Volume II, North-Holland, Amsterdam, 1994, pp.180-187.

[26]   Schäfer, S.: Qualitätssicherung im User Interface Design, Chancen und Grenzen, Diplomarbeit, Universität Bonn, Bonn, 1994.

[27]   Wandke, H., Hüttner, J., Rätz, A., Vogler, K.: Broschüre des Informations und Beratungssystems INRA (Version 1.0), Humboldt Universität Berlin, FB Psychologie, Mai 1991 (unveröffentlicht)

Frank Bachmann[1], Michael Porschen[1], Manfred Ramm[1], Dr. Harald Reiterer[1+2], Stefan Schäfer[1], Helmut Simm[1]

1   Gesellschaft für Mathematik und Datenverarbeitung (GMD)
     Schloß Birlinghoven, D-53731 Sankt Augustin
2   Universität Wien, Institut für Angewandte Informatik und Informationssysteme
     Liebiggasse 3/3-4, A-1010 Wien

SW-ergonomische Normen —
Versuch einer betrieblichen Umsetzung

Astrid Beck, Stuttgart

Manfred Scheifele,
Fraunhofer-Institut für Arbeitswirtschaft und Organisation (IAO), Stuttgart

Zusammenfassung

Verstärkt wird propagiert: die (nicht mehr so ganz) neue EU-Richtlinie zu Bildschirmarbeitsplätzen bietet erweiterte Handlungsmöglichkeiten für die Betriebsräte (DGB Technologieberatung 1993, Wanke et al. 1993, Döbele-Martin, Martin 1993). Doch wie sieht es in der Praxis bei der einer annäherungsweisen Umsetzung und in einer Einigungsstelle tatsächlich aus? Wie schwierig die Umsetzung der EU-Richtlinie und arbeitswissenschaftlicher Erkenntnisse in komplexen Software-Systemen tatsächlich ist (Stichworte: technische Probleme, terminliche Restriktionen, ungenügende Kenntnisse der Beteiligten) und wie Betriebsräte bestimmte Mindestpositionen unter widrigen Bedingungen doch durchsetzen können, soll im folgenden Praxisbeispiel geschildert werden. Daran schließen sich Erkenntnisse und Schlußfolgerungen für eine Systementwicklung an, die auf eine höhere Qualität der Anwendungswerkzeuge abzielen.

1. Ziele und Anforderungen der Arbeitnehmervertretungen

Sahen in der Vergangenheit Gewerkschaften und Betriebsräte in der EDV in erster Linie den Job-Killer und das Kontrollinstrument gegen die Beschäftigten, wird in der Gewerkschaftsdiskussion seit einiger Zeit die Software verstärkt als „Werkzeug" betrachtet, das es aktiv mitzugestalten gilt (Becker-Töpfer 1991, Klotz 1994). Neben den Zielen, die Arbeitsplätze zu sichern und jegliche Art von Kontrolle möglichst zu vermeiden, kommt der betrieblichen Interessenvertretung daher die neue Aufgabe zu, die Partizipation der Betroffenen zu fördern und auf die Qualität des Werkzeugs zu achten. Software–Ergonomie in dieser erweiterten Sichtweise bedeutet nicht nur die Gestaltung von Bildschirmoberflächen sondern auch die Gestaltung der Arbeitsabläufe und der EDV-Arbeitsplätze unter der aktiven Beteiligung der zu qualifizierenden Mitarbeiter.

Bei der Arbeits- und Softwaregestaltung operiert der Betriebsrat (BR) auf der rechtlichen Grundlage des Betriebsverfassungsgesetzes (BetrVG)[1]. Darin sind umfassende Unterrichtungs- und Beratungsrechte (§ 90) festgelegt, wenn Maßnahmen zur Gestaltung des Arbeitsplatzes und von Arbeitsabläufen betroffen sind. Dabei sollen „die gesicherten arbeitswissenschaftlichen Erkenntnisse über die menschen-

---

1   Im öffentlichen Dienst gilt das Bundes- bzw. Landespersonalvertretungsgesetz mit entsprechenden Regelungen. Die Möglichkeiten der Personalräte zur Durchsetzung sind jedoch etwas schlechter.

gerechte Gestaltung der Arbeit"[2] berücksichtigt werden. Ein korrigierendes Mitbestimmungsrecht hat der Betriebsrat (§ 91 BetrVG) bei geplanten oder schon realisierten Maßnahmen dann, wenn gegen gesicherte arbeitswissenschaftliche Erkenntnisse *offensichtlich* verstoßen wird. Bei Systemen, die zur Leistungs- und Verhaltenskontrolle geeignet sind — das trifft fast immer zu, wenn personenbezogene Daten verarbeitet werden — ist seine Zustimmung zur Einführung erforderlich (§ 87(1)6 BetrVG). Bei Regelungen zum Gesundheitsschutz im Rahmen der gesetzlichen Vorschriften oder der Unfallverhütungsvorschriften[3] hat der BR ebenfalls Mitbestimmungsrechte (§ 87(1)7 BetrVG).

Große Erwartungen wurden mit der „EU-Richtlinie" über die Mindestvorschriften bezüglich der Sicherheit und des Gesundheitsschutzes bei der Arbeit an Bildschirmgeräten (90/270/EWG) geweckt. Wesentliche Festlegungen sind:

- die Richtlinie gilt als Voraussetzung für Sicherheit und Gesundheitsschutz
- der Arbeitgeber wird verpflichtet, eine Analyse der Arbeitsplätze durchzuführen um körperliche und psychische Belastungen beurteilen zu können
- die Software muß der Tätigkeit und den Benutzern angepaßt sein
- Grundsätze der Ergonomie sind anzuwenden.

Die Richtlinie sollte bis zum 31.12.1992 als nationales Recht umgesetzt sein, was bisher nicht geschehen ist[4]. Von Arbeitgeberseite wird argumentiert, solange die EU-Richtlinie noch nicht in nationales Recht umgesetzt sei, habe sie noch keine Gültigkeit. In der EDV-Branche werden Befürchtungen vor den Folgen vor allem für laufende Systeme geäußert (vgl. Computerwoche vom 21.01.1994). Von Seiten der Gewerkschaften wird hingegen darauf verwiesen, daß die Bundesregierung ihrer Pflicht zur Umsetzung nicht nachgekommen sei und deshalb „das Recht richtlinienkonform ausgelegt werden muß" (Wanke, Nienstedt, Groß 1993).

Unabhängig vom rechtlichen Stellenwert muß die EU-Richtlinie als gesetzliche Vorschrift im Sinne von § 87 Abs. 1 Ziff. 7 BetrVG gewertet werden, so daß Be-

---

[2] Zu den gesicherten Erkenntnissen zählen u.a. DIN-Normen, die Sicherheitsregeln für Büro–Arbeitsplätze, (ZH 1/535) und die Sicherheitsregeln für Bildschirm–Arbeitsplätze im Bürobereich (ZH 1/618) vom Oktober 1980. Die Sicherheitsregeln entsprechen allerdings nicht mehr dem neuesten Stand. Im Fachausschuß ´Verwaltung´ der Verwaltungs-Berufsgenossenschaft wird noch an ihrer Aktualisierung gearbeitet.

[3] DIN-Normen und die Sicherheitsregeln der Berufsgenossenschaften sind nach einem BAG-Urteil vom 6.12.83 keine gesetzlichen Vorschriften im Sinne von § 87(1)7 BetrVG. Unfallverhütungsvorschriften (UVVen, u.a. Allgemeine Vorschriften, VBG 101) werden von den Berufsgenossenschaften herausgegeben. Eine UVV für die Arbeit an Bildschirmgeräten (VBG 104) ist in Arbeit (Krafft 1993).

[4] Das Bundesarbeitsministerium hat — mit einigem Verzug — Ende 1993 den Entwurf einer Verordnung zur Bildschirmarbeit vorgelegt, der die EU-Richtlinie umsetzen soll. Es war geplant, die „Bildschirmrichtlinie" dann zu verabschieden, wenn das von der Bundesregierung im Herbst 1993 vorgelegte Arbeitsschutzrahmengesetz (ASRG) das Parlament passiert hat. Im Juni 1994 jedoch wurde der Entwurf des Arbeitsschutzrahmengesetzes und damit auch die Bildschirmrichtlinie zurückgezogen.

triebsräte über die Umsetzung der EU-Richtlinie mitzubestimmen haben. Damit hat der BR jetzt „mehr in der Hand" als beispielsweise mit der DIN 66 234 (Teil 8), die „nur" als zu berücksichtigende gesicherte arbeitswissenschaftliche Erkenntnis gewertet werden kann.

Wie schwierig die Umsetzung arbeitswissenschaftlicher Erkenntnisse in komplexen Software-Systemen tatsächlich ist und wie Betriebsräte bestimmte Mindestpositionen unter widrigen Bedingungen doch durchsetzen können, soll anhand des Praxisbeispiels MAPIS geschildert werden.

2. Praxisbeispiel MAPIS (Multifunktionales Administrations- und Projekt-Informationssystem)

Das Unternehmen, in dem rund 7.000 Mitarbeiter beschäftigt sind, bietet durch seine etwa 40 Betriebe Dienstleistungen für private und öffentliche Auftraggeber an. Die Betriebe werden wie Profit-Center geführt. Sie akquirieren und bearbeiten selbständig Projekte. Die Unternehmenszentrale unterstützt die administrative und kostenmäßige Abwicklung der Projekte und disponiert die projektunabhängige der Grundfinanzierung für die Betriebe. Ab Mitte der siebziger Jahre wurde ein EDV-System zur Unterstützung dieser Aufgaben aufgebaut, das ab Mitte der achtziger zu einem Verbund der Betriebe erweitert worden ist. Allerdings werden zunehmend Defizite und Begrenzungen der historisch gewachsenen EDV-Landschaft spürbar: verschiedene Betriebssysteme, immer höherer Wartungsaufwand, die Weiterentwicklung stößt auf Grenzen, Performance-Probleme und Zeitverzögerungen, die durch die Schnittstellen zentral/dezentral und die Brüche zwischen verschiedenen Anwendungssystemen weiter verstärkt werden.

Ende 1989 wurde im Vorstand auf der Basis von Vorstudien die Entscheidung für eine grundsätzliche Neugestaltung der administrativen Systeme getroffen: das Multifunktionale Administrations- und Projekt-Informationssystem (MAPIS). Wesentlich war die Festlegung auf eine Eigenentwicklung eines Systems mit Client/Server-Architektur unter UNIX mit der Datenbank ORACLE, weil die spezifischen Gegebenheiten des Unternehmens den Einsatz kommerziell verfügbarer Systeme als nicht angeraten erscheinen ließen. Maßgaben waren u.a. die Entwicklung einer identischen Anwendungssoftware für die Zentrale und alle Betriebe, einheitliche Hardware, unternehmenseinheitliches Datenmodell. Die Bedeutung der "frühzeitigen Einbeziehung aller erfolgsbeeinflussenden Personen" wurde erkannt. MAPIS soll aus Anwendungen für Haushaltsplanung, administrative Projektabwicklung (incl. Buchungen), Einkauf und Personalverwaltung bestehen.

2.1 Projektstruktur /-organisation

Das Vorgehensmodell des MAPIS-Projekts läßt sich anhand der geplanten Arbeits-
pakete veranschaulichen (Abb. 1). Ziel war es, zunächst die Benutzerhandbücher
komplett mit allen Masken zu entwickeln, bevor mit dem DV-Konzept begonnen
werden sollte. Dafür wurden 35 Prozent des Projektaufwandes veranschlagt (Implementierung 25 Prozent und Test 15 Prozent). Die Entwicklung sollte linear
durchgeführt werden, d.h. Prototyping, oder Entwicklungszyklen waren explizit
nicht vorgesehen.

Bemerkenswert ist die Projekt–Aufbauorganisation, die neben einem Steuerungsgremium aus Vertretern des Managements und des Vorstands, der Gesamtprojekt-
leitung, technischer sowie fachlicher Projektleitung auch Teams für Qualitätssicherung, Systemeinführung, Projektmanagement vorsieht. Desweiteren gibt es zu
allen Anwendungsprojekten eigene Review-Boards sowie Fachberatungsteams, in
denen die Benutzervertreter aus Zentrale und Betrieben ihre Anforderungen einbringen sollen. Der Betriebsrat kommt in der offiziellen Projektorganisation nicht
vor— auch nicht in Fachberatungsteams.

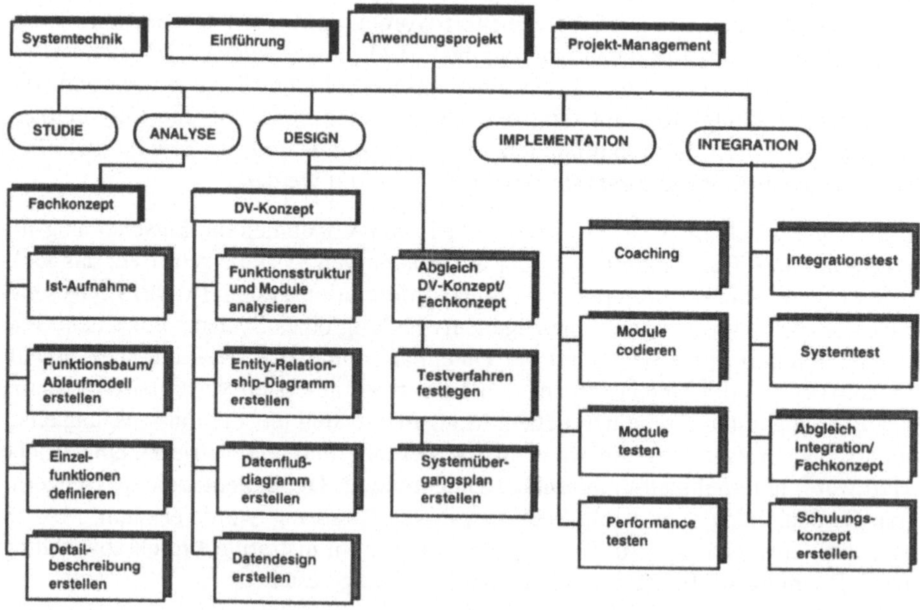

Abb. 1: Phasen/Arbeitspakete des MAPIS-Projekts

2.2 Akteure und Interessen:
die nicht explizit formulierte Betriebspolitik

Auch in diesem Unternehmen wird das überproportionale Wachstum der Zentrale von den *Betrieben* eher mißtrauisch verfolgt. Andererseits disponiert der Vorstand die Grundfinanzierung für die Betriebe, so daß die unterschwelligen Konflikte nicht offen zutage treten. Die Betriebe wollen auch zukünftig ihr Eigenleben behalten und sich nicht zu stark hineinregieren lassen. An aktuelleren Informationen zur Unterstützung ihrer Aufgaben besteht beim jeweiligen Betriebsmanagement generell großes Interesse, wobei es aufgrund der Profit-Center-Orientierung auch Konkurrenz zwischen Betrieben oder sogar zwischen Betriebsteilen gibt. Die Wissenschaftler wollen ihr spezifisches „Prozeßwissen" aus Projektbearbeitung und Projektakquisition nicht der Zentrale oder gar Konkurrenten offenbaren. Sie sehen in MAPIS eher ein (teures) System zur verstärkten Steuerung und Kontrolle.

Die einzelnen *Verwaltungen* der Betriebe bearbeiten Finanzierungsangaben, Kalkulationen, Bestell- und Buchungsvorgänge. Für diese Aufgaben sind die Altsysteme aufgrund der Systembrüche und der unzureichenden Benutzungsoberfläche nicht ausreichend aktuell. Die Kommunikation mit der Zentrale dauert zu lange. Dieses mag erklären, warum die Verwaltungen der einzelnen Betriebe vom Projekt des Vorstands aufgeschlossen gegenüber stehen.

Dem *Vorstand* und den Promotoren geht es dagegen um eine Vereinheitlichung der Organisationsstrukturen der Betriebe, um die Stärkung des zentralen Controllings und der strategischen Planung, während das Massengeschäft (Buchhaltung, Einkauf, Personalverwaltung) eher auf die Betriebe verlagert werden soll. Dies wiederum ruft Befürchtungen bei der Zentralverwaltung hervor, aus Sorge um erworbene Positionen und Befugnisse.

Diese verschiedenen Sichtweisen und Interessenlagen, die Betriebspolitik, wurde nie explizit formuliert. Ganz grob läßt sich in dem vielschichtigen Verhältnis von Zentralisierung zu Dezentralisierung eine Tendenz festmachen: Verlagerung von operativen Aufgaben in die Betriebe, während dispositive Aufgaben eher in die Zentrale kommen.

Die *Entwickler* arbeiten an einem anspruchsvollen technischen Projekt. Sie führen zusammen mit Anwendern aus den Verwaltungen die Systemanalyse durch, haben aber meist keine Erfahrung aus der betrieblichen Arbeit, die betriebspolitische Dimension ist ihnen unbekannt. Wer andere als technische Interessen artikuliert, muß ihnen eher als Störenfried und Verzögerer erscheinen.

Betriebsräte und der Gesamtbetriebsrat (GBR) — vertreten durch seinen EDV–Fachausschuß — sahen in MAPIS zunächst die Gefahr stärkerer Kontrolle der Beschäftigten. Doch in Kenntnis anderer Zentralisierungstendenzen erwartete der GBR mit MAPIS auch Machtverschiebungen zwischen Betrieben und der Zentrale. Das Interesse der Wissenschaftler an eigenbestimmten Bereichen machte sich auch der GBR zu eigen, schließlich ist er auch deren Repräsentant. Der Akteur GBR wollte sich wegen dem überproportional hohen Anteil der Wissenschaftler an der Belegschaft nicht auf einen rein defensiven reaktiven Part beschränken. In den Entwicklungsprozeß mischt er sich explizit und exponiert mit Stellungnahmen und Gestaltungsanforderungen ein.

3. Von den Mühen der betrieblichen Umsetzung ergonomischer Normen

Obwohl in allen offiziellen Dokumenten der Software-Ergonomie sehr hohe Bedeutung beigemessen wurde, zeigte sich im Verlauf des Projektes, daß andere Ziele und Nebenbedingungen höheres Gewicht hatten. Gerade der GBR bzw. sein Fachausschuß setzte sich beharrlich für ein qualitativ hochwertiges und flexibles Werkzeug ein, das auch die Belange der Betriebe berücksichtigen sollte. Die konstruktiven Vorschläge, die mit von der wissenschaftlichen Arbeit einiger Betriebsräte gespeist waren, stießen jedoch häufig wenn nicht auf offene Ablehnung so doch auf sehr reservierte Aufnahme. Die Projektverantwortlichen wiesen dem GBR die traditionelle Gegnerrolle zu, obwohl sie immer wieder Kooperationsbereitschaft einforderten. Andere Gremien des Unternehmens hatten sich bei weitem nicht so ausführlich mit MAPIS auseinandergesetzt. Da es also an einem gemeinsamen Grundverständnis über den Entwicklungsprozeß von MAPIS fehlte, wurde der GBR auf die Durchsetzung rechtlich abgesicherter Positionen zurückgeworfen.

3.1 Rolle der Betriebsräte

Der Gesamtbetriebsrat hat bereits nachdem er die ersten Studien gesichtet hatte, seine Mitbestimmungsrechte für das gesamte Projekt geltend gemacht: im wesentlichen aufgrund der Möglichkeit zur Leistungs- und Verhaltenskontrolle und aufgrund diverser bestehender Betriebsvereinbarungen. Der GBR zeigte sich der Chance aufgeschlossen, die tägliche Arbeit mit benutzungsfreundlichen Programmen und mit aktuelleren Informationen zu erleichtern. Allerdings wollte der GBR und sein zuständiger EDV–Ausschuß mit dem Vorstand zu einem Konsens über die Rahmenbedingungen kommen: minimale maschinelle Kontrolle einzelner Personen, Anerkennung der "Autonomie" von Arbeitsgruppen und Abteilungen, nicht alle "Basis"-Informationen müssen der Zentrale bekannt sein, Service für die Projektarbeit, Ergonomie, abwechslungsreiche Tätigkeiten, Verlagerung von Tätigkeiten unter Wahrung der Interessen der Beschäftigten. Den Entwicklungsprozeß sowie entsprechende Freigabeverfahren wollte der GBR in einer Betriebsvereinbarung geregelt wissen. Darin sollten auch die Modalitäten für eine tatsächliche

Partizipation aller Nutzergruppen niedergelegt werden. Das Teilnahmerecht von GBR-Vertretern an den MAPIS-Gremien sollte ebenfalls sichergestellt werden.

Zu einer verbindlichen Einbeziehung des GBR bzw. seines zuständigen Ausschusses in den Gremien konnte sich der Vorstand jedoch nicht durchringen. Der GBR kommt in der Aufbauorganisation des Projekts gar nicht erst vor (vgl. 2.1). Der Abstimmungsprozeß mit dem GBR war deshalb auch nicht in den Terminplan aufgenommen worden. Eine Betriebsvereinbarung, deren Notwendigkeit zwar unstrittig war, sollte erst abgeschlossen werden, nachdem die Spezifikationen der einzelnen Teilprojekte in den Anwendungshandbüchern (mit Maskenbeschreibungen) niedergeschrieben worden sind. Eine flexible Ausgestaltung des festgelegten Zeitplans für GBR-Abstimmungen wollte der Vorstand auch nicht in Kauf nehmen.

Die Projektverantwortlichen unterrichteten den EDV-Ausschuß des GBR laufend mit den Protokollen aller Gremien und mit Systemunterlagen. In regelmäßigen Abständen gab es Gespräche zwischen dem EDV-Ausschuß und den Projektverantwortlichen. Hauptanliegen der Betriebsräte war zunächst, die Voraussetzungen zu schaffen, damit MAPIS überhaupt regelbar werden konnte: ein Zugriffsschutzkonzept, das einen differenzierten Zugriffsschutz gestattet. Von Anfang an legten sie aber auch Wert auf die Qualität des zu entwickelten Systems.

3.2 Projektplanung und terminliche Restriktionen

Im Rahmen der sehr detaillierten Terminplanung (u.a. Verfolgung der Aktivitäten bis auf Stundenebene) wurde bereits frühzeitig der Einführungszeitpunkt von MAPIS auf Anfang 1995 festgelegt. Vorher sollte ausreichend Zeit für eine Piloterprobung jedes Teilprojektes sein. Das Steuerungsgremium wurde laufend über die plangenaue Entwicklung der Meilensteine unterrichtet. Die ersten unerwarteten Verzögerungen sollten noch durch Produktivitätssteigerungen wettgemacht werden können, die die hinteren Phasen als abkürzbar erscheinen ließen. Allmählich sind die Phasen Test, Integrationstest und Piloterprobung zu Puffern geraten, die immer mehr in Anspruch genommen worden sind, da am Einführungstermin nicht gerüttelt werden durfte.

Mit den (auch technisch bedingten) Verzögerungen (s.a. 3.7) waren die Entwickler immer höherem Druck ausgesetzt. Der GBR mußte mit seinen Anliegen ihnen geradezu als Störenfried erscheinen, der nur noch zusätzliche Schwierigkeiten bereitet und Terminverschiebungen verursacht.

Schon in den Anfängen von MAPIS wurde von den Projektverantwortlichen der Qualitätssicherung (QS) eine hohe Bedeutung zugewiesen, für die zunächst ein entsprechend hoher Ressourcenaufwand veranschlagt wurde. Allmählich mußte dieser Aufwand reduziert werden, um nicht die Gesamtkosten des Projekts überproportional steigen zu lassen. Als zudem noch die Funktion Qualitätssicherung durch höhere Fluktuation auffiel, wurde die Not zur Tugend erhoben und die QS-Aufgaben verschiedenen Personen(gruppen) zugeteilt: fachlich den Fachberatungsteams, technisch der (technischen) Projektleitung und den Entwicklern selbst. Durch diese "integrierte Qualitätssicherung" erwartete sich die Projektleitung zudem eine Verkürzung der Testphasen.

Mitte 1994 wurde der Projektleitung immer deutlicher, daß nach jahrelanger Entwicklung den Benutzern eine alphanumerische Benutzungsoberfläche nicht mehr angeboten werden konnte. Immer lauter wurden die Forderungen nach einer graphischen Benutzungsoberfläche, wie sie den Benutzern von PCs her vertraut ist. Das neue Entwicklungswerkzeug, endlich verfügbar, erforderte einen wesentlich größeren Umstellungsaufwand als ursprünglich angenommen (s. 3.7). Zudem zeigten sich "überplanmäßige Rückschleifen" von den verschiedenen Stellen der QS zu den Entwicklern, so daß sich auch deshalb der Beginn der Pilotphase weiter verzögerte. Auch für die Entwicklung von MAPIS trifft der Befund von Weltz/Ortmann (1992) zu: "Viele Regelungen der Qualitätssicherung, die auf dem Papier überzeugend und einleuchtend aussahen, waren allein schon aus Zeitgründen unmöglich einzuhalten ... Die Qualitätsbeauftragten hatten durchweg zu wenig Zeit, um ihrer Aufgabe nur halbwegs gerecht zu werden. Den Softwareentwicklern fehlte die Zeit, die Vorschläge und Korrekturhinweise der Qualitätssicherer umzusetzen." (S. 99/100).

3.3 Maskengestaltung und Styleguide

Für umfangreiche Projekte ist es Aufgabe der Projektleitung, frühzeitig konkrete Programmier- und Dokumentationsrichtlinien für das – oder besser noch mit dem – Entwicklungs-Team zu erarbeiten. Auch existieren bereits firmenspezifische Entwicklungsrichtlinien in Softwarehäusern und DV-Abteilungen, die sich konkret auf die Benutzungsoberfläche beziehen, sogenannte (Firmen-)Styleguides (Ilg 1993).

Auch in MAPIS wurde ein sehr umfangreicher Styleguide entwickelt, der bis hin zu Funktionstasten und Druckgestaltung entsprechende Regelungen formuliert. Damit einher ging der Entwurf der Bildschirmmasken für das gesamte System.

Problematisch am Styleguide waren die von gängigen Normen und Standards (DIN, CUA, OSF/Motif etc.) mitunter stark abweichenden Regelungen — ein Umstand der sich auch bei anderen Firmen-Style-Guides findet. Der Style Guide wurde mehr oder weniger „von oben verordnet", eine Diskussion mit den Beteiligten oder mit Ergonomie-Experten fand nicht statt. Zudem trat ein weiteres bekanntes Problem auf: je detaillierter bestimmte Regelungen getroffen werden, desto eher werden die Entwickler von diesen abweichen.

Als für alle Beteiligten sehr vorteilhaft erwies sich die Vorgehensweise, frühzeitig konkrete Maskenentwürfe vorzulegen. Somit konnten sich Benutzer und Betriebsräte vor der eigentlichen Entwicklung ein umfangreiches Bild vom zukünftigen System machen, da auch bereits vollständige Handbücher mit Masken- und Feldbeschreibungen vorlagen. Allerdings geht bei soviel Detailinformation (ca. 25 Handbücher) oft der Überblick fürs Ganze verloren oder der Blick wird zu früh auf unwesentliches gelenkt. Arbeitsabläufe werden nur noch anhand von Masken diskutiert. Auch wird sofort ein weiterer Nachteil klar: jede noch so geringfügige Änderung an Layout oder Dokumentation im Handbuch kann umfangreiche Anpassungsarbeiten nach sich ziehen und ist somit unerwünscht. Kurzum: wie die MAPIS-Masken einmal festgelegt waren, sollten sie möglichst auch bleiben. Den Fachberatungsteams wurden die fertigen Handbücher zwar vorgelegt, aber Änderungen und neue Vorschläge wurden nicht motiviert. Alternativen wurden nicht aufgezeigt und entsprechende Vorschläge waren von den Benutzern oder dem Betriebsrat eher unerwünscht. Vorschläge zur Software–Ergonomie wurden nur dann aufgegriffen, wenn sie mit wenig Aufwand zu realisieren waren. Neu- und Umgestaltungen kamen nicht (mehr) in Frage, software-ergonomische Beratung wurde somit auf „Bildschirmkosmetik" reduziert.

3.4 Aufgabenanalyse

Ebensowenig wie an der Masken- und Style-Guide-Entwicklung war der Betriebsrat auch nicht an der Ist-Aufnahme beteiligt. Eine Dokumentation, die beschreibt, wie sich die Masken in den Arbeitsablauf einbetten sollen, gibt es nicht. Daher mußte der Betriebsrat sich bei Fragen zu Aufgaben- und Benutzerangemessenheit auf die Aussagen der Entwickler verlassen, die oft behaupteten, „die Benutzer wollen das genau so und nicht anders", z.B. wenn der Betriebsrat auf die offenkundig überladenen Maskenentwürfe hinwies.

Belegen konnten die Entwickler ihre Aussagen nicht: es gibt keinerlei Dokumentation über typische Abläufe und Alternativen von Arbeitsaufgaben, die Notwendigkeit von Maskenanordnungen ist nicht belegbar, ja es fehlen sogar Mengengerüste, wie Häufigkeiten von (Teil-)Aufgaben, Abarbeitungsreihenfolgen von Daten etc. So stand es quasi „Aussage gegen Aussage", wenn der GBR Vorschläge machte, und die Entwickler diese mit Hinweis auf die angeblichen Benutzerwünsche zurückwiesen. Auf die zuvor beteiligten Benutzer konnte nach abgeschlossenem Fachkonzept nicht (oder kaum) mehr zurückgegriffen werden, um möglicherweise neue Gegebenheiten oder Vorschläge zu diskutieren. Zudem standen viele Vorschläge der Benutzer in software-ergonomischem Widerspruch (wie z.B. umfassende Darstellung vs. übersichtliche Bildschirmgestaltung), sicher auch ein von vielen anderen (Beteiligungs-)Projekten bekanntes Problem.

3.5 Kooperation mit Benutzern

Die vorherigen Ausführungen deuten bereits darauf hin: die Beteiligung der Benutzer kann nur als unzureichend beurteilt werden, was in eklatantem Widerspruch zur Devise der Unternehmensleitung steht, die Beteiligung der Benutzer als kritischen Erfolgsfaktor charakterisierte.

Zwar gibt es die zuvor bereits erwähnten Fachberatungsteams, in die ca. 80 Mitarbeiter mehr oder weniger stark eingebunden waren, was aber gerade mal gut ein Prozent der Belegschaft ausmacht. Beteiligt waren zudem nicht alle Benutzergruppen, Wissenschaftler waren z.B. so gut wie ausgeklammert. Nach Fertigstellung des Fachkonzepts sind in den Gremien gerade noch zwei ständige Benutzervertreter.

Da die Mitarbeit am Projekt nicht besonders angeregt wurde und anfallende Kosten für Dienstreisen und Arbeitszeitausfall von den beteiligten Betrieben getragen werden mußten, war das Engagement sehr unterschiedlich. Dabei blieben nur die, die sich etwas von dem neuen System versprachen, und sich dann konsequent für ihre — oftmals sehr spezifischen Belange — einsetzen konnten. Der Kontakt „zur Basis", und zu eher repräsentativen Anforderungsprofilen ging verloren — wenn er überhaupt je vorhanden war.

Viele der schon früher berichteten Defizite (Beck 1993) bei der Organisation und Durchführung von Benutzerpartizipation fanden sich auch bei MAPIS, wie z.B. unrealistische zeitliche Regelungen, wenig Kooperation und Kommunikation der Benutzer untereinander, fehlende Abstimmungsprozesse unter Berücksichtigung aller Akteure. Es gab z.B. auch in MAPIS keine Evaluation und/oder Benutzertests von Entwürfen.

Die Entscheidungen der Fachberatungsteams wurden als Legitimation für die (schlechte) Maskengestaltung genommen. Entsprechende Qualifizierung, insbesondere zu Software–Ergonomie, Arbeitsgestaltung oder gar zu Richtlinien und Normen hatten die Teilnehmer der Fachberatungsteams jedoch nicht erhalten.

3.6 Modernes System — schwieriger Datenschutz

Wichtiges Anliegen des GBR war, die personenbezogenen und personenbeziehbaren Daten zu schützen (BDSG Anlage zu § 10) und die Möglichkeiten zur automatischen Kontrolle einzuschränken (§87(1)6 BetrVG). Die gewählte Systemarchitektur und die Tatsache, daß die Software identisch in der Zentrale und allen Betrieben vorliegen soll, ließ den Schutz auf Feldebene als nicht handhabbar erscheinen. Stattdessen wurden die Beschränkungen in die Oberfläche verlagert. Für jede Funktion (=Maske) muß eine eigene Berechtigung erteilt werden. Zusätzlich können für die gleiche Funktion auch Projektdaten der einen Abteilung gegenüber dem Zugriff einer anderen Abteilung geschützt werden. Höhere Stellen in der Hierarchie (Betriebsleitung, Vorstand) haben jedoch weiter auf diese Daten Zugriff. Schon dieser Ansatz dürfte zu einem hohen Administrationsaufwand für den Zugriffsschutz führen. Der vom GBR darüber hinaus für manche Funktionsbereiche verlangte „antihierarchische" Schutz, der einen Zugriff "von oben" unterbindet, wurde nur in sehr wenigen Ausnahmen verwirklicht. Derartige Schutzfunktionen müssen aber im Programmcode realisiert werden und sind somit für Überprüfungen durch den GBR/BR nicht unbedingt transparent.

Die Erfahrung der Betriebsräte als den faktischen Datenschutzbeauftragten war, daß Datenschutz mühsam nicht nur technisch sondern auch „in den Köpfen" der Entwickler implementiert werden mußte. Häufig gehörte Ansichten waren: "Wozu denn löschen, wir haben genug Speicherplatz", "Es ist doch klar, daß die gespeicherten Daten zu einer Kontrolle überhaupt nicht taugen. Wer hätte denn daran Interesse". Der Erfahrungshintergrund der Betriebsräte ist den Entwicklern fremd.

3.7 Veränderung der Entwicklungsumgebung

Wie bei großen Software-Entwicklungsprojekten fast schon üblich, änderten sich auch in diesem Projekt Rahmenbedingungen, die zu Beginn nicht kalkuliert wurden bzw. konkret nicht kalkulierbar waren. Hier ist vor allem die Umstellung des Entwicklungswerkzeugs (Oracle Forms 3.0) auf eine neue Version (Oracle Forms 4.0) hervorzuheben. Den Projektverlauf beeinflußten auch die — über die lange Projektlaufzeit durchaus übliche — Mitarbeiterfluktuation verbunden mit notwendigem Einarbeitungsaufwand für neue Entwickler.

Während mit Oracle Forms 3.0 großrechner-orientierte alphanumerische Bildschirme entworfen werden konnten, bietet Forms 4.0 die Möglichkeit, moderne, graphische Benutzungsoberflächen zu entwickeln. Somit könnten mit diesem neuen Entwicklungswerkzeug eine Reihe von Forderungen des Betriebsrats bezüglich der Benutzungsoberflächengestaltung realisiert werden. Doch zunächst mußten sich die Entwickler mit ganz anderen Fragestellungen beschäftigen. Von Problemen, die bei einem Versionswechsel auftreten, kann eigentlich jeder Software-Entwickler ein Lied singen: die neue Version läuft instabil, bewährte Funktionen gibt es nicht mehr, Portierungswerkzeuge gibt es nicht oder funktionieren nur teilweise, es wird jede Menge Neuprogrammierung notwendig, Funktionen laufen nicht wie im Handbuch beschrieben, vieles muß erst wieder mühsam ausgetestet werden und und und — die Liste ließe sich beliebig fortsetzen. Wie sich dies auf die Motivation der Entwickler auswirkt, kann man sich leicht vorstellen, insbesondere dann, wenn außerdem noch starker Termindruck dazukommt.

Die Chance zur Verwirklichung von software-ergonomischen Forderungen mit einem modernen Oberflächenwerkzeug wurde bisher noch zu wenig genutzt. Die

Entwickler haben die geschilderten Probleme mit der Realisierung der geplanten Masken, was dazu führt, daß sie auf zusätzlichen Komfort (z.B. Nutzung von Klapplisten statt Pop-Ups) verzichten, weil sie weitere Probleme und Zeitverzug befürchtet werden.

Ein kurzes Fazit zu den Mühen der betrieblichen Umsetzung: es gibt hoch gesteckte Ziele bei allen Beteiligten, nicht nur beim Betriebsrat, aber Schwierigkeiten in der Planung und Ausführung lassen viele dieser Ziele als (noch) nicht verwirklichbar erscheinen.

4. Gestaltungsversuche der Betriebsräte unter widrigen Bedingungen

Zu einer verbindlichen Einbeziehung des GBR bzw. seines zuständigen Ausschusses in den Gremien konnte sich der Vorstand nicht durchringen. Ein Konsens über die Rahmenbedingungen der MAPIS-Entwicklung ist nicht zustande gekommen (vgl. 3.1). Der EDV–Ausschuß des GBR wurde jedoch ausführlich mit Informationen versorgt. Er wurde in den Verteiler aller Protokolle aufgenommen. Von ihm wurde — wie von anderen auch — erwartet, daß er zu den erstellten Systemspezifikationen Stellung mit ziemlich knapper Terminsetzung bezieht. Der Ausschuß machte jedoch die Erfahrung, daß seine Bemühungen um eine frühzeitige Verständigung zu Kernpositionen häufig übergangen wurden. In regelmäßigen Abständen gab es Gespräche zwischen dem EDV-Ausschuß und den Projektverantwortlichen. Statt einer produktiven Auseinandersetzung wurde dem GBR bedeutet, das seien Themen für die Einigungsstelle. Die Bemühungen um Zugriffsschutz oder Software-Ergonomie führten zu viel Papier, der Umfang der Umsetzung zeigte sich erst sehr viel später bei der Implementierung.

Positiv für die Argumentation des GBR wirkte sich aus, daß der Vorstand vom Nutzen interner Sachverständiger (in moderatem Umfang) überzeugt werden konnte. Damit wurde dem GBR eine kompetente Beschäftigung mit dem Datenmodell und der Software-Ergonomie erleichtert. Er konnte zusammen mit den Sachverständigen begründete Alternativen skizzieren.

Nachdem der EDV–Ausschuß des GBR erfahren mußte, welch geringe Bedeutung seine Vorschläge genossen, hat sich der GBR zur Herausgabe eines Informationsblattes entschlossen. Damit sollte zu einer unternehmensweiten Debatte um die Ziele und die vermuteten Veränderungen der Betriebspolitik von MAPIS in Gang kommen. Es wurde über das Projekt MAPIS berichtet (von dem viele Beschäftigte bis dahin noch gar nichts wußten) und die Kritikpunkte des GBR wurden erläutert.

Insgesamt hat sich für den GBR ergeben, zu "klassischen" Verhaltensweisen zurückgreifen zu müssen: Der zentrale Ansatz Leistungs- oder Verhaltenskontrolle wurde ausgebaut um dann in Verhandlungen auch zu Kompromissen über Einschluß der Arbeitsbedingungen zu kommen. Mit der Ablehnung oder der Verzögerung zumindest von Teilen von MAPIS mußte gedroht werden. Die gesicherten arbeitswissenschaftlichen Erkenntnisse mußten präzisiert werden. Die Projektverantwortlichen gaben einer gütlichen Einigung mit dem GBR keine Chance mehr. Nach Marathon-Verhandlungsrunden provozierten sie beim GBR mit Maximalpositionen die Ablehnung, so daß die Betriebsvereinbarung für die allgemeinen Regelungen und die einzelnen Teilprojekte einer Einigungsstelle vorbehalten waren.

Eine Einigungsstelle wird paritätisch von Arbeitgeber und Betriebsrat besetzt. Sie hat einen Vorsitzenden, auf den sich beide einigen oder der von Arbeitsgericht eingesetzt wird. Betriebsräte nehmen i.d.R. einen Rechtsanwalt auf ihre Seite und häufig einen (oder mehrere) Sachverständigen. Der Vorsitzende ist gehalten, einen Ausgleich der unterschiedlichen Interessen herbeizuführen. Wo das nicht gelingt, muß er sich letztendlich der Auffassung einer Seite anschließen. Dabei darf er die Grenzen des Ermessensspielraums nicht überschreiten, weil er sonst die Anfechtung des Beschlusses der unterlegenen Partei vor dem Arbeitsgericht riskiert. Diese Ausführungen in Verbindung mit der einleitenden Darstellung der rechtlichen Situation sollen erläutern, weshalb die Umsetzung von konkreten Vorgaben zur Ergonomie jeweils genau geprüft wird. Einfach die EU-Richtlinie als für das Unternehmen und das strittige System für gültig zu erklären, ist bei gegensätzlichen Auffassungen der beiden Parteien im Einigungsstellenverfahren nicht zu erwarten. Nebenbei wäre für einen BR auch nicht alles erledigt, wenn die Gültigkeit der Richtlinie postuliert würde: er müßte dann u.U. nachweisen können, daß die Software nicht den arbeitswissenschaftlichen Erkenntnissen entspricht.

Derzeit, solange die nationale Umsetzung noch aussteht, kann der BR sich rechtlich abgesichert auf § 87(1)7 BetrVG stützen: Danach hat er über die Anwendung der Richtlinie mitzubestimmen. Gesicherte arbeitswissenschaftliche Erkenntnisse sind die bereits verabschiedeten DIN-EN-29241 Normen (Regelungen u.a. zu Zeichengröße, Aufgabenangemessenheit, Individualisierbarkeit...). Die Arbeitsplatzanalyse kann auch pragmatisch angegangen werden: anhand von Checklisten, die schon im Eigeninteresse des BR noch mit vertretbarem Aufwand — gute Schulung vorausgesetzt — zu bewältigen sind. Insgesamt hat ein BR bei guter Vorbereitung in einer Einigungsstelle nichts zu verlieren.

5.   Forderungen und konkrete Schritte

Trotz vieler wissenschaftlicher Erkenntnisse der Software-Ergonomie und die Notwendigkeit der breiten Nutzerbeteiligung werden diese in Entwicklungsprojekten noch zu wenig umgesetzt. Dies mag um so mehr erstaunen, weil allenthalben die großen Chancen der neuen Managementkonzepte für den „menschlichen Faktor" herausgestellt werden. Immer noch mangelt es an einer „Kultur", die eine transparente Auseinandersetzung um Ziele und Alternativen zuläßt. (Große) Softwaresysteme müssen in definierter Zeit zum Einsatz kommen, wurde doch mit einem „genauen" Termin- und Kostenplan bei der Bewilligung operiert. Am Fallbeispiel wurde gezeigt, wie diesem Diktat inhaltliche Positionen geopfert werden.

Weltz/Ortmann (1992) fordern in ihrer Untersuchung über das Projektmanagement in Softwareprojekten die „Entmythologisierung der Softwareentwicklung als rein ´ingenieurmäßiger´ Entwicklungsprozeß, durch die ihr Doppelcharakter als Prozeß technischer und ´politischer´ Auseinandersetzung deutlich gemacht wird". Notwendig erscheint ihnen „schließlich die Verfügbarkeit organisatorischer Modelle, die eine tragfähige Grundlage für die Verarbeitung von Interessengegensätzen und Konfliktpotentialen liefern, die die Thematisierung der betriebspolitischen Diskussion erleichtern. Dies gilt nicht zuletzt auch für Ansätze zur Aktivierung der Nutzerpartizipation. Schwierigkeiten bei deren Durchsetzung sind nicht zuletzt auf eine mangelnde Berücksichtigung der ´politischen´ Dimension zurückzuführen." (Weltz, Ortmann, S.131). In diesem Sinne beschließen wir den Beitrag mit Forderungen an die unterschiedlichen betriebliche Akteure.

5.1 Management

Das Management sollte kontinuierliche Lernprozesse zwischen Entwicklern und Anwendern etablieren. Dafür sind beide Gruppen zu qualifizieren, den Anwendern müssen alternative Gestaltungsmöglichkeiten der Software vorgestellt werden, damit ihre Beteiligung nicht nur rein legitimatorischen Zwecken für bereits festgelegte Benutzungsoberflächen dient. Eine frühzeitiger Konsens über die Projektziele erleichtert den Projektfortschritt. Was aus überzogenen Ansprüchen geworden ist, die letztlich auf die umfassende Kontrolle dezentral operierenden Organisationseinheiten und Personen gerichtet ist, belegen die Erfahrungen aus zahlreichen gescheiterten EDV-Großprojekten. Ein zu datengläubiges Management ist vor Datenfriedhöfen nicht gefeit. Eine kontinuierliche Konsensfindung muß Teil des Projektablaufs werden.

Damit die Software-Ergonomie nicht dem Termindiktat geopfert wird, ist auf eine realistische Planung Wert zu legen. Starre Modelle des Projektmanagements erweisen sich indessen als immer weniger wirkungsvoll. Diese sich ausbreitende wissenschaftliche Erkenntnis kollidiert in der Praxis aber häufig noch mit den Interessen eines konservativen Managements. Alternativlösungen sollten möglichst lange offen gehalten werden. Eine realistische Zeitschätzung müßte diese ebenso berücksichtigen wie auch gerade bei Großprojekten die Fluktuation erfahrenen Personals und Umstellungsschwierigkeiten beim Wechsel von Hardware oder Software. Eine

formalisierte und rigide Projektplanung erweist sich besonders dann als zweifelhaft, wenn innerbetriebliche Konflikte verdrängt werden, die etwa zwischen Be-
wahrern bisheriger Strukturen und Neuerern, aufgrund nicht eingestandener poli-
tischer Budgetierung bzw. Terminfestlegung, aufgrund unzureichender Kompe-
tenzabgrenzung zwischen Aufbau- und Ablauforganisation oder aufgrund von
Konkurrenzverhältnissen auftreten können. Eine verbesserte, flexible Planung kann
diese Art von Problemen zwar nicht aus der Welt schaffen, doch lassen sich die
resultierenden Konsequenzen wie z.B. „doppelte Buchführung" und undurch-
sichtige Kompensationsgeschäfte zwischen verschiedenen Managementfraktionen
vermeiden.

5.2 Entwickler

„Zwischen Identifikation und Überforderung" ordnen Weltz/Ortmann (1992) die
Situation der Entwickler ein. In diesem Spannungsfeld übersehen sie leicht die
anderen Interessen der unterschiedlichen betrieblichen Akteure, die den Entwicklungsprozeß auch beeinflussen. Die geforderte Auseinandersetzung übersteigt indessen meist die rein technikbezogene Ausbildung zum Softwarespezialisten.
Gleichwohl sollten sich die Software-Entwickler ein Verständnis für die „betriebspolitische" Dimension des Projektes aneignen, um vor unliebsamen Überraschun-
gen besser gefeit zu sein.

Darüber hinaus sollten SW–Entwickler über ein solides Wissen über Ergonomie,
Arbeitsgestaltung und Datenschutz verfügen. Firmen- oder projektspezifische
Styleguides und Handbücher sollten mit allen Beteiligten frühzeitig abgestimmt
und von Experten überprüft werden. Dabei sind aufgabenbezogene Darstellungsformen für die Benutzerkooperation zu verwenden, nicht nur Maskendarstellungen.

SW–Entwicklern muß nähergebracht werden, was wirkliche Kundenorientierung
auch für sie bedeutet. In diesem Projekt dagegen haben wir einmal mehr erfahren,
daß SW-Entwickler immer noch eher die nächste Deadline oder einen erfolgreichen Modultest als den fernen Benutzer bei ihrem Tun vor Augen haben.

5.3 Benutzer

Alle Benutzergruppen müssen in ausreichender Zahl, über mehrere Abstimmungs-
runden und projektbegleitend beteiligt werden. Organisatorische Rahmenbedingungen mit dem Ziel echter Beteiligung müssen zuvor abgeklärt werden. Wichtig
dabei: Qualifizierung für Partizipation muß Ergonomie, Hinweise über Spielräume
der Systemgestaltung und das organisatorische Umfeld mit einbeziehen. Kritische
Mitarbeit mit Wille zur Entwicklung von Alternativen muß angeregt und gefördert
werden. Benutzer müssen begreifen und die Chance wahrnehmen, daß es um die
Ausgestaltung ihrer Arbeitsbedingungen geht.

5.4 Betriebsrat

Betriebsräte müssen sich frühzeitig mit Systementwicklungen auseinandersetzen, schon bei Analyse und Design, um in den Prozeß Alternativen einbringen zu können. Sie müssen versuchen, die Rahmenbedingungen für die Partizipation der Nutzer so zu beeinflussen, daß sich unterschiedliche Interessen artikulieren können. Benutzerpartizipation im Softwaregestaltungs- und -auswahlprozeß sollte in jedem Falle vom Betriebsrat angestrebt und aktiv mitbegleitet werden. Andererseits müssen Betriebsräte ihre eigene Kompetenz erhöhen, um im Entwicklungsprozeß als hartnäckiger, aber kompetenter Akteur mitwirken zu können. Der § 80(3) BetrVG, der die Hinzuziehung von Sachverständigen ermöglicht, sollte noch stärker in Anspruch genommen werden. Diese können bei der Information, Qualifizierung, Durchführung von Gestaltungsmaßnahmen und vor allem bei der Erstellung von Gutachten unterstützen. Diese Maßnahmen erleichtern auch die Kommunikation mit den Entwicklern.

Wichtig ist auch, die eigene Position in der Betriebsöffentlichkeit darzulegen — auch wenn die Betriebsräte unterschiedliche Positionen zu koordinieren haben (bessere Unterstützung für die einen kann stärkere Kontrolle der anderen bedeuten; Aufwertung von Tätigkeiten ist Wegfall von einfachen....).

Das Recht auf Mitbestimmung bei der Durchführung von und die Teilnahme an Bildungsveranstaltungen (§98 BetrVG) hilft dem Betriebsrat, an der Bildungskonzeption mitzubestimmen und aktiv bei der Auswahl einer geeigneten Qualifizierung mitwirken zu können, die den Beschäftigten Umgang mit der Technik, Überblick und Aufstiegsmöglichkeiten vermittelt.

Dabei darf der "klassische" Ansatz, d.h. die technische Leistungs- und Verhaltenskontrolle zu beschränken (§87(1)6 BetrVG) bei der Auswahl von Standardsoftware (aber natürlich auch bei der Neugestaltung) nicht vernachlässigt werden. Gerade hier sind die entscheidenden Ansatzpunkte für einen BR gegeben.

Wichtig für die Handlungsfähigkeit eines BR ist, eine angemessenes Verhältnis zu dem System zu bekommen, d.h. nicht am Anfang zu starke Ängste oder Vorbehalte zu hegen und während des (langen) Realisierungszeitraums allmählich das Interesse zu verlieren und der Einschätzung aufzusitzen, es komme ja sowieso nichts heraus. Zumindest eine Arbeitsgruppe des BR sollte den Systementwicklungsprozeß beharrlich verfolgen.

5.5 Forschung

Die hier skizzierten Forderungen schließen mit der Forderung an die Software-Ergonomieforschung ab: es müssen noch mehr Techniken entwickelt werden, die im Betrieb unmittelbar eingesetzt werden können. Beispielsweise mangelt es immer noch ein praktikablen Kommunikationsformen für SW–Entwickler und Benutzer. Analyse und Evaluation von Aufgaben- und Benutzeraspekten könnten durch ein handhabbares Methodenrepertoire feste Größe für Projektplanung und -durchführung werden.

Literatur

90/270/EWG (1990): Richtlinie des Rates vom 29. Mai 1990 über die Mindestvorschriften bezüglich der Sicherheit und des Gesundheitsschutzes bei der Arbeit an Bildschirmgeräten (Fünfte Einzelrichtline im Sinne von Artikel 16 Absatz 1 der (Rahmen-)Richtlinie 89/391/EWG)

Beck, A. (1993): Benutzerpartizipation aus Sicht von SW–Entwicklern und Benutzern. Eine Untersuchung von beteiligungsorientierten SW–Entwicklungsprojekten, In: Rödiger, K.–H. (Hrsg., 1993): Software–Ergonomie '93. Von der Benutzungsoberfläche zur Arbeitsgestaltung, Stuttgart: Teubner, 263–274

Becker-Töpfer, E. (1991): Beteiligungsorientierte Formen der Arbeitsgestaltung aus Sicht der betrieblichen und gewerkschaftlichen Interessenvertretung. In: Brödner, P; Simonis, G.; Paul, H. (Hrsg.): Arbeitsgestaltung und partizipative Systementwicklung, Opladen: Leske + Budrich, 147–155

DGB Technologieberatung (1993): Bildschirmarbeit human gestalten, 4. Aufl.

DIN 66 234 Teil 8 (1988): Bildschirmarbeitsplätze: Grundsätze ergonomischer Dialoggestaltung, Beuth-Verlag

DIN EN 29 241, Teile 1-3 (ISO 9241-1/2/3) (1993): Ergonomische Anforderungen für Büro-tätigkeiten mit Bildschirmgeräten, Teil 1: Allgemeine Einführung, Teil 2: Anforderungen an die Arbeitsaufgaben — Leitsätze, Teil 3: Anforderungen an visuelle Anzeigen, Beuth-Verlag

Döbele-Martin, C. ; Martin, P. (1993): Ergonomie–Prüfer. Handlungshilfe zur ergonomischen Arbeits- und Technikgestaltung, Oberhausen

Ilg, R. (1993): Styleguides. In: Ziegler, J.; Ilg, R. (Hrsg.): Benutzergerechte Software-Gestaltung - Standards, Methoden und Werkzeuge. München, Wien: Oldenbourg, S. 25-38

Klotz, U. (1994): Werkzeug statt Maschine — Neue Leitbilder für den Computereinsatz, Frankfurter Zeitung 19.4.1994

Krafft, H.G. (1993): Zukünftige Regelungen für die Arbeit an Bildschirmgeräten. In: Moderne Unfallverhütung, 37, 26-28

VBG 104 (1992): Unfallverhütungsvorschrift "Arbeit an Bildschirmgeräten" der Verwaltungs–Berufsgenossenschaft, Teilvorentwurf November 1992

Wanke, H.-R.; Nienstedt, Ch.; Groß, W. (1993): Die EG–Richtlinie zur Bildschirmarbeit, Schriftenreihe des Technologie-Beratungs-Systems Bremen

Weltz, F.; Ortmann, G. (1992): Das Softwareprojekt - Projektmanagement in der Praxis. Frankfurt/New York: Campus Verlag

ZH 1/535 (1976): Sicherheitsregeln für Büro–Arbeitsplätze, Verwaltungs–Berufsgenossenschaft

ZH 1/618 (1980): Sicherheitsregeln für Bildschirm–Arbeitsplätze im Bürobereich, Verwaltungs–Berufsgenossenschaft

Direkte Manipulation von akustischen Objekten durch blinde Rechnerbenutzer

Ludger Bölke und Peter Gorny
Carl von Ossietzky Universität Oldenburg

Zusammenfassung:

Es gibt verschiedene Systeme, die es sehgeschädigten Benutzern gestatten, mit akustischen Repräsentationen von grafischen Benutzungsschnittstellen zu arbeiten (Graphical User Interface - GUI). Diese Anpassungen von GUIs an die Bedürfnisse von Blinden bieten jedoch kein hinreichendes Feedback, um ihnen eine effiziente Benutzung der Maus zur Cursor-Steuerung zu ermöglichen.

In diesem Beitrag werden die Möglichkeiten zur direkten Manipulation von akustischen Objekten, den Hearcons, beschrieben. Dazu wird ein für die Steuerung der Mausbewegungen unabdingbares akustisches Feedback erzeugt, das die Auswahl und Bearbeitung hörbarer Objekte erlaubt. Die Grundlage für diese Ohr-Hand-Koordination ist eine Kombination von permanent klingenden aktiven Hearcons und einem akustischen Feedback, das die relative Position des Maus-Cursors zu einem Zielobjekt liefert. Durch die Ausnutzung des räumlichen Hörvermögens können die Benutzer auf der technischen Basis der Kunstkopf-Stereophonie die Hearcons nach eigenem Bedarf im Hörraum positionieren.

Über den geschilderten adaptiven Ansatz hinausgehend wird vorgeschlagen, für Sehbehinderte spezielle Benutzungsschnittstellen auf der Basis der skizzierten Technik zu entwickeln. Das Konzept für das "assistive user interface" SPUI-B (StereoPhonic User Interface for Blind) wird vorgestellt.

Schlüsselworte: Akustisches Feedback, Hearcon, Direkte Manipulation, Blinde Benutzer, Ohr-Hand-Koordination, Akustische Benutzungsschnittstelle

1. Einleitung

Blinde Menschen müssen ihre Sehschädigung durch taktile und akustische Wahrnehmung kompensieren. Um ihnen den Zugang zum Rechner zu ermöglichen, sind die Sprachausgabe (vor allem in den USA) und die Braille-Ausgabe (vor allem in Deutschland) entwickelt worden. Diese adaptiven Techniken basieren auf dem sogenannten Screenreader, der die Ausgaben der Anwendungsprogramme direkt dem Bildschirmspeicher entnehmen. In den textbasierten Systemen können die auf dem Bildschirm ausgegebenen Zeichen im ASCII-Code in einem Textpuffer abgelegt und anschließend direkt in Braille-Zeichen oder Sprache transformiert werden. In grafischen Systemen wird die Bildschirmausgabe dagegen als Bitmap gespeichert. Ein Bit enthält nun die Information, ob ein Pixel auf dem Bildschirm gesetzt wird oder nicht. Das aus dieser Bitmap entstehende Pixelmuster wird dann vom Betrachter als Text oder Grafik wahrgenommen, der nicht direkt in taktile oder akustische Form umgewandelt werden kann.

Die grafischen Systeme bilden die technische Basis für grafische Benutzungsoberflächen (GUIs), die eine intuitive Benutzung ermöglichen sollen. Objekte und

Funktionen werden in textueller oder bildlicher Form direkt auf dem Bildschirm dargestellt; sie werden referenziert, indem sie mit einem Mausklick selektiert oder mit der Maus bewegt werden (drag and drop). Um dem Pixelmuster die textuelle Information zu entnehmen, sind spezielle zeit- und rechenintensive Mustererkennungsprogramme nötig, die jedoch (noch) nicht beliebige Zeichensätze erkennen können. Dies war lange Zeit mit ein Grund dafür, daß Blinde von der Benutzung grafischer Oberflächen ausgeschlossen waren.

Eine Lösung dieses Problems wird durch das "off-screen-model" gegeben [16]. Es fängt alle Funktionen ab, die sich auf die Bildschirmausgabe beziehen, und speichert die Informationen in ASCII-Format in einer Datenbank. Um den blinden Benutzern die Arbeit mit einer grafischen Oberfläche zu ermöglichen, werden außerdem alle Cursor-Bewegungen verfolgt und die textuelle Information über das Objekt, welches gerade vom Cursor berührt wird, an die Sprachausgabe oder die Braille-Zeile weitergegeben.

Das folgende Kapitel beschreibt drei spezielle Probleme, mit denen blinde Benutzer konfrontiert sind, wenn sie mit grafischen Benutzungsoberflächen arbeiten. Dabei wird deutlich werden, daß sie mit den gegebenen Adaptionen von Anwendungsprogrammen nicht ebenso effektiv und effizient arbeiten können wie sehende Benutzer. Im Anschluß daran wird die direkte Manipulation von akustischen Objekten erläutert. Diese Technik ermöglicht blinden Benutzern eine effiziente direkte Manipulation von Objekten mit der Maus und sie bildet die Basis für das "assistive" System SPUI-B (StereoPhonic User Interface for Blind).

2. Grafische Benutzungsoberflächen und blinde Benutzer

2.1 GUIs und ihre Barrieren für blinde Benutzer

Die Vorteile von GUIs (gegenüber textuellen Schnittstellen mit kommando- und menüorientierter Steuerung) sind bereits auf breiter Ebene in den letzten Jahren diskutiert worden. Deshalb werden hier nur die vier, im Zusammenhang mit diesem Thema wichtigsten Vorteile genannt:
1. Die Metapher gewohnter Umgebungselemente, die mit Icons grafisch dargestellt werden; diese Icons sind Referenzen auf Datenstrukturen, Programme, Funktionen usw.
2. Die direkte Manipulation der sichtbaren Objekte mit unmittelbarem visuellen Feedback für alle Aktionen.
3. Die einfache Steuerung parallel zugänglicher bzw. ablaufender Programme und Funktionen.
4. Die topographische Strukturierung und Organisation der sichtbaren Objekte auf dem Bildschirm durch Zuordnen von Semantik zu den Positionen.

So wie GUIs auf der einen Seite weniger geübten Benutzern die Arbeit am Rechner erleichtern, so erschweren sie auf der anderen Seite sehgeschädigten Benutzern die Arbeit und verursachen sogar neue Hindernisse für sie.

Boyd et al. [2] geben folgende drei Problembereiche an, die die Ursache dafür bilden, daß blinde Benutzer nicht ebenso effektiv und effizient mit GUIs arbeiten können wie Sehende:

1. Die *Pixel-Barriere* bezieht sich auf das Problem, daß die Bildschirmausgaben im Pixelformat abgespeichert werden, welches von einem gewöhnlichen Screenreader nicht gelesen werden kann. Eine Lösung für dieses *technische Problem* ist das bereits erwähnte *off-screen-model*.

2. Die *Maus-Barriere* zielt auf das Problem, daß blinde Benutzer die Maus nicht effektiv als Eingabegerät nutzen können. Dieses *motorische Problem* resultiert daraus, daß ein geeignetes Feedback zur Maussteuerung fehlt. Sehende sind aufgrund der *Auge-Hand-Koordination* in der Lage, die Maus zu benutzen. Sie *sehen* Cursor und Zielobjekt und sie *sehen*, wie sich die relative Lage des Maus-Cursors zu den Objekten verändert, wenn die Maus bewegt wird. Blinde erhalten keine Rückmeldung über die relative Lage des Cursors zu den Objekten und von daher können sie GUIs nicht in derselben Art und Weise benutzen wie Sehende.

3. Die *Grafik-Barriere* stellt das schwierigste Problem dar, da es ein *semantisches Problem* ist. Mehr und mehr Informationen werden grafisch präsentiert; zudem geben Topographie und Topologie der Objekte zusätzliche Hinweise über die Objekte selbst und ihre Beziehungen zueinander. Eine Transformation dieser grafischen Darstellungen in Ton- und Sprachausgabe ist immer mit einem Verlust von Informationen verbunden. Textuelle Beschreibungen von Grafiken sind häufig sehr lang, ungenau, kompliziert und umständlich ("Ein Bild sagt mehr als tausend Worte"). Eine technische Lösung des Problems ist das zweidimensionale taktile Display ("Stuttgarter Stiftplatte"); die Nachteile sind jedoch die geringe Auflösung (8 dpi) und die hohen Anschaffungskosten.

Wie bereits erwähnt, ist die Grafik-Barriere ein semantisches Problem und es ist fraglich, ob es je zufriedenstellend gelöst werden kann. Grafische Symbole werden verwendet, um ein Objekt in offensichtlicher Weise zu repräsentieren. Icons z.B. sind häufig vereinfachte Bilder von Objekten des alltäglichen Lebens, z.B. Mülleimer oder Radiergummi. Werden solche Icons in eine taktile Ausgabe mit einer wesentlich geringeren Auflösung als der Bildschirm transformiert, so ist es sehr zweifelhaft, ob blinde Benutzer die Objekte ebenso leicht erkennen können, da sie die Objekte nur vom Ertasten in Originalgröße kennen. Die verkleinerte taktile Figur wird nicht eindeutig zu erkennen sein und sie verliert vollständig den metaphorischen Bezug zur realen Umgebung.

Blinde Benutzer müssen also die Bedeutung der taktilen Icons als abstrakte Darstellungen erlernen und dieser Lernprozeß steht in krassem Widerspruch zu dem Ziel der GUIs, intuitiv und offensichtlich benutzbar zu sein. Dieser Nachteil kann auch nicht durch die Entwicklung kostengünstigerer zweidimensionaler taktiler Displays mit einer höheren Auflösung als die zur Zeit erhältlichen Displays beseitigt werden [8]. Nichtsdestotrotz ist eine Weiterentwicklung solcher Displays wichtig und wird sicherlich die Möglichkeiten der interaktiven Arbeit mit schematischen Darstellungen von Objekten erweitern.

Es gibt einige Projekte, die versuchen, Grafiken in akustische Ausgaben zu transformieren [17], oder die Bilder für Blinde beschreiben [12], aber es existieren zur

Zeit noch keine praktischen Lösungen, um beliebige Grafiken akustisch leichtver-
ständlich darzustellen.

2.2 Adaptionen existierender GUIs für blinde Benutzer

Zur Zeit existierende Adaptionen überwinden nicht die drei eben aufgeführten
Barrieren. Im allgemeinen lösen sie die Pixel-Barriere, indem sie das off-screen-
model verwenden und visuelle Objekte auf akustische Objekte im
dreidimensionalen Hörraum abbilden. Häufig können blinde Benutzer zum
Navigieren auf der Benutzungsoberfläche nur die Cursor-Tasten verwenden.

Im *GUIB*-Projekt (Textual and Graphical User Interfaces for Blind People) [12]
wird z.B. der Bildschirminhalt von *MS-Windows* zum einen in eine taktile Form
übersetzt, wobei die Topographie der Objekte beibehalten wird; zum anderen wird
die taktile Ausgabe um die akustische Ausgabe mit Raumklang erweitert. Jedem
Objekt der GUI - auch dem Cursor - ist ein Geräusch zugeordnet. Die Geräusch-
quellen der Objekte erscheinen im Hörraum an den gleichen Positionen wie die
sichtbaren Objekte auf dem Bildschirm. Bewegt sich der Cursor auf dem
Bildschirm, so bewegt sich die Geräuschquelle in derselben Richtung im
akustischen Raum. Auf diese Art erhält der Benutzer die Information über die
absolute Position des Cursors auf dem Bildschirm; die übrigen Objekte ertönen
erst, wenn sie vom Cursor berührt werden. Mit diesem akustischen Feedback ist
jedoch ein direktes Selektieren eines Objektes nicht möglich, da kein Feedback
über die relative Lage des Cursors zum Zielobjekt gegeben wird. Diese Situation ist
für Sehende vergleichbar mit einem schwarzen Bildschirm, auf dem nur der Cursor
zu sehen ist und die Objekte erst aufleuchten, wenn sie vom Cursor berührt werden.

Der *SonicFinder* [9, 10, 11], eine Anpassung der Finder-Oberfläche des Macintosh
für sehbehinderte Benutzer, arbeitet nach demselben Prinzip. Die Objekte sind als
akustisch passive Objekte implementiert, die nur dann ertönen, wenn ein
bestimmtes Ereignis eingetreten ist. In einer späteren Version des SonicFinders
werden sogenannte *Soundholder* eingeführt, die an ein Objekt gebunden werden
können und die akustisch Auskunft über ihre relative Position zum Cursor geben.
Sie tönen permanent und ihre Lautstärke fällt mit steigender Entfernung zum
Cursor, d.h. entfernt sich der Cursor von einem Objekt, so ertönt der diesem Objekt
zugeordnete Soundholder leiser, und bei Annäherung wird er lauter. Da kein
Raumklang verwendet wird, können die Soundholder nicht lokalisiert werden, um
z.B. die Maus sofort in die richtige Richtung zu bewegen. Zudem bleibt das
Grundproblem erhalten: wie finde ich ein Objekt, bevor ihm ein Soundholder
zugeordnet ist?

Im *Soundtrack*-Projekt [6, 7] werden den absoluten Positionen der Objekte verschiedene Töne zugeordnet (Keyboard-Metapher). Tests ergaben jedoch, daß nur
ein Musiker in der Lage war, den Tönen die entsprechende Positionsinformation zu
entnehmen. Alle übrigen sehgeschädigten Testpersonen zählten die bereits vom
Cursor passierten Objekte, um zum Zielobjekt zu gelangen.

Im *Mercator*-Projekt [13] wird blinden Benutzern der Zugang zu XWindows unter
Unix ermöglicht. In diesem Projekt wird das Navigationsproblem dadurch gelöst,

daß die grafische Anordnung durch eine baumartige Hierarchie ersetzt wird. Um ein Objekt zu selektieren, muß die Baumstruktur mit Hilfe der Cursor-Tasten durchsucht werden. Die Objekte ertönen wiederum nur dann, wenn ein bestimmtes Ereignis eingetreten ist.

Zusammenfassend kann festgestellt werden, daß die bekannten Adaptionen grafischer Benutzungsoberflächen nicht alle drei Barrieren überwinden. Sie ermöglichen es Blinden nicht, GUIs in derselben Art und Weise wie die Sehenden mit der Maus zu benutzen. Mit Hilfe des off-screen-models wird das Pixel-Problem gelöst, jedoch ist die graphische Benutzungsoberfläche häufig auf eine textuelle menü-/kommandoorientierte Schnittstelle reduziert, deren Objekte mit den Cursor-Tasten selektiert werden können. Die topographische Anordnung der Objekte hat keinerlei Bedeutung (Ausnahme GUIB-Projekt); eine direkte Manipulation der Objekte ist nicht möglich, da ein ausreichendes akustisches Feedback nicht gegeben wird. Dieses Feedback muß direkt und nicht indirekt, wie etwa bei der Keyboard-Metapher, wahrnehmbar sein.

3. Direkte Manipulation akustischer Objekte

Die Basis für die akustische direkte Manipulation wird zum einen gebildet durch akustisch *aktive* Objekte, den *Hearcons*, die im 3-D-Raum positioniert werden, und zum anderen durch ein akustisches Feedback für die *Ohr-Hand-Koordination* zur Steuerung der Cursor-Bewegungen mit der 3-D-Maus oder anderen geeigneten 3-D-Zeigegeräten.

3.1 Hearcons

In unserem System soll jedes relevante Objekt der Benutzungsoberfläche durch ein sogenanntes Hearcon repräsentiert werden. Ein Hearcon ist charakterisiert durch:
- das Geräusch, welches das Objekt repräsentiert,
- seine Lautstärke,
- seine Positionskoordinaten im Raum,
- seine räumliche Ausdehnung.

Die verwendeten Geräusche werden dabei nicht auf bestimmte Geräuschklassen eingeschränkt, wie etwa bei den *auditory icons* in Gaver's *SonicFinder* [9], bei dem nur Umweltgeräusche verwendet werden, oder bei Blattner's *earcons* [1], die aus synthetischen Tonfolgen zusammengesetzt sind. Den Hearcons kann jedes Geräusch zugeordnet werden, welches zu dem gegebenen Objekt paßt. Die Koordinaten geben die aktuelle Position im Raum an; die räumliche Ausdehnung ist ein künstliches Attribut für eine Geräuschquelle, die eigentlich als punktförmig angenommen wird. Sie wird jedoch benötigt, um den Raum abzugrenzen, in dem ein Hearcon als vom Cursor selektiert gilt.

Im Unterschied zu bisherigen Adaptionssystemen werden Hearcons als akustisch aktive Objekte realisiert, die permanent klingen und nicht nur dann, wenn ein bestimmtes Ereignis eingetreten ist. Sie bilden die Grundlage für das akustische Feedback, um blinden Anwendern die effiziente Benutzung der Maus zu ermöglichen. Die Verwendung von Raumklang (Kunstkopf-Stereophonie) erlaubt ein beliebiges

Positionieren der Hearcons im Raum und die topographische Anordnung der Objekte kann dadurch wahrgenommen werden, daß alle Hearcons parallel klingen.

Weiterhin ist ein Hearcon als das *aktuelle* Hearcon ausgezeichnet. Nur die Ausgaben oder der Inhalt des Objekts, welches durch das aktuelle Hearcon repräsentiert wird, können vom Benutzer mit Hilfe eines Ausgabegerätes seiner Wahl (Sprachausgabe, Braille-Zeile, Bildschirm) wahrgenommen werden. Alle anderen Hearcons repräsentieren nur ihre Objekte. Ebenso beziehen sich alle Eingaben nur auf das Objekt des aktuellen Hearcons. Dies ist vergleichbar mit einem Fenster-System, bei dem alle Fenster, bis auf das aktuelle als Icons repräsentiert werden und nur der Inhalt des aktuellen Fensters auf dem Bildschirm dargestellt wird. Die Vorgehensweise bei der Eingabe verhindert nebenbei das 'unselected-window'-Problem, da der Benutzer nie im Zweifel darüber ist, welches Fenster gerade die Eingabe akzeptiert.

3.2 Das Selektieren eines Hearcons

Der Ansatz, die Hearcons als akustisch aktive Objekte zu realisieren, ermöglicht es dem Benutzer, ein Zielhearcon direkt zu selektieren, ohne den gesamten Raum danach zu durchsuchen. Der Benutzer hört die topographische Anordnung der Objekte und damit auch die Position seines Zielobjekts und er erkennt, in welche Richtung die Maus bewegt werden muß, um dorthin zu gelangen.
Um die Maus-Barriere zu überwinden, muß ein geeignetes Feedback zur Ohr-Hand-Koordination, die die Handbewegungen zum Positionieren des Cursors steuert, realisiert werden. Der Benutzer benötigt ein permanentes und unmittelbares Feedback, inwiefern sich die relative Position des Cursors gegenüber dem Zielobjekt verändert, wenn die Maus bewegt wird. Aus diesem Grund ändert sich die akustische Ausgabe aller Hearcons, wenn der Maus-Cursor bewegt wird.

Es werden im Projekt folgende Alternativen für das akustische Feedback untersucht:
1. Jede Veränderung der Position des stillen Cursors resultiert in einer Änderung der akustischen Parameter (z.B. Tonhöhe, Lautstärke, Timbre) der Hearcons. Beispielsweise erklingt ein Hearcon lauter oder im Ton klarer, wenn sich der Cursor nähert, und es klingt leiser oder im Ton dumpfer, wenn sich der Cursor entfernt. (Analogie: das Cursor-Icon ist unsichtbar, der Helligkeitsgrad der anderen Icons gibt die Nähe zum Cursor an.)
2. Dem Cursor selbst ist ein Hearcon zugeordnet und die Geräuschquelle bewegt sich mit dem Cursor im Raum. Alle übrigen Hearcons klingen unverändert weiter.
3. Eine Bewegung des klingenden Cursors führt dazu, daß sich die Geräuschquelle des Cursors im Hörraum entsprechend mitbewegt und daß sich die akustischen Parameter der Hearcons ändern (Kombination der Alternativen 1 + 2).
4. Der Cursor ist an einer festen Position und eine Bewegung der Maus resultiert in einer Bewegung aller Hearcons. (Analogie: der Benutzer bewegt sich zusammen mit dem Cursor durch eine Artificial Reality.)
Welche Alternative am besten geeignet ist, um die Ohr-Hand-Koordination zu steuern und welche am meisten das Selektieren eines Hearcons in einer gegebenen

Anwendung unterstützt, muß in Versuchen gezeigt werden, die im Projekt in enger Zusammenarbeit mit den Fachbereichen Physik (Akustik) und Psychologie im Rahmen des gemeinsamen Graduiertenkollegs Psychoakustik erfolgen.

Das Berühren eines Hearcons mit dem Cursor muß unmittelbar angezeigt werden. Das System *AudioWindows* [4, 5] benutzt dazu den sogenannten "spotlight". Die Idee hinter dem Spotlight ist der "gerade noch erkennbare Unterschied" (Just noticeable difference, kurz *jnd*), einem wichtigen Aspekt der Psychophysik: es ist die kleinste erkennbare Unterscheidung in der Wahrnehmung (akustisch, visuell, taktil) zwischen zwei Objekten [15]. Unterhalb dieses Schwellwertes werden nur ein Objekt oder zwei identische Objekte erkannt. Eine andere Lösung ist ein zusätzlicher Ton, der beim Berühren eines Hearcons ertönt.

Dasselbe Problem entsteht, wenn ein gefundenes Hearcon selektiert wird. Entweder ändern sich die akustischen Parameter des Hearcons ein wenig, oder aber ein weiterer Ton wird eingeführt, um die Selektierung des Hearcons anzuzeigen.

3.3 Die Manipulation der Hearcons

Im folgenden werden einige weitere Funktionen zur Manipulation der Hearcons kurz vorgestellt.

Verschieben eines Hearcons

Das Verschieben eines Hearcons erfolgt in der gleichen Art und Weise wie bei GUIs. Ein Hearcon muß gefunden und selektiert werden, bevor es zu seiner Zielposition gezogen werden kann. Während des Verschiebens muß ein unmittelbares Feedback darüber erfolgen, wo sich das Hearcon gerade befindet. Dazu bewegt sich die Geräuschquelle entsprechend der Cursor-Position.

Erzeugen eines Hearcons

Bevor ein Hearcon irgendwo plaziert werden kann, muß es natürlich erzeugt werden. Dazu ruft der Benutzer die Funktion 'erzeuge_hearcon' entweder durch Maus-Doppelklick auf, wenn sich der Cursor auf keinem Hearcon befindet, oder er drückt eine Funktionstaste. Die Position des neuen Hearcons kann entweder
- jeweils eine feste Position im Raum (z.B. direkt vor dem Benutzer) sein, oder
- die Position des Maus-Cursors, als der Vorgang gestartet worden ist.

Nach dem Erzeugen eines neuen Hearcons befindet sich der Cursor implizit auf diesem Hearcon, um dem Benutzer die Möglichkeit zu geben, das neue Hearcon unverzüglich an seine Zielposition zu bewegen, ohne es noch einmal selektieren zu müssen.

Löschen eines Hearcons

Um ein Hearcon zu löschen, muß es zunächst gefunden und selektiert werden, bevor die 'lösche_hearcon'-Funktion aufgerufen wird:
- Die erste Alternative ist, das Hearcon physikalisch zu einem Mülleimer-Hearcon zu ziehen, und
- die zweite Alternative ist, die Funktion über eine Funktionstaste aufzurufen.

Ein Verändern der Größe eines Hearcons ('shrinking' in GUIs) ist bei diesem
Ansatz nicht von weiterem Interesse, da die Hearcons die Objekte nur
repräsentieren und keine Informationen über den Inhalt des Objekts ausgeben. Das
Vergrößern eines Hearcons hätte nur den Effekt, daß die Fläche zum Finden dieses
Hearcons wächst.

4. Einschränkungen des Ansatzes: Parallel klingende Hearcons

Dieser Ansatz ist natürlich einigen Einschränkungen unterlegen und löst nicht alle
aufgezählten Probleme der Blinden im Umgang mit GUIs. Ebensowenig können
gegebene GUIs derart adaptiert werden, daß jedem Objekt der Oberfläche ein
aktives Hearcon zugeordnet wird.

Das menschliche Gehör kann nur eine um einige Größenordnungen kleinere
Anzahl von Informationen gleichzeitig bewußt wahrnehmen wie die Augen. Als
Konsequenz können nur viel weniger aktive Hearcons gleichzeitig ertönen als
visuelle Icons gleichzeitig auf dem Bildschirm erscheinen können, wenn sie noch
bewußt wahrgenommen werden sollen. Ertönen zu viele Hearcons, ist es sehr
zweifelhaft, ob ein einzelnes Hearcon separat herausgefiltert und lokalisiert werden
kann - trotz des sogenannten 'Cocktail-Party-Effekts' (die Fähigkeit, innerhalb eines
'Volksgemurmels' einen ganz bestimmten räumlichen Bereich akustisch wahrzunehmen).

Es werden im Projekt folgende Lösungen näher untersucht:
1. Nicht alle Hearcons sind aktive Objekte und klingen permanent, sondern nur
    die in der Nähe des Cursors (siehe Gaver's Lösung im SonicFinder, inkl. der
    Soundholder).
2. Homogene Hearcons werden so gruppiert, daß sie wie *ein* Hearcon klingen. Ein
    akustisches Gruppieren von Geräuschen kann mit ähnlichen Geräuschen
    (gleiches Timbre) oder mit Tönen, die in der Tonhöhe dicht beieinander liegen,
    realisiert werden [3].
3. Hearcons können beliebig vom Benutzer gruppiert und die Gruppen im Klang
    gedämpft ("in den Hintergrund geschickt") werden.
Baumartige Strukturen (siehe Mercator-Projekt) werden nicht weiter verfolgt, da
sie die räumliche Metapher verlassen.

Der Nachteil der ersten Lösung ist, daß nicht mehr alle Hearcons direkt selektiert
werden können, da der Benutzer sie nicht alle hört. Soll ein Hearcon selektiert
werden, welches nicht klingt, so muß der Benutzer entweder die Position des
Hearcons kennen, um die Maus in die richtige Richtung zu bewegen, oder im
schlimmsten Fall muß der ganze Raum abgesucht werden, bis das Zielobjekt gefunden ist. Eine Vorbedingung für Lösung 2 ist, daß überhaupt ähnliche Hearcons
existieren, die dann gruppiert werden können. Die dritte Lösung schließlich erlaubt
dem Benutzer beliebig zu gruppieren und der Gruppierung eine Bedeutung beizu-
messen, allerdings ist dieser Vorgang dann ein zusätzlicher Arbeitsschritt.

Es wird im Projekt experimentell untersucht, wieviel Hearcons parallel klingen und vor allem bewußt wahrgenommen werden können und welche Lösung am besten geeignet ist, wenn viele Hearcons parallel klingen sollen.

Als Schlußfolgerung kann festgehalten werden, daß Hearcons und die erläuterte Ohr-Hand-Koordination nicht in jedem Fall zur Adaptierung gegebener GUIs geeignet sind, da ernsthafte Probleme auftauchen, wenn zu viele Hearcons gleichzeitig präsentiert werden müssen. Statt der Suche nach einer Adaption, die den blinden Benutzern alle Möglichkeiten und Vorteile von GUIs eröffnet, wird im nächsten Kapitel das "assistive interface" SPUI-B (StereoPhonic User Interface for Blind) vorgestellt, welches blinden Benutzern die erwähnten Vorteile der Fenster-Systeme und die akustische direkte Manipulation eröffnet.

5. SPUI-B - eine assistive Technik

Zuerst ist die Frage zu beantworten, warum eine neue Spezialumgebung (in diesem Sinne assistive Technik) nur für blinde Benutzer entwickelt werden soll, wo doch bekannt ist, daß Benutzer mit Behinderungen gerade an adaptiven Lösungen, d.h. an Adaptionen der Programme für nicht- (oder weniger) behinderte Benutzer interessiert sind. Die Antwort ist, daß Adaptionen gegebener GUIs immer mit einem Verlust an Informationen verbunden sind (wie z.B. dem Verlust der Information, die in der topographischen Anordnung der Objekte verborgen ist). Ein assistiver Ansatz eröffnet vielleicht eine größere Chance, eine Spezialumgebung für Blinde zu entwickeln, die ihnen Zugang zu den Informationen ermöglicht, die z.B. in der topographischen und/oder der topologischen Anordnung der Objekte enthalten ist, und ihnen dieselben funktionalen Vorteile bietet, wie die GUIs den Sehenden. Daneben soll es natürlich möglich sein, innerhalb der Spezialumgebung mit anderen Adaptionen graphischer Benutzungsoberflächen zu arbeiten, so daß blinde Benutzer trotzdem in der Lage sein werden, mit sehenden Kollegen an einem Bildschirm bzw. Arbeitsplatz zu arbeiten.

5.1 Ziele

Das Hauptziel von SPUI-B ist, blinden Benutzern die Vorteile der Fenster-Systeme zugänglich zu machen, ohne bloß gegebene GUIs zu adaptieren. Das ist erstens die Möglichkeit, Metaphern gewohnter Umgebungen zu realisieren, zweitens mehrere Anwendungen, die parallel laufen, auf einfache Art zu steuern, indem ohne großen Aufwand zwischen ihnen hin- und hergeschaltet werden kann, drittens die direkte Manipulation von Hearcons, um die Benutzungsoberfläche an die eigenen Bedürfnisse anzupassen und viertens mit diesen Mitteln der topographischen Anordnung der akustischen Objekte eine Bedeutung geben.

5.2 Entwurfskonzept

Die Basis von SPUI-B ist ein Fenster-System, welches den Benutzern erlaubt, eine beliebige Anzahl an Fenstern zu erzeugen, in denen beliebige Anwendungen gestartet werden können, genau wie in einem visuellen Fenster-System. Eines der Fenster ist das aktuelle Fenster, welches die Tastatureingabe akzeptiert und welches seine Ausgaben zu dem vom Benutzer gewählten Ausgabegerät schickt. Sehende Benutzer werden den Bildschirm bevorzugen, während blinde Benutzer die Braille-Zeile oder die Sprachausgabe wählen werden. In diesem Sinne ist SPUI-B nicht nur eine Spezialumgebung für Blinde, sondern es kann ebenso von Sehenden benutzt werden.
Jedes nicht-aktuelle Fenster wird durch ein Hearcon repräsentiert. SPUI-B erlaubt die Wahl beliebiger Geräusche, z.B. Musik oder Umweltgeräusche. Als Ausgabe-technik wird die Kunstkopf-Stereophonie verwendet. Die Bewegung der Maus wird durch die Ohr-Hand-Koordination gesteuert. So kann der Benutzer die Hearcons, wie weiter oben erläutert, direkt manipulieren. Die Hearcons können beliebig im Hörraum plaziert werden, so daß die topographische Anordnung für den blinden Benutzer eine Bedeutung erlangen kann, da er die Anordnung selbst gestaltet. Darüber hinaus kann er jedes Fenster direkt selektieren, da das zugehörige Hearcon lokalisiert werden kann. Als Resultat kann der Wechsel zwischen den Anwendungen ganz einfach und ohne großen Aufwand für die Blinden dadurch erfolgen, daß eines der nicht-aktuellen Hearcons angefahren und aktiviert wird.

Weiterhin dienen die Hearcons als Gedächtnisstütze, da der Benutzer hört, wenn eine Anwendung noch läuft. Um die Funktion als Gedächtnisstütze ebenso gut wie grafische Icons zu erfüllen, reicht es jedoch nicht aus, daß erkannt wird, daß noch *irgendeine* Anwendung läuft. Der Benutzer muß dazu auch noch erkennen können, *welche* Anwendung noch läuft.

Die Brauchbarkeit des Systems hängt hier in hohem Maße von der Wahl der Ge-räusche ab, die die entsprechenden Fenster und Anwendungen repräsentieren.

5.3 Die Wahl der Geräusche

Wie bereits erwähnt, ist SPUI-B nicht beschränkt auf eine spezielle Geräuschklasse. Falls es zu einem Objekt oder einer Anwendung ein entsprechendes Objekt im 'realen' Leben gibt, so wird das dazugehörige Geräusch zur Repräsentation des Rechnerobjekts herangezogen. Die Idee dieses Ansatzes ist, die diesen Geräuschen inhärente Information über das Objekt zu nutzen, so daß die Bedeutung dieses Hearcons nicht erst erlernt werden muß. Nicht alle Objekte der Rechnerwelt haben jedoch entsprechende hörbare Pendants im realen Leben. In solchen Fällen muß ein Geräusch gewählt werden, das evtl. nicht in unmittelbarem Zusammenhang mit dem Objekt steht, und die Bedeutung dieses Hearcons muß erlernt werden.

Der Gebrauch von synthetischer Sprache erfolgt nur auf Wunsch des Benutzers, da zum einen die Hearcons akustisch aktive Objekte sind, die permanent klingen, und zum anderen Sprache ein recht langsames Ausgabemedium ist.

Mynatt faßt einige Anforderungen an die den Rechnerobjekten zugeordneten Geräusche zusammen [14]:

- Die konzeptuelle Abbildung: Werden die Objekte offensichtlich
  repräsentiert?
- Die Benutzerempfindung: Klingt das Geräusch angenehm?
- Die physikalischen Parameter: Keine gegenseitige Maskierung der Geräusche, gute Klangqualität,...

Zudem muß es möglich sein, zueinander in Beziehung stehende Objekte durch geringe Unterschiede in ihren akustischen Parametern zu repräsentieren, um die Relationen zu verdeutlichen.

Ausgehend von diesen Anforderungen folgt, daß Hearcons nur dann als Gedächtnisstütze fungieren können, wenn die konzeptuelle Abbildung von der Anwendung zur akustischen Darstellung erfolgreich ist.

Welche Geräusche den Forderungen am besten entsprechen, hängt sowohl von den entsprechenden Anwendungen als auch vom Benutzer selbst ab. Ein Musiker wird ein Hearcon sicherlich anders hören als ein Nicht-Musiker. Deshalb muß den Benutzern die Möglichkeit eingeräumt werden, sich geeignete Geräusche selbst auszusuchen.

5.4 Projektorganisation

Das Projekt begann 1993 und befindet sich noch in der Phase der Erforschung der grundsätzlichen Prinzipien. Seit Beginn 1995 ist es Bestandteil des Graduiertenkollegs Psychoakustik der Carl von Ossietzky Universität Oldenburg.

Als technische Plattform werden DOS-kompatible PCs verwendet, die mit Soundkarten und Braille-Zeile ausgestattet sind.

6. Zusammenfassung

Anhand der Diskussion um die Pixel-, Maus- und Grafik-Barriere wurde gezeigt, daß einige der Vorteile der grafischen Benutzungsoberflächen von Blinden nicht genutzt werden können. Die Grafik-Barriere ist das schwierigste Problem und gibt Anlaß zu der Vermutung, daß Blinde prinzipiell nicht in die Lage versetzt werden können, GUIs ebenso effektiv und effizient wie Sehende zu benutzen.

Gegebene Adaptionen von GUIs lösen das Problem der Pixel-Barriere, jedoch ist die Arbeit mit ihnen häufig auf das Finden und Selektieren von Funktionen und Objekten mit Hilfe der Cursor-Tasten reduziert. Die topographische Anordnung der Objekte ist für die Blinden kaum wahrnehmbar, so daß eine mögliche Informationsquelle verborgen bleibt. Eine akustische Lösung zur effizienten Benutzung der Maus wurde vorgestellt, die auf akustisch aktiven Hearcons und der Ohr-Hand-Koordination mit akustischem Feedback über die relative Position des Cursors zu den Hearcons beruht. Weitere Funktionen zur direkten Manipulation der Hearcons sind erläutert worden.

Zu erforschende Probleme des Ansatzes sind die Auswahl geeigneter Geräusche, um Objekte akustisch offensichtlich zu repräsentieren, und die Anzahl gleichzeitig klingender Hearcons.

Von den allgemeinen Betrachtungen wurde das Konzept einer Spezialumgebung abgeleitet, das zum Entwurf des StereoPhonic User Interface for Blind (SPUI-B) führt.

Literaturverzeichnis

[1] M. M. Blattner, D. A. Sumikawa, R. M. Greenberg: *Earcons and Icons: Their Structure and Common Design Principles,* Human Computer Interaction 1989, Vol.4

[2] L. H. Boyd, W. L. Boyd, G. C. Vanderheiden: *Graphics-Based Computers and the Blind: Riding the Tides of Change,* in: Proceedings of the 6th Annual Conference "Technology and Persons with Disabilities" Los Angeles, 20-23.3.1991

[3] A. S. Bregman: *Auditory Scene Analysis - The Perceptual Organization of Sound,* MIT Press, Cambridge 1990

[4] M. Cohen, L. Ludwig: *Multidimensional audio window management ,* International Journal of Man-Machine Studies, 1991, Vol 34, pp.319-336

[5] M. Cohen: *Throwing, pitching and catching sound: audio windowing models and modes,* International Journal on Man-Machine-Studies, 1993, Vol 39, pp 269-304

[6] A. D. N. Edwards: *Design of Auditory Interfaces for Visually Disabled Users,* in *Human Factors in Computing Systems* Proceedings of the CHI '88, ACM SIGCHI

[7] A. D. N. Edwards: *Soundtrack: An Auditory Interface for Blind Users,* Human Computer Interaction 1989, Vol.4

[8] J. Fricke, H. Bähring: *A graphic input/output tablet for blind computer users,* in W. Zagler (Ed.): *Computers for handicapped Persons,* Proceedings of the 3rd International Conference, Vienna, July 7-9, 1992

[9] W. W. Gaver: *The SonicFinder: An Interface that uses Auditory Icons,* Human Computer Interaction 1989, Vol.4

[10] W. W. Gaver, R. B. Smith: *Auditory Icons in Large-Scale-Collaborative Environments,* in H. Diaper et al. (Eds.): *INTERACT '90,* Proceedings of the 13 3rd Int. Conference on Human-Computer Interaction, Cambridge, U.K.. 27-31 August, 1990

[11] W. W. Gaver, R. B. Smith, T. O´Shea: *Effective Sounds in complex Systems: The Arkola Simulation,* in ´Reaching through technology´, CHI ´91 Proceedings, 1991

[12] P. L. Emiliani (Project Manager): *Publications from the TIDE project GUIB until to September 1993,* GUIB Consortium, 1993

[13] E. D. Mynatt, W. K. Edwards: *Mapping GUIs to Auditory Interfaces,* in Proceedings of the UIST´92, Nov.15-18, 1992, Monterey, CA

[14] E. D. Mynatt: *Designing with Auditory Icons: How Well Do We Identify Auditory Cues,* in: Proceedings of the CHI'94 Conference, Boston, M., April 24-26, 1994

[15] J. G. Roederer: *Physikalische und psychoakustische Grundlagen der Musik,* Springer Verlag Berlin Heidelberg New York, 1977

[16] R. S. Schwerdtfeger: *Making the GUI Talk,* BYTE, December 1991

[17] S. Smith, R. D. Bergeron, G. G. Grinstein: *Stereophonic and surface sound generation for exploratory Data Analysis* in CHI '90 Proceedings April 1990

Die Autoren bedanken sich bei den Mitgliedern des Graduiertenkollegs Psychoakustik der Universität Oldenburg, besonders bei Prof. Dr. Volker Mellert und seiner Abteilung.

Ludger Bölke und Peter Gorny
Carl von Ossietzky Universität Oldenburg - Fachbereich Informatik, D-26111 Oldenburg
EMail: Ludger.Boelke@informatik.uni-oldenburg.de, Gorny@informatik.uni-oldenburg.de

Alltagspraxis der Hypermediagestaltung – Erfahrungen beim Einsatz des World Wide Web und Mosaic in der Lehre

Andreas Brennecke, Reinhard Keil-Slawik

Heinz Nixdorf Institut Universität–GH Paderborn

Zusammenfassung

Hypermediasysteme werden vorwiegend als Lehr- und Präsentationssysteme eingesetzt. Dabei handelt es sich in der Regel um einzelne Anwendungen und isolierte Lerneinheiten, die von Autoren entwickelt und dann nur noch interaktiv „gelesen" werden. Der Beitrag untersucht, inwieweit solche Systeme, speziell das World Wide Web, auch zur kontinuierlichen Erstellung und Aktualisierung von Arbeitsunterlagen geeignet sind. Aus den Erfahrungen der Lehrenden und den Bewertungen der Studierenden werden Konsequenzen abgeleitet, wie die Systemgestaltung und der Einsatz solcher Systeme verbessert werden können.

1   Einleitung

Hypermediasysteme werden – neben reinen Präsentationsaufgaben – vielfach im Bereich der Ausbildung verwendet, weil die Möglichkeit der interaktiven Erschließung von Lehrunterlagen den Lernenden die Möglichkeit gibt, situations- und personenbezogen Sinnzusammenhänge zu erschließen. Umgekehrt bieten die Möglichkeiten des „nicht-sequentiellen Schreibens", als das Ted Nelson Hypertext charakterisierte („By "hypertext" I mean non-sequential writing." [10]), den Lehrenden ein erweitertes Repertoire an Ausdrucksmöglichkeiten, insbesondere unter Einbeziehung von Multimedia-Dokumenten. Ob sich jedoch dieses Potential produktiv entfalten läßt, zeigt sich erst unter den Bedingungen der Alltagspraxis. Der nachfolgende Beitrag faßt unsere Erfahrung mit der Gestaltung und Benutzung eines Hypertextsystems zur Erstellung und Verwaltung von Lehrveranstaltungsunterlagen zusammen.

Neben einer kurzen Einführung in diese spezifischen Rahmenbedingungen des Anwendungskontextes, werden wir zum einen die Möglichkeiten und Probleme der Bereitstellung der Unterlagen behandeln, wie auch die Erfahrung und Reaktionen seitens der Studierenden. In der abschließenden Bewertung fassen wir die wichtigsten Einsichten zusammen und leiten daraus einige wichtige Hinweise zur Gestaltung und zum Einsatz von Hypermediasystemen ab.

2   Einsatzkontext

Seit Ende 1992 gibt es an der Universität-GH Paderborn das Fachgebiet *Informatik und Gesellschaft*. Ziel ist, die Kompetenzen der Bewertung und Gestaltung von Informationstechnologien zu verbinden und die informatikrelevanten Teile der Wechselbeziehung Mensch–Technik–Umwelt zu erforschen. Um dieses auch praxisrelevant durchführen zu können, ist ein Forschungsschwerpunkt der Arbeitsgruppe die Einrichtung eines elektronischen Hörsaals, an und in dem die Rolle von Technik im Bereich der Bildung und Ausbildung untersucht werden soll. Ein entscheidender Ausgangspunkt für diese Untersuchung ist die Sichtweise, Artefakte wie Bücher, Bilder, Ton- oder Filmdokumente ebenso wie interaktive Systeme als externes Gedächtnis zu betrachten (siehe Keil-Slawik [6]), und darüber sowohl die spezifische Qualität der Technik in bezug auf soziale Lernprozesse zu bestimmen und Vorschläge als auch Strategien für eine angemessene Gestaltung zu gewinnen. Praxisrelevant bedeutet dabei, solche Bestimmungen nicht allein auf der Basis solcher Prinzipien theoretisch abzuleiten, sondern das vielfältige Wirkungsgeflecht der Alltagspraxis zur Überprüfung des Stellenwertes der Prinzipien und zur Validierung von Gestaltungsvorschlägen zu erheben (vgl. das Beispiel in Keil-Slawik [7]).

Aus diesem Grund haben wir beschlossen, die Lehrveranstaltungsunterlagen nebst Fragen und Übungsaufgaben zur Vertiefung des Stoffes den Studierenden mittels eines Hypermediasystems auf dem Universitätsrechnernetz zur Verfügung zu stellen. Neben Fragen der Kosten und der allgemeinen Verfügbarkeit der Unterlagen war ein wichtiger zu berücksichtigender Gesichtspunkt, daß die Unterlagen während des Semesters erweitert und verändert werden müssen.

Es gibt mittlerweile eine Reihe von rechnerbasierenden Systemen, die in der Lehre eingesetzt werden. So wird z. B. BWL-Studierenden die HERMES-CD für das Selbststudium angeboten (siehe Schoop [12], [13]). Hierfür werden Einzelteile in einer Autorenphase erstellt und anschließend in einer Redaktionsphase getestet. Erst danach werden die Dokumente in die eigentliche Anwendungsumgebung integriert. MILES ist ein Datenbank-Autorensystem als Träger der gesamten Lehrsammlung eines Studienfaches. Die Datenbank enthält multimediale Komponenten, die über eine (Hypertext-ähnliche) Vernetzung zu Lehreinheiten zusammengestellt werden (siehe Wiemer [15]). MIAS soll das medizinische Curriculum modular und systematisch abdecken. Programm-Moduln werden hierfür von einem technischen Entwicklerteam nach Storyboards erstellt, die von klinischen Experten stammen (siehe Fischer [3]).

Systeme dieser Art sind aufgrund mangelnder Anpaßbarkeit für unsere Zwecke nicht geeignet. Eine weitere Einschränkung ergab sich aus der verfügbaren technischen Infrastruktur.

Das Universitätsrechnernetz besteht überwiegend aus UNIX-Workstations. Es gibt nur wenige PC- und Macintosh-Systeme, für die die überwiegende Zahl von

Hypertextsystemen existiert. Aus diesen Gründen fiel die Wahl auf das World Wide Web (kurz WWW oder Web), für das es eine Reihe Public-Domain-Programme wie z. B. den Viewer Mosaic gibt. Das Web ist ein weltweit verteiltes Hypertextsystem, das auf verschiedenen Plattformen verfügbar ist. Daten werden darin auf lokalen Servern abgelegt und können über das Internet abgerufen werden (siehe z. B. Zores [16]; Aktuelle Informationen über das World Wide Web finden sich im Web selbst z. B. unter der URL-Adresse: http://www.informatik.tu-muenchen.de/about_www.html).

Ein weiterer Vorteil besteht darin, daß sich Multimedia-Dokumente integrieren lassen, die dann jeweils den Aufruf eines Standardwerkzeuges zur Anzeige oder zum Abspielen bewirken. Soweit solche Standardwerkzeuge lokal verfügbar sind, können entsprechende Multimedia-Dokumente benutzt und bearbeitet werden. Obwohl die nachfolgend noch skizzierten Features und Funktionen des Web sehr vielversprechend klingen, zeigte sich bald, daß kleine Beschränkungen bereits gravierende Konsequenzen für die Gestaltung der Unterlagen bedeuten, insbesondere wenn man berücksichtigt, daß die Erstellung von Unterlagen starken Einschränkungen bezüglich Zeit und verfügbarer Resourcen unterliegt.

3   World Wide Web und Mosaic

Zu Beginn der Lehrveranstaltung war an der Universität-Paderborn seit ca. einem halben Jahr ein Web-Server und der Viewer Mosaic installiert, so daß ein lauffähiges System bereits zur Verfügung stand.

Das Web benutzt die Dokumentbeschreibungssprache HTML (HyperText Markup Language), die dem Autor recht wenig Gestaltungsmöglichkeiten bietet. Es gibt – anders als in vielen Systemen wie z. B. Hypercard – keine graphisch gestalteten Seiten, die dem Leser angezeigt werden. Die Dokumente bestehen aus einer sequentiellen Aneinanderreihung von Textelementen und Bildern. Zeilenumbrüche werden durch die Fensterbreite erzeugt oder sind vom Autor vorgegeben. Typische Textelemente sind neben einfachem Text Überschriften, Listen oder Aufzählungen. In die Texte lassen sich Bilder in zwei Bitmap-Formaten (X Bitmap (XBM) und GIF) einbinden.

Abbildung 1 zeigt die Darstellung eines HTML-Dokuments mit dem Viewer Mosaic. Das Dokument wird im inneren Rahmen angezeigt. Rundherum sind die Bedienfelder von Mosaic angeordnet. Die Hypertextstruktur wird über farbige bzw. unterstrichene Textteile markiert, die Verweise (links) in andere Dokumente präsentieren. Weitere Darstellungsformen der Hypertextstruktur – z. B. als Graph – bietet Mosaic nicht. Beim Selektieren der markierten Textteile wird das Dokument geladen, auf das der Verweis zeigt. In einem Fenster kann aber immer nur ein Dokument gleichzeitig dargestellt werden.

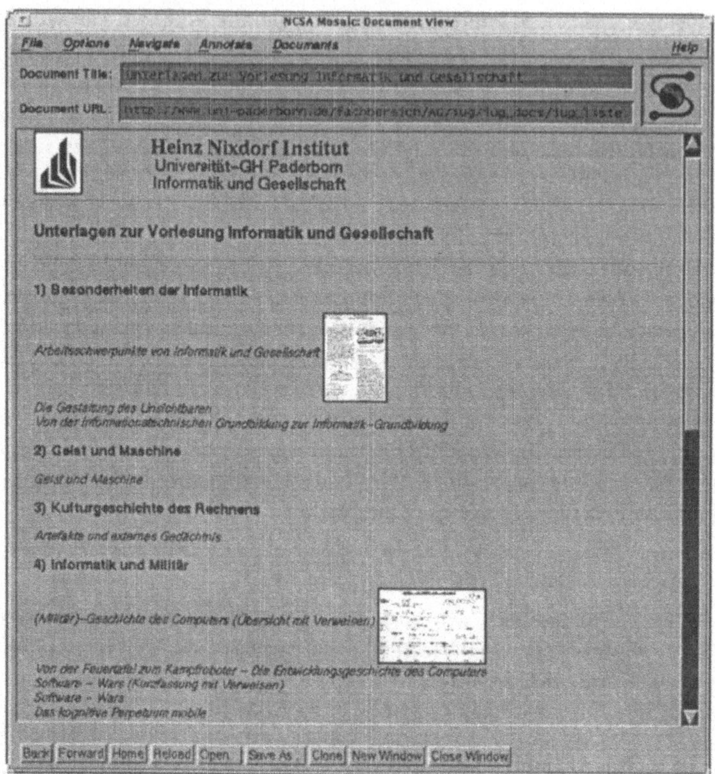

Abb. 1: Die Oberfläche des Viewers Mosaic. Angezeigt wird ein Hypertext mit einigen
         importierten Bildern.

Der gesamte Text wird linksbündig gesetzt, die Schriftgröße hängt von den gewähl-
ten Schriftkonstrukten (Überschrift, Haupttext, Aufzählung, ...) und den
Einstellungen des Benutzers ab, der Schrifttyp (Times, Helvetica, New Century oder
Lucida) und Schriftgröße (klein, mittel, groß) als Option einstellen kann.
Einrückungen werden nur in Aufzählungen vorgenommen. Für das Logo und den
Schriftzug der Universität in Abbildung 1 wurden zwei GIF-Bilder eingebunden.
Der Schriftzug der Universität läßt sich in HTML so nicht gestalten, es gibt keine
Textzentrierung, Tabulatoren oder beliebiges Plazieren. Eingerückter Text läßt sich
nur mit der nicht proportionalen Schrift Courier erstellen. Dieser wird ebenfalls
linksbündig gestzt, allerdings lassen sich buchstabenbreite Leerzeichen einfügen.
Auch die Bilder unter „Besonderheiten der Informatik" und „Informatik und Militär"
lassen sich nicht frei positionieren. Sie verhalten sich wie Buchstaben im Textfluß
und verbreitern somit den Zeilenabstand der gesamten Zeile.

Die freie Gestaltung einer Seite läßt sich nur mit seitenfüllenden importierten Graphiken bewerkstelligen, die mit anderen Programmen erstellt werden müssen und dann in das Dokument eingebunden werden. Für weitere multimediale Dokumente lassen sich externe Präsentationswerkzeuge wie z. B. ein Movie-Player starten.

4 Online-Erstellung von Hypertexten

Von R. Kuhlen werden drei grundsätzlich verschiedene Verfahren einer Autoren- bzw. Konversionskomponente zur Erstellung von Hypertext beschrieben (siehe [8]):

- Es ist möglich, linearen Text mit Hilfe von Konvertierungsprogrammen automatisch in einen Hypertext zu übersetzen. Eine Aufteilung in Einzeltexte erfolgt dabei meist nach Kapiteln und Abschnitten, die Verweisstruktur repräsentiert in der Regel neben der linearen Verkettung das Inhaltsverzeichnis, Fußnoten und ein Indexverzeichnis.

- Hypertext läßt sich auch durch eine Zusammenstellung multimedialer Dokumente und deren Verbindung über Verweise erstellen.

- Durch die Repräsentation von Wissen in einzelnen Knoten, die über eine Verweisstruktur verbunden werden, lassen sich Hypertext-Dokumente direkt erzeugen.

Für das Web gibt es eine Reihe von Werkzeugen, so daß alle drei genannten Verfahren unterstützt werden. Das Angebot an Werkzeugen für das Web ist sehr dynamisch: Fast wöchentlich gibt es neue Konvertierungstools, Editoren für HTML oder Hilfsprogramme. Um hier auf dem laufenden zu bleiben, bedarf es einer ständiger Recherche im Web, in dem Neuerungen laufend als Hypertext-Dokumente angekündigt werden.

Da die Hypertext-Dokumente für die Vorlesungsunterlagen aber online d. h. neben anderen Tätigkeiten im Semester erstellt wurden, konnten nicht alle Werkzeuge installiert und getestet werden. Bei der Installation einiger Programme bedurfte es ohnehin der Unterstützung der Rechnerbetreuungsgruppe, so daß auch nicht alle verfügbaren Programme direkt eingesetzt werden konnten.

Neu geschriebene Texte wurden mit einem Texteditor direkt in HTML geschrieben. Die zu Beginn getesteten HTML-Editoren stellten sich als nicht brauchbar heraus. Bei ihnen ist nur das Einfügen von HTML-Konstrukten nicht jedoch das Ändern und Löschen möglich. Nachfolgend konzentrieren wir uns deshalb auf die automatische Konvertierung von Texten und die Einbindung multimedialer Komponenten.

4.1 Automatische Konvertierung

Ein Teil der in unserem Hypertextsystem integrierten Texte wurde mit einem Konverter von LaTeX nach HTML erzeugt. Wegen der eingeschränkten Zeit, die während des Semesters zur Verfügung stand, konnten keine vollständig neuen

Hypertexte entworfen und geschrieben werden. Ausgangspunkt waren Dokumente, die bereits in der Arbeitsgruppe erstellt worden waren. Ein Teil dieser Dokumente bestand aus in LATEX gesetzten Texten und konnte mit dem Konverter direkt übersetzt werden. Andere lineare Texte oder Textteile wurden mit einem OCR-Scanner eingelesen. Hierbei hat es sich als erfolgreich herausgestellt, diese ASCII-Texte ebenfalls zuerst in LATEX zu setzen und dann den Konverter zu benutzen. Bei einem Vergleichstest konnte in der gleichen Zeit ein doppelt so großer Text über den Umweg LATEX nach HTML konvertiert werden als durch ein direktes Umwandeln mit HTML-Editoren; letztere sind noch recht rudimentär und bieten wenig Funktionalität.

Die vom Konverter generierten HTML-Dokumente erzeugen einen baumförmigen Hypertext, dessen Schichten sich nach Kapiteln, Abschnitten und Unterabschnitten strukturieren. Will man durch Verweise Sinnzusammenhänge ausdrücken, so müssen diese nachträglich manuell eingefügt werden. Die automatische Konvertierung nimmt einem die Umsetzung der Formate von LATEX nach HTML ab, eine abschließende Nachbearbeitung der Dokumente ist aber in der Regel noch erforderlich.

4.2 Einbindung multimedialer Komponenten

Zwar können multimediale Objekte in das Web integriert werden, aber die technischen Möglichkeiten bei der Nutzung vorhandener Hardware und die Online-Erstellung setzten ihrem Gebrauch gewisse Grenzen. Das Web stellt die Daten auf einem zentralen Server zur Verfügung, so daß diese über das Rechnernetz an die einzelnen Arbeitsplätze übertragen werden müssen, was bei Videos, die meist mehrere Megabyte groß sind, Kapazitätsprobleme schafft. Des weiteren stehen an den Arbeitsplätzen der Studierenden keine Video- oder Soundkarten zur Verfügung. Die Erstellung eigener Videos und Animationen erfordert einen sehr großen zeitlichen Aufwand sowie eine entsprechende Kompetenz. Eine andere Beschränkung ergab sich aus den personellen Resourcen, die notwendig gewesen wären, um Bewegtbilder zu erzeugen. Der Einsatz von Videos oder Animationen konnte von uns deswegen noch nicht vorgenommen werden.

So wurden die Hypertexte ausschließlich durch Bilder erweitert. Das Einbinden von Bildern in HTML bereitet keine Schwierigkeiten, solange diese in einem standardisierten Bitmap-Format vorliegen. Die meisten Bilder wurden mit einem Scanner eingelesen, der Rest mit Zeichenprogrammen auf dem Rechner erstellt. Hierbei wurde ein vektororientiertes Zeichenprogramm eingesetzt, das mehr Möglichkeiten z. B. bei der Skalierung bietet als ein pixelorientiertes Malprogramm. Die Bilder wurden dann nachträglich in das erforderliche Bitmap-Format konvertiert. Beim Erzeugen der Bitmap-Dateien mußte bei der Wahl der Auflösung zwischen einer guten Lesbarkeit und einer annehmbaren Netzübertragungszeit abgewogen werden.

Ab der zweiten Semesterhälfte stand das Programm Imagemap zur Verfügung, mit dem Verweise aus Graphiken heraus angelegt werden können. Hiermit ergaben sich weitergehende Gestaltungsmöglichkeiten, da HTML selbst keine graphischen Anordnungen unterstützt. Nun konnten Zusammenhänge erstmals in einer räumlichen Struktur dargestellt werden, von der aus Verweise zu näheren Erklärungen führen.

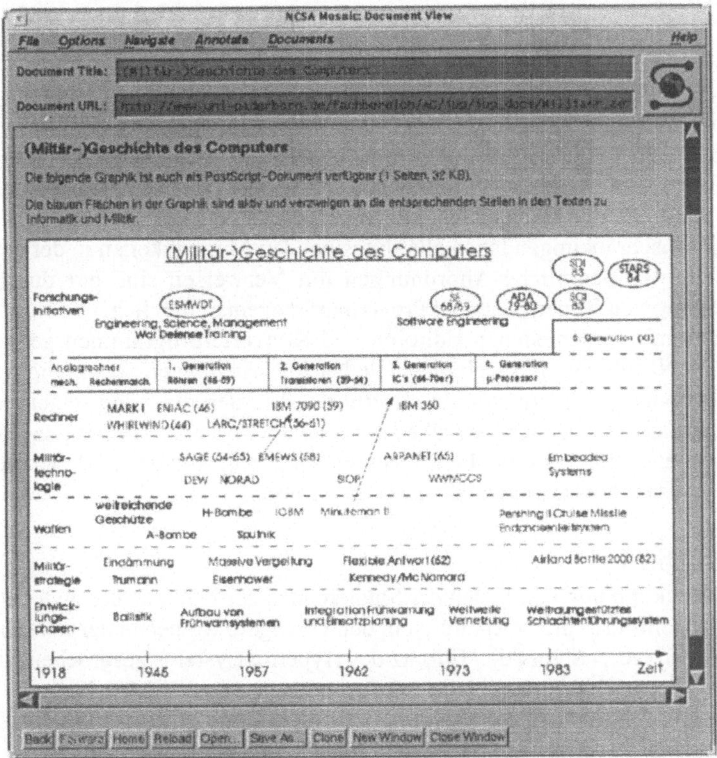

Abb. 2: Graphische Darstellung von Zusammenhängen im Web durch Einlesen eines Bildes.
Verweisen kann direkt aus dem Bild heraus nachgegangen werden.

Der Zeitaufwand für die Erstellung solcher Dokumente ist allerdings recht hoch. Neben der graphischen Gestaltung müssen für die Verweisstruktur in verschiedenen Dokumenten Ankerpunkte und die Verweise auf die entsprechenden Ankerpunkte gesetzt werden. Hierzu gibt es für das Web noch keine geeigneten Werkzeuge. Die Verweise werden in einer Datei textuell beschrieben. Es ist nicht möglich, sie interaktiv mit der Maus zu setzen.

In Abbildung 2 wird ein Bild aus den Lehrveranstaltungsunterlagen gezeigt, das in einer graphischen Übersicht die Militärgeschichte des Computers darstellt. Darin wurde der in der Vorlesung behandelte Stoff zeitlich angeordnet und so noch einmal der inhaltliche Zusammenhang verdeutlicht. Nähere Erklärungen zu einzelnen Schlagworten und Abkürzungen können über Verweise aufgerufen werden, indem mit der Maus der entsprechende Begriff im Bild ausgewählt wird. Für welche Begriffe dies möglich ist, wird durch eine farbige bzw. helle Markierung angezeigt. Anders als in den HTML-Texten ist es aber nicht möglich, die schon einmal besuchten Verweise zu markieren.

4.3 Fazit

Das Web ist ohne allzu großen Einarbeitungsaufwand zur Erstellung von Lehrveranstaltungsunterlagen einsetzbar. Die ersten Texte waren schon nach einem Nachmittag Arbeit in das Web eingebunden. Allerdings sind die für das Web zur Verfügung stehenden Werkzeuge noch recht rudimentär. WYSIWYG gibt es bei der Erstellung noch nicht und Verweise können nicht graphisch gesetzt werden.
Die größten Einschränkungen lagen allerdings bei den Möglichkeiten in der graphischen Gestaltung. Räumliche Anordnungen mit Verweisen sind nur durch die Nachbearbeitung von Bildern mit dem Programm Imagemap möglich.
Dadurch, daß mit verschiedenen Editoren und Konverterprogrammen gearbeitet wurde und im Nachhinein oft noch manuelle Nachbearbeitungen notwendig waren, ist es sehr schwierig, Änderungen durchzuführen. Werden die Änderungen in den Ursprungsdokumenten gemacht, so gehen die Daten der Nachbearbeitung nach einer erneuten Konvertierung verloren. Führt man die Änderungen in den Zieldokumenten aus, so sind diese inkonsistent zu den Ursprungsdaten.

5 Evaluation

Eine erste Rückkopplung von Seiten der Studierenden zur Nutzung des angebotenen Hyermediasystems erfolgte wöchentlich in den Übungen, wo neben der inhaltlichen Diskussion auch laufend über den Einsatz des Hypertextsystems gesprochen wurde. Zur weiteren Beurteilung wurden die Studierenden zu ihren Eindrücken und ihrer Nutzung befragt, indem sie ihre Meinung zu Beginn und am Ende des Semesters jeweils in einem Fragebogen zumeist durch freie Nennungen äußern konnten.
An der Veranstaltung nahmen ca. 80 Studierende teil. Der Nutzungsgrad des Systems als Medium zum Lesen von Texten stellte sich im Nachhinein als sehr niedrig heraus – Gründe hierfür werden später noch erläutert. Dennoch läßt sich aus den Nennungen der Studierenden einiges über ihre Probleme beim Gebrauch des Systems ablesen.
Die Veranstaltung wurde von ca. 70 % Diplominformatikern und je 15 % Wirtschaftsinformatikern und Lehramtstudierenden besucht. Alle hatten bereits mehrjährige Rechnererfahrung, mit einem Hypertextsystem hatten jedoch 56 % noch nie gearbeitet. Nur 8 % gaben an, ein Hypertextsystem häufig zu nutzen.

Die Einstellung der Studierenden zu der Idee, Lehrveranstaltungsunterlagen auf dem Rechnernetz zur Verfügung zu stellen, war überwiegend positiv (siehe Tabelle 1). Dabei ist zu berücksichtigen, daß die Studierenden nur beschränkt auf das System angewiesen waren. Zum einen wurde neben dem Hypertext auch eine Kopiervorlage zur Verfügung gestellt, zum anderen bestand die Möglichkeit, einzelne Dokumente oder Teile des Hypertextes auszudrucken.

	sehr gut	gut	geht so	nicht gut	keine Angabe
zu Beginn	6 %	49 %	25 %	7 %	13 %
am Ende	17 %	42 %	22 %	9 %	17 %

Tab. 1: Nennungen der Studierenden auf die Frage: „Wie findest Du die Idee, Unterlagen auf dem Rechnernetz bereitzustellen", zu Beginn und am Ende des Semesters.

Der Nutzungsgrad von Mosaic zum Lesen der Hypertexte war mit 36 % der Studierenden nicht sehr hoch und nahm zum Ende des Semesters sogar auf 7 % ab (siehe Tabelle 2), obwohl die Studierenden die Idee rechnergestützter Unterlagen überwiegend gut fanden. Das bevorzugte Medium zum Lesen war letztendlich doch Papier. In den weiteren Angaben der Studierenden lassen sich dafür einige Gründe ausmachen.

	nicht gelesen	in Mosaic gelesen	fotokopiert	ausgedruckt
zu Beginn	12 %	36 %	42 %	34 %
am Ende	19 %	7 %	52 %	27 %

Tab. 2: Nutzung von Mosaic zum Lesen der Texte zu Beginn und am Ende des Semesters (es wurde jeweils die Nutzung der ersten drei Texte zu Beginn und der letzten drei Texte am Ende der Veranstaltung gemittelt).

Die Qualität der heutigen Bildschirme ist im Vergleich zu Papier noch nicht sehr gut. Dies wurde von ca. 20 % der Studierenden genannt (Tabelle 4). Dieses Problem werden vermutlich erst strahlungsarme, hochauflösende und kontrastreiche LCD-Bildschirme lösen. Erst dann wird vielleicht auch so mancher Probeausdruck eingespart. Daß unser System kein Papier gespart hat, wurde den Studierenden erst im Laufe des Semesters klar, zu Beginn gaben 45 % der Studierenden die Papierersparnis als Grund für den Einsatz rechnergestützter Unterlagen an (siehe Tabelle 3).

angegebene Gründe dafür	zu Beginn	am Ende
Umweltschutz durch Papierersparnis	45 %	15 %
hohe Verfügbarkeit durch Zugang über beliebige Rechner des Netzes der Universität	35 %	41 %
Ausnutzung von Hypertextmöglichkeiten	10 %	11 %
allgemeines Üben des Umgangs mit Rechnern	9 %	2 %
einfaches Auffinden durch automatisierte Suche	7 %	9 %
Drucken bzw. Kopieren für die eigenen Unterlagen kann selektiv, nach einer Auswahl im Hypertextsystem, erfolgen	7 %	4 %
es gibt keine Gründe dafür	1 %	2 %

Tab. 3: Die häufigsten Nennungen auf die Frage: „Welche Gründe sprechen Deiner Meinung dafür, Unterlagen auf dem Rechnernetz bereitzustellen" (zu Beginn und am Ende des Semesters abgefragt).

angegebene Gründe dagegen	zu Beginn	am Ende
Lesbarkeit von Texten ist am Bildschirm schlecht	22 %	20 %
Papier hat Vorteile: Markieren, Randbemerkungen und einfaches Blättern	17 %	28 %
Zugang zum Hypertextsystem nur an der Universität	17 %	7 %
beschränkte Rechneranzahl (Engpässe)	14 %	20 %
der Lärmpegel in den Rechnerräumen an der Universität ermöglichen kein konzentriertes Lesen	10 %	15 %
hoher Einarbeitungsaufwand in das System	6 %	-
es gibt keine Gründe dagegen	7 %	13 %

Tab. 4: Die häufigsten Nennungen auf die Frage: „Welche Gründe sprechen Deiner Meinung dagegen, Unterlagen auf dem Rechnernetz bereitzustellen" (zu Beginn und am Ende des Semesters abgefragt).

Papier bietet im Vergleich zum eingesetzten Hypertextsystem wegen der Unabhängigkeit von der technischen Infrastruktur und neben der besseren Lesbarkeit weitere Vorteile. Die Möglichkeit, Stellen beliebig zu markieren oder Bemerkungen an den Rand zu schreiben, wurde zu Beginn von 17 %, am Ende von 28 % der Studierenden angemerkt (Tabelle 4). Das eingesetzte Hypertextsystem bietet zwar die Möglichkeit eigener Anmerkungen, diese wurde aber von nur 2 % der Studierenden genutzt (Tabelle 6).
Das hat unter anderem auch technische Gründe. Pro Dokument kann z. B. nur eine Anmerkung gesetzt werden, die an den Anfang des Dokuments plaziert wird, auch wenn dieses insgesamt mehrere Seiten umfaßt. Mit einigen technischen Tricks lassen

sich Anmerkungen auch an anderen Stellen plazieren, allerdings nur dort, wo der Autor einen Anker als möglichen Zielpunkt eines Verweises vorgesehen hat. Der Leser hat auch keine Möglichkeiten, eigene Verweise in den Dokumenten zu setzen, d. h. die vom Autor vorgegebene Verweisstruktur zu ändern, um z. B. die für ihn relevanten Stellen in Beziehung zu setzen. Bezüglich der räumlichen Orientierung lassen sich Fotokopien und Bücher auf einem Schreibtisch besser umorganisieren, allerdings ohne das Netz der durch die jeweilige Anordnung verkörperten Beziehungen speichern zu können. Hypertext könnte diese Möglichkeit bereitstellen, allerdings müßte dazu die strikte Trennung von Autor – der die Verweisstruktur vorgibt – und Leser – der einer fertigen Verweisstruktur folgt – aufgehoben werden. Der Leser muß Anmerkungen schreiben können, die ja nichts anderes als Dokumententext sind, und Verweise ziehen können, d. h. er muß Leser und Autor sein.

Ein großes Problem sahen zu Beginn 14 % und am Ende 20 % der Studierenden in der beschränkten Anzahl an Arbeitsplatzrechnern an der Universität (siehe Tabelle 4). Gerade im Semester sind tagsüber nur wenig freie Rechner zu finden, die zudem mit guten Bildschirmen ausgestattet sind. Weiterhin wurde die hohe Lautstärke in den Rechnerräumen zu Beginn von 10 % und am Ende von 15 % der Studierenden beanstandet (siehe Tabelle 4), wo andere z. B. in Gruppenarbeit Programmieraufgaben lösen. Hier könnte bei einem größeren Einsatz von rechnergestützten Unterlagen die Einrichtung eines rechnerbestückten Leseraums Abhilfe schaffen. Eine andere Möglichkeit wäre, den Studierenden das Hypertextsystem und die Dokumente auf ihrem eigenen Rechner zur Verfügung zu stellen. Dies ist mit einem Public-Domain-Programm wie Mosaic kein Problem, da es auf verschiedenen Plattformen verfügbar ist; die Unterlagen müßten dann an die unterschiedlichen Rechner (hauptsächlich PC und Macintosh) angepaßt werden.

Bei den Gründen für den Einsatz rechnergestützter Unterlagen stand bei den Studierenden die hohe Verfügbarkeit im Vordergrund. Dies wurde zu Beginn von 35 % und am Ende von 41 % der Studierenden genannt (siehe Tabelle 3). Der Zugriff auf die Daten ist von jedem Arbeitsplatz des Universitätsrechnernetzes aus möglich. Die Ausnutzung von Hypertextmöglichkeiten und Suchfunktion sahen jeweils ca. 10 % der Studierenden als ein Argument für den Rechnereinsatz (siehe Tabelle 3). Damit Hypertext aber Vorteile bietet, müssen die Texte gut aufbereitet sein. Ein automatisches Konvertieren eines linearen Textes in eine Hypertext-Baumstruktur bringt dem Leser wenig Vorteile. Zum Aufbau sinnvoller Hypertexte müssen graphischer Aufbau, Einsatz von Verweisen und inhaltliche Gestaltung einhergehen. Hier weisen unsere Unterlagen sicherlich noch Defizite in der Aufbereitung des Stoffes auf.

Gestaltungsaspekt	Bewertungsskala	Mittelwert
Lesbarkeit der Texte	(3 gut – -3 schlecht)	0
Größe der Einzeltexte	(3 zu klein– -3 zu groß)	-2
Anzahl der Verweise	(3 zu viele – -3 zu wenige)	0
Auffindbarkeit wichtiger Information	(3 gut – -3 schlecht)	-1
Wiederauffindbarkeit von Information	(3 gut – -3 schlecht)	0
Orientierung im System	(3 gut – -3 schlecht)	0

Tab. 5: Bewertung einiger Gestaltungsaspekte auf einer Skala von -3 bis 3 (Mittelwerte durch Medianberechnung).

In der Befragung nach der Gestaltung schnitt unser System mittelmäßig, bei der Größe der Einzeltexte und der Aufindbarkeit wichtiger Information eher schlecht ab (siehe Tabelle 5).
Der schlechte Nutzungsgrad des Systems liefert auch einen Indikator dafür, warum in bezug auf die Nutzung der Systemfunktion (siehe Tabelle 6) das Speichern und Ausdrucken von Dokumenten den großen Stellenwert hatte.

genutzte Funktionalität	
Speichern von Dokumenten im eigenen Verzeichnis oder auf Diskette	42 %
Speichern von Einstiegsverweisen auf Texte (Hotlist)	37 %
Ausdrucken von Texten oder Textteilen	24 %
Suchen nach Stichworten im Text	11 %
Verfassen persönlicher Anmerkungen	2 %

Tab. 6: Häufigste Nennungen auf die Frage: „Welche Funktionalität hast Du bei der Arbeit mit Mosaic ausgenutzt"

Aus der Tabelle läßt sich auch ablesen, daß der individualisierte Zugang zu den Dokumenten einen höheren Stellenwert hatte als beispielsweise klassische Datenbank- und Informationssystemfunktionen wie z. B. das Suchen nach Stichworten. Dies entspricht auch den Zielvorstellungen, die die Begründer der Hypertext-Idee Vannevar Bush und Ted Nelson hatten, indem sie die Selektion von Informationseinheiten aufgrund individuell angelegter Verweise und Einstiegspunkte als zentrales Konzept von Hypertext bezeichneten ([1],[9]). Allerdings gilt hier festzustellen, daß eine Verallgemeinerung dieser Aussage unzulässig ist, weil die tägliche Nutzung des Systems zu gering war.
Als zusätzliche Information zu der Befragung wurden bei Diskussionen in den Übungen alleinstehende Texte als schlecht bewertet, da sie besser auf Papier gelesen werden können. Für gut befunden wurden graphische Darstellungen, die auch als Träger von Verweisstrukturen dienen (siehe Abbildung 2). Diese erläutern die

Zusammenhänge und ermöglichen ein schnelles Auffinden der wesentlichen Information über die Verweise.

Lehrveranstaltungsunterlagen als gut aufbereiteten Hypertext zur Verfügung zu stellen läßt sich nur schwer während anderer Tätigkeiten im Semester realisieren. Es bedarf hierfür neben dem hohen Zeitaufwand und der Beherrschung des Systems sowohl einer inhaltlichen als auch einer gestalterischen Kompetenz zur Aufbereitung der Unterlagen.

6 Fazit

Viele Gestaltungsaspekte von interaktiven Systemen im allgemeinen und Hypermediasystemen im besonderen lassen sich nicht universell und kontextfrei bestimmen. Unter den Bedingungen der Alltagspraxis, d. h. Zeit- und Ressourcenknappheit, zeigt sich erst, ob Designkonflikte angemessen von den Entwicklern gelöst worden sind. Designkonflikte treten immer dann auf, wenn eine oder mehrere sich partiell widersprechende Anforderungen nur auf Kosten von anderen verwirklicht werden können. Beispielsweise steht die Frage der Bildqualität (Auflösung) grundsätzlich im Konflikt mit Speicherbedarf und Übertragungszeit oder die Einfachheit der Hypertexterstellung wird mit einer Einschränkung der Gestaltungsmöglichkeiten erkauft. Unter den Bedingungen der Alltagspraxis erweist sich erst, welche Bedeutung einzelnen Gestaltungselementen zukommt.

Dabei wird auch deutlich, daß die häufig geforderte Trennung von Applikation und Interaktion im Hinblick auf die Bewertung von Handhabbarkeit und Durchschaubarkeit ebenso wenig rein interaktionsbezogen evaluiert werden kann wie die Aufgabenangemessenheit eines Systems ohne Bezug zum Arbeitskontext. So sind die Länge und der Aufbau von Texten erheblich entscheidender gewesen für die schlechte Nutzung unseres Systems als beispielsweise die Einschränkungen der textuellen Gestaltungsmöglichkeiten. Daß beispielsweise Texte wie auch Bilder immer nur linksbündig bezogen auf den Seitenrand bzw. die aktuelle Textposition gesetzt werden können und nicht frei plazierbar sind, ist keinem der von uns angesprochenen Web-Nutzer aufgefallen, umso stärker aber den Dokumentgestaltern.

Um Tabellen setzen zu können, muß man beim Edieren des HTML-Dokuments das Textattribut „fixed" angeben. Das Resultat: Tabellen werden im Schrifttyp Courier angezeigt und der umgebende Text in der Schriftart, die bei der Benutzung gesetzt wird. Nur so ist es möglich, mit Tabulatoren zu arbeiten (räumliche Textgestaltung). Allerdings ist allen Beteiligten die beschränkte Auswahl an zur Verfügung stehenden Schriften (Fonts) aufgefallen; doch spielte dies kaum eine Rolle.

Die Tatsache, daß heutige Bildschirme ein ungeeignetes Medium zum Lesen und Betrachten größerer Dokumente sind, muß unter gegenwärtigen Bedingungen in Kauf genommen werden. Technisch sind hier in nächster Zeit entscheidende Verbesserungen zu erreichen. Das gilt natürlich auch für die Gestaltung des Arbeitsumfelds (Leseraum).

Gravierender und grundlegender für die Systemgestaltung sind dagegen die Fragen der Verzahnung von Autor- und Leser-Zyklen. Hypermediasysteme sind vorwiegend als Autoren- und Präsentationssysteme konzipiert und erlauben daher den Nutzern nicht, die vorgegebenen Strukturen ihren Anforderungen gemäß umzuändern oder zu erweitern. Dazu wäre, wie der Autor von HERMES, Eric Schoop, feststellt „ein gänzlich neuer Systemansatz erforderlich" ([13] S. 165). Zwar gibt es auch von vornherein auf Kooperationsunterstützung angelegte Systeme wie z. B. SEPIA (siehe Hannemann et al. [4]), doch wäre hier noch zu überprüfen, inwieweit das in diesem System realisierte kognitive Modell des kooperativen Schreibens auch für die Verwaltung und Fortschreibung von Lehrveranstaltungsunterlagen geeignet wäre, bei denen zwar eine starke Verzahnung von Autor-Leser-Zyklen erforderlich ist, die Erstellung dieser Dokumente aber selten kooperativ erfolgt.
Neben dem Web gibt es das weltweit verteilte Hypermedia-System Hyper-G (Kappe, Maurer [5]). Es bietet mehr Möglichkeiten als das Web, z. B. werden die Verweise von den Texten getrennt gespeichert. Dies ermöglicht es dem Leser eine „persönliche" Verweisstruktur abzulegen oder Anmerkungen an beliebige Textstellen zu schreiben. Hyper-G hat ebenfalls eine Trennung von Dokumentbeschreibungssprache und Navigationswerkzeug und besitzt wie das Web nur wenige graphische Gestaltungsmöglichkeiten. Zur Zeit untersuchen wir, ob die prinzipiellen Vorteile von Hyper-G sich auch mit den bisher verfügbaren Werkzeugen bei der Erstellung und dem Einsatz rechnergestützter Lehrveranstaltungsunterlagen ausnutzen lassen und ob es evtl. sinnvoll ist, einen ganz anderen Systemansatz zu wählen.
Im Moment hat das Web Vorteile in der Standardisierung sowie der hohen Verfügbarkeit und Weiterentwicklung von Werkzeugen. Sowohl das Web als auch Hyper-G bedürfen jedoch grundlegender funktionaler und softwareergonomischer Verbesserungen.
Diese betreffen in erster Linie die Tatsache, daß beide hochgradig textbasiert sind. Verweis- und Dokumentstrukturen sind mehrdimensional und lassen sich kaum oder nur sehr schwer über eindimensionale (sequentielle) textuelle Strukturen erfassen und veranschaulichen. Insofern kommt der räumlichen Anordnung und Gestaltung multimedialer Bildschirmobjekte eine besondere Bedeutung zu. Zwar kann man mit Hilfe von zusätzlichen Werkzeugen unter Mosaic Verweisstrukturen an Bildelemente knüpfen, doch geschieht dies sozusagen durch die Hintertür. Neben dem bereits erwähnten hohen Aufwand bei Revisionen tritt das Problem auf, daß Verweise, die in solchen Bildstrukturen selektiert worden sind, nicht – wie bei allen anderen Verweisen – ihr Farbattribut ändern, um anzuzeigen, daß der entsprechende Knoten bereits besichtigt worden ist.
Darüber hinaus ist es im Web auch nicht möglich, eine graphische Struktur, die den Navigationsraum visualisiert, zu benutzen, um anzuzeigen, wo man sich gerade befindet. Dieses Defizit der unzureichenden Rückmeldung läuft damit wesentlichen softwareergonomischen Kriterien der Navigationsunterstützung (Nievergelt [11]),

der direkten Manipulierbarkeit (Shneiderman [14]) oder der Präsentation des Handlungsabschlusses (Keil-Slawik [7]) zuwider. Aufgrund dieser fehlenden Rückmeldemöglichkeiten ist es auch nicht möglich, das Prinzip der interreferentiellen Ein-/Ausgabe (Draper [2]) zu verwirklichen oder mit „Drag & Drop"-Mechanismen Objekt- und Verweisstrukturen zu manipulieren.

Insofern fallen sowohl das Web als auch Hyper-G trotz ihrer Hypermedia-Eigenschaften hinter den Stand der Kunst in der Software-Ergonomie zurück und erweisen sich als nur begrenzt tauglich zur Bereitstellung von Lehrveranstaltungsunterlagen.

7   Ausblick

Die hier geschilderten Erfahrungen haben gezeigt, daß neuartige Möglichkeiten der Technik und aktueller Kenntnisstand im Bereich der Software-Ergonomie nicht notwendigerweise gleichermaßen dem Stand der Kunst entsprechen. Sowohl auf der Seite der Gestaltung wie auch auf der Seite der Nutzung treten Designkonflikte auf und müssen Kompromisse geschlossen werden.

Die beiden wesentlichen hier aufgezeigten softwareergonomischen Defizite von Web bzw. Mosaic, nämlich unzureichendes Feedback und unzureichende räumliche Plazierbarkeit von Bildschirmobjekten könnten durch eine erweiterte Dokumentbeschreibungssprache z. B. HTML+ vermindert werden. (Einen Entwurf von HTML+ findet man im World Wide Web unter der URL-Adresse: „http://www. informatik.tu-muenchen.de/tum.informatik/admin/html_doc.html") Damit können insbesondere die Anforderungen an die Textgestaltung realisiert werden.

Andere Probleme lassen sich mit dem Web nicht lösen. Durch die Trennung von Dokumentenbeschreibungssprache (HTML) und Navigationswerkzeug (z. B. Mosaic) ist es zwar möglich, neue Werkzeuge für das Web zu entwickeln, die auf HTML aufsetzen, aber eine stärkere Verzahnung von Autor-Leser-Zyklen ist nicht ohne weiteres realisierbar, weil dann eine Trennung von Dokument und Verweisstruktur erforderlich wäre. Hyper-G als „Hypertext-System der 2. Generation" bietet diese Trennung von Dokument und Verweisstruktur, basiert aber im wesentlichen auch auf einer textbasierten Dokumentbeschreibungssprache mit wenig graphischen Gestaltungsmöglichkeiten.

Es gilt weiterhin zu untersuchen, welchen Stellenwert einzelne Aspekte wie weltweite Vernetzung, offene Systeme, benutzereigene Verweise oder räumliche Navigation für den jeweiligen Einsatz haben und wie diese durch verschiedene Systeme abgebildet werden können. Hier müssen die unterschiedlichen Vor- und Nachteile gegeneinander abgewogen werden, da bisher keines der vorgestellten Systeme alle Anforderungen gleichermaßen erfüllt. Die entscheidende Frage, die zu beantworten ist, lautet: Auf welcher Grundlage lassen sich die hier dargestellten Anforderungen durch neue Versionen und Werkzeuge umsetzen, oder müssen dazu gänzlich neue Systeme entwickelt werden?

Literatur:

[1] Bush, V.: As we may think. Atlantic Monthly 176, S. 101 - 108, July, 1945

[2] Draper, S.W.: Display Managers as the Basics for User-Machine Communication. In: Norman, D.A., Draper, S.W. (Eds.): User Centered System Design – New Perspectives on Human-Computer Interaction. Lawrence Earlbaum: Hillsdale London, 1986

[2] Fischer, M.: MIAS – Medizinisches Informations- und Ausbildungssystem. In: Glowalla, U. und Schoop, E. (Hrsg.): Hypertext und Multimedia – Neue Wege in der computergestützten Aus- und Weiterbildung. S. 145 - 148, Springer Verlag, 1992

[4] Hannemann, J., Thüring, M.: Das Hypermedia-Autorensystem SEPIA. In: Glowalla, U. und Schoop, E. (Hrsg.): Hypertext und Multimedia – Neue Wege in der computergestützten Aus- und Weiterbildung. S. 118 - 136, Springer Verlag, 1992

[5] Kappe, F., Maurer, H.: Hyper-G – a large universal hypermedia system and some spin-offs. ACM Computer Graphics, experimantal special online issue. Available by anonymous ftp from siggraph.org in directory publications/May_93_online/Kappe.Maurer May, 1993

[6] Keil-Slawik, R.: Konstruktives Design. Ein ökologischer Ansatz zur Gestaltung interaktiver Systeme. Habilitation, Forschungsberichte des Fachbereichs Informatik, Bericht Nr. 90-14, TU Berlin, 1990

[7] Keil–Slawik, R.: Telepresence with Time Delays. Designing the User Interface of a Telemedicine Workstation Under Real Life Constraints. Proc. of IMAGINA '93, Monte Carlo, Febr., 1993

[8] Kuhlen, R.: Hypertext – ein nichtlineares Medium zwischen Buch und Wissensbank. Springer Verlag 1991

[9] Nelson, T. H.: As we will think. In: Online 72: Conference Proceedings of the International Conference on Online Interactive Computing. Online Computer Ltd. Uxbridge (UK) S. 439 - 454, 1973

[10] Nelson, T.H.: Computer Lib. Dream Machines. Tempus Books of Microsoft Press Redmond, 1987

[11] Nievergelt, J.: Die Gestaltung der Mensch-Maschine-Schnittstelle. In: Kupka S. 41 - 50, 1983

[12] Schoop, E., Pohl, C. und Sonntag, R.: Die Hermes-CD – Betriebswirtschaftslehre als Hypermedia-Informationssystem. In: Dette, K. und Pahl, P. J. (Hrsg.): Multimedia, Vernetzung und Software für die Lehre – Mikrocomputer-Forum für Bildung und Wissenschaft 4. S. 33 - 40, Springer Verlag, 1991

[13] Schoop, E.: Benutzernavigation im Hypermedia Lehr-/Lernsystem HERMES. In: Glowalla, U. und Schoop, E. (Hrsg.): Hypertext und Multimedia – Neue Wege in der computergestützten Aus- und Weiterbildung. S. 149 - 166, Springer Verlag, 1992

[14] Shneiderman, B.: Direct Manipulation – A Step Beyond Programming Languages. IEEE Computer; Vol. 16, No. 8; S. 57 - 69, August, 1983

[15] Wiemer, W. Heuser, J., Kaak, D. und Schmidtmann, M.: MILES/Studienmodell Physiologie – Ein multimediales PC-Datenbanksystem als Universalträger der Lehrsammlung eines Studienfaches. In: Dette, K. und Pahl, P. J. (Hrsg.): Multimedia, Vernetzung und Software für die Lehre – Mikrocomputer-Forum für Bildung und Wissenschaft 4. S. 28 - 32, Springer Verlag, 1991

[16] Zores, R.: Weltweites Hypermediasystem WWW. In: SUGinfo 2/94 Herausgeber: SUN User Group Deutschland e. V. S. 3 - 6, 1994

URL-Adressen:

Im folgenden werden noch einige URL-Adressen angegeben, unter denen man im World Wide Web weitere Informationen findet.

Überblick über WWW:
http://www.informatik.tu-muenchen.de/about_www.html

Lehrstuhlbeschreibung der Arbeitsgruppe „Informatik und Gesellschaft" an der Universität-GH Paderborn:
http://www.uni-paderborn.de/fachbereich/AG/iug/iug.html

Lehrveranstaltungsunterlagen der Arbeitsgruppe:
http://www.uni-paderborn.de/fachbereich/AG/iug/iug_docs/iug_liste.html

WWW '94 Conference Workshop: Teaching & Learning with the Web:
http://tecfa.unige.ch/edu-ws94/ws.html

Auflistung einiger Projekte, die das WWW ebenfalls für die Lehre einsetzen:
http://www.uni-konstanz.de/misc/lehre.html

Andreas Brennecke, Reinhard Keil-Slawik
Heinz Nixdorf Institut
Universität–GH Paderborn
Warburger Str. 100
33098 Paderborn

Tel. 05251/602064, 05251/602066
Fax 05251/603427
Email anbr@uni-paderborn.de, rks@uni-paderborn.de

OASE: Eine Arbeitsplatzumgebung für komplexe Anwendungssysteme

Edmund Eberleh und Falco Meinke
SAP AG, Walldorf

Zusammenfassung

Für sehr viele Endanwender der betriebswirtschaftlichen Anwendungen des SAP R/3-Systems stellt das R/3-System ihre eigentliche Arbeitsumgebung dar, in der sie sich die meiste Zeit bewegen, während die Benutzungsoberfläche des Betriebssystems oft gar nicht oder nur sehr selten benötigt wird. Für die Anwender des SAP-Systems ist somit eine für sie zugeschnittene Arbeitsplatzumgebung bereitzustellen, die Ihnen eine effiziente und ergonomische Erledigung ihrer Tätigkeiten ermöglicht. Die dargestellte Arbeitsplatzumgebung OASE stellt einen Prototypen hierfür dar, der in einigen Aspekten bereits im aktuellen R/3-Release eingesetzt ist.

Der SAP-Arbeitsplatz OASE ist als ein kompaktes Fenster konzipiert, aus dem heraus die gesamte Funktionalität des R/3-Systems zugreifbar und kontrollierbar ist. OASE besteht aus vier Bereichen: (1) einem Standardmenü, welches die Gesamtheit aller SAP-Anwendungen in graphischer Form standardmäßig strukturiert zugänglich macht, (2) einem Benutzermenü, welches eine benutzerindividuell zusammengestellte Untermenge aller Anwendungen enthält, (3) einer temporären und längerfristigen Objektablage mit e-mail Eingangskorb, und (4) einer Diensteleiste mit anwendungsübergreifenden Funktionen und Devices. Die Größe der einzelnen Bereiche kann je nach Bedarf individuell verändert werden.

1   Problemstellung

Das R/3-System von SAP ist ein Softwarepaket für integrierte betriebswirtschaftliche Standardanwendungen. Es ist modular aufgebaut und deckt den gesamten Bereich betrieblicher Funktionen ab. Das R/3-System basiert auf einer Client/Server-Architektur und läuft auf den Betriebssystemen UNIX, Windows, OS/2 und MacOS. Es stellt sich für den Endbenutzer als eine Anwendung innerhalb seiner anderen Desktop-Programme dar, und präsentiert sich im Look & Feel der jeweiligen Plattform. Während das R/3-System nach außen somit als eine abgeschlossene Einheit im Desktop des Front-End-Geräts erscheint, stellt es nach innen ein sehr komplexes Paket vieler miteinander vernetzter Anwendungen und Daten dar, mit eigener Systemverwaltung und Entwicklungs- und Kommunikationswerkzeugen. Für den Benutzer wiederholen sich damit im R/3-System strukturell viele Tätigkeiten aus der übergeordneten

Betriebssystemoberfläche: Anwendungen finden und aufrufen, Dateien öffnen und ablegen, das System einstellen und verwalten.

Für sehr viele Endanwender stellt dabei das R/3-System ihre eigentliche Arbeitsumgebung dar, in der sie sich die meiste Zeit bewegen, während die Benutzungsoberfläche des Betriebssystems oft gar nicht oder nur sehr selten benötigt wird. Für die Anwender des SAP-Systems ist somit eine für sie zugeschnittene Arbeitsplatzumgebung bereitzustellen, die Ihnen eine effiziente und ergonomische Erledigung ihrer Tätigkeiten ermöglicht. Die im folgenden dargestellte Arbeitsplatzumgebung OASE stellt einen Prototypen hierfür dar, der in Teilen bereits im aktuellen R/3-Release eingesetzt ist. OASE stellt den zentralen Ausgangspunkt für die Bearbeitung aller Anwendungs- und Systemobjekte dar. Die Arbeitsplatzumgebung OASE erlaubt alternativ und parallel einen **O**bjekt-, **A**nwendungs-, **S**tandard- und **E**reignisorientierten Zugriff auf alle R/3-Systemfunktionalität aus einem kompakten und flexiblen Kontroll-Fenster heraus.

Bei der Konzeption und Gestaltung von OASE wurden die aktuellen und zukünftigen Entwicklungen im Bereich der Desktop-Manager analysiert und berücksichtigt. OASE stellt einen Versuch dar, einige der dort auftretenden Probleme zu vermeiden bzw. bestehende Lösungsansätze zu übertragen und zu integrieren. Als Problembereiche stellten sich für SAP insbesondere die schnelle Zugreifbarkeit auf individuell unterschiedliche Funktionen, die Angemessenheit der Objektorientierung, die effiziente Verwaltung und Verknüpfung mehrerer amodaler Fenster sowie die effiziente Nutzung des beschränkten Bildschirmplatzes dar.

2 Bestehende Desktop-Lösungen und deren Benutzbarkeitsprobleme

2.1 Apple Macintosh Finder

Der Apple Macintosh Finder stellt die klassische graphisch-direktmanipulative Benutzungsoberfläche für ein PC-Betriebssystem dar [1]. Anwendungen und Objekte werden als Icons oder in Fenster dargestellt, und können mit der Maus manipuliert werden. In Verbindung mit der Bürometapher und dem WYSIWYG-Prinzip war dieses graphische Fenstersystem ein wesentlicher Fortschritt in der Benutzbarkeit eines Computersystems.

Es zeigten sich jedoch bald Grenzen bzw. Unzulänglichkeiten eines derartigen Ansatzes. So war die Verwaltung eines großen Dateisystems nur anhand von geschachtelten Ordnern bzw. Fenstern ziemlich unübersichtlich, da kein schneller Blick in die Tiefe der Hierarchie möglich war. Mit der Baumdarstellung des Dateisystems in späteren Betriebssystemversionen sollte dieses Problem gelöst werden. Das schnelle Hin- und Herschalten zwischen parallel laufenden Anwendungen nur mittels der ständig sichtbaren geöffneten Fenster erzeugte bei zunehmender Anwendungszahl ebenfalls Benutzbarkeitsprobleme. Ein Menü mit einer Liste der laufenden Anwendungen war die Konsequenz dieser Problematik.

Und schließlich war die Benutzungsoberfläche in vielen Fällen nicht wirklich objektorientiert, sondern basierte auf Anwendungen, die Objekte bearbeiteten. Derartige Anwendungen wurden in den verschiedenen Systemversionen graduell zu einer immer direkteren Manipulation der eigentlichen Objekte erweitert. In der neuesten Systemversion stellt der Finder somit eine evolutionär gewachsene Betriebsystemoberfläche dar, in der viele einzelne Ideen und Entwicklungen miteinander verwoben sind.

2.2   IBM CUA workplace

Mit der Version OS/2 2.0 präsentierte IBM eine Benutzungsoberfläche für ihr PC-Betriebsysystem, welches dem IBM-CUA Standard von 1991 entsprach [4]. Diese Oberfläche war von Grund auf gemäß dem Prinzip der Objektorientierung aufgebaut, d.h. daß Icons wirklich Objekte repräsentieren, die in verschiedenen Sichten angeschaut werden können.
Die Nutzung des Programms offenbart jedoch in deutlicher Weise den elementaren Nachteil der prinzipiell scheinbar angemessenen Objektorientierung: indem alle Funktionen nur über ein Objekt zugänglich sind, gestaltet sich das Arbeiten mit dem System als permanente Suche nach dem richtigen Objekt und der richtigen Sicht des Objektes. Scheinbar einfache Aufgaben wie das Öffnen der Dateisystemübersicht werden so für ungeübte Nutzer fast unlösbar, wenn er die Eigenschaften der einzelnen Objekte nicht kennt bzw. die Systemobjekte nicht mit den mentalen Objekten korrespondieren. Das Öffnen einer neuen Ansicht eines Objektes in jeweils einem neuen Fenster führt darüberhinaus in kürzester Zeit zu einer Unmenge geöffneter Fenster.

2.3 NeXTStep

Die Besonderheit des innovativen NeXTSTEP-Workplace zeigt sich u. a. anhand des File Viewers, mit dem das UNIX-Dateisystem verwaltet wird [10]. Die Hierarchie des Dateisystems wird mittels einer Reihe von Spalten in einem Browser angezeigt. Dabei enthält jede Spalte jeweils die Einträge eines Verzeichnisses. Der Name des Verzeichnis wird über der Spalte mit Icon in der Pfadsicht (Icon Path) angezeigt. Wird ein Eintrag in einer Spalte ausgewählt und repräsentiert dieser wiederum ein Verzeichnis, so wird in der rechten Spalte daneben der Inhalt des Unterverzeichnisses angezeigt. Der Browser kombiniert eine parallele Sichtbarkeit von Einträgen auf verschiedenen Hierarchieebenen mit einer parallelen Sichtbarkeit eines gesamten Hierarchiepfades.

Um auf oft benötigte Verzeichnisse direkt Zugriff zu haben, können diese aus dem Icon Path direkt per Drag & Drop auf eine Ablage (Shelf) oberhalb des Icon Path deponiert werden. Ein Mausklick auf ein Symbol im Shelf zeigt den dazugehörigen Pfad im Browser an. Dieses Browser-Shelf Konzept erlaubt damit eine sehr schnelle Navigation in großen Hierarchien ohne Orientierungsverlust. Außer auf dem Shelf können Objekte in einem „Dock" genannten Bereich am rechten Rand des Workplace abgelegt werden. Ins Dock können allerdings nur Programmobjekte eingestellt werden, und die Kapazität ist auf 13 Elemente beschränkt. Das Dock erlaubt ein schnelles Starten der individuell am häufigsten benötigten Anwendungen.

Mittels der Hide-Funktion können alle Fenster einer Anwendung unsichtbar gemacht werden. Klick auf das Icon der Anwendung im Dock macht die verborgenen Fenster wieder sichtbar. Dieser Mechanismus der virtuellen Bildschirme erlaubt eine sehr effiziente Verwaltung der vielen Fenster der Next-Anwendungen.

2.4 Microsoft Chicago

Die Benutzungsoberfläche von Microsoft Windows 4.0 (Chicago) stellt eine Kombination aus den bisherigen Ansätzen dar [7]. Sie kombiniert eine teilweise Objektorientierung mit prozeduralen Aspekten über das sog. Start-Menü, welches den Benutzer gezielt zu den Objekten bzw. Programmen führt. Damit sollen die bei OS/2 auftretenden Orientierungsverluste vermieden werden. Die dem Next-Dock ähnliche individualisierbare Task Bar dient der Visualisierung und dem schnellen Zugriff auf laufende Anwendungen. Zuletzt genutzte Objekte werden im Start-Menü ebenfalls zum schnellen Zugriff gesondert gespeichert und angezeigt.

2.5   Desktop-Erweiterungen

Die Unzulänglichkeiten der klassischen Betriebssystem-Oberflächen haben zu einer Vielzahl von Drittprodukten geführt, die eine alternative Programm- und Dateiverwaltung darstellen sollen [8]. Ein derartiges Programm ist z.B. Dashboard von Hewlett-Packard, welches den Programmanager von Windows ersetzen soll [5]. Es verwendet die Metapher eines Armaturenbrettes, und nimmt ohne geöffnete Fenster nur etwa ein Achtel der Desktopoberfläche in Anspruch. Die hervorstechenden Eigenschaften dieser Windows Shell liegen in der Möglichkeit der Schachtelung von Programmgruppen, der Verwendung erweiterter virtueller Bildschirme zur Eindämmung der Fensterflut, schneller Zugreifbarkeit von Diensten und Funktionen über Schaltflächen sowie den umfangreichen Individualisierungs-möglichkeiten durch den Benutzer.

2.6   Forschungsprototypen

Neuere Ansätze versuchen die Begrenzungen durch die klassische zweidimensionale Desktop-Oberfläche zu überwinden. „Workscape" stellt eine konsequente Erweiterung des Desktops auf eine 3D-Büroumgebung dar, in der aber noch relativ ähnlich zu bisherigen Desktops gearbeitet wird [6].
„Pad ++" verläßt die klassische Ordner-Metapher und betrachtet die Objekte als beliebig tiefe Hierarchie, die durch kontinuierliches zoomen zugänglich gemacht werden können [2].

„GLOBE" abstrahiert von der Büroumwelt und organisiert die Funktionalität entsprechend der mentalen Repräsentation in drei Dimensionen (Arbeitsablauf, Handlungsspielraum, Handlungssequenz) [3]. Als allgemeinere Metapher zur Abbildung dieser Dimensionen auf die Benutzungsoberfläche dient ein dreidimensionaler Globus.

Obwohl derartige Arbeiten sicherlich interessante Lösungen für die geschilderten Probleme darstellen, setzen sie viel an Verarbeitungsleistung und Grafikfähigkeit der Hardware voraus. Für den industriellen Einsatz kommen sie daher vorerst kaum in Betracht.

3   Der SAP-Workplace „OASE"

3.1 Organisation der aktuellen R/3-Funktionalität

Die insgesamt über 1000 Anwendungen des SAP-Systems mit ihren vielen
untergeordneten Detailfunktionen sind entsprechend der betriebswirtschaftlichen
Zusammengehörigkeit hierarchisch gruppiert, um dem Anwender ein schnelles
Auffinden der gesuchten Funktion zu ermöglichen. Im Sinne einer arbeitsplatz-
und aufgabenangemessenen Gestaltung soll eine Benutzerin all die Arbeitsobjekte
und -funktionen, die sie für ihre tägliche Arbeit benötigt, mühelos auffinden und
auswählen können. Die Abbildung dieser Anwendungshierarchie auf die
Benutzungsoberfläche geschieht zur Zeit (R/3 Release 2.2) in hierarchischen Pull-
down-Menüs (s. Abb. 1). Dieser Einstieg ähnelt damit sehr der von MS Chicago
gewählten Lösung der.

Darüberhinaus wurde eine objektbezogene Gruppierung der Funktionen angestrebt.
Die gesamte Anwendungshierarchie ist aus diesem Grunde auf drei Hauptebenen
verteilt (s. Abb. 1): Die zentrale Navigationsebene stellt die mittlere sog.
"Arbeitsgebietsebene" dar. In der Menüleiste dieser Ebene sind all die
Arbeitsobjekttypen aufgeführt, die an einem Arbeitsplatz üblichererweise
bearbeitet werden. In den untergeordneten Kaskadenmenüs stehen die auf diese
Objektklassen anwendbaren Funktionen. Durch Wählen einer dieser
Funktionenwird die entsprechende Anwendung aufgerufen, in der dann
verschiedenen Objekte aus der gewählten Objektklasse nacheinander bearbeitet
werden können. Beenden der gewählten Objektklassenbearbeitung läßt den
Benutzer auf sein Arbeitsgebietsmenü zurückkehren. Beendet der Benutzer auch
dieses Menü, so gelangt er auf die erste Hauptebene des Anwendungsbaumes.
Diese Ebene existiert nur einmal im gesamten R/3-System und dient ausschließlich
zum
Verzweigen in die einzelnen Arbeitsgebietsmenüs, von denen es ca. 100 gibt.

Neben dieser hierarchischen Navigation gibt es mehrere Abkürzungen und
Direktaufrufe: (1) Die Anwendungen selbst sind untereinander verknüpft, soweit
für den Arbeitsablauf nötig. (2) Der Benutzer kann durch eine Systemeinstellung
direkt das Startmenü setzen, so daß er die erste Hauptebene nicht mehr manuell
durchlaufen muß. (3) Direkteingabe eines Transaktionscodes in ein Befehlsfeld ruft
sofort eine Anwendung aus einem beliebigen Zweig des Menünetzes auf.

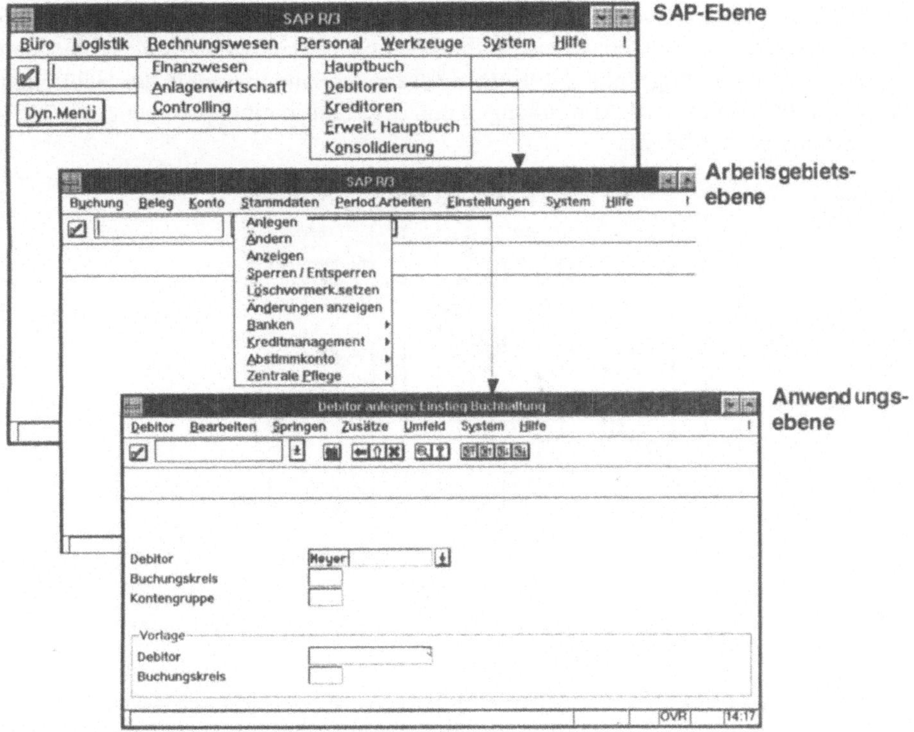

Abb. 1: Die R/3-Menüstruktur (Release 2.2)

Obwohl das jetzige Menüsystem somit alle in der ISO-Norm geforderten
ergonomischen Eigenschaften erfüllt, könnten einige Aspekte möglicherweise noch
effizienter gestaltet werden: es enthält nur das von SAP vorkonfigurierte
Standardmenü und erlaubt im Standard keine benutzerspezifischen
Arbeitsplatzmenüs, es bildet nur indirekt eine Objekt- bzw. Vorgangsorientierung
ab, es erlaubt keine direkte Manipulation bei System- bzw. Dateioperationen, und
es existiert nur ein gemeinsames Primärfenster für Einstiegsmenü und Anwendung.

3.2 Nutzungscharakteristiken von SAP-Anwendungen

Um die genannten möglichen Defizite auf ihre Bedeutung für Endbenutzer
einschätzen zu können, wurde im Rahmen einer Aufgabenanalyse im Bereich
Logistik-Einkauf u.a die tägliche Nutzungshäufigkeit von SAP-Anwendungen
erhoben und mit Benutzerparametern in Beziehung gesetzt [9]. In der Studie

zeigten sich prägnant unterschiedliche Nutzergruppen und Nutzungsprofile von SAP-Software. In der untersuchten Stichprobe von 18 Personen aus verschiedenen Firmen besteht ein umgekehrt u-förmiger Zusammenhang zwischen der Häufigkeit insgesamt bekannter SAP-Anwendungen und der Häufigkeit täglich benutzer SAP-Anwendungen (s. Abb. 2).

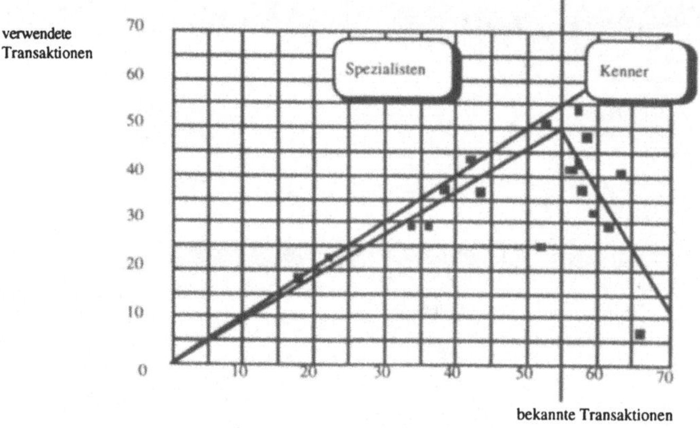

Abb 2: Zusammenhang zwischen der Zahl bekannter und verwendeter Transaktionen (Einkauf) mit Regressionsgeraden

Dieser Verlauf läßt auf wenigsten zwei unterschiedliche Gruppen von Benutzern schließen: die Spezialisten, die die relativ geringe Zahl ihnen bekannter Anwendungen auch alle täglich nutzen (bis maximal 50 Anwendungen), und die Generalisten, die von der größeren Zahl ihnen bekannter Anwendungen nur wenige (kleiner 50) in der täglichen Arbeit benutzen. Es liegt die Vermutung nahe, daß diese gelegentlichen Benutzer eine Vorgesetztenfunktion haben. Beide Gruppen nutzen darüber hinaus unterschiedliche Anwendungen des gleichen Arbeitsgebietes, bzw. nutzen die gleiche Anwendung unterschiedlich häufig.

Aus diesen Befunden ergeben sich zwei Forderungen für die Systemgestaltung: (1) Benutzer müssen einen schnellen Zugriff auf eine begrenzte Menge von Anwendungen haben, und (2) Benutzer müssen sich diese von ihnen genutzte Anwendungsmenge individuell zusammenstellen können.

3.3 Neudesign des SAP-Workplace

Vor dem Hintergrund der geschilderten Probleme wurde ein Prototyp einer SAP-Arbeitsplatzumgebung konzipiert und unter Nextstep entwickelt. Abb. 3 stellt

schematisch die verschiedenen Bereiche des SAP-Workplace „OASE" dar. Jeder
Bereich verkörpert jeweils einen zentralen Benutzbarkeitsaspekt, so daß in der
Gesamtheit eine flexible Arbeitsumgebung vorliegt, die je nach Interesse
individuell unterschiedlich genutzt werden kann.

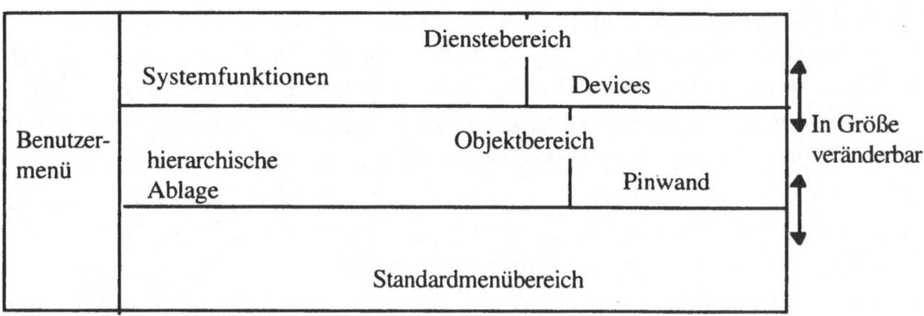

Abb. 3: Aufteilung des SAP-Workplacefensters

Standardmenü

Der Standardmenübereich dient zur strukturierten Aufnahme der gesamten
Standardfunktionalität des R/3-Systems. Er stellt ein Abbild des bisherigen R/3-
Menüsystems der SAP- und Arbeitsgebietsebene dar. Die Aufteilung in SAP- und
Arbeitsgebietsebene bleibt erhalten. Die SAP-Ebene soll lediglich als
Präsentations- und Einstiegsebene fungieren, während die Arbeitsgebietsebene den
Kern des SAP-Workplace darstellt. Die Anzeige der Arbeitsgebietsebene geschieht
durch Zoomen eines Anwendungsbereiches auf der SAP-Ebene. In Anlehnung an
das SAP-Logo sind die einzelnen Objektklassen, die mit dem System bearbeitbar
sind, als Rauten dargestellt, die durch Mausklick im Primärfenster geöffnet werden
(s. Abb. 4). Die Objektklassen sind hierarchisch um einen zentralen Ring
gegliedert, und je nach Benutzerinteresse können einzelne Äste durch Klick auf
den inneren Ring geöffnet oder geschlossen werden. Der Benutzer kann sich damit
individuell das gesamte Systemangebot auf seine aktuellen Bedürfnisse einstellen.

Benutzermenü - Anwendungsorientierung

Das Benutzermenü stellt ein Hauptelement von OASE dar. Es verknüpft die klassische und oftmals sinnvolle Anwendungsorientierung mit Individualisierungsaspekten. Durch das Benutzermenü können alle Anwendungen, die ein Anwender benötigt, aus der Vielfalt des gesamten Standardmenüs individuell herausgezogen und gruppiert werden. Diese Anwendungen sollten parallel neben dem Hauptfenster sichtbar und durch Doppelklick mit der Maus aufrufbar sein. Um diese Eigenschaft zu erfüllen, ist für das Benutzermenü eine Spalte am Fensterrand vorgesehen, die sich dem Benutzer wie ein Regal darstellt. In dieses Regal können die gewünschten Anwendungen vom Endbenutzer einsortiert werden, indem eine Anwendung (Raute) aus dem Standardmenü mit der Maus auf einen freien Regalplatz gezogen wird. Der Anwender hat nun noch die Möglichkeit zur individuellen Bezeichnung der Anwendung sowie zur direktmanipulativen Vertauschung der Reihenfolge der Anwendungen in den Regalplätzen, so daß etwa eine bestimmte Arbeitsreihenfolge abgebildet werden kann. Darüberhinaus gibt es als Strukturierungsmittel der Anwendungen noch das Konzept der mehrfachen Regalbretter. Anwendungen aus verschiedenen Arbeitsbereichen können auf verschiedene Regalbretter verteilt werden, so daß immer nur die für einen aktuellen Arbeitskontext sinnvollen direkt sichtbar sind.

Ein R/3 Anwender arbeitet in der Regel mit einer großen Anzahl von Objekten. Dabei benötigt er bestimmte Objekte nur ein- oder wenige Male (z.B. ein Erfassungsbeleg), während er andere Objekte vielleicht mehrmals täglich benutzt (z.B. Übersichtslisten über Bestände). Zweck des Objektbereiches ist es nun, die Arbeitsobjekte eines Benutzers aus beiden Kategorien für ihn schnell zugreifbar zu machen.

Der Ablagebereich ist direkt unterhalb des Dienstebereichs angeordnet. Er ist ebenfalls in zwei Bereiche aufgeteilt, einer hierarchischen und einer listenartigen Ablage (Pinwand). Letztere ist dafür vorgesehen, Objekte aufzunehmen, auf die der Anwender unmittelbaren Zugriff haben möchte, weil er sie häufiger benötigt. Sie dient der kurzfristigen Ablage weniger, aktueller Objekte. Dagegen ist der hierarchische Teil des Objektbereichs eher für die strukturierte Ablage einer größeren Objektzahl gedacht. Objekte können zwischen beiden Bereichen per Drag&Drop ausgetauscht werden. Außerdem lassen sich Objekte auf ein Device im Dienstebereich ziehen. Ein Doppelklick auf ein Objekt oder Ziehen desselben in ein geöffnetes Anwendungsfenster bringt das Objekt im Primärfenster der zugehörigen Anwendung zur Anzeige.

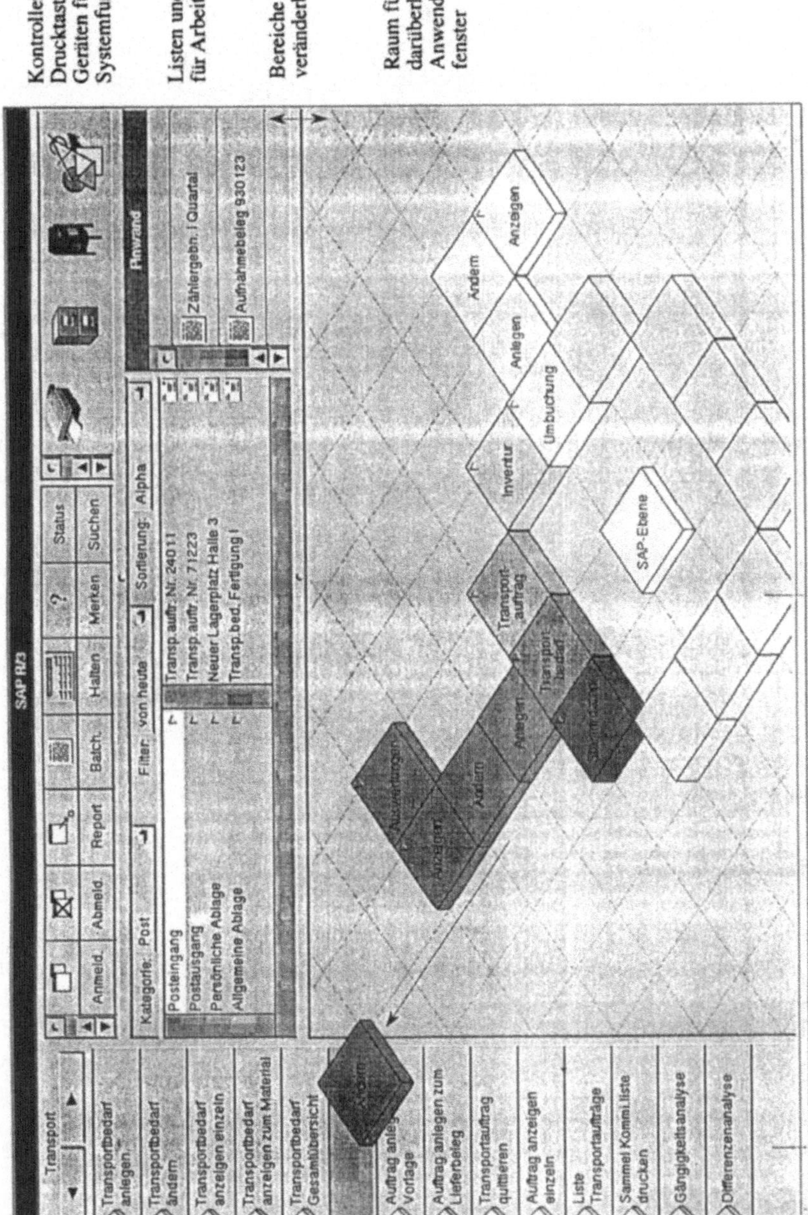

Abb. 4: SAP-Workplace OASE mit Standardmenü, Benutzermenü, Ablagecontainer und Werkzeugleiste

Ablagebereich - Objekt-/Vorgangsorientierung

Eine Kategorie innerhalb der hierarchischen Ablage, die ein Teil des Mailsystems des R/3 abbildet, ist mit „Post" benannt. Hier kann ein- und ausgehende Post verwaltet werden, indem Dokumente per Drag&Drop in den Postausgang gezogen werden oder eingehende Post per Doppelklick gelesen wird. Wenn eine Nachricht in einem bestimmten Format vorliegt, kann auch gleich eine verbundene Anwendung mit dem Objekt gestartet werden. So können beispielsweise Belegobjekte zur Weiterverarbeitung an andere Mitarbeiter versendet werden. Durch diese Möglichkeit wird der Arbeitsfluß (workflow) zwischen Arbeitsplätzen unterstützt und beschleunigt.

Dienstebereich - Ereignisorientierung

Der Dienstebereich ist als Leiste am oberen Fensterrand plaziert. Er teilt sich in den Funktionsbereich und den Devices-Bereich. Der Funktionsbereich ist als eine Matrix von Drucktasten realisiert, die die wichtigsten Funktionen zur Verwaltung des R/3 Arbeitsplatzes enthält und häufig benötigte anwendungsübergreifende Dienste zur Verfügung stellt.

In der rechten Hälfte des Dienstebereiches sind die Devices angeordnet, wie Drucker, Papierkorb, lokaler Massenspeicher, Postkasten und Grafik. Sobald der Anwender ein gültiges Objekt mit der Maus über ein Device zieht, gibt letzteres dem Benutzer durch Animation des Piktogramms ein Feedback, daß das Device das Objekt verarbeiten kann. Nach Loslassen des Objektes über einem Device wird das Objekt entsprechend verarbeitet. Über eine Blätterleiste können weitere Drucktastenzeilen mit weniger oft benötigten Funktionen oder weitere Devices eingeblendet werden.

Gesamtstruktur

Der SAP-Workplace ist als ein kompaktes Kontroll-Fenster konzipiert, daß im Normalfall größer als das Anwendungsfenster ist. Nach Aufruf einer Anwendung wird das neue Anwendungsfenster über dem Standardmenübereich des Workplacefensters plaziert. Dadurch bleiben das Benutzermenü, der Dienstebereich und auch Teile des Objektbereiches sichtbar und unmittelbar zugreifbar, was für effizientes Arbeiten mit dem Workplace sehr wichtig ist (s. Abb. 5).

Der Benutzer kann die Aufteilung des Workplacefensters an seine Aufgaben und persönlichen Präferenzen anpassen, indem er die Größe von Dienste-, Objekt- und Standardmenübereich verändert. Dies erfolgt durch Ziehen der Trennbalken (sash) zwischen den Bereichen mit der Maus. Wenn der Anwender z.B. alle benötigten Transaktionen in sein Benuztermenü kopiert hat, kann er den Standardmenübereich ausblenden, indem er den Trennbalken zwischen Objekt- und Standardmenübereich ganz nach unten zieht. Dadurch kann im Objekt- oder Dienstebereich mehr Information angezeigt werden.

Um ein Objekt, das in einem Anwendungsfenster angezeigt ist, im Objektbereich ablegen zu können oder auf ein Device zu übertragen, ist in der Titelleiste des Anwendungsfensters ein Dokument-Icon vorgesehen. Dieses Mini Icon symbolisiert das im Anwendungsfenster angezeigte Objekt. Das Mini Icon kann mit der Maus sowohl in die hierarchische Ablage als auch auf die Pinwand oder ein Device gezogen werden.

3.4   Aktueller R/3-Entwicklungsstand

Einige der dargestellten Eigenschaften von OASE sind in Ausschnitten bereits im aktuellen R/3-System realisiert (s. Abb. 6). So existiert das Benutzermenü als zweites amodales Fenster neben dem Primärfenster der Anwendungen, und im Rahmen der Workflow-Komponente existiert ein gemeinsamer Eingangskorb der Arbeitsobjekte bzw. -vorgänge und der persönlichen elektronischen Post. Wird ein Eintrag des Benutzermenüs ausgewählt, so kann das Anwendungsicon mit der Maus auf ein Primärfenster einer Anwendung gezogen werden. Die gewählte Anwendung wird anschließend in diesem Fenster gestartet. Alternativ zur Listendarstellung kann im Benutzermenü eine hierarchische Baumdarstellung gewählt werden.

Zur effizienten Fensterverwaltung mehrerer parallel laufender Anwendungen wurde eine Fensterleiste entwickelt (s. Abb. 6). Diese Fensterleiste wird vom System angezeigt, sobald der Benutzer zwei Anwendungsfenster übereinanderschiebt. Dieses bewirkt eine Stapelfunktion, die die beiden Fenster übereinanderlegt, so daß nur noch das oberste Fenster des Stapels sichtbar ist. Die Titel der gestapelten Fenster werden als Drucktaste in der Fensterleiste angezeigt. Klick auf eine derartige Fenstertaste bringt das entsprechende Fenster nach oben und verbirgt alle Fenster der vorher aktiven Anwendung.

Die bisherigen Rückmeldungen der SAP-Benutzer zum Benutzermenü sind positiv. Es wird an einer schrittweisen Erweiterung des aktuellen Systems um die weiteren

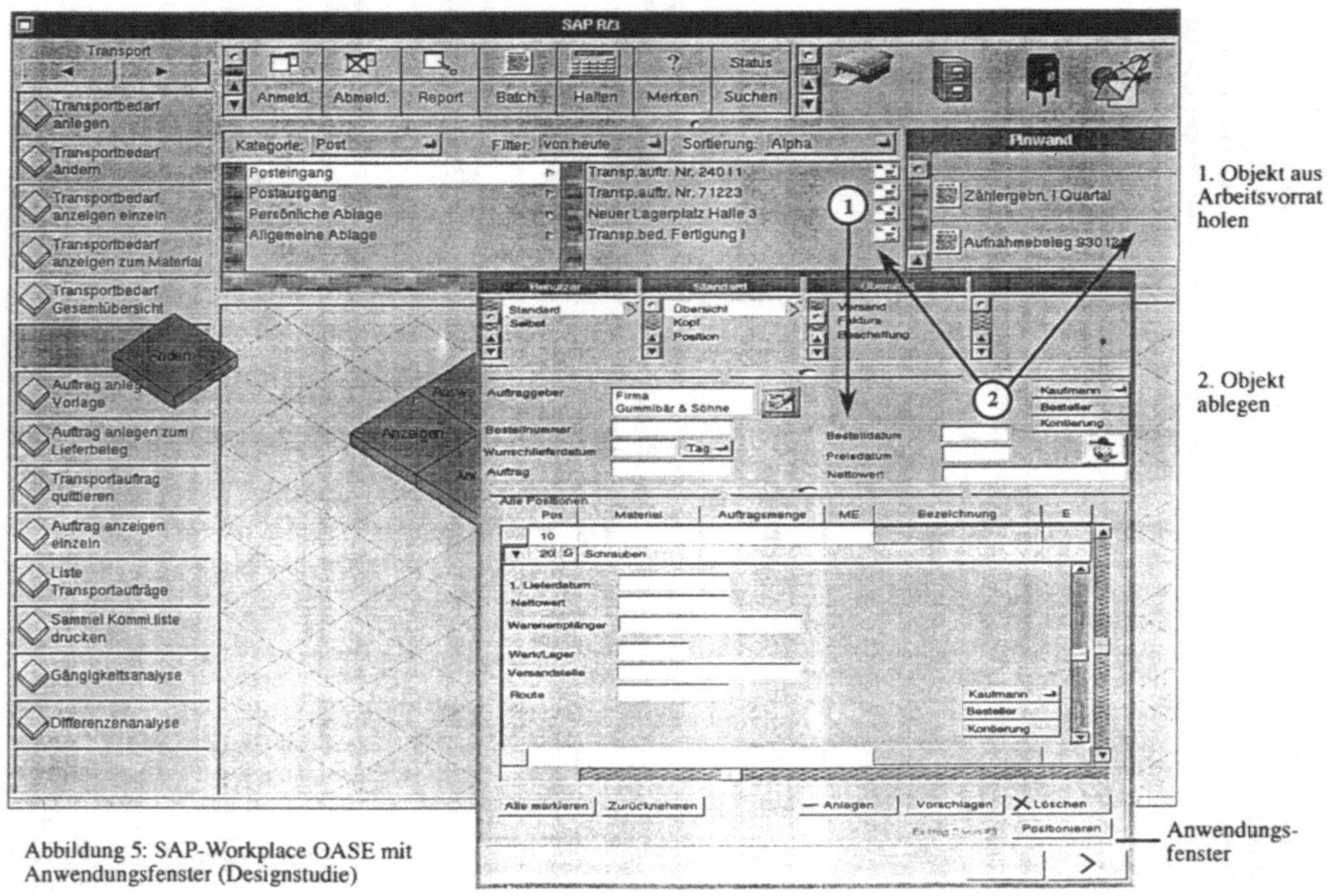

Abbildung 5: SAP-Workplace OASE mit Anwendungsfenster (Designstudie)

Aspekte von OASE gearbeitet. Dieses soll vor dem Hintergrund weiterer Diskussionen mit SAP-Endbenutzern geschehen, da nur deren Einschätzung bzw. Arbeitserleichterung ein Maß für die Sinnhaftigkeit und Notwendigkeit derartiger Entwicklungen darstellen kann.

Für die ergonomische Forschung läßt sich als Erkenntnis aus den bisherigen Arbeiten folgendes festhalten:
- eine reine Objektorientierung auf der Oberfläche wird vielen Benutzern bzw. Arbeitsplätzen nicht gerecht
- es sollte verstärkte Aufmerksamkeit auf Probleme der Fensterver-waltung und deren Optimierung gerichtet werden
- der gesamte Arbeitskontext und -ablauf sollte stärker in den Mittelpunkt des Design rücken. Effiziente Navigations- und Überwachungshilfsmittel sind bei komplexen Anwendungssystem ein ebenso wichtiger Faktor der Benutzbarkeit wie die ergonomische Gestaltung einer singulären Anwendung.

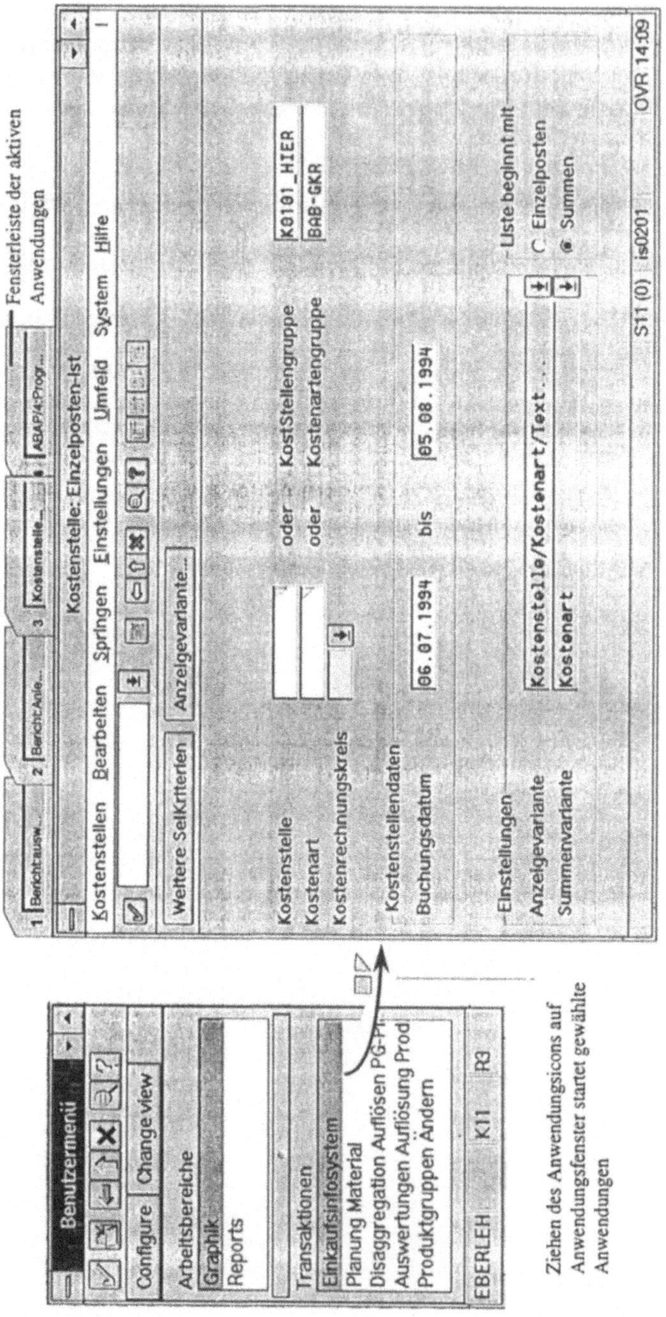

Abb. 6: Benutzermenü zum Aufruf von Anwendungen und Fensterleiste
zur Fensterverwaltung laufender Anwendungen (SAP R/3 Release 3.x)

Literaturverzeichnis

[1]   Apple Computer, Inc.: Macintosh Human Interface Guidelines, Reading, Mass.: Addison-Wesley 1992.

[2]   Bederson, B., Stead, L., Hollan, J.: Pad++: Advances in multiscale interfaces, in: C. Plaisant (Ed.), CHI ,94 conference companion, Boston, Mass. 1994, S. 315-316

[3]   Eberleh, E.: Browsing cognitive task spaces instead of working on the desktop, in: H.-J. Bullinger (Hrsg.), Human aspects in computing, Amsterdam: Elsevier 1991, S. 419-423

[4]   IBM Corp.: SAA/CUA guide to user interface design, IBM Report SC 34-4289-00, Cary, North-Carolina 1991

[5]   Leckebusch, J.: Alternativen für den Windows-Desktop, in: c't 1993, Heft 9, S. 74-78

[6]   Lucas, P, Schneider, L.: Workscape: a scriptable document management environment, in: C. Plaisant (Ed.), CHI ,94 conference companion, Boston, Mass. 1994, S. 9-10

[7]   Microsoft Corp.: An application design guide for microsoft windows (Chicago). Preliminary release, Microsoft 1993

[8]   Miller, R.: 20 ways to remodel your desktop, in: PC Magazine, September 29, 1992, pp. 111-179

[9]   Neugebauer, C.: Arbeitsorganisation und mentale Repräsentation betriebswirtschaftlicher Software. Unveröffentlichte Diplomarbeit, RWTH Aachen 1991

[10]   Next Computer, Inc.: Nextstep user interface guidelines, Reading, Mass.: Addisson-Wesley 1992

Dr. Edmund Eberleh
SAP AG
Neurottstr. 16
D-69190 Walldorf

Dipl. Wirtsch.-Inf. Falco Meinke
Bau Software Unternehmen GmbH
Wietzeaue 72
D-39896 Bissendorf/Hannover

Arbeitsplatzintegration und Medienintegration: Mensch-Computer-Interaktion in kooperativen Anwendungen

Hans-Werner Gellersen und Max Mühlhäuser
Telecooperation Office (TecO) an der Universität Karlsruhe

Zusammenfassung

Kooperationsunterstützung und Medienintegration sind die großen Herausforderungen bei dem Ziel, menschliches Handeln in seiner ganzen Komplexität informationstechnisch zu unterstützen. Wir betrachten beide Aspekte mit Bezug auf die Mensch-Computer-Interaktion und identifizieren hier Arbeitsplatz- und Medienintegration als Kernprobleme, die in der heutigen Benutzerschnittstellenentwicklung ungenügend reflektiert werden. Wir stellen schließlich eine Vorgehensweise zur Entwicklung arbeitsplatz- und medienintegrierter Benutzerschnittstellen vor, die Teil einer Entwicklungsumgebung für kooperative Anwendungen ist. Zentrale Konzepte unseres Ansatzes sind *Arbeitsszenarien* zur Kooperationsbeschreibung und *Modalitätenplanung* basierend auf problemnahen Charakterisierungen von Interaktionen und Medien.

1 Einleitung

Traditionell unterstützen Computer einzelne Anwender bei der Bearbeitung isolierter Aufgaben. Menschliche Aktivität besteht jedoch nicht aus isolierten Handlungen, sondern aus einem komplexen Geflecht von Interaktionen mit anderen Menschen und Interaktionen mit verschiedenen Aktivitäten. In jüngster Zeit findet sich nun eine Neuorientierung in der Informatik, die stärker auf den Menschen als Gestalter und Betroffenen informationstechnischer Prozesse eingeht. So beschreibt Rafael Capurro als Aufgabe der Informatik die "technische Gestaltung menschlicher Interaktionen mit der Welt" [1]. Die großen Herausforderungen sind dabei die umfassende Unterstützung von Kooperation, da menschliches Handeln größtenteils in Kooperation erfolgt, und Medienintegration, um dem enormen Kommunikationsbedarf bei der "Interaktion mit der Welt" Rechnung zu tragen.
Wir betrachten in diesem Artikel Kooperationsunterstützung und Medienintegration bezogen auf die Mensch-Computer-Interaktion. In Abschnitt 2 identifizieren wir Arbeitsplatz- und Medienintegration als zentrale Herausforderungen an die Mensch-Computer-Interaktion. Daran schließt sich eine kurze Kritik der heutigen Benutzerschnittstellenentwicklung an. In Abschnitt 4 stellen wir schließlich die

Entwicklung arbeitsplatz- und medienintegrierter Benutzerschnittstellen in ITEMS, einer Entwicklungsumgebung für kooperative medienintegrierte Anwendungen, vor.

2 Herausforderungen an die Mensch-Computer-Interaktion in kooperativen Anwendungen

2.1 Kooperationsunterstützung

2.1.1 Das Koordinationsproblem

Kernproblem der Kooperation ist die Koordination. Je komplexer oder wahrscheinlichkeitsbehafteter die Ziele und Aufgaben einer Kooperation sind, je größer ist der Informationsverarbeitungsaufwand für deren Koordination. Es gibt im wesentlichen zwei Ansätze für die Handhabung des Koordinationsaufwands: *formale Strukturen* und *Kommunikationspfade* [14]. Der erste Ansatz reduziert die Komplexität und Ungenauigkeit von Kooperationen mittels formaler Strukturen. Ein Beispiel hierfür ist Software Engineering, wo die Koordination einer sehr komplexen Kooperation auf formalen Spezifikationen aufbaut, wodurch der Kommunikationsaufwand für die Koordination verringert wird. Der zweite und grundlegend unterschiedliche Ansatz zur Handhabung der Koordination basiert auf einer Erhöhung der Informationsverarbeitungskapazitäten durch Bereitstellung von Kommunikationspfaden, wie z.B. electronic mail. Als Ergebnis dieser beiden grundlegend unterschiedlichen Ansätze zur Koordination finden sich heute zwei noch weitgehend isolierte Welten der Kooperationsunterstützung, die als *Groupware* und *Workflow Management* bezeichnet werden [7].

2.1.2 Groupware und Workflow Management.

Groupware ist mit der Unterstützung vergleichsweise kleiner Arbeitsgruppen befaßt, in denen eng gekoppelt (meist synchron) an einem gemeinsamen Problem gearbeitet wird. Groupware-Anwendungen sind z.B. Multimedia-Konferenzen und gemeinsames Editieren. Koordination wird hier durch Kommunikation unterstützt aber im wesentlichen den Teilnehmern überlassen, d.h. die Koordination basiert auf dem sozialen Verhalten der Gruppe. Der Computer wird hier mehr als Medium denn als Werkzeug verstanden. Im Gegensatz zur Groupware liegt dem Workflow Management das Kooperationsmodell der Aufteilung und Lösung von Teilproblemen zugrunde. Die Teilnehmer und Aktivitäten in einem Workflow (d.h. in einer Vorgangsbearbeitung) sind vergleichsweise lose gekoppelt, die Interaktion erfolgt vorwiegend asynchron. Die Koordination ist wohlstrukturiert und im Gegensatz zur Groupware unter Systemkontrolle. Einzelne Kooperationsteilnehmer sind sich der

Koordination oft nur teilweise bewußt, genauso wie Mitarbeiter in einer größeren Organisation deren Funktionsweise oft nur partiell verstehen. Anwendungen für Workflow Management sind wohldefinierte Arbeitsabläufe in den verschiedensten Bereichen.

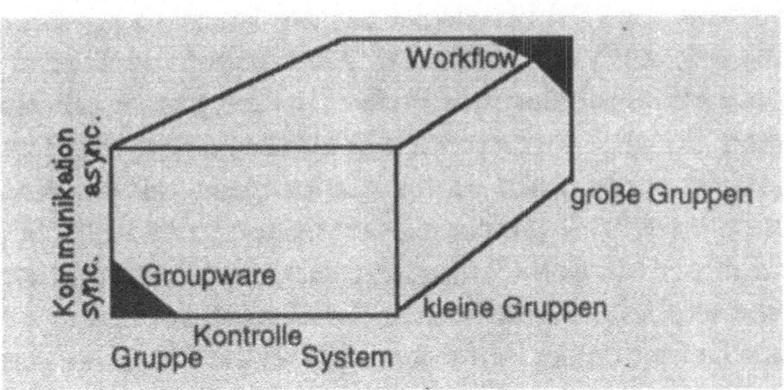

Abb. 1: Ein Entwurfsraum für Kooperationsunterstützung

Groupware und Workflow Management unterstützen zwei Extreme menschlicher Zusammenarbeit, wie in Abb. 1 skizziert. Eine Harmonisierung der beiden Ansätze verspricht Unterstützung für ein weitaus größeres Spektrum menschlicher Zusammenarbeit.

2.1.3 Mensch-Computer-Interaktion und Kooperation

Traditionell hat sich die Mensch-Computer-Interaktion mit Einbenutzer-Schnittstellen zu monolithischen Anwendungen befaßt. Im Kontext kooperativer Anwendungen entstehen neue Herausforderungen für die Mensch-Computer-Interaktion:

- Mehrbenutzerschnittstellen im Groupware-Kontext
- Arbeitsplatzintegration: Zugang zu verschiedenen Aktivitäten am Arbeitsplatz

Mehrbenutzerschnittstellen. Kooperation basiert generell auf einem gemeinsamen Kontext, der im allgemeinen durch gemeinsame Daten realisiert wird. Bei synchroner Kooperation leitet sich hieraus das Problem ab, über Mehrbenutzer-Schnittstellen konsistente Sichten auf die gemeinsamen Daten zu schaffen. Grundlage hierfür muß eine Trennung von Modell und Sichten sein, die zwar ohnehin gute Entwurfpraxis ist, aber bei Mehrbenutzer-Schnittstellen zwingend wird, um einzelne Sichten mit Kooperationssemantik anreichern zu können [11]. Derartige Kooperationssemantik kann z.B. die Granularität der Aktualisierung von Sichten betreffen.

Arbeitsplatzintegration. Die Integration von Benutzerschnittstellen zu vormals isoliert betrachteten Anwendungen, die im Rahmen der Arbeitsplatzintegration als integrale Bestandteile einer Arbeitsumgebung gesehen werden, ist eine weitere Herausforderung für die Mensch-Computer-Interaktion. Medes kommt bei der Untersuchung einer integrierten Projektunterstützung zu dem Schluß, daß der Benutzerschnittstelle eine Schlüsselrolle bei der Integration zufällt [10]. Er fordert uniformen Zugang zu verschiedenen Aktivitäten und zum Austausch mit Kooperationspartnern. Grundlegende Problemstellung ist in diesem Zusammenhang die Heterogenität der zu integrierenden Werkzeuge. Arbeitsplatzintegration wird natürlich keinesfalls allein durch uniformen Umgang mit einzelnen Komponenten einer Organisation erreicht. Vielmehr müssen auch organisatorische Zusammenhänge einbezogen und in der Mensch-Computer-Interaktion reflektiert werden. Grundlage hierfür muß eine explizite Modellierung der Kooperationen in einer Organisation sein, die sich zur Definition der Anforderungen an die Mensch-Computer-Interaktion eignet.

2.2 Medienintegration

Mit der zunehmenden Leistung von Arbeitsplatzrechnern und der Verfügbarkeit erschwinglich gewordener Multimedia-Erweiterungen hält Multimedia am Arbeitsplatz einzug. Die Integration von Multimedia ist jedoch noch nicht sehr weit fortgeschritten, wie in den folgenden Abschnitten diskutiert wird.

Vorausgeschickt werden soll aber zunächst eine Klärung der im folgenden verwendeten Begriffe *Medien, Modalitäten, Multimedia* und *Multimodalität.* Den Begriff Medium verwenden wir bezogen auf die Übertragung oder Darstellung von Information. Modalität bezieht sich im Gegensatz zu Medium auf die Dynamik der Interaktion und legt die Art und Weise, in der man Information erhält, transportiert oder vermittelt, fest. Die Begriffe Multimedia und Multimodalität werden verwendet, wenn mehrere Kommunikationskanäle vorliegen. Im Gegensatz zu Multimedia wird dabei im Fall der Multimodalität auf mehreren Kanälen Bedeutung aus der kommunizierten Information abstrahiert. Danach ist ein Voice-Mail-System als multimedial zu bezeichnen, wenn Sprache nur aufgezeichnet und abgespielt werden kann. Wird die Sprache jedoch ausgewertet, liegt ein multimodales System vor.

2.2.1 Medienintegration an der Bedienoberfläche

Im Alltag benutzen Menschen viele verschiedene Medien und Modalitäten zur Kommunikation basierend auf der Erfahrung, daß einige Medien und Modalitäten in gewissen Kontexten effektiver als andere sind. Es ist offensichtlich, daß auch die Mensch-Computer-Interaktion wesentlich verbessert werden kann, wenn verschiedene Medien zur multimedialen Darstellung zur Information und verschiedene Modalitäten zur multimodalen Interaktion kombiniert werden.

Medien- und Modalitätenplanung. Eine große Herausforderung für die Entwicklungsunterstützung von Benutzerschnittstellen ist ein besseres Verständnis verschiedener Medien/Modalitäten, um sie geeignet einsetzen zu können. Existierende Ansätze zur Klassifikation von Medien/Modalitäten sind jedoch technologieorientiert, wie etwa in [2]. Derartige Klassifikationen erweisen sich als ungeeignet für die Abbildung von Medien und Modalitäten auf die Anforderungen der Mensch-Computer-Interaktion, da sie sich nicht am Inhalt der Interaktion orientieren

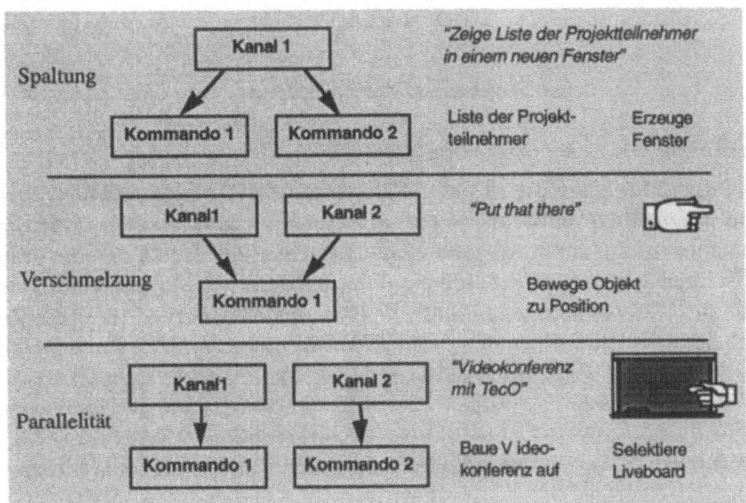

Abb. 2: Verarbeitung multimodaler Information

Verarbeitung multimodaler Information. Die Kombination von Modalitäten zur multimodalen Interaktion ist eine weitere gravierende Herausforderung für die Mensch-Computer-Interaktion. Die Verarbeitung multimodaler Information ist äußerst komplex, da die Informationsverarbeitung auf den involvierten Kommuni-kationskanälen koordiniert werden muß. Dabei kann es erforderlich sein, daß Informationseinheiten aufgespalten, verschmolzen oder parallel verarbeitet werden müssen. Die Spaltung von Information kann z.B. erforderlich sein, um Kommandos für unterschiedliche Verarbeitungskontexte zu extrahieren. Umgekehrt kann Verschmelzung erforderlich sein, um ein sinnvolles Kommando zu gewinnen. Im parallelen Fall müssen verschiedene Informationseinheiten unterschiedlichen

Verarbeitungskontexten zugeleitet werden. Abb. 2 illustriert diese verschiedenen Anforderungen mit Beispielen. Die eigentliche Komplexität der multimodalen Informationsverarbeitung liegt darin, daß a priori nicht bekannt ist, wann Information zu spalten, zu verschmelzen oder parallel zu verarbeiten ist.

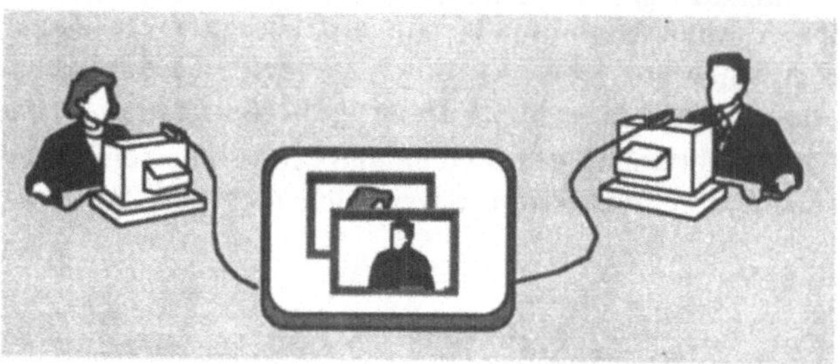

Abb. 3: Ein Videokonferenz-Szenario

2.2.2 Anwendungsweite Medienintegration

Medienintegration darf nicht auf die Benutzeroberfläche beschränkt sein. Der Integrationsanspruch muß viel weiter gefaßt und auf alle Aspekte einer Software ausgedehnt werden, wie im folgenden am Beispiel einer Videokonferenz illustriert wird. Nehmen wir folgendes Videokonferenz-Szenario: in einer Videokonferenz werden die Videos der verschiedenen Kooperationspartner in verschiedenen Fenstern auf dem Bildschirm dargestellt. Werden diese Fenster ganz oder teilweise verdeckt, so kommt Videoinformation, die über das Netzwerk transportiert wird und dabei wertvolle Bandbreite allokiert nie zur Darstellung (Abb. 3). Ein und dasselbe Medienobjekt, das Video eines Konferenzteilnehmers, wird vom Netzwerktransportsystem und vom Fenstersystem isoliert und nicht integriert manipuliert.

2.2.3 Orthogonale Medienunterstützung

Eine Grundvoraussetzung für die Integration verschiedener Medien und Modalitäten ist einfache und gleichartige Handhabung. In existierenden Multimedia-Anwendungen wird nicht ausreichend von den Unterschieden der Medienhandhabung auf Betriebssystem- und Netzwerkebene abstrahiert. Auf diesen Ebenen werden Medien aufgrund unterschiedlicher Speicher- und Transportanforderungen getrennt bearbeitet. Die mangelnde Abstraktion von diesen Ebenen resultiert in was Laurel *Medienmodalität* nennt: Medien wie z.B. Video erscheinen als grobgranulare abgeschlossene Blöcke, die sich schwer handhaben und mit anderen Medien integrieren lassen [9]. Schlüssel zu einer orthogonalen Unterstützung sind also die Abstraktion von technoligieorientierten Darstellungen und die Schaffung einer geeigneten Mediengranularität. Einige Medien, z.B. Text, haben eine sehr feine Granularität (hier: Zeichen) die es sehr einfach macht, orthogonale Unterstützung für z.B. das Editieren, Ordnen und Zusammenfügen zu realisieren. Andere Medien, wie z.B. Video, haben eine Granularität, die sich nicht am Inhalt orientiert (z.B. Frames) und sich daher nicht als Grundlage für orthogonale Unterstützung eignet. Eine geeignete inhaltsbezogene Granularität ist auch Voraussetzung für natürliche Interaktion mit Multimedia und für die Koordination multimodaler Interaktion.

Die Etablierung einer inhaltsbezogenen Granularität setzt das Verstehen von Medien voraus. Hier gibt es schon sehr vielversprechende Ansätze aus dem Forschungsbereich der Multimedia-Indexierung, so z.B. Wordspotting, mit denen sich vormals als Blöcke betrachtete Medienobjekte, z.B. die Aufzeichnung einer Audio/Video-Konferenz, strukturieren lassen. Die Mensch-Comuter-Interaktion ist gefordert, derartige Konzepte in die Entwicklung von Benutzerschnittstellen zu integrieren.

2.3 Zusammenfassung der Herausforderungen an die Mensch-Computer-Interaktion

Wir haben im Kontext der Kooperationsunterstützung und Medienintegration eine Reihe von Herausforderungen an die Mensch-Computer-Interaktion identifiziert:

- Mehrbenutzerschnittstellen
- Arbeitsplatzintegration
- Medien- und Modalitätenplanung
- Verarbeitung multimodaler Information
- Anwendungsweite Medienintegration
- Orthogonale Medienunterstützung

Im folgenden konzentrieren wir die weitere Diskussion auf die Herausforderungen, die wir primär im Bereich der Mensch-Computer-Interaktion angesiedelt sehen: Arbeitsplatzintegration, Medien- und Modalitätenplanung, sowie Verarbeitung multimodaler Information. Die anderen Herausforderungen müssen in einem größeren Kontext, der über die Mensch-Computer-Interaktion hinausgeht, gelöst werden [3]; auf einen solchen integrativen Lösungsansatz werden wir aus Platzgründen nur kurz eingehen (vgl. 4.1). Wenn wir im folgenden von Medienintegration sprechen, beziehen wir uns auf Integration an der Bedienoberfläche mit den Aspekten der Modalitätenplanung sowie der Verarbeitung multimodaler Information.

3 Existierende Ansätze der Benutzerschnittstellenentwicklung

In diesem Abschnitt wird kurz reflektiert, inwieweit heutige Ansätze zur Entwicklung von Benutzerschnittstellen geeignet sind, Arbeitsplatz- und Medienintegration zu unterstützen.

3.1 Existierende Ansätze

Toolkits. Der Ansatz, Benutzerschnittstellen basierend auf Toolkits, d.h. Bibliotheken von Bausteinen für die Implementierung der Bedienoberfläche, zu entwickeln, ist derzeit dominierend. Die Nachteile dieser Vorgehensweise sind jedoch vielfältig [7,14]:

- Die Bibliotheken sind präsentationsorientiert und vernachlässigen semantische Aspekte der Interaktion (mangelnde Integration mit semantischen Objekten).
- Es fehlt jegliche methodische Unterstützung zur Förderung guter Entwurfspraxis bei der Anwendung von Toolkits.
- Toolkits sind weitgehend an konkrete Interaktionstechniken gebunden; selbst von Plattformen wird nur ansatzweise abstrahiert (in sog. *Portability Toolkits*).

Generell erweisen sich toolkitbasierte Benutzerschnittstellen als sehr schwer wart- oder portierbar, da verschiedenartige Entwurfsentscheidungen (z.B. Dialogstruk-tur) nicht explizit modelliert werden.

Application Frameworks. In Application Frameworks ist der Fokus im Gegensatz zu Toolkits nicht allein auf der Bedienoberfläche. Application Frameworks bieten weitergehende Bibliotheken und Werkzeuge für die Beschreibung aller Teile einer Anwendung. Genau darin ist aber auch der Hauptnachteil dieser Entwicklungs-umgebungen begründet: alle Teile der Anwendung müssen in der monolithischen Umgebung realisiert werden. Wenn es auch Ansätze gibt, die Multimedia und Kooperation berücksichtigen (z.B. RENDEZVOUS [11]), so sind Application Frameworks insgesamt doch unzureichend, da sie von einer abgeschlossenen, homogenen Anwendungswelt ausgehen und die am Arbeitsplatz inhärent gegebene Heterogenität von Werkzeugen, Aktivitäten und Ressourcen nicht unterstützen.

User Interface Management Systeme (UIMS). UIMS generieren und kontrollieren Benutzerschnittstellen basierend auf Modellen interaktiver Anwendungen, wobei semantische Aspekte der Interaktion betont werden. ACE stellt z.B. semantik-orientierte Dialogbausteine anstatt präsentationsorientierter Toolkitobjekte zur Verfügung [7]. In ähnlicher Weise erlauben viele UIMS ein gewisses Maß an Abstraktion von Interaktionstechniken, allerdings orientieren sich die Dialog-modelle doch sehr stark an der Benutzerschnittstellenstruktur heute dominierender grafischer Bedienoberflächen. Eine Unterstützung verschiedener Medien und Mo-dalitäten findet sich in existieren UIMS noch nicht. Während UIMS allgemein eine Trennung von Benutzerschnittstelle und Anwendung zugrundeliegt, gibt es doch nur wenige Ansätze, die darauf aufbauend integrierte Benutzerschnittstellen zu mehreren Anwendungen unterstützen. Zu nennen sind hier SCENARIOO [13] und Chiron-1 [14], die allerdings auch nur die Realisierungsbasis für integrierte Schnittstellen schaffen und nicht etwa auf Kooperationsmodellen aufbauen.

3.2 Integration von Werkzeugen und Methoden

In der heutigen Benutzerschnittstellenentwicklung finden wir zwei wesentliche Schwachpunkte: mangelnde Integration verschiedener Medien und Modalitäten sowie fehlende methodische Unterstützung für eine gute Entwurfspraxis. Heutige Werkzeuge bieten zwar ausgefeilte Bibliotheken grafischer Interaktionsobjekte und fortgeschrittene Konstruktionsumgebungen, sind aber doch sehr eingeschränkt was die Schnittstellengestaltung betrifft. Um fortgeschrittene Benutzerschnittstellen, in denen mehrere Benutzer in verschiedenen Modalitäten interagieren, entwickeln zu können, müssen die verschiedenen spezialiserten Werkzeuge integriert werden.

Eine weitere Voraussetzung für fortgeschrittene Benutzerschnittstellenentwicklung ist methodische Entwurfsunterstützung. Es gibt in der Mensch-Computer-Interak-tion eine Vielzahl von Methoden für Analyse, Entwurf und Evaluierung, die sich jedoch kaum etablieren, da auch sie recht eingeschränkt sind und Teilaspekte des Entwurfsproblems isoliert betrachten. Um die Benutzerschnittstellenentwicklung entscheidend zu verbessern, müssen verschiedene Werkzeuge und Methoden stärker integriert werden. Die Benutzerschnittstellenentwicklung selbst ist ebenfalls stärker in die Anwendungsentwicklung einzubinden, da nur dann komplexere Probleme, wie z.B. die anwendungsweite Medienintegration, gelöst werden können.

4 Entwicklung arbeitsplatz- und medienintegrierter Benutzerschnittstellen in ITEMS

4.1 Softwaretechnik für medienintegrierte Anwendungen

Im Projekt ITEMS arbeiten wir an der Softwaretechnik für kooperative medienintegrierte Anwendungen. Ein Kernaspekt von ITEMS ist die grafische Modellierung solcher Anwendungen. Sie wird motiviert durch die Komplexität der betrachteten Anwendungen und die Notwendigkeit, Entwicklern und Nutzern ein Verständnis dieser Anwendungen zu vermitteln. Kooperative medienintegrierte Anwendungen involvieren verschiedene Nutzer, bestehen aus einer größeren Zahl verteilt ablaufender Prozesse und integrieren in größerem Umfang bereits existierende (*legacy*) Software. Softwareentwürfe werden daher in ITEMS primär als Verständnishilfen und Kommunikationsmittel aufgefaßt, erst nachrangig als maschinenlesbare Grundlage für die automatische Erzeugung von Programm-Code oder für Überprüfung und Verifikation.

Dabei soll ein und derselbe (modulare) Softwareentwurf den Ausgangspunkt bilden für die Entwicklung der kompletten komplexen Anwendung, insbesondere für die Softwareentwicklung in den drei Bereichen *Synergy*, *Modality* und *Ubiquity* [4].

Synergy bezeichnet dabei die Kooperation zwischen den Nutzern der Anwendung, i.S.v. Teamarbeit (Groupware) und Vorgangsbearbeitung (Workflows), unter Einbeziehung von Mehrbenutzer-Schnittstellen (vgl. 2.1) und Dokument-bearbeitung.

Modality umreißt den Bereich der Arbeitsplatz- und Medienintegration, wie er in diesem Artikel näher beschrieben ist.

Der Begriff Ubiquity rührt von der Tatsache her, daß die zunehmende Computer-nutzung für alle Tätigkeitsbereiche einer Organisation (i.S. des Anspruches kooperativer medienintegrierter Anwendungen) zur Folge hat, daß den Nutzern praktisch ständig der Zugang zu Computer-Ressourcen gewährt werden muß, nicht nur dann, wenn sie sich an einem festen, angestammten Arbeitsplatz befinden. Es ist daher das Anliegen des Schwerpunktes Ubiquity, mobile Benutzer von kooperativen medienintegrierten Anwendungen so durch mobile Software und mobile Computerressourcen intelligent zu komplementieren, daß die Nutzung der Anwendung weitgehend ubiquitär ermöglicht wird.

4.1.1 Modellierungskonzept

Um derart komplexe Sachverhalte, wie solche Anwendungen sie offensichtlich darstellen, unter softwareergonomischen Gesichtspunkten an Entwickler und Nutzer kommunizieren zu können, wird das Modellierungskonzept von ITEMS nach folgenden Grundprinzipien entwickelt:

- Aufteilung in verschiedene, konsistent miteinander verbundene grafische Sichten

- Ermöglichen unvollständiger, ausschnittsweiser Entwürfe vor allem in frühen Entwurfsphasen bzw. für Entwürfe, die der Kommunikation mit dem Nutzer dienen; Basis ist hier die Unterstützung sog. Szenarien.

- Aufgabe strenger Kategorisierung der Form "Sachverhalt XYZ kann ausschließlich in der Form xyz modelliert werden" und konsistenter Detaillierungsgrade zugunsten intuitiver Modellierung.

- Konzentration auf Szenariengraphen und Higraphen als wichtigste Grundformen der grafischen Modellierung. Als Szenariengraphen bezeichnen wir spezielle ungerichtete Graphen, deren Notation die Formulierung von optionalen bzw. (dynamisch veriierenden) Mengen von Knoten und Kanten erlauben. Diese Eigenschaft wird im Kontext von verteilten und Mehrbenutzer-Systemen häufig benötigt. Higraphen wurden von [5] übernommen und sind hierarchisch strukturierte Graphen mit Hyperkanten (d.h. solchen mit mehreren Quellen oder Zielen).

- Exemplarische Abfassung der Modellierungsmethode, wobei die angebotene Phasenabfolge (gleichzusetzen mit Abfolge der Bearbeitung von Sichten) nicht als bindend anzusehen ist; dies bedingt natürlich die entsprechende Flexibilität im Modellierungssystem selbst.

4.1.2 Die fünf Sichten des Modells

Die wesentlichen fünf Sichten(-typen), die in ITEMS angeboten werden, werden im folgenden kurz beschrieben (vgl. Abb. 4). Die gewählte Reihenfolge entspricht der Phasenabfolge in der exemplarischen Modellierungsmethode.

Arbeitsszenarien. Der Einstieg in die Modellierung einer Anwendung kann dadurch erfolgen, daß in loser Folge Arbeitszenarien der Benutzung aneinandergereiht werden. In jedem Szenario wird sehr grob skizziert, wie (häufig *mehrere*) Benutzer mit der Anwendung in Beziehung stehen. Zur Modellierung der Benutzer werden sog. *Benutzerzugänge* als Knotentypen angeboten. Ein Benutzerzugang modelliert weder exakt einen Benutzer noch exakt eine Benutzeroberfläche. Er beschreibt die Menge an Verbindungen zu anderen Benutzerzugängen sowie zu Softwarekomponenten, welche ein Benutzer *in einer bestimmten Rolle* zur Verfügung hat. Als weitere Knotentypen werden zwei Arten von Softwarekomponenten unterschieden. Einerseits solche, die Zugang zu persistenter Information verschaffen; hierfür wird die Bezeichnung *Archiv* benutzt, während die semantischen Einheiten der

Information als *Dokumente* bezeichnet werden (wobei dieser Begriff weit gefaßt wird). Andererseits werden sog. *Agenten* ausgezeichnet. Agenten sind selbständige Softwarekomponenten, die Aufgaben *eigenständig* erfüllen. *Nicht* als eigenständige Agenten gekennzeichnet werden beispielsweise Editier- oder Tabellenkalkulationsprogramme (Zugänge zu Archiven) oder Kontrollmodule zur Verwaltung von Beziehungen zwischen Knoten eines Szenarios (z.B. zum Auf- und Abbau von Videophonie-Verbindungen). In Abschnitt 4.3 werden Arbeitsszenarien an einem Beispiel illustriert.

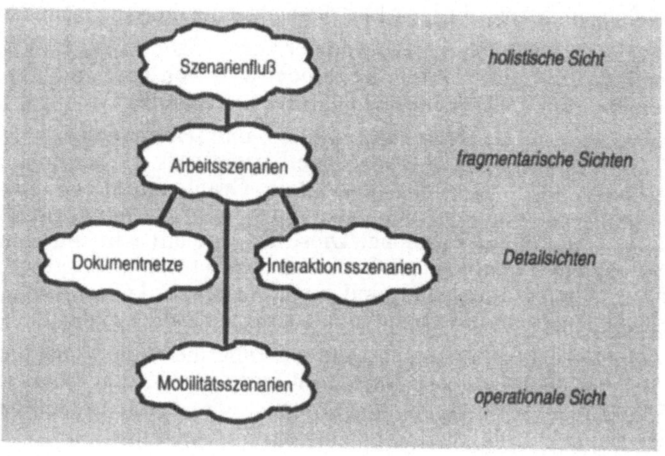

Abb. 4: Sichten auf kooperative medienintegrierte Anwendungen im ITEMS-Modell

Szenarienfluß. In Szenarienflüssen 'schrumpfen' die Arbeitsszenarien zu einzelnen Knoten. Es wird festgelegt, welche Szenarien Teilaspekte ein und desselben Benutzungs-Schnappschusses sind (also theoretisch in einem einzigen Szenario beschrieben werden könnten), welche in enger Kopplung durch Kommandos oder Ereignisse ineinander übergehen können (solche Szenarien gehören häufig i.S. von Groupware zusammen), und welche durch Dokumente oder Meilensteine voneinander getrennt sind (i.S. der Vorgangsbearbeitung).

Dokumentnetze. Die ggf. aus Arbeitsszenarien und Szenarienfluß vordefinierten Dokumente werden hier genauer strukturiert, allerdings auf semantischer Ebene, d.h. ohne Festlegung spezifischer Layouts und ggf. unter Offenhalten der zu wählenden Medien (im Multimedia-Sinne). Dokumente werden als Graphen aus Informationseinheiten aufgefaßt (im Hypertext-Sinne), wobei Dokumentnetze beispielsweise die Zahl der Instanziierungen von Knoten und Kanten (z.B. Anzahl der Kapitel, Anzahl der Querverweise, etc.) offenlassen und stattdessen eine Art Dokument-*Stil* beschreiben. Knotenarten unterscheiden vor allem Art und Grad der *Editierbarkeit* (fix, vorgefüllt, blanko, transient).

Interaktionsszenarien. Ausgehend von den Arbeitsszenarien werden hier spezifischere Festlegungen zu den Benutzerzugängen getroffen. Dabei werden die mit einem Benutzerzugang verbundenen Knoten (andere Benutzerzugänge, Archive, Agenten), das notwendigen Management für die jeweiligen Verbindungen, sowie die vom Benutzerzugang aus initiierbaren Szenarienübergänge gemäß Szenarienfluß berücksichtigt (im Sinne der flexiblen Modellierungsmethode ist hier natürlich die Einschränkung gegeben, daß mindestens ein Arbeitsszenario existieren muß, bevor ein Interaktionsszenario entworfen werden kann). Die Rolle der Interaktionsszenarien wird in Abschnitt 4.2 deutlicher.

Mobilitätsszenarien. Diese fünfte Klasse von Sichten ist vornehmlich für die Abbildung einer Anwendung auf konkrete operationale Umgebungen verantwortlich und wird nur sehr eingeschränkt in der Kommunikation mit dem Nutzer eingesetzt. Hierin werden Randbedingungen und Informationen festgelegt, welche der eingangs des Abschnittes genannten dynamischen Komplementierung mobiler Nutzer durch mobile Software und mobile Computerressourcen dient. Dazu können zunächst für alle Knoten der Arbeitsszenarien sog. *Fragmente* definiert werden, wodurch sich z.B. unterschiedliche (ggf. verteilt kooperierende) Module der Benutzerzugänge, Agenten und (ggf. replizierten) Archive beschreiben lassen. Weiterhin werden Geräte und Gerätefragmente (z.B. PCMCIA-Speicher als mobile Geräteteile und gleichzeitige Informationsträger) sowie Teilnetze in Mobilitäts-szenarien miteinander in Beziehung gesetzt. Auf der Basis eines mobilitätsspezi-fischen Quality-of-Service-Modelles werden alle Fragmente mit QoS-Parametern versehen, so daß die Laufzeitunterstützung i.S. von Migration und Fernzugriff, Replikation und Rekonziliation etc. vorgeplant werden kann.

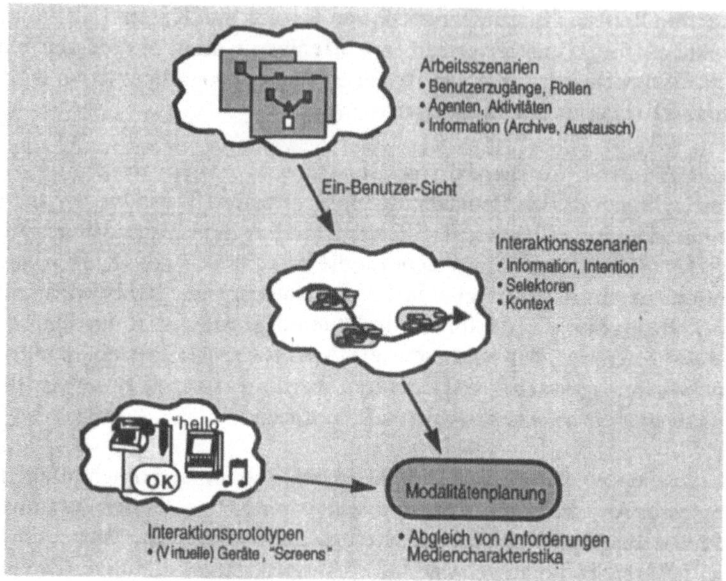

Abb. 5: Entwicklung arbeitsplatz- und medienintegrierter Benutzerschnittstellen in ITEMS

4.2 Entwicklung von Benutzerschnittstellen in ITEMS

Die Entwicklung von Benutzerschnittstellen in ITEMS wird in Abb. 5 skizziert. Ausgangspunkt der Entwicklung sind Arbeitsszenarien, in denen die Anforderungen an die Mensch-Computer-Interaktion gesammelt werden. Auf dieser Ebene herrscht eine Mehrbenutzer-Sicht, d.h. alle Interaktionen werden zusammengefaßt modelliert und nicht etwa auf einzelne Benutzer bezogen. In einem zweiten Schritt wird davon abstrahiert und eine Einbenutzer-Sicht auf die Arbeitsszenarien eingenommen. Darauf basierend werden alle Interaktionen, die einen Arbeitsplatz betreffen, gesammelt und in *Interaktionsszenarien* beschrieben. Diese Beschreibung abstrahiert von etwaigen Interaktionstechniken, d.h. von Medien und Modalitäten. In einem weiteren Entwurfsschritt ist diese Beschreibung Grundlage für eine automatisierte Modalitätenplanung, die primär auf einem Abgleich von Interaktionsanforderungen und Medieneigenschaften basiert, aber auch Interaktionsprototypen berücksichtigt. Interaktionsprototypen sind Fragmente für die Schnittstellenrealisierung, die z.B. mit Schnittstelleneditoren erstellt werden können. Die Modalitätenplanung ist für die Auswahl von Modalitäten zuständig, aber auch für die

Generierung von Koordinationsinformation, die die Grundlage multimodaler Informationsverarbeitung darstellt.

4.3 Arbeitsszenarien: Modellierung der Anforderungen

4.3.1 Szenarien als Entwurfswerkzeug

Insbesondere in der Mensch-Computer-Interaktion erweisen sich traditionelle Beschreibungsmethoden des Requirements Engineering als unbefriedigend, da sie sich nicht für die kompakte Beschreibung menschlicher Arbeit (schon gar nicht *Zusammen*arbeit) eignen. Mit zunehmendem Interesse werden Szenarien untersucht, die sich vor allem durch zwei Merkmale als Werkzeug für das Requirements Engineering anbieten [9]:

- Szenarien können sowohl Situationen als auch Abläufe darstellen
- Szenarien vermitteln einen größeren Kontext für dargestellte Situationen (Ressourcen, Intentionen)

Prinzipiell sind Szenarien als Ausschnitte zu verstehen, die Aspekte eines größeren Sachverhalts unter Abstraktion von anderen Gesichtspunkten untersuchen. Der Ausschnitt kann dabei zeitlich ausgedehnt sein oder aber auch einer Momentaufnahme entsprechen. Innerhalb eines Szenarios wird eine umfassende Beschreibung angestrebt: involvierte Personen, Informationsflüsse, Werkzeuge, etc. Ein Szenario ist jedoch nicht geeignet, einen Teil einer Anwendung abgeschlossen zu modellieren; es dient vielmehr zur Sammlung von Anforderungen für ein solches Modell.

4.3.2 Arbeitsszenarien in ITEMS

Die ITEMS-Arbeitsszenarien basieren, wie in Abschnitt 4.1 beschrieben, aus Benutzerzugängen, Archiven und Agenten. Im folgenden wird die Modellierung von Arbeitsszenarien am Beispiel *Teleteaching* illustriert.

Abb. 6 stellt die Arbeitsszeanrien Vorlesung und Gruppenarbeit dar, die etwa in einen Workflow eingebettet sein könnten. In dem Vorlesungsszenario interagieren mehrere Studenten (*-Notation) mit einem Dozenten (1-Notation) über den Vorlesungsstoff ('Folien'). Einzelne Studenten können in eine direkte Interaktion (*Konversation*) mit dem Dozenten treten. Da 'Redner' eine weitere Rolle in der zugrundeliegenden Kooperation ist, wird dafür ein expliziter Benutzerzugang modelliert. Benutzerzugänge können also sowohl für organisatorische Rollen (Dozent/Student) als auch für aufgabenbezogene Rollen geschaffen werden. Unter aufgabenbezogenen Rollen verstehen wir solche, die auf einen konkreten

Interaktionskontext bezogen sind (Redner/Zuhörer). Organisatorische Rollen implizieren dabei oft ein gewisses Verhalten in einem konkreten Interaktionskontext (z.B. Student=Zuhörer). Das Gruppenarbeitsszenario modelliert die Inetraktion mehrerer Studenten mit einem Intelligenten Tutor, der basierend auf einer Aufgabenbeschreibung die Gruppenarbeit organisiert (z.B. auf Lernfortschritt, Fehlkonzeptionen etc. reagiert). Ein Dozent kann sich in diese Interaktion einschalten, um etwa besondere Fragen zu klären. Die Optionalität der Teilnahme des Dozenten wird durch die 0-Notation angedeutet.

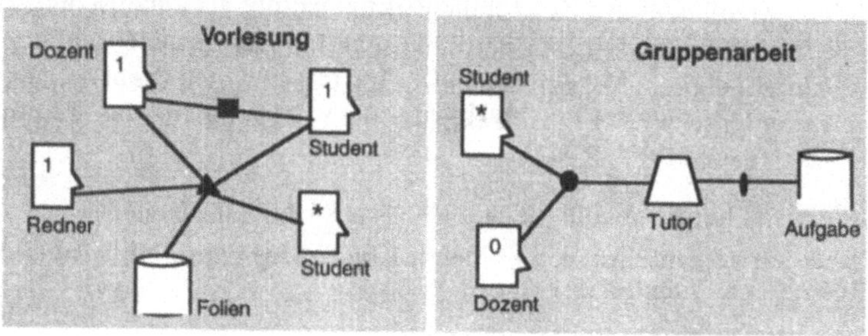

Abb. 6: Teleteaching-Arbeitsszenarien: a) Vorlesung, b) Gruppenarbeit

4.4 Interaktionsszenarien: medienneutrale Interaktionsbeschreibung

4.4.1 Rollenbasierte Sammlung der Interaktionsanforderungen

Zur Beschreibung eines integrierten Arbeitsplatzes wird eine Einbenutzer-Sicht auf die Arbeitsszenarien eingenommen. Dazu wird in Abstraktion von einem tatsächlichen Benutzer eine Arbeitsplatz-Rolle eingeführt, für die dann in einem ersten Schritt zu definieren ist, welche aufgabenbezogenen Rollen sie einnehmen kann. Eine Arbeitsplatz-Rolle kann auch mit einer oder mehreren organisatorischen Rollen assoziiert werden, die in der Regel Befugnisse modellieren und aufgabenbezogenes Rollenverhalten implizieren. Wurde die Arbeitsplatz-Rolle wie beschrieben definiert, lassen sich alle für den Arbeitsplatz relevanten Verbindungen zusammenfassen. Die Verbindungen werden dabei gemäß der involvierten Knoten klassifiziert:

- Konversation: Interaktion zwischen Benutzerzugängen
- Editieren/Abfragen: Interaktion mit Archiven
- Dialog: Interaktion mit Agenten

Für jede dieser Verbindung sind zwei verschiedene Interaktionsanforderungen zu beachten. Primär ist natürlich die Interaktion mit den verbundenen Knoten zu beschreiben. Daneben sind aber auch Interaktionen, die zur Handhabung von Verbindungen (etwa dem Auf- und Abbau) erforderlich sind, zu erfassen.

Über die Betrachtung der einzelnen Verbindungen hinaus müssen schließlich auch noch die Interaktionen modelliert werden, die Übergänge zwischen Szenarien ermöglichen. Bezogen auf die oben vorgestellten Teleteaching-Szenarien könnte für einen studentischen Arbeitsplatz z.B. der Zugang zur Gruppenarbeit mit der Vorbedingung der Vorlesungsteilnahme versehen werden. Dieser Modellierungsaspekt ist ganz wesentlich für die Arbeitsplatzintegration und muß mit szenarienübergreifenden organisatorischen Zusammenhängen abgestimmt werden.

4.4.2 Medienneutrale Beschreibung von Interaktionen

An die Beschreibung der Interaktionen an einem Arbeitsplatz werden zwei wichtige Forderungen gestellt: sie muß *problemnah* und *mediengerecht* sein. Typischerweise sind Interaktionsbeschreibungen sehr stark an die spätere technische Realisierung durch präsentationsorientierte Bausteine angelehnt. Eine problemnahe Beschreibung sollte hingegen Inhalt, Intention und Randbedingungen von Interaktionen modellieren. Hinter der Forderung nach mediengerechter Beschreibung verbergen sich zwei Anliegen. Erstens sollte eine die Beschreibung medienneutral sein, also Medien weder implizieren noch a priori ausschließen, um so eine Basis für Medienintegration zu schaffen. Zweitens muß eine Interaktionsbeschreibung so aufgebaut sein, daß sie Rückschlüsse für die Auswahl von Medien und Modalitäten unterstützt.

Die Interaktionsbeschreibung in ITEMS basiert auf *Selektoren, Constraints* und *Informationscharakterisierung*. Die Verwendung von *Selektoren* zur Beschreibung von benutzerinitiierten Interaktionen wurde weitgehend von [7] übernommen. Alle Aktionen eines Benutzers lassen sich dabei auf eine Auswahl (von Daten oder von weiteren Aktionen) abbilden, die durch Selektoren modelliert werden. Selektoren werden je nach Art der Auswahl (1-n, m-n, etc.) orthogonal zum Inhalt der Auswahl klassifiziert. *Constraints* erlauben die Beschreibung von Abhängigkeiten zwischen einzelnen Interaktionen, z.B. bei der Aktivierung von Selektoren. Durch Constraints werden implizit Modi eingeführt. Diese Modi sind Folge der Interaktionssemantik und nicht zu verwechseln mit Modi, die aus Erwägungen der Benutzbarkeit eingeführt werden. Als Alternative zu der Verwendung schwer wartbarer Constraints untersuchen wir in ITEMS Vor- und Nachbedingungen für die Interaktionsbeschreibung.

Den Forderungen nach Problemnähe und Mediengerechtigkeit wird durch *Informationscharakterisierung* Rechnung getragen. Die in der Mensch-Computer-Interaktion kommunizierte Information wird typischerweise basierend auf implementierungsorientierten Datentypen beschrieben (Text, Video, etc.). In ITEMS entwickeln wir dahingegen eine Charakterisierung von Information, die sich problemnah an Intentionen orientiert und die mit Medieneigenschaften abgeglichen werden kann. Derartige Charakteristika wären z.B. *Kausalbeziehung, Prozeduren, Vergleich*, etc. Entscheidend ist hier, daß die Charakteristika medienunabhängig sind, aber sehr wohl die Eigenschaften der Information beschreiben, die für deren Darstellung relevant ist.

4.5 Modalitätenplanung und Generierung von Koordinationsinformation für die multimodale Informationsverarbeitung

Die Modalitätenplanung realisiert die Auswahl und Kombination geeigneter Medien und Modalitäten für ein gegebenes Interaktionsszenario. Grundlage dafür ist in Analogie zu der oben beschriebenen Informationscharakterisierung eine Charakterisierung von Medien und Modalitäten, die sich auf deren Ausdrucksfähigkeiten bezieht. Die Charakterisierung muß sich dabei an den verschiedensten Anforderun-gen, die es in einer Mensch-Computer-Interaktion geben kann, anlehnen ('Übermittlung exakter Werte', 'Vergleich auf einen Blick', etc.). Mediencharakteristika schaffen die Möglichkeit, die Intentionen einer Interaktion getrennt von Interaktionsform und zu vermittelnder Information zu beschreiben: Medien können z.B. *warnend* oder *anweisend* sein.

Auf Basis von Mediencharakterisierung kann eine zweistufige Modalitätenplanung realisiert werden. Dabei werden Informationstypen in zwei Stufen zunächst auf Informationscharakteristika und dann auf Mediencharakteristika abgebildet. Ein Beispiel soll dieses verdeutlichen: Datenpaare wie z.B. geografische Koordinaten, haben zweidimensionalen Charakter (erste Stufe). Dieser kann durch Medien/ Modalitäten zum Ausdruck gebracht werden, die zweidimensionalität ausdrücken können (zweite Stufe): z.B. Graphen, Tabellen, Karten. Durch Auswerten weiterer Charakteristika kann schließlich die am besten geeignete Modalität ermittelt werden.

Bei der Modalitätenplanung handelt es sich um ein Zusammenspiel von verschiedenen Medien und Modalitäten. Somit tritt ein allgemeines Koordinierungsproblem auf, bei dem neben dem zeitlichen und dem räumlichen Aspekt auch die Konsistenz der einzelnen Medien und Modalitäten zu beachten ist. Neben der Auswahl der einzelnen Modalitäten muß die Modalitätenplanung somit auch Koordinationsinformnation generieren. Hierfür konzipieren wir sog. *ModalityScripts*, die die Synchronisation von Modalitäten beschreiben. Eine Laufzeitkomponente kann auf dieser Basis Entscheidungen über das Spalten, Verschmelzen und Parallelisieren von Informationsflüssen treffen.

5 Zusammenfassung

In diesem Artikel haben wir Arbeitsplatz- und Medienintegration als Kernprobleme der Mensch-Computer-Interaktion in kooperativen Anwendungen diskutiert. Wir haben einen Ansatz für die Entwicklung arbeitsplatz- und medienintegrierter Benutzerschnittstellen vorgestellt, der in eine Entwicklungsumgebung für kooperative Anwendungen eingebettet ist. Ausgangspunkt sind dabei Arbeitsszenarien, die eine intuitive Beschreibung menschlicher Zusammenarbeit unterstützen, und aus denen sich Interaktionsanforderungen für einzelne Arbeitsplätze ableiten lassen. Kern der Beschreibung der einzelnen Interaktionen in Interaktionsszenarien sind problemnahe und medienneutrale Charakterisierungen der zu kommunizierenden Information. Weiter haben wir eine Modalitätenplanung vorgestellt, die basierend auf Mediencharakterisierung eine Auswahl von Medien und Modalitäten für eine gegebene Interaktion durchführt und Koordinationsinformation für die multimodale Informationsverarbeitung generiert.

Unser Ansatz wird intensiv durch grafische Editoren unterstützt. Auf dieser Basis haben wir bereits umfassende und sehr positive Erfahrung mit Arbeitsszenarien gesammelt. Interaktionsszenarien und Modalitätenplanung befinden sich hingegen noch in der Phase der Evaluierung und Verfeinerung.

Literatur

[1] Capurro, R. Ethik und Informatik: Die Herausforderung der Informatik für die praktische Philosophie. Informatik-Spektrum, Jg. 13, 1990, S. 311-320.

[2] Card, S.K., Mackinlay, J.D., Robertson, G.G. The Design Space of Input Devices. Blattner, M., Dannenberg, R. (Ed.): Multimedia Interface Design, Addison-Wesley, 1992.

[3] Gellersen, H.-W. Support of User Interface Design Aspects in a Framework for Distributed Cooperative Applications. Taylor, R. (Ed.): Software Engineering and Human Computer Interaction., Springer LNCS, 1994.

[4] Gellersen, H.-W., Mühlhäuser, M., Frick, O. Multi-user and Multimodal Aspects of Multimedia. Proc. of 1st EuroGraphics Symposium on Multimedia/Hypermedia in Open Distributed Environments, Graz, Juni 1994, Springer, S. 278-297.

[5] Harel, D. On Visual Formalisms. Communications of the ACM, 31(5), S. 514-530, Mai 1988.

[6] Hasenkamp, U., Kirn, S., Syring, M. CSCW — Computer Supported Cooperative Work. Addison-Wesley, 1994.

[7] Johnson, J. Selectors: Going Beyond User-Interface Widgets. Proc. of CHI '92, Monterey, CA, Mai 1992, S. 273-279.

[8]   Kuutti, K. A common vocabulary for talking abpout work processes: sceanrios in HCI & SE
      design. Preprints of Workshop on Software Engineering and Human-Computer Interaction:
      Joint Research Issues, Sorrento, Mai 1994, S. 237-241.

[9]   Laurel, B., Oren, B., Don, A. Issues in Multimedia Interface Design: Media Integration and
      Interface Agents. Proc. of CHI '90, Seattle, April 1990, S. 133-139.

[10] Medes, J. The development of human-computer interface descriptions from organisational
      models. Bullinger, H.-J. (Ed.): Human Aspects in Computing, Proc. of the 4th Conf. on HCI,
      Stuttgart, Sept. 1991, Elsevier, S.572-576.

[11] Patterson, J. Comparing the Programming Demands of Single-User and Multi-User
      Applications. Proc. of UIST '91, Hilton Head, SC, Nov. 1991, S. 87-94.

[12] Pickering, J., Grinter, R. Software Engineering and CSCW: A Common Research Ground. In
      Taylor, R. (Ed.) Software Engineering and Human Computer Interaction., Springer LNCS,
      1994.

[13] Roudaud, B., Lavigne, V., Lagneau, O., Minor, E. SCENARIOO: A New generation UIMS.
      Proc. of INTERACT '90, Aug. 1990, Elsevier Science Publ., S. 607-612.

[14] Taylor, R., Johnson, G. Separation of Concerns in the Chiron-1 User Interface Development
      and management System. proc. of INTERCHI '93, Amsterdam, April 1993, S. 367-374.

Hans-Werner Gellersen und Max Mühlhäuser
Telecooperation Office (TecO) an der Universität Karlsruhe
Vincenz-Prießnitz-Str. 1
76131 Karlsruhe
Tel. (0721) 6902-39
hwg@tk.telematik.informatik.uni-karlsruhe.de

Software-ergonomische Evaluation von Kiosksystemen im Museum

Andreas M.Heinecke, Sigrid Bumann, Thomas Kerstan
Universität Hamburg, FB Informatik / ANT

Zusammenfassung.

Der Einsatz von Kiosksystemen im Museum steht aufgrund der damit verbundenen Kosten unter einem Rechtfertigungsdruck, der eine Evaluation der eingesetzten Systeme erforderlich macht. Diese Evaluation darf ebenfalls aus Kostengründen nicht zu aufwendig sein, soll aber Aufschluß über die Benutzung der Systeme durch die Besucher[1] geben. Benutzungsprotokolle (Logfiles) sind, gegebenenfalls in Kombination mit Befragungen, eine Evaluationsmethode, die mit vertretbarem Aufwand wichtige Daten zur Nutzung des Systems liefern kann. Im Deutschen Salzmuseum wird diese Methode zur Evaluation zweier Anwendungen im Rahmen eines Modellversuchs eingesetzt.

Schlüsselwörter.

Kiosk-Systeme (Points of Information), Hypermedia, Evaluation, Benutzungsprotokolle (Logfiles)

1 Einsatz von Kiosksystemen in Museen

Der Einsatz von Multimedia- und Hypermedia-Anwendungen in Museen, die als Kiosksysteme von den Besuchern selbst allein genutzt werden, nimmt ständig zu. Akzeptanz und Nutzung solcher Systeme durch das Publikum sind dagegen bisher kaum untersucht oder beschrieben worden.

1.1 Vorhandene Anwendungen

Kiosksysteme im Museum laufen in der Regel in speziellen Informationsterminals mit Bedienung über eine Rollkugel (Trackball) oder einen berührungssensitiven Bildschirm (Touchscreen). Leider gibt es zur Zeit noch keinen Überblick über derartige Anwendungen in der Literatur, so daß Interessierte auf einzelne Veröffentlichungen in einschlägigen Konferenzbänden, Zeitschriften oder sogar der Tagespresse angewiesen sind. Die dort beschriebenen Anwendungen lassen sich grob in drei Klassen einteilen.

---

[1] Aus Gründen der leichteren Lesbarkeit werden Personenbezeichnungen nur im grammatisch maskulinen Genus benutzt. Diese Bezeichnungsweise soll sowohl für das weibliche als auch für das männliche natürliche Genus gelten.

Einzelanwendungen vermitteln zu einen bestimmten Objekt oder Thema zusätzliche Informationen. Als Beispiel für diese Klasse sei die im Museum für das Fürstentum Lüneburg laufende Anwendung zur Erschließung einer mittelalterlichen Weltkarte genannt [16].

Bei *verteilten Anwendungen* liefern einzelne Terminals innerhalb einer Ausstellung oder eines Themenbereichs jeweils verschiedene zusätzliche Informationen. So wurden bei der Austellung "Pompeji wiederentdeckt" siebzehn PCs zur Vermittlung von Informationen über die Exponate der Ausstellung und deren Umfeld eingesetzt, wobei jedes Terminal im Prinzip eine Einzelanwendung darstellte [7].

Das Wassermuseum Aquarius in Mülheim/Ruhr ist dagegen ein Beispiel für eine *integrierte Anwendung*, bei der der Computer-Einsatz sich nicht auf Informationsterminals beschränkt, sondern das ganze Ausstellungskonzept durchdringt. Dort werden 30 Multimedia-Stationen eingesetzt, die neben Bildplattenspielern und digitalem Video zur Informationsausgabe auch Dioramen, Schaukästen und Buchstabenlaufbänder direkt ansteuern [14].

1.2 Fragen der Akzeptanz und Nutzung

Für Museen stellt sich die Frage nach dem Einsatz von Hypermedia-Anwendungen auch unter Kosten-/Nutzen-Gesichtspunkten: "lohnt" sich die Entwicklung einer museumsspezifischen Hypermedia-Anwendung und die Beschaffung der dafür notwendigen Hardware in dem Sinne, daß dadurch das Museum attraktiver für Besucher wird? Diese Frage mündet unmittelbar in die Frage nach der Akzeptanz des installierten Systems und nach seiner Benutzbarkeit. Für entsprechende Untersuchungen ergeben sich eine Reihe von Fragestellungen bezüglich des Navigationsverhaltens und des Vorgehens der Benutzer, der Systembeherrschung, der Präferenz bestimmter Medien, der inhaltlichen Schwerpunkte, des Informationsgehaltes und der Informationsvermittlung.

Die genannten Fragestellungen spannen einen weiten Rahmen für die Evaluation von Computersystemen im Austellungsbereich von Museen auf, der von der Gestaltung der Benutzungsoberfläche über die Museumspädagogik und -didaktik bis zur Lernpsychologie reicht. Bei unseren weiteren Überlegungen wollen wir uns hauptsächlich auf die Evaluierung nach Gesichtspunkten der Software-Ergonomie zur Verbesserung der Benutzbarkeit der Systeme konzentrieren.

2 Evaluation von Kiosksystemen

Von den Evaluationsmethoden der Software-Ergonomie sind für Kiosksysteme im Museum nicht alle gleichermaßen geeignet. Das häufig ungünstige Verhältnis zwischen Evaluationsaufwand und daraus erwachsendem Nutzen für das Museum insbesondere bei Einzelanwendungen dürfte der Hauptgrund dafür sein, daß auch die für dieses Gebiet geeigneten methodischen Ansätze nur äußerst selten genutzt werden.

2.1 Methodische Ansätze

Ein umfassender Ansatz, der entsprechend ISO 9241 Teil 2 [8] von der Arbeitsaufgabe des Benutzers ausgeht, läßt sich hier nur schwer verfolgen, da es eine solche Arbeitsaufgabe im üblichen Sinne nicht gibt. Die Besucher wollen herausgefordert werden, aber nichts Bestimmtes lernen ("educational browsing", [1]). Insofern ist es sinnvoll, bei der Evaluation hauptsächlich auf das tatsächliche Benutzerverhalten sowie auf die Einschätzung des Systems durch die Benutzer zu achten.

Subjektive Methoden (Einteilung der Evaluationsmethoden nach [13]) erscheinen auf diesem Gebiet sehr wichtig, da es ja vor allem auf die Akzeptanz der Systeme durch die Besucher ankommt. Mit Fragebögen und Interviews läßt sich die Einschätzung der Systeme bezüglich Benutzbarkeit, Informationsgehalt und Gestaltung ermitteln. Fragebögen haben den Vorteil, daß sie bei der Verteilung nahezu keinen und bei der Auswertung einen relativ geringen Aufwand erzeugen. Interviews erfordern dagegen einen deutlich höheren Personalaufwand und sind deshalb meist nur in geringem Umfang möglich.

Objektive Methoden können ebenfalls nur dann genutzt werden, wenn sie mit geringem Aufwand durchführbar sind. Eine Beobachtung, die zusätzliches Personal erfordert, scheidet also normalerweise aus. Daher ist in erster Linie die automatische Protokollierung der Benutzeraktionen (Logfile-Recording) geeignet, vor allem, wenn die Aktionen so aufgezeichnet werden, daß eine automatische oder halbautomatische Auswertung möglich ist. Sie kann gegebenenfalls dadurch ergänzt werden, daß ohnehin vorhandenes Aufsichtspersonal einige wenige zusätzliche Informationen notiert, etwa zur Person des jeweiligen Benutzers.

Leitfadenorientierte Methoden sind in diesem Anwendungsgebiet meist nicht durchführbar. So dürfte die Bewertung durch ein Expertenurteil nur bei integrierten oder großen verteilten Anwendungen möglich sein, vor allem, wenn aufgrund der Größe des Projektes Spezialisten für Software-Ergonomie dem Entwicklungsteam angehören oder zur Verfügung stehen.

Experimentelle Methoden scheiden schon wegen des Aufwands im allgemeinen aus. Außerdem geht es bei diesem Anwendungsbereich meist nicht darum, bestimmte Hypothesen zu testen, sondern eine Bewertung des vorhandenen Systems zu erreichen.

2.2 Bisherige Untersuchungen

Die bisher einzige ausführlich dokumentierte Studie zur Akzeptanz und Nutzung eines Multi-Media-Systems in einem deutschen Museum ist die von Noschka-Roos und Lewalter [14] durchgeführte Untersuchung des Touchscreen-Systems 'Erneuerbare Energien' in der gleichnamigen Abteilung des Deutschen Museums in München. Die Nutzung des Touchscreen-Systems wurde untersucht mit Interviews, die mit Benutzern nach Benutzung des Systems geführt wurden, und mit einer systematischen verdeckten Beobachtung während der Nutzung des Systems. Erwähnenswert ist, daß die verdeckte Beobachtung aufwendig über einen zweiten Bildschirm erfolgte und per Hand protokolliert wurde. Möglichkeiten der technischen Unterstützung durch das untersuchte System wurden also nicht genutzt.

Auch außerhalb des deutschen Sprachraums hat es bisher nur wenige Versuche zur Evaluation von Kiosksystemen gegeben. Wanning [17] berichtet über die Evaluation eines mit Touchscreen gesteuerten interaktiven Video-Systems im Dänischen Widerstandsmuseum in Kopenhagen. Fragestellungen und Vorgehen sind nahezu identisch mit der Münchener Untersuchung. Fahy, Poulter und Sargent [5] haben das Hypermedia-Front-End eines Museums-Informationssystems evaluiert nach den Kriterien: Effektivität des Layouts, Qualität der Bildschirm-Information, Klarheit der Terminologie und Funktionalität der Steuerknöpfe. Nebenzahl [11] berichtet über die bisher einzige Untersuchung, bei der die Möglichkeit der Protokollierung von Benutzer-Aktionen durch das System selbst genutzt wurde. Protokolliert wurde die Zahl der Berührungen der einzelnen Bedienknöpfe und die in den angewählten Bereichen verbrachte Zeit.

3   Untersuchte Museumsanwendungen

Das Deutsche Salzmuseum hat im Rahmen eines vom Niedersächsischen Ministerium für Wissenschaft und Kultur geförderten Modellversuchs "Museum der Zukunft - Möglichkeiten und Grenzen moderner Museumspräsentation" verschiedene Formen der Präsentation in Sonderausstellungen verwirklicht. Eine diese Formen ist die Nutzung von Computern, wofür zwei Anwendungen aus verschiedenen Themenkreisen erstellt wurden.

3.1 Lüneburg - eine Großstadt um 1600

Diese Anwendung vermittelt mit Texten und Bildern auf der Basis eines Plans der Stadt Lüneburg von 1574 Informationen über die Stadt Lüneburg um 1600 als Ganzes sowie in Planausschnitten über markante Gebäude und Orte. Bei beiden Darstellungsformen befindet sich rechts ein Textfeld, in dem die zu der jeweiligen Ansicht gehörenden Informationstexte einzeln dargestellt werden können. Zum Gesamtplan gibt es 35 solche Texte, zu den Planausschnitten 73. In beiden Ansichten sind die

Texte mit Hilfe jeweils eines hierarchisch gegliederten Inhaltsverzeichnisses zugreifbar. Zwischen den Texten kann mit Hilfe von Aktionssworten (Hypertext-Links) sowie in der Sequenz des Inhaltsverzeichnisses (nächster / vorheriger) gewechselt werden.

Es gibt ungefähr 280 Aktionsworte, die zu einem Wechsel zwischen Informationstexten führen. Weitere etwa 80 Aktionsworte erlauben das Einblenden zusätzlicher Information (Text oder Grafik) in die jeweilige Darstellung. Zu den meisten Texten sind ein oder mehrere Bilder vorhanden, die über Schaltflächen (Buttons) mit einem stilisierten Photoapparat abgerufen werden können. Weitere Buttons dienen der Navigation und Orientierung. In den Planausschnitten können Objekte direkt angeklickt werden, um die zugehörigen Texte zu erhalten. Alle Eingaben erfolgen über eine Rollkugel.

3.2 Salz. Eine Hypermedia-Präsentation des Deutschen Salzmuseums

Diese Anwendung vermittelt Informationen zu Salzmineralien (Entstehung, Aufbau, Aussehen, Fundorte etc.) mit Hilfe von Texten, Fotos, Grafiken und Animationen / Simulationen. Die Bedienung ist auf den Einsatz eines Touchscreens zugeschnitten, es gibt nur wenige Bedienelemente pro Bildschirmseite. Jede Bildschirmseite enthält eine Kapitel-Überschrift sowie am unteren Rand Bedienelemente für die Standard-Funktionen (Aufruf der Orientierungsseite / Inhaltsverzeichnis, Neustart, Rücksprung zum übergeordneten Thema, Zurückgehen zum zuletzt angezeigten Kapitel). Es gibt 107 Bildschirmseiten zu 9 verschiedenen Kapiteln, die in einer Hierarchie so angeordnet sind, daß sie über zwei bis drei Auswahlschritte mit zwei bis drei Wahlmöglichkeiten aufgerufen werden können (vgl. Abb. 1). Diese Informationsseiten der Kapitel zeigen immer einen Text zusammen mit einem zugehörigen Foto oder einer Grafik oder Animation. Jedes Kapitel enthält dabei zwischen 2 und 48 solcher Kombinationen, zwischen denen mit Schaltflächen, die grafische Symbole tragen (zum Anfang, rückwärts, vorwärts etc.), navigiert werden kann. 65 Aktionsworte erlauben als Verweise den direkten Wechsel zwischen den Informationsseiten.

4   Vorgehensweise bei der Evaluation der Systeme

Bei der Evaluation der im Deutschen Salzmuseum installierten Systeme wird versucht, mit möglichst geringem Aufwand eine Bewertung und Verbesserung der Systeme zu erreichen. Hierzu werden informelle Beobachtungen, Protokollierungen der Benutzeraktionen und Fragebögen eingesetzt.

4.1 Informelle Beobachtungen und Gespräche

Die Systementwicklung erfolgte in interdisziplinärer Zusammenarbeit zwischen Museumsfachleuten und Informatikern auf dem Wege des evolutionären Prototypings (vgl. [9]). Bei Vorführungen der Anwendungen für Museumsfachleute und während der Testphase der Prototypen im Publikumseinsatz ließen sich im Gespräch und durch informelle Beobachtung bereits Schwachpunkte beziehungsweise Fehler der Systeme feststellen, die während der Entwicklung nicht bemerkt worden waren.

Neben der Behebung von Fehlern wurde auch eine Überarbeitung der Bedienkonzepte durchgeführt, indem in Abkehr von WINDOWS-Konventionen die Auswahl in der Liste des Inhaltsverzeichnisses mit einfachen Klick ermöglicht wurde und alle Schaltflächen vom Button-Up- auf das Button-Down-Prinzip umgestellt wurden, so daß bereits beim Niederdrücken der Taste unmittelbar mit der optischen Rückmeldung auch die zugehörige Aktion ausgeführt wird. Insgesamt bildete die Testphase einen Schnittpunkt zwischen Prototyping und Evaluation.

4.2 Benutzungsprotokolle

Die beiden Anwendungen "Salz" und "Lüneburg um 1600" erzeugen im Betrieb Benutzungsprotokolle (Logfiles), in denen alle Benutzungsaktionen detailliert mit Kontext und Uhrzeit festgehalten werden. Protokolliert wird dabei das Ereignis (z.B. Klick, Doppelklick, längeres Gedrückthalten der Taste), das betroffene Objekt (z.B. Schaltfläche, Aktionswort, Zeichenobjekt), die vor dem Ereignis aktuelle Bildschirmseite, die nach der Aktion aktuelle Bildschirmseite und die Zeit (in Sekunden).

Durch diese automatische Protokollierung läßt sich unter anderem die Häufigkeit des Aufrufs von und der Verweildauer bei bestimmten Informationen, die Nutzung der einzelnen Navigationsmöglichkeiten sowie die Verteilung der Klicks auf funktionslose Teile der Darstellung ermitteln. Hieraus lassen sich Schlüsse auf das Interesse der Benutzer, auf ihre Vorgehensweise und auf ihre Beherrschung des Systems ziehen. Beschränkt auf die ersten beiden Fragen (Aufrufhäufigkeit und Verweildauer bei bestimmten Informationen) werden ähnliche Protokollierungsverfahren mittlerweile auch bei kommerziellen Kiosk-Systemen eingesetzt [15].

4.3 Fragebögen

Bei allen Ausstellungen des Deutschen Salzmuseums, die im Rahmen des Modellversuchs "Museum der Zukunft" stattfanden, wurden Befragungen der Besucher durchgeführt. Hierzu wurden Fragebögen erstellt, die Fragen zu der jeweiligen Ausstellung, zu Museumsbesuchen allgemein und zur Person des Besuchers enthielten. Die Fragebögen wurden vom Museumspersonal an die Besucher beim Verlassen der Ausstellung ausgegeben. Als Anreiz zur Beantwortung wurden unter den zurückgegebenen oder später zurückgesandten Fragebögen Bücher verlost. Außerdem erhielten alle Besucher, die den Fragebogen direkt im Museum abgaben, einen kleinen Steinsalzkristall.

Die Anwendung "Salz" war Teil der Sonderausstellung "unter/übertage - Salzminera-
lien aus aller Welt", die mit einem solchen Fragebogen bewertet wurde. Von den 24
Fragen des Fragebogens bezogen sich dabei drei unmittelbar auf die Anwendung
"Salz". Gefragt wurde, wie die Besucher mit dem System zurechtkämen, welche
Probleme sie ggf. mit der Benutzung hätten und wie sie die Inhalte des Programms
bewerteten. Die Anwendung "Lüneburg um 1600" ist Teil der Dauerausstellung. Um
sie mitbewerten und mit der Anwendung "Salz" vergleichen zu können, wurden im
Fragebogen hierzu die gleichen drei Fragen gestellt mit der Vorbedingung "Wenn Sie
die Dauerausstellung besucht haben". Daran schloß sich noch die Frage an, welches
der Systeme den Besuchern besser gefalle und warum.

5 Untersuchungsergebnisse

Für die Evaluation der Systeme im Deutschen Salzmuseum wurden die Benutzungsprotokolle der Anwendung "Salz." und der Anwendung "Lüneburg um 1600"
sowie die Fragebögen zur Sonderausstellung "unter/übertage" herangezogen. In der
folgenden Darstellung steht die Anwendung "Salz." im Vordergrund. Die Anwendung "Lüneburg um 1600" wird nur in geringem Maße zum Vergleich herangezogen. Die Auswertung der Benutzungsprotokolle von "Lüneburg um 1600" [3] kann
hier aus Platzgründen nicht dargestellt werden.

5.1 Auswertung der Benutzungsprotokolle des Systems "Salz"

Für die Untersuchung wurden alle Protokolle des sechswöchigen Zeitraums vom
12.08.1994 bis zum 22.09.1994 ausgewertet. Das System war während dieser Zeit
etwa 320 Stunden in Betrieb. In diesem Zeitraum besuchten etwa 1350 Personen die
Ausstellung. Es wurden 27077 Benutzeraktionen protokolliert. Um eine Abschät-
zung zu erhalten, wieviele Besucher das Kiosksystem benutzten, haben wir zunächst
die Ereignisse danach eingeteilt, wie lang die auf das Ereignis folgende Pause ist. Es
zeigt sich, daß bei 65 % der Ereignisse das nächste Ereignis schon nach weniger als
5 Sekunden folgt. Unterhalb einer Pausendauer von 15 Sekunden liegen schon
92 % der Ereignisse, unterhalb 30 Sekunden 97 % und unterhalb 1 Minute 98 %.
Ab 2 Minuten Länge sind die Pausen etwa gleich verteilt. Daher gehen wir davon
aus, daß bei Pausen von zwei Minuten und mehr niemand mehr am Gerät ist, mithin
nach jeweils mehr als 2 Minuten Pause ein Benutzerwechsel stattgefunden hat. Unter
dieser Prämisse haben in dem genannten Zeitraum 479 Personen das System benutzt.

Ereignisse				Anteil an 2707
mit Wirkung			Anteil an 1746	65 %
Seitenwechsel-Buttons		Anteil an 7789	45%	29%
vorwärts	6339	81%	36%	23%
rückwärts	686	9%	4%	3%
Kapitelanfang	255	3%	1%	1%
schnell vorwärts	69	1%	0%	0%
schnell rückwärts	291	4%	2%	1%
zuletzt gezeigte	149	2%	1%	1%
Parameter-Einstellung		Anteil an 3225	18%	12%
Schwelle hoch	1000	31%	6%	4%
Schwelle runter	896	28%	5%	3%
Wasserspiegel hoch	689	21%	4%	3%
Wasserspiegel runt.	640	20%	4%	2%
Themenwahl		Anteil an 2989	17%	11%
Leitseiten u. Orient.	2655	89%	15%	10%
Mineralienwahl	334	11%	2%	1%
Aufruf der Orientierung		631	4%	2%
Neustart		704	4%	3%
Übergeordnetes Kapitel		805	5%	3%
Zurück (Kapitel)		1128	6%	4%
Sprung mit Hotwords		196	1%	1%
ohne Wirkung			Anteil an 9610	35%
Überreaktionen		Anteil an 4712	49%	17%
Doppelklick	3235	69%	34%	12%
langes Drücken	1477	31%	15%	5%
Falsche Taste		1943	20%	7%
Fehlklicks		2955	31%	11%

Die Gesamtnutzungsdauer des Systems betrug etwa 34,5 Stunden, was lediglich 11 % der Öffnungszeit der Ausstellung entspricht. Allerdings gibt es keine Statistik darüber, wie sich die Besucher auf die Öffnungszeiten verteilt haben. Untersucht man die Benutzungsdauer bezogen auf die einzelnen Benutzer, so ergibt sich eine maximale Benutzungsdauer von 37 Minuten (möglicherweise haben hier Wechsel unterhalb der 2-Minuten-Grenze stattgefunden), eine minimale Benutzungsdauer von 0 Sekunden (isoliertes Ereignis mit nachfolgender Pause von mehr als 2 Minuten), ein arithmetisches Mittel von 4 Minuten 20 Sekunden und ein Median von 2 Minuten 36 Sekunden. Ordnet man die protokollierten Ereignisse den 479 Benutzern zu, so ergibt sich ein Maximum von 904 Ereignissen, ein Minimum von 1 Ereignis, ein arithmetisches Mittel von 57 und ein Median von 27 Ereignissen.

Tab. 1: Häufigkeit der möglichen Benutzeraktionen

Die inhaltliche Aufteilung der Benutzerereignisse zeigt Tab. 1, wobei zu beachten ist, daß viele der aufgeführten Benutzungsmöglichkeiten nur auf bestimmten Bildschirmseiten vorhanden sind. Die hohe Zahl der Parameter-Einstellungen für die Simulation der Entstehung von Salzlagerstätten erklärt sich daraus, daß die Höhe des Wasserspiegels und der Bodenschwelle nur schrittweise eingestellt werden kann und jeder Schritt als ein Ereignis gezählt wurde. Da noch kein funktionsfähiger Touchscreen zur Verfügung stand, wurde die Anwendung über einen Trackball mit drei Tasten bedient, von denen die linke mit einer Beschriftung als Auswahltaste markiert war.

Die Verteilung der Benutzeraktivitäten auf die einzelnen Seiten der Anwendung ist sehr unterschiedlich. In Abb. 1 ist für die Auswahlseiten und jeweils die ersten beiden und die letzte Seite der 9 Kapitel angegeben, wie oft sie direkt mit Hilfe von

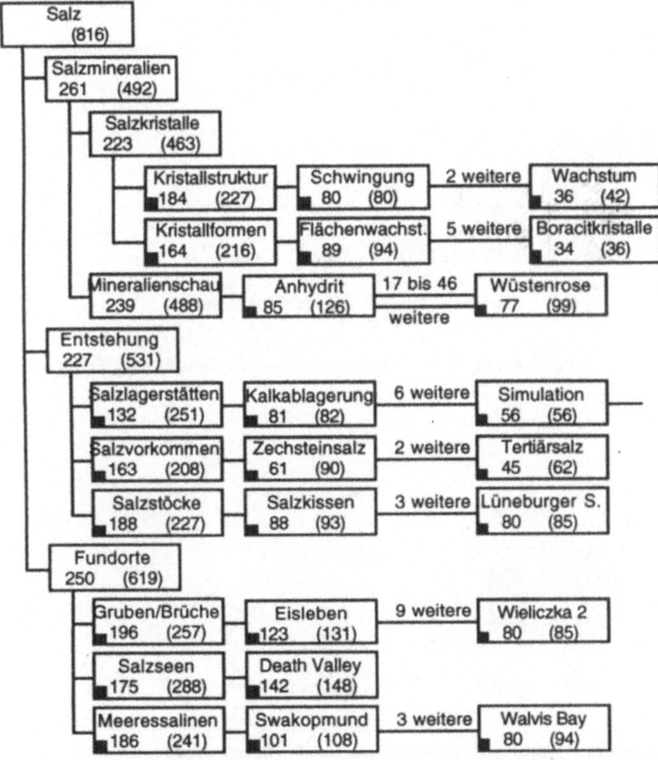

Abb. 1:   Aufrufhäufigkeiten der Seiten. Die räumliche Anordnung entspricht der Anordnung der Auswahl-Schaltflächen und gibt die Auswahl-Hierarchie bis zu den schwarz markierten Informationsseiten wieder.

Themenauswahl und Vorwärts-Buttons aufgerufen wurden. In Klammern ist angegeben, wie oft die Seiten insgesamt betreten wurden. Auf die Auswertung der Verweildauern auf den Seiten kann hier aus Platzgründen nicht eingegangen werden.

5.2 Daten aus den Fragebögen

Ausgewertet wurden die Fragebögen, die nach Abschluß der System-Testphase ab Mitte Juli abgegeben wurden. Diese Fragebögen wurden von 51 männlichen und 57 weiblichen Personen ausgefüllt. Die einzelnen Altersgruppen sind dabei etwa nach der Besucherstruktur vertreten. Familien / Paare haben z.T. nur einen gemeinsamen Bogen abgegeben, so daß sich eine Gesamtzahl von 100 ergibt. Abb. 2 zeigt die Bewertung der Anwendung "Salz.", Abb. 3 die der Anwendung "Lüneburg um 1600". Abb. 4 gibt eine vergleichende Bewertung der beiden Anwendungen. Die hohe Zahl der Enthaltungen bei Abb. 3 und Abb. 4 erklärt sich z.T. daraus, daß nicht alle Besucher der Sonderausstellung auch die Dauerausstellung mit dieser Anwendung besuchten.

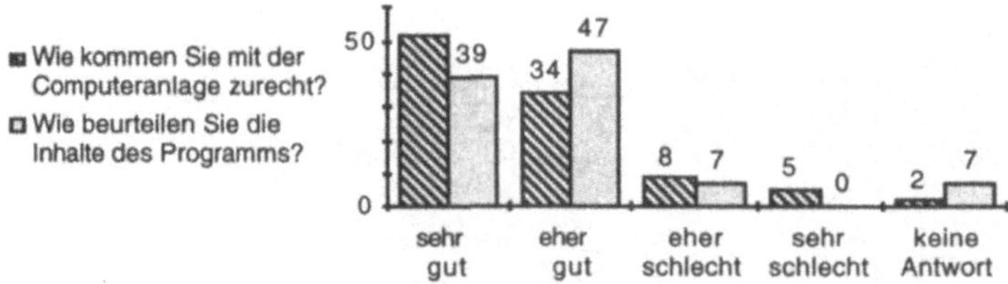

Abb. 2:   Bewertung der Anwendung "Salz." durch die Besucher der Sonderausstellung

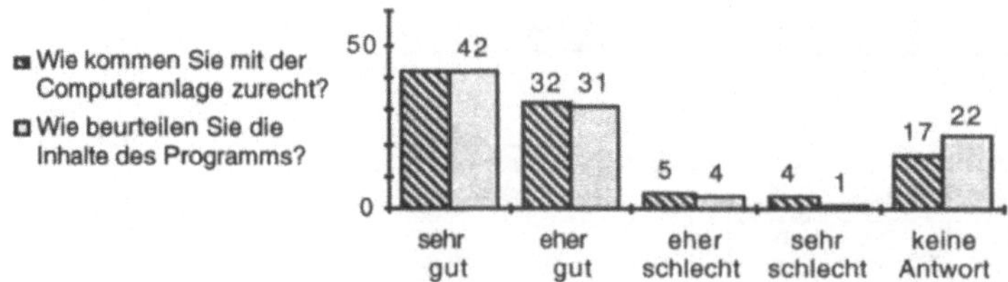

Abb. 3:   Bewertung der Anwendung "Lüneburg um 1600" durch die Besucher der Sonderausstellung.

Welches Programm gefällt Ihnen besser?

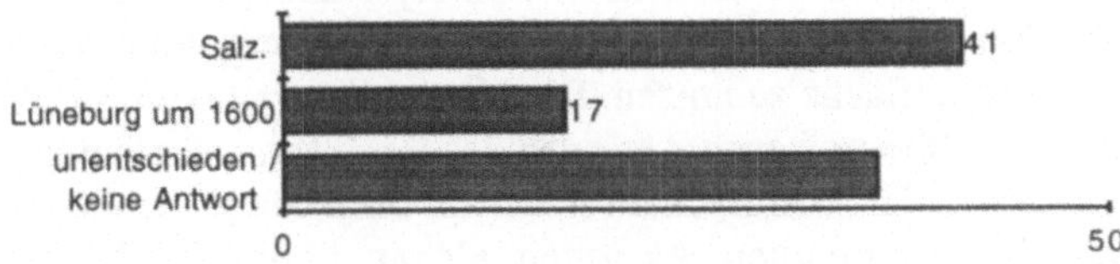

Abb. 4:   Vergleichende Bewertung beider Anwendungen durch die Besucher der Sonderausstellung.

Die frei zu beantwortenden Fragen nach Problemen bei der Benutzung wurden in der Regel nicht oder mit "keine" beantwortet. Lediglich die Bedienung mit Hilfe der Rollkugel und die Zuordnung der Rollkugel-Tasten wurde zweimal bemängelt. Die freie Frage nach der Begründung für die Entscheidung zwischen den Systemen wurde hingegen 42 mal beantwortet. Hierbei wurde in erster Linie "interessanter" angegeben (14 mal für "Salz." und 5 mal für "Lüneburg um 1600"). 5 mal wurde "Salz." bevorzugt, weil es übersichtlicher / einfacher zu verstehen sei, jedoch auch einmal "Lüneburg um 1600", weil es komplexer und vielfältiger sei.

5.3 Analyse von Benutzerverhalten und Benutzerbewertung

Die Hälfte der Benutzer bleibt weniger als 2,5 Minuten am Gerät. Alle 4,6 Sekunden findet im Mittel eine Benutzeraktion statt. Von den erfolgreichen Aktionen entfallen 36 % (immerhin ungefähr ein Viertel aller Aktionen) auf das Betätigen des Vorwärts-Buttons. Zusammen erweckt dies den Eindruck, daß hier lediglich geblättert wird, ohne die Information zu betrachten, oder um nach einem kurzen Blick zu erkennen, daß die dargestellte Information nicht interessiert.

Andererseits ist die Bewertung der Inhalte wie auch der Bedienung durch die Benutzer sehr gut. Lediglich ein Befragter sprach sich generell gegen den Einsatz von Computern in einer Museumsausstellung aus. Mehrere Besucher fragten beim Museumspersonal nach, ob die Anwendungen käuflich zu erwerben seien.

Möglicherweise teilen sich die Benutzer des Systems in zwei Gruppen: Die einen klicken schnell ein wenig herum, hauptsächlich um zu sehen, wie das System reagiert, ohne inhaltlich interessiert zu sein. Die anderen nehmen sich Zeit, um die dargestellten Inhalte anzusehen und nutzen die Möglichkeiten des Systems bewußt. Vermutlich sind die Angehörigen der zweiten Gruppe insgesamt mehr interessiert und auch eher bereit sich in einem Fragebogen zu äußern. Diese Personen würden die eigentliche Zielgruppe für die Entwicklung solcher Kiosksysteme bilden, und ihr Benutzerverhalten wäre besonders interessant. Es sollte deshalb versucht werden, speziell die Daten dieser Gruppe zu untersuchen.

Obwohl die Fragebögen wenig Anlaß und praktisch keine Anhaltspunkte zur Verbesserung der Systeme bieten, kann versucht werden, eine Analyse der Benutzeraktionen ohne Wirkung (immerhin ein Drittel aller Aktionen) zu einer ergonomischen Verbesserung der Benutzungsoberfläche zu nutzen. Die Hälfte dieser wirkungslosen Aktionen besteht aus sozusagen übertriebenen Aktionen des Benutzers, nämlich langanhaltendem Drücken der Taste oder mehrfachem Drücken. Dies kommt sowohl in Verbindung mit eigentlich erfolgreichen Aktionen als auch bei Klicks auf funktionslose Bereiche vor und ist eigentlich nicht als Fehler zu werten.

Ein Fünftel der wirkungslosen Aktionen besteht aus dem Drücken der falschen Taste der Rollkugel. Diese Fehlerquelle scheidet beim Einsatz eines Touchscreens aus und läßt sich zwischenzeitlich durch eine bessere Kennzeichnung der Auslösetaste und / oder eine Gleichbehandlung der Tasten in der Software (nur bedingt möglich) minimieren.

Als eigentliche Fehlklicks, also Klicks auf funktionslose Bereiche, sind 11 % aller Benutzerereignisse zu anzusehen. Dies ist ein relativ hoher Anteil, wenn man bedenkt, daß der Cursor über allen Objekten mit Funktion eine andere Form annimmt als über funktionslosen. Betrachtet man die Fehlklicks genauer, so liegen etwa ein Fünftel von ihnen auf der gerade angezeigten Grafik, der Rest verteilt sich etwa gleichmäßig auf Textblöcke, Farbfelder und Texte von Legenden, die unmittelbare Umgebung von Buttons, Plätze, an denen auf der vorhergehenden Seite Buttons lagen, und völlig leere Bereiche.

Die Ereignisse der vorletzten Kategorie entstehen hauptsächlich beim hektischen Weiterblättern, indem auf der jeweils letzten Seite an der Stelle des dort nicht mehr vorhanden Vorwärts-Buttons geklickt wird. Eine Verbesserung durch Änderung der Benutzungsoberfläche erscheint hierfür ebenso wie für die Klicks auf völlig leere Bereiche kaum möglich. Das gleiche gilt für die "Zielfehler" in unmittelbarer Umgebung von Buttons, die wohl auf motorischen Problemen im Umgang mit der Rollkugel beruhen und beim Einsatz eines Touchscreens aufhören sollten.

Ansatzpunkte zur ergonomischen Verbesserung bieten somit lediglich die etwa 4 % der Gesamtereignisse, die aus Klicks auf Objekte bestehen, die keine Funktion tragen. In der überarbeiteten Version des Systems wird deshalb versucht, die Bedienelemente noch deutlicher hervorzuheben und die Legenden der Grafiken so zu gestalten, daß sie sich deutlicher von Schaltflächen unterscheiden. Zusätzlich wird in die Orientierung, deren Aufrufzahl die Anzahl der Benutzer deutlich übersteigt, eine Bildschirmseite aufgenommen, auf der erklärt wird, wie Schaltflächen und Aktionsworte aussehen und was sie bewirken. Möglicherweise läßt sich hierdurch die Systembeherrschung verbessern.

Bezüglich der Navigation ist interessant, daß nur sehr selten mit Hilfe von Aktionsworten navigiert wird (1 % der Benutzeraktionen). Dies muß allerdings in Relation zu der Zahl der vorhandenen Hotwords gesetzt werden. Untersucht man, über welche Navigationsaktionen die einzelnen Seiten verlassen wurden, so läßt sich feststellen, daß bei Seiten mit Aktionsworten die Nutzung eines jeden Aktionswortes zwischen 3 und 10 % der Weiternavigation umfaßte. Benutzer, die sich intensiver mit der Anwendung beschäftigen, nehmen diese Möglichkeit also durchaus wahr. In der überarbeiteten Version des Systems werden die Aktionsworte zusätzlich durch Farbe hervorgehoben, und es wird in der Orientierung auf die Möglichkeit der Nutzung solcher Verweise hingewiesen. Es bleibt abzuwarten, ob sich deren Nutzung dadurch weiter erhöht.

Betrachtet man die Aufrufhäufigkeit der einzelnen Seiten, so zeigt sich, daß allem Anschein nach weder die räumliche Anordnung der Auswahlschaltflächen noch die Hierarchiestufe die Häufigkeit der Wahl eines Themas beeinflußt (vgl. Abb. 1). Die jeweils erste Informationsseite eines jeden Kapitels wird größenordnungsmäßig gleich oft aufgerufen. Das Minimum beträgt hier 132 für Salzlagerstätten (Aufrufreihenfolge: mittlerer Button - oberer Button), das Maximum 196 für Gruben und Brüche (unterer Button - oberer Button).

Interessant ist der deutliche Unterschied in der Aufrufhäufigkeit zwischen der jeweils ersten Informationsseite und den folgenden bei den Kapiteln Kristallstruktur, Kristallformen, Salzlagerstätten, Salzvorkommen und Salzstöcke (Rückgang auf 61 bis 37 %). Bei diesen Kapiteln bestehen die Seiten aus einer Kombination von Text und Grafik. Bei den Kapiteln Gruben / Brüche, Salzseen und Meeressalinen, die eine Kombination aus Foto und Text enthalten, ist der Rückgang nicht ganz so stark (auf 81 bis 54 %). Möglicherweise wirkt sich hier eine leichte Präferenz des Mediums Foto aus.

6   Schlußfolgerungen

Die Auswertung der Benutzungsprotokolle hat gezeigt, daß diese ein wichtiges Hilfsmittel für die Evaluation von Kiosksystemen im Museum sein können. Sie ermöglichen nicht nur Aussagen zum Benutzerverhalten, sondern geben auch konkrete Hinweise für ergonomische Verbesserungen der Benutzungsoberfläche, selbst bei Anwendungen, die in der subjektiven Bewertung durch die Benutzer ein gutes Urteil erhalten. In einem zweiten Schritt läßt sich dann untersuchen, ob die Änderungen den gewünschten Erfolg erzielt haben.

Wenn die Protokollierung mit Hinblick auf eine spätere Auswertung in einer Datenbank konzipiert wird, lassen sich mit relativ geringen Aufwand Fragen zur ergonomischen und inhaltlichen Gestaltung beziehungsweise Verbesserung von Kiosksystemen im Museum beantworten. Auch für andere Kiosksysteme dürfte die Evaluation mit Hilfe einer solchen Protokollierung sinnvoll sein. Auch weitergehende Fragen wie die der Medienpräferenz oder des Einflusses der räumlich-hierarchischen Gliederung lassen sich untersuchen, wenn verschiedene Varianten einer Anwendung einander gegenübergestellt werden.

Neben der reinen Protokollauswertung können Fragebögen und Beobachtungen sinnvoll sein, insbesondere, wenn eine Untersuchung mit Bezug zu bestimmten Benutzergruppen erfolgen soll. Für die weitere Untersuchung der Systeme ist geplant, die ohnehin vorhandenen Überwachungskameras für eine verdeckte Beobachtung zu benutzen, die es ermöglicht, eine Differenzierung der Benutzer etwa nach Alter und Geschlecht vorzunehmen oder danach, ob das System von einer Einzelperson oder gemeinsam von einer Gruppe genutzt wird. Ebenso läßt sich dabei feststellen, welche Besucher nicht an das System herangehen. Für spezielle Fragen lassen sich dann zusätzlich kurze Fragebögen verwenden, bei denen ein besserer Rücklauf zu erwarten ist.

7 Literatur

[1]   Allison, D. K., Gwaltney, T. "How People Use Electronic Interactives...". In [2].

[2]   Bearman, D. (Ed.); 1991: "Archives and Museums Informatics Technical Report No.14 - Hypermedia & Interactivity in Museums, Proceedings of an International Conference, October 14-16, 1991, Pittsburgh, Pennsylvania".

[3]   Bumann, S. und Kerstan, T.; 1994: "Methoden zur software-ergonomischen Evaluation von informationsvermittelnden Kiosksystemen auf Hypermedia-Basis in Museen". Studienarbeit, Fachbereich Informatik, Universität Hamburg.

[4]   Eberleh, E., Oberquelle, H. und Oppermann, R. (Hrsg.); 1994: "Einführung in die Software-Ergonomie, 2. Auflage". de Gruyter, Berlin.

[5]   Fahy, Poulter, Sargent; 1993: "Hypermuse: A Prototype Hypermedia Front-End for Museum Information Systems". In [10].

[6]   Gloor, P.A. und Streitz, N.A.; 1990: "Hypertext und Hypermedia - von theoretischen konzepten zur praktischen Anwendung". Springer-Verlag, Berlin.

[7]   Hamburger Abendblatt; 1993: "Mit dem Computer in die Römerzeit". Hamburger Abendblatt, 24./25. Juli 1993.

[8]   ISO 9241 Teil 2: "Ergonomic requirements for office work with visual display terminals (VDTs): Guidance on task requirements". International Standards Organization, Genf.

[9]   Kieback, A. u.a.; 1992: "Prototyping in industriellen Software-Projekten, Erfahrungen und Analysen". Informatik Spektrum, Band 15, Heft 2.

[10] Lees, D. (Ed.); 1993: "Archives & Museums Informatics Technical Report No. 20 - Museums and Interactive Multimedia, Proceedings of an International Conference held in Cambridge, England, 20-24 September 1993". Larman Printers, Cambridge.

[11] Nebenzahl, L. A.; 1993: "Evaluating Interface Design throgh User Data Collection". In [10].

[12] Noschka-Roos, A. und Lewalter, D.;1993: "Untersuchungsbericht: Akzeptanz und Nutzung des Touch-Screen-Systems 'Erneuerbare Energien'. Eine Studie in der Abteilung Neue Energietechniken des Deutschen Museums". Deutsches Museum, München.

[13] Oppermann, R. und Reiterer, H.; 1994: "Software-ergonomische Evaluation". In [4].

[14] Steinhau, H.; 1993: "Mit allen Wasser gewaschen". Screen Multimedia 1/1993.

[15] Vichr, A.; 1994: "Kottan ermittelt. Was ist Protokollierung?" Screen Multimedia 6/1994.

[16] Warnke, M.: "Das Thema ist die ganze Welt: Hypertext im Museum". In [6].

[17] Wanning, T.; 1991: "Evaluating Museum Visitors' Use of Interactive Video". In [2].

Dr. Andreas M. Heinecke
Kerstan
hi.soft Systemberatung
Liebermannstr. 50
22605 Hamburg
Tel. 040 8804967
Fax 040 8804967

Sigrid Bumann, Thomas

Fachbereich Informatik / ANT
Vogt-Kölln-Str. 30
22527 Hamburg
Tel. 040 54715-436
Fax 040 54715-552

Ein wissensbasiertes System zur Unterstützung des Benutzers bei der ergonomischen Farbzusammenstellung für Dialogmasken

Peter Heintzen, Volker Kruschinski, Helmut Balzert
Lehrstuhl für Software-Technik, Ruhr-Universität Bochum

Zusammenfassung
Schon für viele Bereiche der Gestaltung von Mensch-Computer-Dialogen, wie Auswahl und Anordnung von Dialogelementen, wurden Style-Guides geschaffen, die ein Regelwerk für die ergonomische Maskenerstellung darstellen. Die Aussagen der Style-Guides über die farbliche Gestaltung von Dialogmasken sind jedoch sehr allgemein. Die vorliegende Arbeit befaßt sich mit der Schaffung eines Regelwerkes zur ergonomischen Farbgestaltung beim Entwurf von Dialogmasken. Das Regelwerk wird dazu benutzt, den Benutzer bei der Zusammenstellung von Dialogelement-Farben durch Farbvorschläge zu unterstützen. Die Regeln werten Farb-Charakteristika (Farbfamilie, Helligkeit, Sättigung) und Farb-Beziehungen (Kontrast-, Äquivalenz- und Synonymfarbe) zur Bestimmung der Vorschläge aus. Die Basis aller Regeln bilden das biologische und psychologische Wissen über das Farbempfinden des Menschen. Das entstandene wissensbasierte System ist eine Teilkomponente des JANUS-Systems, das am Lehrstuhl für Software-Technik an der Ruhr-Universität Bochum entwickelt wird.

1   Einführung

Schon lange sind die biologischen Grundlagen für das Farb- und Helligkeitsempfinden des menschlichen Auges bekannt.Trotzdem fehlen bisher Systeme, die den Ergonomen bei der Zuordnung von Farben zu den entsprechenden Dialogelementen unterstützen. Bisher wurde dieser Farb-Auswahlvorgang dem subjektivem Empfinden des Dialogmasken-Entwicklers überlassen. Doch gerade, wo in der heutigen Zeit die Zahl der Berufstätigen immer weiter zunimmt, die während ihres gesamten Arbeitstages vor einem Computer-Bildschirm sitzen, wird der Bedarf einer ergonomischen Farbzusammenstellung immer größer. Eine ergonomische Farbzusammenstellung muß gewährleisten, daß ein Benutzer auch nach mehrstündiger Arbeit die Farben der Masken noch als angenehm empfindet.

Bei der Automatisierung der Farb-Vorauswahl können psychologische Einflußfaktoren für die Farbwahl nicht berücksichtigt werden. Psychologische Einflußfaktoren sind die synästhetische Wirkung (ausgelöste Gefühle und Stimmungen) und die Alltagsbedeutung einzelner Farben. Alltagsbedeutungen können sowohl allgemeingültig als auch berufsspezifisch sein. Allgemeine Farbbedeutungen werden schon von frühester Jugend an, z.B. durch den

Farbeinsatz bei der Regelung des Straßenverkehrs, anerzogen. Deshalb assoziiert der Betrachter bestimmte Farbtöne direkt mit bestimmten Bedeutungen [7]:

* Rot   -   Stop, Feuer, heiß, Gefahr
* Gelb   -   Vorsicht, langsam, Test
* Grün   -   Weitergehen, in Ordnung, Sicherheit, Pflanzen
* Blau   -   kalt, Wasser, Ruhe, Himmel, neutral
* Grau   -   neutral

Am Beispiel der Farbe Blau werden ihre verschiedenen berufsspezifischen Bedeutungen gezeigt [7]:

* Für Finanzmanager   - Fähigkeit oder Zuverlässigkeit einer
                                 Gemeinschaft
* Für Mediziner   - Tod
* Für Kernreaktor-Überwacher   - Kälte oder Wasser

Da ein Farbexpertensystem kein Hintergrundwissen über die fallspezifische semantische Bedeutung der jeweiligen Farbwahl hat, ist dies keine Randbedingung bei der Zusammenstellung der Farbvorschläge.
Technische Einflußfaktoren wie Lichtverhältnisse am Arbeitsplatz und verwendete Hardware können aus Gründen fehlendem Wissens über den Zielarbeitsplatz nicht berücksichtigt werden.
Das angewendete Wissen bezieht sich auf die Farbwahrnehmung und die physiologischen Einschränkungen des Auges. Die physiologischen Einschränkungen sind ein Sammelbegriff für Farbenblindheit, Farbenschwäche, chromatische Aberration und Verteilung der Farbrezeptoren auf der Netzhaut.

Um Kriterien für die Auswahl von Farben zu finden, müssen erst einmal Merkmale gefunden werden, mit der sich eine Farbe charakterisieren läßt. Nicht nur die Farbe als solche, sondern das Zusammenspiel aller verwendeten Farben innerhalb einer Maske spielt eine große Rolle bei der Auswahl von geeigneten Regeln. Bei der Regelfindung wird deutlich, daß es zwei unterschiedliche Problembereiche gibt, die aufeinander aufbauen:

. Charakterisierung und die damit verbundene Unterscheidungsmöglichkeit von Farben bzw. Farbtönen anhand verschiedener Farbeigenschafen.
. Auswertung von Farbabhängigkeiten zwischen Dialogelementen bzw. Dialogelementteilen einer Maske.

2   Farbklassifikation

Farbtöne auf Computerbildschirmen lassen sich durch die numerischen RGB-Werte (Rot-Grün-Blau-Werte) der drei Elektronenkanonen des Monitors in eindeutiger Weise voneinander abgrenzen. Diese absoluten Farbabstufungen stehen aber kaum im Verhältnis zum individuellen Farbempfinden des menschlichen Auges. Die Gesamtheit aller meßbaren Farben liegt bei ungefähr 7,5 Millionen.

Das menschliche Auge dagegen ist nur in der Lage 160 Farbtöne und 600.000 Farbnuancen [3] voneinander zu unterscheiden. Die auf einem Computer darstellbaren Farben werden durch ihre RGB- bzw. HSV-Werte (Hue (Farbwert), Saturation (Sättigung), Value(Helligkeitswert)) repräsentiert. Deshalb bleibt nur die Möglichkeit, Wissen über das menschliche Farbempfinden durch numerische Werte auszudrücken, um so eine Farbe mittels ihrer RGB-Zusammensetzung bewerten und charakterisieren zu können.

Betrachtet man eine Farbe, so entsteht das Problem, demjenigen, der diese nicht sehen kann, durch Zuordnung charakteristischer Eigenschaften eine möglichst genaue Beschreibung zu geben, so daß er sich diese Farbe aufgrund dessen vergegenwärtigen kann. Charakteristische Eigenschaften werden besonders dann wichtig, wenn man zwei Farbtöne miteinander vergleichen will. Selbst wenn sich zwei Farbtöne gleichen, möchte man sie sprachlich unterscheiden können.

Eine leichte Unterscheidungsmöglichkeit besteht, wenn der Betrachter sagen kann: „Dieser Farbton ist rot, der andere ist blau". Hieraus wird deutlich, daß Farbtöne Namen haben. Doch welche Farben können als Grundfarben, die Farben, aus denen sich alle anderen Farben zusammensetzen bezeichnet werden ? Eine Antwort gibt der Aufbau der Netzhaut (Retina) des menschlichen Auges. Die Netzhaut ist die lichtempfindliche Fläche des Auges. Sie beinhaltet zwei Arten von Rezeptoren. Die Stäbchen und die Zäpfchen. Beide „übersetzen" das einfallende Licht in Nervenimpulse. Die Stäbchen haben kein Farbempfinden und sind mehr für das Sehen bei geringen Helligkeitsniveaus, zum Beispiel bei Nacht, verantwortlich. Zäpfchen reagieren erst ab einem höheren Helligkeitsniveau. Sie sind für das Erkennen von Farben verantwortlich. Auf der Retina gibt es drei unterschiedliche Populationen dieser Farbrezeptoren, die für jeweils unterschiedliche Frequenzbereiche des einfallenden Lichtes sensitiv sind. Ungefähr 64 % der Zäpfchen sind bevorzugt sensitiv für langwelliges Licht mit der maximalen Empfindlichkeit bei 575 nm Wellenlänge. Sie werden als rot-empfindliche Rezeptoren bezeichnet, obwohl ihre maximale Empfindlichkeit im Frequenzbereich für Wahrnehmung der Farbe Gelb liegt. Ungefähr 32 % der Zäpfchen haben ihre maximale Empfindlichkeit bei 535 nm. Diese werden als grün-empfindliche Rezeptoren bezeichnet. Die restlichen 2 % haben ihre maximale Empfindlichkeit bei 445 nm. Sie werden deshalb als blau-empfindlich bezeichnet.

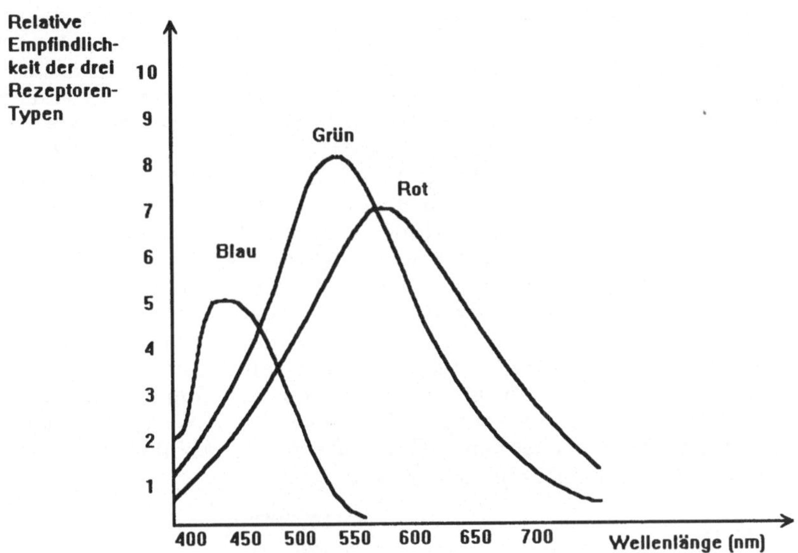

Abb. 1: Empfindlichkeit der Rezeptoren in Abhängigkeit zur Wellenlänge [10]

Die Farbrezeptoren des Auges legen die Farben Rot, Grün und Blau als Grundfarben für das menschliche Farbempfinden fest. Alle übrigen Farbtöne werden durch additive Mischung der drei Grundfarben erzeugt. Die Empfindung eines Farbtons wird durch die Frequenz des einfallenden Lichtes bestimmt. Die Interpretation der Zusammensetzung des Farbtons aus Grundfarben hängt von der Empfindlichkeit der jeweiligen Rezeptoren-Typen bei dieser Frequenz ab. Die Entstehung eines Farbeindrucks setzt ein bestimmtes Helligkeitsniveau voraus, ab dem die Zäpfchen erst sensibel für Licht werden.

Das menschliche Auge ist in der Lage, aus einer Mischfarbe die Grundfarben herauszuinterpretieren. Ab einem bestimmten Mischungsverhältnis besteht nicht mehr die Möglichkeit, die dominante Grundfarbe eindeutig zu erkennen. Dieser Interpretationsvorgang läßt sich als eine Art Kippzustand bezeichnen. Es ist keine Entscheidung möglich, ob z.B. die Mischfarbe Türkis der Farbe Blau oder der Farbe Grün zugeordnet wird. Die Farbempfindungen von Blau und Grün sind annähernd gleich. Ein gleiches Verhalten läßt sich auch bei der Mischung von zwei anderen Grundfarben beobachten. Eine Ausnahme bildet das aus Rot und Grün gemischte Gelb. In der von Hering [10] entwickelten Vierfarbentheorie kommt Gelb trotz ihrer Eigenschaft als Mischfarbe eine besondere Bedeutung zu. Die Theorie stützt sich auf der Tatsache, daß Gelb in vielen psychologischen Experimenten gleiche Eigenschaften wie eine Grundfarbe aufweist. Physiologisch wird diese Theorie durch die Funktionsweise der den Zapfen nachgeschalteten neuronalen Einheiten bestätigt. Die neuronalen Mechanismen dieser Einheiten vergleichen die Aktivität der Zapfen miteinander. Es gibt zwei Einheiten, auch Detektoren genannt, die für die Farbempfindung verantwortlich sind. Der erste Detektor vergleicht die Aktivität der rotempfindlichen Zapfen mit der Aktivität der grünempfindlichen Zapfen. Der zweite Detektor vergleicht die Aktivität der blauempfindlichen Zapfen mit der aufsummierten Aktivität der rot- und grünempfindlichen Zapfen. Die aufsummierte Aktivität der rot- und grünempfindlichen Zapfen wird als Empfindung des Farbtons Gelb interpretiert. Das Farberleben wird somit durch die Kombination der Ausgangssignale eines Rot/Grün- und eines Gelb/Blau-Detektors bestimmt. Ein Eingangsignal pro Detektor veringert, das andere verstärkt sein Ausgangssignal zum Gehirn. Sind beide Eingangssignale gleich stark, kann sich der Detektor auf keine Farbe festlegen. Er sendet sein Ruhe-Ausgangssignal, als wenn gar keine Farb-Bewertung stattgefunden hätte. Es ist deshalb niemals die Wahrnehmung von Rot und Grün bzw. von Gelb und Blau an einem Ort möglich. Paare von Rot und Grün bzw. von Gelb und Blau werden darum als Gegenfarben bezeichnet. Liegen die Ausgangssignale beider Detektoren auf dem Ruhepotential, so wird die einfallende Strahlung an diesem Punkt als „unbunt" (Grau) empfunden. Abbildung 2 zeigt die schematische Darstellung der Rezeptorenauswertung.

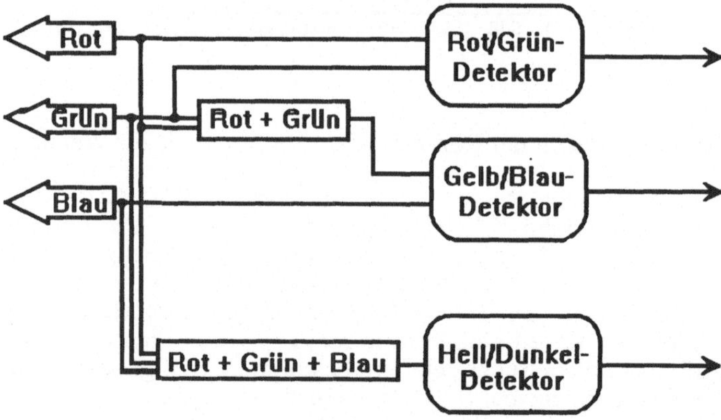

Abb. 2: Schematische Darstellung der Farb-Rezeptorenauswertung [3]

Jeder Farbton, der in einem speziellen Farbbereich liegt, wird durch den im Rahmen dieser Arbeit entstandenen Farbexpertensystems einer bestimmten Farbfamilie zugeordnet. Eine Farbfamilie ist die Menge aller Farbtöne, die in einem bestimmten Farbbereich liegen. Insgesamt lassen sich neun Farbfamilien, bestehend aus Grund- und Mischfarben herausarbeiten. In Tabelle 1 ist zu erkennen, daß das RGB-Mischungsverhältnis konform zur Stellung der Frequenz auf dem sichtbaren Frequenzband ist. Die einzige Ausnahme stellt Violett dar. Eine andere Darstellungsart in Verbindung mit der Gegenfarbtheorie ist der in Abbildung 3 gezeigte Farbkreis aus den vier „Urfarben" Rot, Grün, Gelb und Blau

(Roche Lexikon Medizin). Hier schließt sich an Violett direkt wieder Rot an. Das sichtbare Farbspektrum wird quasi zu einem Kreis zusammengebogen. Die entsprechenden Gegenfarbfamilien liegen jeweils auf der gegenüberliegenden Seite des Kreises. Grau kommt nicht als Farbfamilie vor. Grau ist ein Lichtgemisch ohne eine oder mehrerer dominanter Grundfarben. Auch in der Malerei wird Grau nicht als Farbe angesehen. Bei der Farbgebung einer Bildschirm-Dialogmaske spielen die Elemente von Grau als neutraler Farbfamilie eine wichtige Rolle, besonders dann, wenn die Wahl von Farbvorschlägen durch Gegenfarb-Unverträglichkeiten eingeschränkt wird.

Farbfamilie	RGB-Zusammenstellung	maximales Farbempfinden bei Wellenlänge in nm	maximales Farb-Helligkeits-Verhältnis (10 für Weiß)
Rot	Rot	700	4,9
Orange	Gelb + Rot	600	(7,6+4,9)/2 = 6,25
Gelb	Rot + Grün	570	7,6
Gelb-Grün	Gelb + Grün	535	(7,6+7,1)/2 = 7,35
Grün	Grün	500	7,1
Türkis	Blau + Grün	493	7,4
Blau	Blau	470	2,7
Violett	Rot + Blau	400	3,7
Grau	Rot + Grün + Blau	-	10

Tabelle 1: Die Farbfamilien geordnet nach dem maximalen Farbempfinden entlang des sichtbarem Farbspektrums

Die Farbbereichs-Intervallgrenzen der Farbfamilien wurden subjektiv mit Hilfe des Farbwertes aus dem HSV-Farbsystem festgelegt. Der Wertebereich des Farbwertes aus diesem System stellt praktisch einen Umlauf auf dem Farbkreis von Rot ausgehend im Uhrzeigersinn dar.

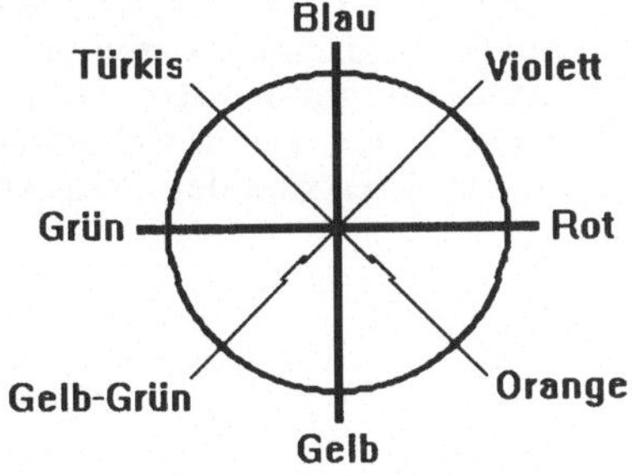

Abb. 3: Der Farbkreis

Neben der Empfindung eines Farbtons gibt es noch zwei weitere Dimensionen des Farberlebens.

Die eine ist die der *Sättigung*. Sie beschreibt die Intensität des Farberlebnisses. Zum Beispiel ist Weiß ungesättigt, Rosa zum Teil gesättigt und ein leuchtendes Rot ist zutiefst gesättigt. Der Sättigungsgrad hängt von dem Anteil an weißem Licht des einfallenden Strahls ab. Ein besonders starkes oder ein besonders schwaches Ausgangssignal des Rot/Grün- bzw. des Gelb/Blau-Detektors deutet auf eine starke Farbsättigung hin. Bei „blassen" Farben bewegt sich das Ausgangssignal in der Nähe des Ruhesignals.

Die andere Dimension des Farberlebens ist die der *Helligkeit*. Physikalisch gesehen wird die Helligkeitsempfindung durch die Amplitude der Lichtwellen bestimmt. Das Erleben von Schwarz bis Weiß findet in einem den Zapfen nachgeschalteten Hell/Dunkel-Detektor statt, der die gesamte Aktivität der rot-, grün- und blauempfindlichen Farbrezeptoren bewertet. Die sehr viel geringere Anzahl der blauempfindlichen Zapfen gegenüber die der anderen, ist der Hauptgrund für den nur geringen Einfluß von blauem Licht am Helligkeitsempfinden.

Durch die entsprechend größere Anzahl und die größere Empfindlichkeit von rot- und grünempfindlichen Farbrezeptoren wird Gelb am intensivsten empfunden. Wird ein gelber und ein blauer Lichtstrahl gleicher Energiedichte miteinander verglichen, so erscheint der gelbe mehr als doppelt so hell. Diese Erscheinung wird als Helmholtz-Kohlrausch-Effekt bezeichnet. Murch [7] ermittelte 1984 eine Farb/Helligkeits-Skala (siehe oben Tabelle 1). Er vergleicht die Helligkeitsempfindung von Weiß, festgelegt als 10 cd/m2, mit der anderer Farben. Nicht bekannte Werte wurden der Einfachheit halber aus den Daten der direkt benachbarten Farbfamilien durch Mittelwertsbildung ermittelt. Die Sättigung eines Farbtons hat also einen großen Einfluß auf die Helligkeitsempfindung des Auges. Ein stark gesättigter Farbton (besonders im Gelb/Grün-Bereich) wird häufig als heller empfunden, als einer mit größerem Weißlicht-Anteil. Dieser Effekt wird z.B. in Waschmittelwerbungen eingesetzt. Die stark gesättigten Farben der gezeigten Kleidungsstücke werden in der Werbung als „leuchtende" Farben bezeichnet. Gerade „leuchtende" Farbtöne ziehen besonders das Augenmerk des Betrachters auf sich.

Nach dieser Betrachtung der biologischen und psychologischen Grundlagen für das Farbempfinden, entsteht das Problem der Umsetzung auf die Charakterisierung von RGB-Werten. Die vom Computer-Bildschirm erzeugten Farben bestehen aus bestimmten Anteilen der Monitor-Grundfarben Rot, Grün und Blau (RGB). Die Wirkung des erzeugten Farbtons setzt sich, wie in Abbildung 4 gezeigt, im Prinzip aus drei Einflußfaktoren zusammen. Dies sind der Grau-Anteil, der Misch-Anteil und der Basis-Farbanteil.

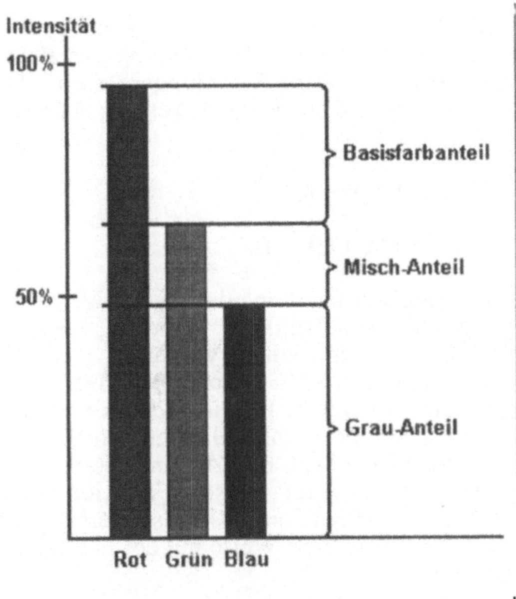

Abb. 4: Zusammensetzung eines Farbtons am Beispiel von RGB-Werten

Der Grau-Anteil ist der Anteil eines Farbtons, den alle drei Monitor-Grundfarben in ihrer Intensität gemeinsam haben. Er wird bestimmt durch den Farbanteil der Monitor-Grundfarbe mit der geringsten Intensität. Dieser Anteil bestimmt wesentlich die Helligkeit einer Farbe. Er ist äquivalent zu der oben erwähnten Zumischung von Weißlicht.
Der Mischanteil des Farbtons ist der Anteil, den die beiden intensivsten Monitor-Grundfarben oberhalb des Grau-Niveaus gemeinsam haben. Er wird bestimmt durch die Farbintensität der zweitstärksten Monitor-Grundfarbe.
Alles was oberhalb des Misch-Anteils liegt, wird als Basis-Farbanteil bezeichnet. Er wird, wie zu sehen ist, durch die Monitor-Grundfarbe mit der stärksten Farbintensität bestimmt.

Das Verhältnis von Basis-Farbanteil und Misch-Anteil bestimmt den Farbwert des HSV-Systems und damit die Farbfamilie. Ist die Summe aus Basis-Farbanteil und Misch-Anteil kleiner gleich 3 % bezogen auf 100 % Maximalintensität, so kann nicht mehr entschieden werden, welcher Farbfamilie man diesen Farbton zuordnen soll. Die neuralen Farbdetektoren senden ihr Ruhe-Ausgangssignal. Darum werden alle zutreffenden Werte-Kombinationen der neutralen Farbfamilie Grau zugeordnet.

Da der Helligkeitseindruck von der Sättigung des Farbtons als Funktion der Farbfamilie abhängt, soll nun eine neue Sättigungs-Definition auf der Grundlage der Farb/Helligkeits-Skala von March gemacht werden. Der Sättigungswert des HSV-Systems ist unabhängig von den zwei anderen Größen und liefert deshalb unbefriedigende Ergebnisse. Sättigung ist eine subjektive Bewertung des Farbeindrucks und kein absoluter Wert. Die Summe aus Basis-Farbanteil und Mischanteil wird im folgenden als Sättigungsreferenz bezeichnet. Dem Farbexpertensystem reichen drei Zustände zur Bewertung der Sättigung aus:

- gering   (blasser Farbton)
- mittel
- stark   (leuchtender Farbton)

Die Zustands-Übergangswerte werden durch die zugehörige Farbfamilie bestimmt. Der als untere Referenz empirisch ermittelte Übergangswert von „mittel" nach „stark" für die Farbfamilie Gelb, mit einem Verhältnis von 7,6 cd/m2, liegt bei einem Sättigungs-Referenzwert von 82 % bezogen auf 100 % Maximalwert. Über Dreisatzrechnung lassen sich die Werte für die anderen Farbfamilien berechnen. Wie zu erwarten war, erreichen die Farbfamilien an den Rändern des sichtbaren Spektrums (Blau, Violett und Rot) nicht den Zustand „stark". Selbst das Element mit der größten Farbreinheit jeder dieser Familien erreicht nicht die Leuchtwirkung, um hell zu erscheinen. Die Zustands-Übergangswerte von „mittel" nach „gering" lassen sich auf die gleiche Weise mit der Farbfamilie Blau mit 70 % als oberer Referenz ermitteln.

Für die Charakterisierung der Helligkeit wird eine mittlere Helligkeit berechnet, d.h. es wird der Mittelwert aus dem Rot-, dem Grün- und dem Blauwert gebildet. Das Ergebnis wird einem der folgenden drei *Helligkeitszustände* zugeordnet:

- dunkel
- normal
- hell

Für das Helligkeitsempfinden ist im wesentlichen der Grau-Anteil verantwortlich. Darum wird für die subjektive Einstellung der Übergangsgrenzwerte ein Grauton verwendet, um sich nicht durch den Sättigungseinfluß der Helligkeitsempfindung beeinflussen zu lassen. Gute Ergebnisse lassen sich mit dem Zustandsübergang der mittleren Helligkeit von normal nach hell bei 75 % bzw. von normal nach dunkel bei 21 % bezogen auf 100 % Maximalwert erzielen.

Das Farbexpertensystem bewertet einen Farbton nach den drei Kriterien *Farbfamilie*, *Sättigung* und *Helligkeit*. Die drei Kriterien sind wie folgt unterteilt:

Farbfamilie : Rot, Grün, Blau, Gelb, Gelb-Grün, Orange, Türkis, Violett, Grau
Sättigung   : gering, normal, stark
Helligkeit   : dunkel, normal, hell

Mehr Kriterien oder eine schärfere Unterteilung sind nicht notwendig, da das Ziel ist, herauszufinden, ob Farben bei festgelegten Randbedingungen zueinander passen. Die Parameter der Farbbewertungsregeln werden, wie gezeigt, durch subjektive Auswertung von Farbbetrachtungen festgelegt. Diese, speziell die

Intervallgrenzen, bieten Ansatz zur Kritik, da Randentscheidungen immer vom Farbempfinden des jeweiligen Betrachters abhängen. Der Benutzer ist deshalb in der Lage, das Farbexpertensystem in der automatischen Bewertung eines Farbtons nach den genannten Kriterien zu überstimmen.

Die Ergebnis-Ausgabe der Farb-Bewertung des Expertensystems ist integriert in dem in Abbildung 5 gezeigten Benutzerdialog. Sie wird repräsentiert durch die drei Drop-Down-List-Boxen auf der linken Seite.

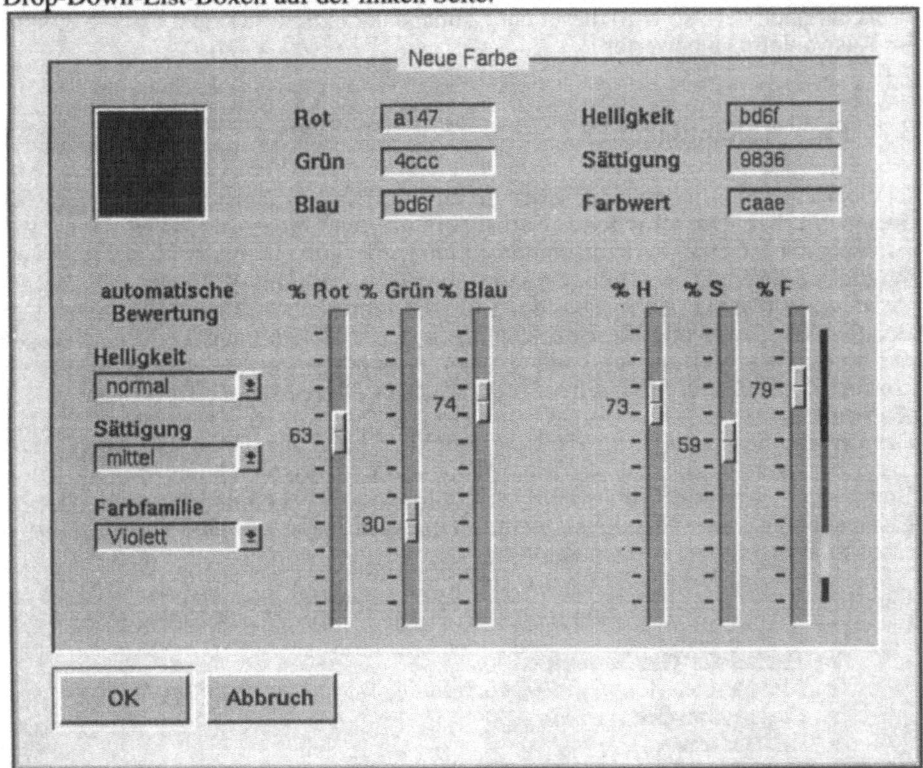

Abb. 5: Der Farb-Bewertungsdialog

Mit Hilfe zweier Farbsysteme (RGB und HSV) stellt der Benutzer den gewünschten Farbton, angezeigt in der linken oberen Ecke, ein. Die Schieberegler und Wert-Eingabefelder sind zueinander synchronisiert. Verschiebt der Benutzer einen Regler oder ändert er einen der Werte in den Eingabefeldern, passen die Schieberegler und Wert-Eingabefelder des anderen Farbsystems automatisch ihre Werte an. Die Bewertung durch das Farbexpertensystem wird repräsentiert durch die Drop-Down-List-Boxen auf der linken Seite der Dialogmaske. Bei jeder Wert-Veränderung des gerade bearbeiteten Farbtons wird der Auswertungs-Mechanismus angestoßen. Das Ergebnis sind die aktuell angezeigten Zustände. Ist

der Benutzer mit der automatischen Bewertung nicht einverstanden, so wählt er aus
den in den Auswahl-Listen angebotenen möglichen Zuständen einen anderen aus.
Durch Aktivierung des OK-Buttons wird die neu kreierte Farbe mit der aktuellen
Bewertung in die Liste der zur Verfügung stehenden Farben aufgenommen. Durch
die Überstimmungs-Möglichkeit des Farbexpertensystems durch den Benutzer
besteht auch die Möglichkeit, bei der anschließenden Auswertung der
Farbabhängigkeiten unsinnige Vorschläge zu erhalten. Hat der Benutzer z.B. den
Farbton Schwarz als hellen, stark gesättigten Vertreter der Farbfamilie Gelb
nachcharakterisiert, so wird dieser auch anhand der extrem falschen Charakteristika
im Auswahlprozeß bewertet.

3   Farb-Abhängigkeiten

Ist die Menge aller möglicher Farben durch den Benutzer festgelegt, so kann er mit
diesen Farben verschiedene Farbschemata festlegen und verwalten. Ein
Farbschema ist eine Farbzusammenstellung, die auf ein festgelegtes Beispiel-
Widgetset (Widget = Dialogelement) bezogen ist. Die Widgets aus diesem
Referenz-Widgetset (siehe Abbildung 6) sind repräsentativ für alle Widgets. Das
Beispiel-Widgetset und die Bezeichnungen der Dialogelement-Teile sind wegen
der weiten Verbreitung und der großen Akzeptanz von Microsoft-Windows,
konform zu denen des Einstellungsdialogs der Microsoft-Windows 3.x
-Farbpalette.
Die Farben, die man z.B. für ein Eingabefeld und den dazugehörigen Eingabetext
festgelegt, werden auch für die Einfärbung von List-Boxen, Combo-Boxen, Spin-
Buttons usw. verwendet. Insgesamt sind 15 repräsentative Dialogelement-Teile, die
Komponenten ganzer Dialogelemente, herausgearbeitet worden, für die jeweils
eine Farbe zugeordnet werden kann:

      Dialogelement-Teile:
- Titelleiste, Titelleistentext
- Menüleiste, Menütext, Hervorhebung, Text aktiv, Text deaktiviert
- Programmarbeitsbereich
- Bildlaufleiste
- Hintergrund
- Fenstertext
- Schaltfläche, Schaltflächentext
- Eingabefeld, Eingabetext

Es existieren acht unterschiedliche Dialogelement-Typen (jede Zeile in der Liste
„Dialogelement-Teile" stellt im Prinzip ein Dialogelement dar) von deren Farbwahl
man alle anderen, auch zusammengesetzte Typen ableiten kann.

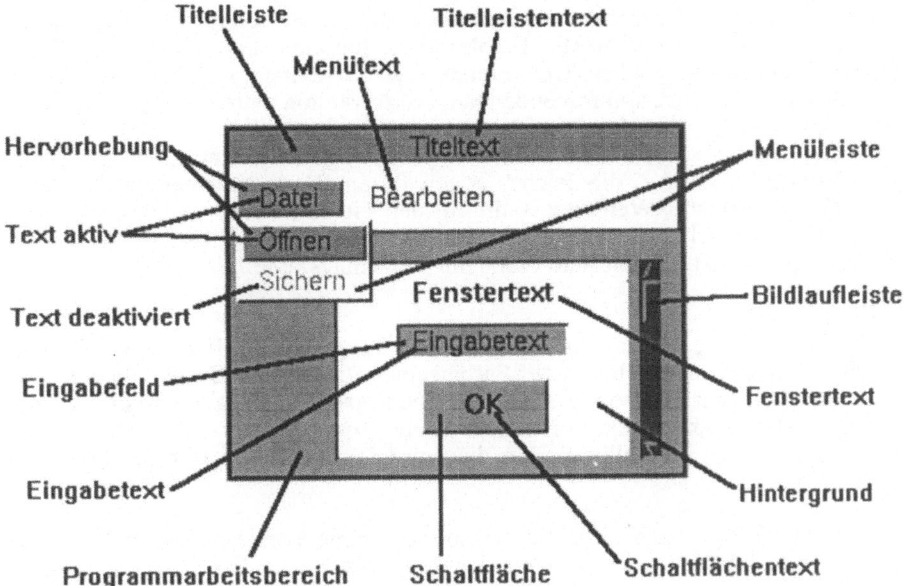

Abb. 6: Das Beispiel-Widgetset

Das Zusammenspiel der Dialogelement-Teile erfordert bestimmte Farb-Beziehungen bzw. Farb-Abhängigkeiten zwischen aneinander angrenzenden Teilen. Die Farbe eines Textes muß sich zum Beispiel deutlich von der Farbe der ihn umgebenden Hintergrundfläche abheben, um die Lesbarkeit zu erhöhen. Eine andere Beziehung besteht zwischen aneinander angrenzenden Flächen. Es können zwei Fälle unterschieden werden. Beim ersten Fall ist es nicht wichtig, daß sich die Farben unterscheiden, ob überhaupt eine Grenze wahrgenommen werden muß oder eine Grenze sowieso schon durch einen deutlichen Rahmen vorgegeben ist. Ein Beispiel hierfür ist der Push-Button, der normalerweise auf dem Fenstersystem des Apple-Macintosh nicht standardmäßig farblich hervorgehoben wird. Beim anderen Fall soll gereade eine Fläche farblich hervorgehoben werden. Bei einer Menüleiste möchte man z.B. das gerade aktive Menü und den Eintrag, auf dem momentan der Mauszeiger steht, farblich besonders kennzeichnen (siehe „Hervorhebung" des Beispiel-Widgetsets in Abbildung 6).
Hieraus lassen sich drei Beziehungen, die eine Farbe eines Dialogelement-Teils zu einer ihr angrenzenden Farbe eines anderen Dialogelement-Teils haben kann, herausarbeiten:

- Kontrastfarbe
- Äquivalenzfarbe

• Synonymfarbe

Bei der Präsentation der Regeln zur Beschreibung von Beziehungen zwischen den Farben werden die Attributwerte (Farbfamilie, Helligkeit und Sättigung) zweier Farben miteinander verglichen. Um unterscheiden zu können, welcher Attributwert zu welcher Farbe gehört, soll folgende Namenskonvention gelten:

> Die Farbe, deren Tauglichkeit gerade getestet wird, heißt *Vorschlagsfarbe*, die Farbe, gegen deren Werte getestet wird, *Vergleichsfarbe*. Werden z.B. die für den Titelleistentext passenden Farben gesucht, so ist die Farbe der Titelleiste die Vergleichsfarbe, die gerade aktuell getestete Farbe die Vorschlagsfarbe.

Die Kontrastfarbe
Eine Farbe wird als Kontrastfarbe bezeichnet, wenn sich die Farbe eines Textes oder eines Symbols deutlich von der Farbe seiner Hintergrundfläche abheben soll. Die Lesbarkeit wird besonders durch einen großen Helligkeitsunterschied verbessert, also dunkler Text vor hellem Hintergrund (Positiv-Darstellung) bzw. umgekehrt (Negativ-Darstellung). Aus diesem Kriterium lassen sich die folgenden Regeln ableiten:

• Ist die Vergleichsfarbe dunkel, so müssen die Vorschlagsfarben hell bzw. jedoch mindestens von normaler Helligkeit und starker Sättigung sein.
• Ist die Vergleichsfarbe hell, so müssen die Vorschlagsfarben dunkel bzw. jedoch höchstens von normaler Helligkeit und geringer Sättigung sein.
• Hat die Vergleichsfarbe eine normale Helligkeit und eine starke Sättigung, so müssen die Vorschlagsfarben dunkel sein.
• Hat die Vergleichsfarbe eine normale Helligkeit und eine geringe Sättigung, so müssen die Vorschlagsfarben hell sein.

Da man einen dunkelgrünen Text auf einem hellgrünen Hintergrund schlechter erkennt, als wenn der Text z.B. dunkelblau wäre, so ist es sinnvoll die eigene Farbfamilie und alle Farbfamilien, in denen die dominierenden Farben auch dominant sind, nicht zu berücksichtigen. Werden z.B. alle Farben der Farbfamilie Grün aus der Vorschlagsliste gestrichen, so werden auch die direkten Nachbarn im Farbkreis, Gelb-Grün und Türkis, gestrichen.

Gehört die Vergleichsfarbe einer der Grundfarbfamilien (Rot, Grün, Blau und Gelb) an, so müssen alle Vorschlagsfarben, die der Gegenfarbfamilie angehören, gelöscht werden. Dies ist sinnvoll, da die neuralen Farb-Detektoren (Rot/Grün bzw. Gelb/Blau) eine Farbe und ihre Gegenfarbe an demselben Ort nicht oder nur schwer unterscheiden können (siehe 2. Farbklassifikation). Da auch bei den Mischfarbfamilien die Farbanteile der Grundfarben denen der Gegengrundfarben auf der auf dem Farbkreis gegenüber liegenden Mischfarbe gleichen, werden auch hier die Vorschlagsfarben der Gegenfarbfamilie gelöscht.

Bei der Text- oder Symboldarstellung dürfen Rot und Blau niemals zusammen eingesetzt werden. Das langwellige Rot benötigt einen anderen Brennpunkt, als das kurzwellige Blau. Die erforderliche unterschiedliche Linsenakkommodation (Einstellung des Brennpunktes durch Linsenkrümmung) führt zu Unschärfen und zu Ermüdungserscheinungen durch häufige Akkommodation. Das Auge kann nur für eine der beiden Farben scharfgestellt werden; die andere erscheint dementsprechend unscharf.

Vorschlagsfarben aus der Farbfamilie Grau werden wegen ihrer Farbfamilien-Zugehörigkeit nie gestrichen, denn Grau gilt als neutrale Farbfamilie.

Die Äquivalenzfarbe
Eine Farbe wird als Äquivalenzfarbe bezeichnet, wenn bei aneinander grenzenden Flächen ein Eindruck der Zusammengehörigkeit vermittelt werden soll. Diesen Eindruck erreicht man vor allem dadurch, daß die Vorschlagsfarbe der gleichen oder einer verwandten Farbfamilie der Vergleichsfarbe angehört. Darum werden Vorschlagsfarben, die der Gegenfarbfamilie oder einem der beiden direkten Nachbarn auf dem Farbkreis angehören, aus der Vorschlagsliste gestrichen. Gehört die Vergleichsfarbe zur Familie Gelb, so werden mögliche Vorschläge der Familien Blau, Türkis und Violett gestrichen.
Elemente der Farbfamilie Grau vertragen sich mit jeder Farbfamilie.

Die Synonymfarbe
Eine Farbe wird als Synonymfarbe bezeichnet, wenn bei angrenzenden Flächen ein Eindruck der Zusammengehörigkeit vermittelt werden soll, sie sich aber trotzdem, zumindest geringfügig, farblich voneinander abheben sollen. Bei der Eingrenzung der Farbvorschläge gilt das gleiche, wie bei der Äquivalenzfarbe. Zusätzlich wird aber die Vergleichsfarbe selbst aus der Liste der möglichen Farbvorschläge entfernt. Auf diese Weise wird eine Unterscheidung erzwungen.
Neben den Regeln zu den Farbbeziehungen gibt es auch noch zwei Regeln, welche die Vorschlagsauswahl durch den Dialogelement-Teil selbst einschränken.

Die Flächenfarb-Einschränkung
Für große Flächen ohne Funktion, wie der Programmarbeitsbereich und der Hintergrund, werden Vorschlagsfarben mit starker Sättigung aus der Vorschlagsliste entfernt. Zum einen lenken die stark gesättigten Farben vom eigentlichen Benutzerdialog ab, zum anderen werden die Zäpfchen im Auge durch die Reizüberflutung überlastet, so daß farbige Nachbilder entstehen können (farbiger Sukzessivkontrast)[10].

Die Textfarb-Einschränkung
Blau ist als Farbfamilie für Textfarben ungeeignet. Der Grund liegt in der Verteilung der Zapfen auf und um den fovealen Bereich (gelber Fleck), der Stelle des schärfsten Sehens auf der Netzhaut. Im zentralen fovealen Bereich gibt es kaum blauempfindliche Farbrezeptoren, so daß dieser Bereich fast unempfindlich gegenüber blauem Licht ist.Fixiert das Auge ein kleines blaues Objekt z.B. einen Buchstaben, so kann dieser, zumindest bei kleiner Schriftgröße, nur schwer wahrgenommen werden [9][10].

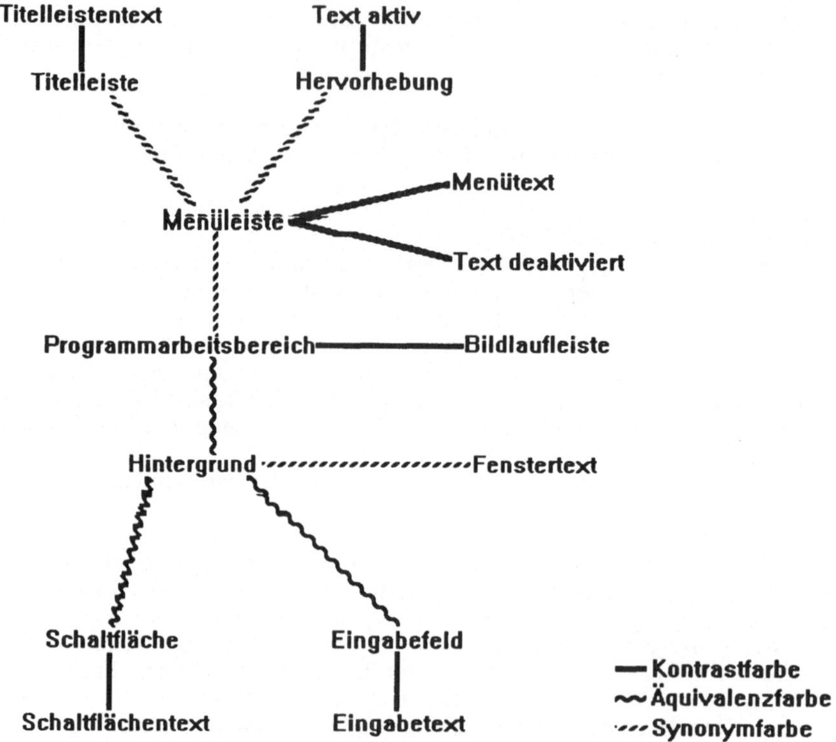

Abb. 7: Kontrast-, Äquivalenz- und Synonymfarb-Beziehungen der
Dialogelement-Teile untereinander

Alle die zuvor vorgestellten Beziehungen, Einschränkungen und Farbbewertungen
bilden, in konkreten Regeln ausgedrückt, die Wissensbasis des
Farbexpertensystems. Die Regeln stützen sich auf Fakten, die innerhalb der
physiologischen und psychologischen Grundlagenforschung ermittelt wurden. Die
Randbedingungen bei der Lösungsfindung werden im Dialog mit dem Benutzer
festgelegt.

4   Zusammenstellung eines Farbschematas

Das Ziel ist, Farbzusammenstellungen der Dialogelement-Teile zu erstellen, die
allen Regeln entsprechen. Zu diesem Zweck wird der Benutzer bei der Farbauswahl

durch Vorschläge des Farbexpertensystems unterstützt. Abbildung 8 zeigt den Auswahldialog.

Dem Benutzer stehen 28 fest vordefinierte und 7 selbstdefinierbare Farben pro Farbschema zur Verfügung. Die eigenen Zusatz-Farben können mit Hilfe des im 2. Abschnitt vorgestellten Farbbewertungsdialoges definiert werden. Die „Paletten"-Piktogramme unterhalb der eigenen Farben aktivieren den Definitions-Dialog. Da die Anzahl der verwendeten Maskenfarben sieben nie überschreiten sollte, reicht die Anzahl der sebstdefinierbaren Farben selbst dann aus, wenn der Benutzer keine Farbe aus der Grundfarbpalette verwenden möchte.

Um den Auswahlmechanismus zu starten, können die Dialogelement-Teile aus der Beispiel-Maske direkt angeklickt werden.

Vor Beginn der Regelabarbeitung werden alle aktiven Farben zu möglichen Farbvorschlägen erklärt. Aktive Farben sind Farben, deren Attributwerte (Farbfamilie, Helligkeit und Sättigung) bekannt sind. Alle 28 Farben der Grundfarbpalette sind aktive Farben. Die Zusatzfarben sind bis zu ihrer Definition durch den Benutzer passive Farben, da ihre Attributwerte unbekannt sind. Durch die Definition einer Zusatzfarbe werden nicht nur ihre RGB-Werte, sondern auch ihre Farbfamilie, ihre Helligkeit und ihre Sättigung automatisch oder durch den Benutzer mitbestimmt. Da die Attributwerte der Zusatzfarbe diese für die Regelauswertung ausreichend klassifizieren, wird sie zu einer aktiven Farbe. Es können somit maximal 35 Farben zu möglichen Farbvorschlägen erklärt werden. Die Zahl 35 ist nur auf den vorgestellten Auswahldialog in Abbildung 8 bezogen. Das bewertende Expertensystem ist für eine unbegrenzte Anzahl von Farben ausgelegt.
Bei der Suche passender Farben für ein Dialogelement-Teil werden nacheinander alle Farbbeziehungen zu anderen Dialogelement-Teilen mit deren jeweiligen Vergleichsfarbe abgearbeitet. Danach folgen die Regeln, die die Auswahl durch den Teil-Typ selbst einschränken. Die Regelpakete arbeiten destruktiv, d.h. pro Regellauf (jede Beziehung steht für einen Regellauf) werden alle Farben aus der Liste der möglichen Farbvorschläge entfernt, die die Regelanforderungen nicht erfüllen können. Alle Farben, die übrigbleiben erfüllen alle Farbbeziehungen zwischen den Dialogelement-Teilen und zu sich selbst. Sie werden nach Abschluß der Beziehungsabarbeitung dem Benutzer als Entscheidungshilfe präsentiert. Er kann dann eine von den vorgeschlagenen Farben (gekennzeichnet durch einen schwarzen Rahmen) zu dem aktuell betrachteten Dialogelement-Teil auswählen. Die aktuelle Farbe des Dialogelement-Teils ist durch ein Kreuz auf der Farbfläche markiert. Eine Wahl einer anderen Farbe durch den Benutzer ist dennoch möglich. Nur kann dann keine Garantie für eine ergonomische Farbwahl übernommen werden.

Vor dem Verlassen des Dialoges ist ein kompletter Test aller Farbbeziehungen und Abhängigkeiten möglich. Mit Hilfe eines Informationsfensters wird in textueller Form auf Farbunverträglichkeiten einzelner Dialogelement-Teile hingewiesen und vor einer zu großen Anzahl verwendeter Farben (größer sieben) gewarnt.

Fertiggestellte Farbschemata können mit Hilfe der hinterlegten objektorientierten Datenbank gespeichert und verwaltet werden.

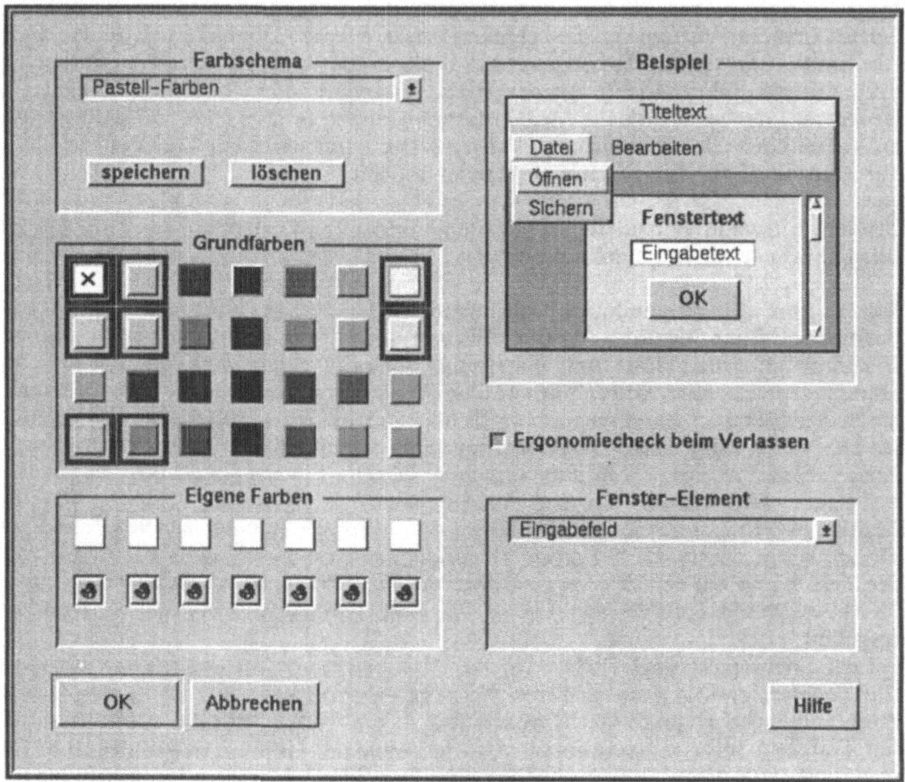

Abb. 8: Der Auswahl Dialog

5   Ausblick

Nicht nur für den Einsatz von Farben bei der Erstellung von Dialogmasken gibt es auf physiologischen und psychologischen Grundlagen basiertes Wissen. Auch für die problembezogene Auswahl und die Positionierung von Dialogelementen lassen sich Regeln z.B. aus Style-Guides ableiten. Mit Hilfe dieser Regeln kann der gesamte Prozeß der Maskengenerierung unterstützt bzw. sogar automatisiert werden.

Eine der ersten Realisierungen auf dem Gebiet der automatischen Maskengenerierung durch Anwendung von Ergonomie-Wissen ist das JANUS-System [1], [2], welches am Lehrstuhl für Softwaretechnik an der Ruhr-Universität Bochum entwickelt wird. Aus den Informationen eines vom Benutzer eingegebenen OO-Modells werden durch Dialog- und Layout-Expertensysteme

Bildschirm-Masken erzeugt, die dem aktuellen Wissensstand auf dem Gebiet der Ergonomie entsprechen.

Die vorliegende Arbeit ist eine Teilkomponente des JANUS-Systems. Sie ermöglicht es dem Benutzer, den Bereich der Farbgebung bei der Maskengenerierung zu konfigurieren.

6   Literatur

[1]   Balzert H., *Der JANUS-Dialogexperte: Vom Fachkonzept zur Dialogstruktur in: Softwaretechnik Trends, August 1993 S. 62 - 72* , Proceedings der GI-Fachtagung Softwaretechnik in Dortmund, 1993

[2]   Balzert H., *Das JANUS-System: Automatisierte, wissensbasierte Generierung von Mensch-Computer-Schnittstellen in: Informatik - Forschung und Entwicklung S. 22-35*, Springer-Verlag, 9/1994

[3]   Bibliographisches Institut Mannheim/Wien/Zürich, *Schüler-Duden Die Biologie*, Duden, 1976

[4]   Bourne L., Ekstrand B., *Einführung in die Psychologie*, Dietmar Klotz, 1992

[5]   Herczeg M., *Software-Ergonomie*, Addison-Wesley, 1994

[6]   Mayhew D., *Principles and Guidelines in Software User Interface Design*, Prentice-Hall Englewood Cliffs, 1992

[7]   Rubin T., *User Interface Design for Computer Systems*, Ellis Horwood Limited, Halsed Press 1988

[8]   Urban & Schwarzenberg, *Lexikon Medizin*, Roche, 1987

[9]   Wandmacher J., *Software-Ergonomie*, Walter de Gruyter, Berlin, New York 1992

[10] Ziegler J., Ilg R. (Hrsg), *Benutzergerechte Software-Gestaltung*, Oldenbourg, 1993

Visuelle Programmierung graphischer Benutzungsoberflächen: Ein wissensbasierter, generischer Ansatz

Jürgen Herczeg
Universität Stuttgart

Zusammenfassung

Durch die Verbreitung graphischer Benutzungsoberflächen-Editoren setzt sich visuelle Programmierung bei der Entwicklung von Benutzungsschnittstellen immer mehr durch. Die meisten der verfügbaren Werkzeuge schöpfen jedoch das Potential interaktiver visueller Programmierung nur unzureichend aus und besitzen in bezug auf Funktionsumfang und Flexibilität zahlreiche Schwächen. Diese liegen oftmals bereits in den verwendeten Toolkits begründet. Der vorliegende Beitrag zeigt Probleme bestehender Werkzeuge auf. Es wird ein Ansatz vorgestellt, mit dem visuelle Programmierwerkzeuge durch Repräsentation von Wissen auf Toolkit-Ebene teilweise automatisch generiert werden können. Diese bilden ein „Metasystem", mit dem Benutzungsschnittstellen auch zur Anwendungslaufzeit erstellt und modifiziert werden können, und ermöglichen so eine sinnvolle Kombination konventioneller und visueller Programmierung. Die generische Modellierung der Werkzeuge gewährleistet einfache Erweiterbarkeit. Darüberhinaus ermöglicht der gewählte wissensbasierte Ansatz auch die Unterstützung des Benutzers, beispielsweise bei der Berücksichtigung ergonomischer Entwurfsrichtlinien.

1 Visuelle Programmierung graphischer Benutzungsoberflächen

Die Möglichkeit, graphische Benutzungsoberflächen durch *visuelle Programmierung* zu erstellen, wurde bereits mit großem Interesse verfolgt, bevor diese einem größerem Anwenderkreis zugänglich waren. PERIDOT [10] war ein Vorläufer solcher *visuellen Programmierwerkzeuge*. Durch die Verbreitung graphischer Benutzungsoberflächen und den steigenden Bedarf an effizienten Entwicklungsmethoden und -werkzeugen haben sich graphische *Benutzungsoberflächen-Editoren (Interface Builder)* als Bestandteile von *User Interface Management Systemen (UIMS)* mittlerweile etabliert; Beispiele dafür sind der NEXT Interface Builder [19] und DEVGUIDE [16]. Diese erlauben es, Teile einer Benutzungsschnittstelle graphisch interaktiv ohne den direkten Einsatz einer konventionellen Programmiersprache zu erstellen. Sie arbeiten üblicherweise nach dem Baukastenprinzip unter Verwendung direkter Manipulation. Wesentliche Unterschiede bestehen allerdings darin,

- welche Aspekte *visuell* erstellt werden können: statisches Layout, Dynamik, Anwendungsanbindung;

- welche Interaktionstechniken zur Verfügung stehen: nur vordefinierte „Standard-Bausteine" oder beliebige durch den Benutzer definierbare *Interaktionsobjekte*;

- welche Implementierungsplattformen unterstützt werden: Programmiersprachen, Fenstersysteme, Baukästen (*Toolkits*).

Kommerzielle Werkzeuge konzentrieren sich im Hinblick auf „Standard-Anwendungen" mehr und mehr auf letzteren Aspekt mit dem Ziel plattformunabhängiger *portabler Benutzungsoberflächen*. Mit der Bindung an vorgefertigte Bausteine und strenge Entwurfsrichtlinien (*User Interface Style-Guides*) schränken sie jedoch notwendigerweise das Spektrum erstellbarer Benutzungsoberflächen – und damit auch die Kreativität des Entwicklers – stark ein. Systeme, die im Bereich der Forschung entwickelt und eingesetzt werden, wie z. B. GARNET [13] oder die im folgenden vorgestellten Entwicklungsumgebung XIT, decken ein wesentlich breiteres Spektrum von Anwendungen ab (z. B. direkte Manipulation oder interaktive Graphik) sowie Entwurfsmöglichkeiten, die deutlich über statisches Layout hinausgehen.

1.1   Schwächen visueller Programmierwerkzeuge

Visuelle Programmierung bietet sich auf natürliche Weise an, wenn die zu erstellenden Konstrukte visueller Art sind. Der Anwendung des WYSIWYG-Paradigmas ist es zuzuschreiben, daß visuelle Programmierung bei graphischen Benutzungsschnittstellen wesentlich erfolgreicher ist als bei interner Anwendungslogik. Dennoch werden Benutzungsschnittstellen trotz einer Vielzahl verfügbarer Werkzeuge größtenteils *nicht* durch visuelle, sondern durch konventionelle Programmierung – meist unter Einsatz von Baukästen (*Toolkits*) – erstellt [12]. Verschiedene Gründe sind dafür verantwortlich:

- Graphische Editoren beschränken sich auf die Bearbeitung der visuellen Aspekte einer Benutzungsschnittstelle, wie etwa Darstellungsattribute (Farben, Fonts, etc.) und Layout. Die Beschreibung der Dynamik und die Anbindung an die Anwendung – die weitaus schwierigeren Aufgaben – werden nur selten graphisch unterstützt. Durch *demonstratives Programmieren*, das in Systemen wie DRUID [14] oder MARQUISE [11] eingesetzt wird, ist es lediglich möglich, graphisch visualisierte, nicht jedoch interne dynamische Aspekte zu beschreiben.

- Zur Spezifikation der Dynamik werden werkzeugspezifischen Dialogbeschreibungssprachen oder konventionellen Programmiersprachen verwendet, die meist regelbasiert oder prozedural sind. Dies führt zu einem Bruch mit dem objektbasierten visuellen Programmierparadigma.

- Viele Unzulänglichkeiten graphischer Werkzeuge, wie beispielsweise unzureichende Mechanismen zur Anbindung an die Anwendung, liegen bereits in den verwendeten Toolkits begründet [4]. Der Mangel an Ausdrucksmöglichkeiten wird durch die visuelle Programmierschnittstelle zwangsläufig noch verstärkt. So wird etwa die auf Toolkit-Ebene angebotene Möglichkeit zur Definition neuer Interaktionsobjekte graphisch nicht unterstützt.

- Die meisten Interface Builder sind bezüglich ihrer Benutzungsoberfläche und dem Funktionsumfang statisch und können nicht erweitert werden. Das Spektrum wünschenswerter Erweiterungen reicht von zusätzlichen Interaktionsobjekten bis hin zu anwendungsspezifischer Funktionalität.

- Interaktive Anwendungen werden nach wie vor durch traditionelle Stapelverarbeitung erstellt: Die Benutzungsoberfläche wird mit einem Graphikeditor (statt einem Texteditor) konstruiert, übersetzt und an die Anwendung angebunden. Erst jetzt entsteht ein lauffähiges Programm, mit dem Oberfläche und Anwendung zusammen getestet werden können. Notwendige Modifikationen zwingen dazu, viele solcher Zyklen zu durchlaufen. Simulationskomponenten, die oft zusätzlich angeboten werden, erlauben es lediglich, dynamisches Verhalten auf syntaktischer Ebene zu testen. Semantische Rückmeldungen, wie sie für direktmanipulative Anwendungen charakteristisch sind, lassen sich nicht einbeziehen.

- Als Ausgabe wird Programmcode in einer werkzeugspezifischen oder allgemeinen Programmiersprache erzeugt. Umgekehrt können jedoch Benutzungsschnittstellen, die durch konventionelle Programmierung erstellt wurden, durch visuelle Werkzeuge *nicht* nachbearbeitet oder eingebunden werden. Die Wiederverwendbarkeit vorhandener Module wird somit trotz eines objektorientierten Ansatzes nicht unterstützt.

Der inkrementellen und explorativen Vorgehensweise, die für den Entwurf graphischer Benutzungsschnittstellen typisch ist, werden bestehenden Entwicklungswerkzeuge aufgrund dieser Schwächen nicht gerecht. Das Potential interaktiver visueller Programmierung wird nur ansatzweise ausgeschöpft. Interaktive objektorientierte Programmierumgebungen, wie beispielsweise

SMALLTALK-80, aus denen sich Benutzungsschnittstellen-Entwicklungsumgebungen ursprünglich entwickelt haben, zeichnen sich durch Werkzeuge aus, mit denen jeder Programmaspekt – insbesondere auch zur Laufzeit – inspiziert und manipuliert werden kann. Bei graphischen Benutzungsoberflächen ist dies aufgrund der strikten Trennung zwischen Entwicklungs- und Laufzeitumgebung nur in eingeschränktem Maße möglich: *Window-Manager* ermöglichen es dem Benutzer, Anwendungsfenster, z. B. bezüglich ihrer Position und Größe, zu manipulieren. Mit *Ressource-Editoren* können Attribute wie Farben oder Schriftarten geändert werden. Ebenso einfach sollte es beispielsweise für den Benutzungsschnittstellen-Entwickler oder auch Endbenutzer sein, interaktiv die Einträge eines Anwendungsmenüs umzuordnen, neu zu benennen oder weitere Einträge hinzuzufügen.

1.2   Ein wissensbasierter, generischer Ansatz

In *wissens-* oder *modellbasierten* Systemen, wie z. B. UIDE [2], wird die Benutzungsschnittstelle automatisch aus der Anwendung und einer Beschreibung ihrer Semantik generiert. Das Anwendungsmodell spiegelt sich dabei unmittelbar auf der Benutzungsoberfläche wider. Da das Hauptinteresse auf der Anwendungslogik liegt, entstehen jedoch in aller Regel wenig attraktive Oberflächen. Dies kann durch Einbeziehen visueller Programmiertechniken, wie etwa in HUMANOID [18], verbessert werden.

Im folgenden wird ein wissensbasierter Ansatz vorgestellt, bei dem wesentliche Teile der Benutzungsschnittstelle visueller Programmierwerkzeuge automatisch generiert werden. Das dazu benötigte Wissen wird nicht wie in traditionellen wissensbasierten Systemen explizit in den Werkzeugen, sondern bei den Interaktionsobjekten des Toolkits selbst repräsentiert. Die Werkzeuge liegen bei diesem generischen Ansatz nur als Rahmensysteme vor und werden durch die bearbeiteten Objekte spezialisiert; d. h. nicht der Editor „weiß", wie Benutzungsschnittstellen erstellt und manipuliert werden, sondern die darin enthaltenen Bausteine. Daraus ergeben sich zahlreiche Vorteile [5]:

- Das dem Toolkit zugrunde liegende Modell wird von den visuellen Werkzeugen vollständig wiedergegeben; diese verfügen nahezu über die gesamte interne Funktionalität bezüglich Präsentation, Dynamik und Anwendungsschnittstelle. Modifikationen und Erweiterungen des internen Modells, z. B. neue Interaktionsobjekte, stehen automatisch auch auf der Benutzungsoberfläche zur Verfügung.

- Benutzungsschnittstellen können nicht nur aus den angebotenen Bausteinen konstruiert werden, sondern es lassen sich auch Oberflächen visuell bearbeiten, die auf Toolkit-Ebene programmiert wurden. Visuelle und konventionelle Programmierung können sich also sinnvoll ergänzen.

- Benutzungsschnittstellen können auch zur Programmlaufzeit bearbeitet werden; eine Simulationskomponente wird somit nicht benötigt. Es besteht keine strikte Trennung mehr zwischen Entwicklungs- und Laufzeitumgebung. Daraus ergibt sich auch die Möglichkeit zum Modifizieren oder Individualisieren einer Oberfläche durch den Endbenutzer.

Durch den wissensbasierten Ansatz werden die Werkzeuge zu *Metasystemen* für graphische Benutzungsschnittstellen, sogenannten *Meta-Benutzungsschnittstellen* [6]. Einfache Beispiele dafür wurden bereits in USIT [7] und INTERVIEWS [9] implementiert. Diese waren jedoch statisch und auf allgemeine Eigenschaften von Interaktionsobjekten (z. B. geometrische Attribute) beschränkt. Im folgenden wird ein Metasystem beschrieben, das auch objektspezifische Eigenschaften berücksichtigt.

2   Ein Metasystem für Interaktionsobjekte

Ein generisches Metasystems für Interaktionsobjekte wurde als Bestandteil der Benutzungsschnittstellen-Entwicklungsumgebung XIT realisiert.

2.1   Systemarchitektur

Abb. 1 zeigt die geschichtete Systemarchitektur von XIT [3]. XIT ist in COMMON LISP und CLOS implementiert und benutzt CLX und CLUE [8] als Schnittstellen zum X Window System. Darauf bauen zwei Toolkit-Ebenen auf: XIT_{low}, ein Rahmensystem für Interaktionsobjekte, und XIT_{high} eine erweiterbare Interaktionsobjekt-Bibliothek. Diese objektorientierten Baukästen bilden die Wissensbasis für die visuellen Programmierwerkzeuge des Metasystems XIT_{visual}. Der Editor, Inspektor und Browser aus XIT_{visual} werden im folgenden näher beschrieben.

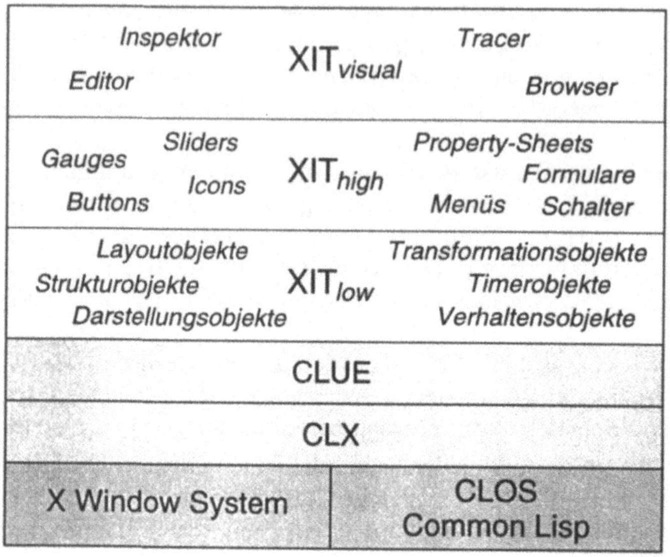

| Inspektor | XIT_visual | Tracer |
| Editor | | Browser |

Abb. 1 Systemarchitektur von XIT

2.2 Interaktionsobjekt-Editor

Zum Erzeugen und Verknüpfen von Interaktionsobjekten kommt die in graphischen Editoren verbreitete Baukasten-Metapher zur Anwendung. Dabei werden Objekte durch Drag&Drop aus einer *Palette* in den *Arbeitsbereich* gezogen. Abb. 2 zeigt beispielsweise, wie mit dem Interaktionsobjekt-Editor ein Fenster zur Darstellung der Eigenschaften eines Rechners erzeugt wird.

Die verfügbaren Interaktionsobjekte sind über verschiedene Paletten verteilt. Es existieren Paletten für die Standard-Bausteine von XIT_{high} (Buttons, Menüs, Property-Sheets etc.) sowie für die in XIT_{low} enthaltenen Grundbausteine (*Darstellungsobjekte* für Text, Bild und Graphik, allgemeine *Strukturobjekte*); aus diesen können neue Interaktionsobjekte zusammengesetzt werden, wie etwa das mit einem Ein/Aus-Schalter versehene graphische Bildschirmobjekt (s. Abb. 2). Die Verteilung der Objekte über verschiedene Paletten ermöglicht es, eine große Anzahl von Bausteinen anzubieten, diese semantisch zu gruppieren und bei Bedarf auch Paletten mit anwendungsspezifischen Bausteinen nachzuladen. Abb. 3 zeigt beispielsweise, wie über eine Palette mit speziellen Icons Rechner und Peripheriegeräte graphisch konfiguriert werden können.

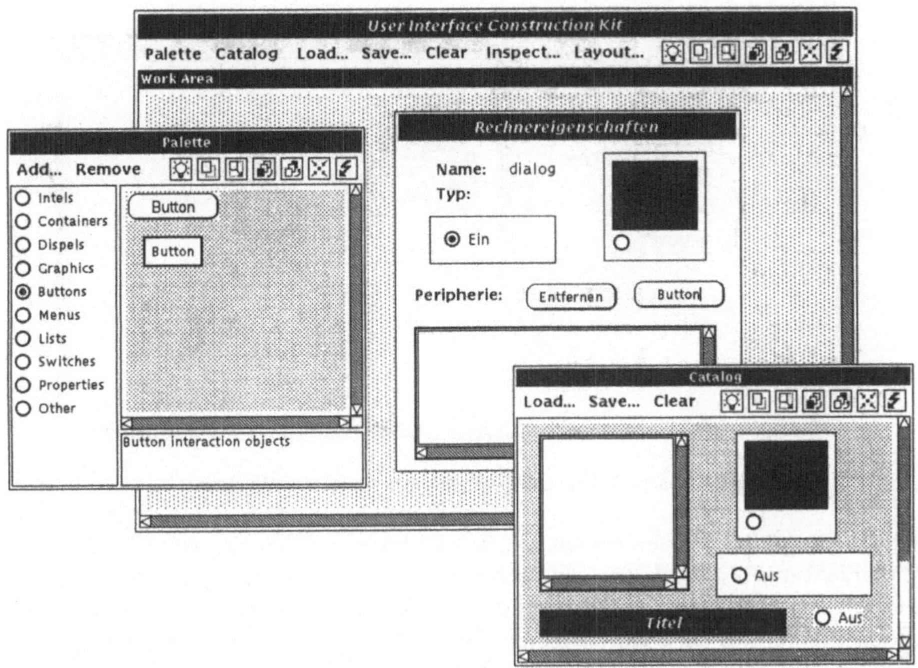

Abb. 2: Interaktionsobjekt-Editor

Paletten werden aus internen *Palettenbeschreibungen* generiert. Diese enthalten für jeden Eintrag die zugehörige graphische Darstellung und die Aktion, die bei dessen Auswahl ausgelöst wird. Im einfachsten Fall wird lediglich auf eine Interaktionsobjekt-Klasse verwiesen, die an der jeweiligen Zielposition instantiiert wird. Allgemein können Paletteneinträge jedoch beliebige Aktionen auslösen. So werden etwa durch die Bausteine der Palette in Abb. 3 zusätzlich zu Interaktionsobjekten auch entsprechende Anwendungsobjekte erzeugt und verknüpft. Der Benutzungsoberflächen-Editor wird auf diese Weise zum Anwendungseditor.

Bei der Auswahl eines Paletteneintrags wird kein graphisches Abbild, sondern das tatsächliche Interaktionsobjekt instantiiert; dieses kann deshalb nicht nur im Arbeitsbereich des Editors, sondern in einem beliebigen Anwendungsfenster abgelegt werden (vgl. Abb. 3). Der gesamte Bildschirm wird somit zum Arbeitsfläche. Die erzeugten Objekte sind voll funktionsfähig; es wird nicht zwischen einem Konstruktions- und einem Testmodus unterschieden. Auf diese Weise können mit dem Editor Anwendungen auch zur Laufzeit – beispielsweise um zusätzliche Buttons oder Menüeinträge – erweitert werden.

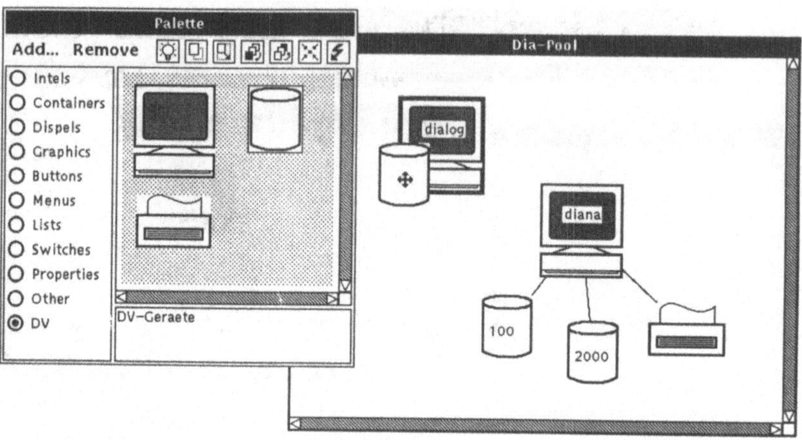

Abb. 3: Anwendungsspezifische Palette zur Konfiguration von Rechnerobjekten

Durch Drag&Drop-Operationen können Objekte – abhängig von Quellobjekt, Zielobjekt und Zielposition – (a) geometrisch *positioniert*, (b) *aggregiert* (d. h. ineinander verschachtelt) oder (c) miteinander *assoziiert* werden. In Abb. 3 werden durch Ziehen des Festplatten-Icons aus der Palette auf ein Rechner-Icon eine Kombination dieser verschiedenen Aktionen ausgeführt: Das entsprechende Icon wird (1) instantiiert, (2) in das umgebende Fenster (in einem vorgegeben Abstand vom Rechner-Icon) eingefügt und (3) mit diesem durch eine Linie verbunden; (4) die zugrundeliegenden Anwendungsobjekte werden miteinander assoziiert. Das Wissen darüber, welche Aktionen durch Drag&Drop ausgelöst werden, ist bei den beteiligten Objekten repräsentiert; z. B. „weiß" ein Platten-Icon, daß es nur auf einem Rechner-Icon abgelegt werden darf; das betreffende Rechner-Icon führt die Verknüpfung der zugeordneten Anwendungsobjekte durch.

Einfacher als das Erzeugen von Interaktionsobjekten aus Bausteinen der Paletten ist oftmals das Kopieren bereits vorhandener, ähnlicher Objekte. Beispielsweise kann ein neuer Menüeintrag durch Kopieren eines bestehenden Eintrags und Modifizieren von Eigenschaften der Kopie (z. B. Text und Aktion) erzeugt werden. Eine generische Kopieroperation erlaubt es, auch komplexe Interaktionsobjekte (z. B. komplette Anwendungsfenster) zu kopieren. In der Objektkopie werden auch die entsprechenden Verweise auf Anwendungsobjekte hergestellt. Das zur Implementierung der Kopieroperation notwendige Wissen ist bei den Interaktionsobjekten definiert. Nach demselben generischen Prinzip arbeitet auch die Code-Generierung, die zum Abspeichern der Objekte im Arbeitsbereich dient. Sie ist deshalb nicht auf die in den Paletten enthaltenen Objekte beschränkt, sondern erzeugt für *jedes* XIT-Interaktionsobjekt den zugehörigen Lisp-Programmcode.

Der *Katalog* (vgl. Abb. 2) ist eine durch den Benutzer interaktiv erweiterbare Palette für komplexere Bausteinen. Er kann als Zwischenspeicher für Copy&Paste-Operationen dienen oder – durch die Möglichkeit, den Inhalt abzuspeichern und zu laden – zum Austausch von Bausteinen über verschiedene Anwendungen hinweg. Das Einfügen und Entnehmen von Objekten ist über die generische Kopieroperation realisiert.

2.3 Interaktionsobjekt-Inspektor und -Browser

Da die Bausteine in herkömmlichen Benutzungsoberflächen-Editoren fest vorgegeben sind, genügen zur Manipulation ihrer Attribute vordefinierte, statische Property-Sheets. Diese sind nur auf die im Arbeitsbereich des Editors enthaltenen Objekte anwendbar, nicht aber auf die entsprechenden Objekte in der späteren Anwendung. In XIT wurde dazu der *Interaktionsobjekt-Inspektor* entwickelt. Er ermöglicht das Inspizieren und Manipulieren der Eigenschaften jedes beliebigen Interaktionsobjektes und den Einstieg in andere Werkzeuge des Metasystems.

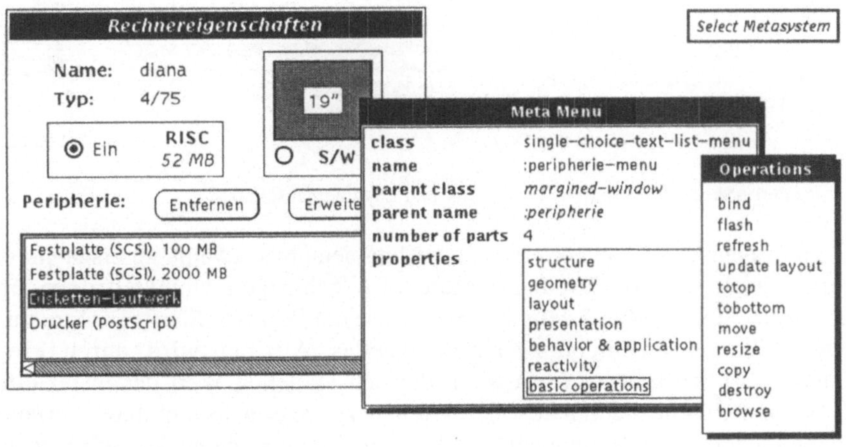

Abb. 4: Aufruf des Metasystems

Der Aufruf des Inspektors erfolgt durch Betätigen des *Metasystem-Buttons* (`Select Metasystem`) und anschließende Auswahl des gewünschten Interaktionsobjekts, in Abb. 4 beispielsweise die Peripherieliste des erstellten Eigenschaftsfensters; für Objekte, die sich im Arbeitsbereich des Editors befinden, genügt ein spezieller Mausklick. Daraufhin öffnet sich das *Metamenü*, das Auskunft über Name und Klasse des betreffenden Interaktionsobjekts sowie über hierarchisch über- und untergeordnete Objekte gibt. Es werden verschiedene Eigenschaftskategorien zur Auswahl gestellt, zu denen spezielle Inspektoren individuelle Objektsichten bieten. Über den Eintrag `basic operations` wird ein Menü geöffnet, das u. a. Operationen zum Bewegen, Kopieren oder Löschen des Objekts anbietet (s. Abb. 4).

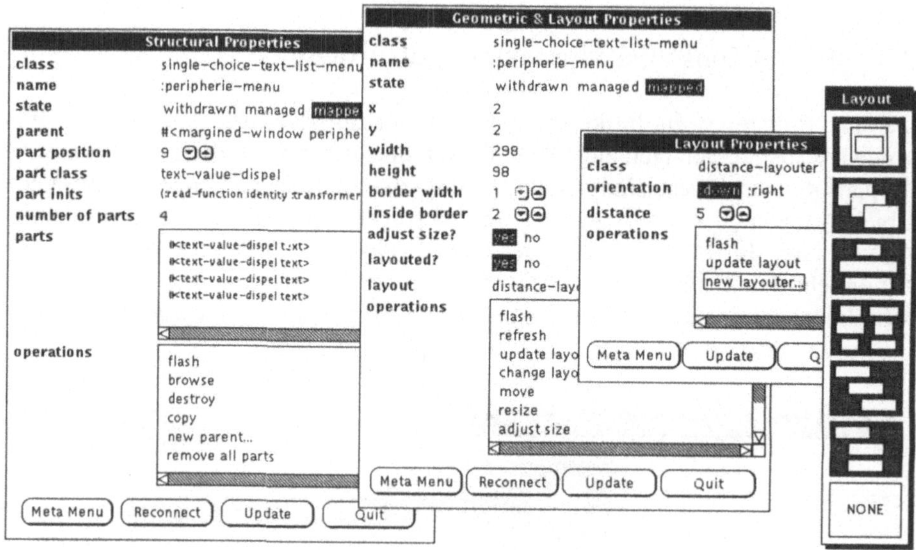

Abb. 5: Property-Sheets für Struktur-, Geometrie- und Layouteigenschaften

Wird eine Eigenschaftskategorie aus dem Metamenü ausgewählt, so erscheint ein Property-Sheet, das die zu dieser Kategorie gehörigen Objektattribute und -operationen enthält. Abb. 5 zeigt beispielsweise die Property-Sheets für Struktur-, Geometrie- und Layouteigenschaften. Jedes Attribut wird durch einen attributspezifischen Eintrag visualisiert, der den aktuellen Wert des inspizierten Objekts darstellt und geeignete Interaktionen zu dessen Manipulation erlaubt. Handelt es sich bei den Attributwerten selbst um komplexere Objekte, wie etwa die zur geometrischen Anordnung von Interaktionsobjekten verwendeten *Layoutobjekte* (vgl. Abb. 5), oder steht eine große Zahl möglicher Attributwerte zur Auswahl, so werden zu deren Visualisierung eigene Property-Sheets und Paletten verwendet. Dies trifft auch auf die zur Beschreibung von Präsentationseigenschaften verwendeten Darstellungs-Ressourcen zu, wie z. B. Farben, Fonts, Bitmaps und Cursor (vgl. Abb. 6). Alle Property-Sheets werden beim Aufruf mit den für das betreffende Objekt spezifischen Einträgen dynamisch generiert und aus Effizienzgründen zur Wiederverwendung zwischengespeichert.

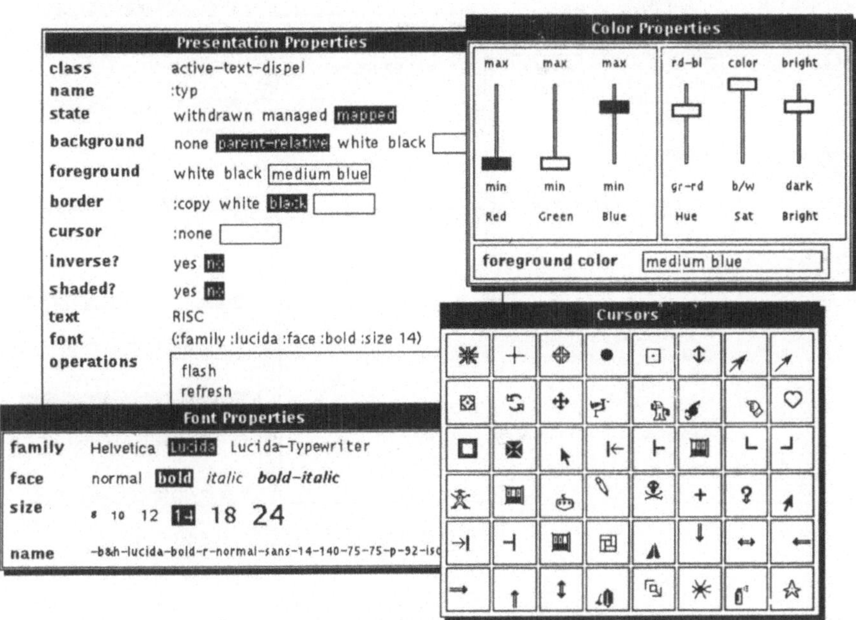

Abb. 6: Property-Sheets und Paletten für Präsentationseigenschaften

Jedes Property-Sheet kann über den `Reconnect`-Button an andere Objekte desselben Typs gebunden werden. Wird dieser gedrückt, so sind alle Bildschirmobjekte dieses Typs sensitiv. Sollen verschiedene Objekte gleiche Eigenschaften erhalten, so ist es einfacher, anstatt dieselben Attributwerte mehrfach einzugeben, diese von einem Objekt zum anderen zu kopieren. Das Kopieren erfolgt aus dem betreffenden Attribut-Eintrag im Property-Sheet durch Zeigen auf das Objekt, von dem der Wert übernommen werden soll; dabei sind diejenigen Bildschirmobjekte sensitiv, die über das entsprechende Attribut verfügen. Kopiert werden können Präsentationsattribute (um Objekten gleiches Aussehen zu geben), geometrische Attribute (um Objekte einheitlich anzuordnen), Strukturattribute (um beispielsweise alle Einträge eines Menüs in ein anderes zu kopieren) oder auch Attribute, die das dynamische Verhalten eines Objekts beschreiben.

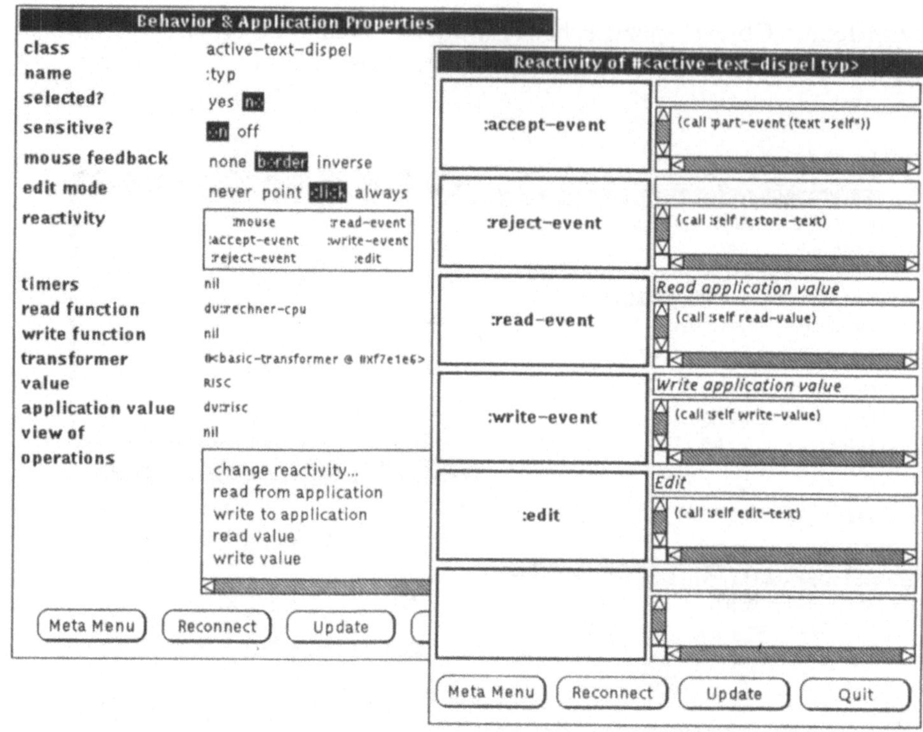

Abb. 7: Property-Sheets für dynamisches Verhalten und Anwendungsanbindung

Da in XIT dynamisches Verhalten und Anwendungsanbindung ebenfalls über Objektattribute spezifiziert werden, können auch diese mit dem Inspektor bearbeitet werden. Abb. 7 zeigt die entsprechenden Property-Sheets für ein editierbares Textfeld. Verweise auf Anwendungsobjekte und -operationen können durch Texteingabe interaktiv eingerichtet und verändert werden. Bei Abhängigkeiten zwischen Interaktionsobjekten werden die entsprechenden Verweise durch Zeigehandlungen angegeben. So wurde etwa das in Abb. 4 dargestellte Eigenschaftsfenster mit einem in Abb. 3 konfigurierten Rechnerobjekt verknüpft, um dessen Eigenschaften darzustellen. Das Erstellen und Verknüpfen der Objekte dieses Anwendungsszenarios kann somit vollständig visuell erfolgen. Komplexere dynamische Abhängigkeiten, die nicht deklarativ über Objektattribute beschreibbar sind, werden durch Editieren von Event-Aktions-Tabellen spezifiziert und unmittelbar auf die betreffenden Objekte übertragen. Events und häufig verwendete Aktionen stehen in Pop-Up-Menüs zur Verfügung.

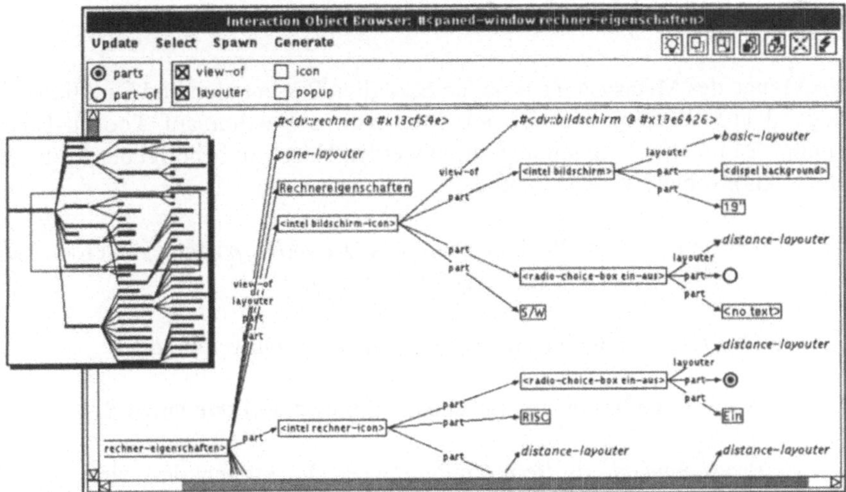

Abb. 8: Interaktionsobjekt-Browser

Der Inspektor erlaubt es, ausgehend von einem Objekt durch die Interaktionsobjekt-Hierarchie zu navigieren. Dazu kann im Property-Sheet für Struktureigenschaften (vgl. Abb. 5) durch Auswahl des hierarchisch übergeordneten Objekts (`parent`) oder eines der untergeordneten Objekte (`parts`) das zugehörige Metamenü aufgerufen werden. Diese einstufige Art der Navigation ist jedoch aufgrund des fehlenden Überblicks für komplexe Interaktionsobjekt-Hierarchien ungeeignet.

Eine graphische Visualisierung von Objektstrukturen bietet der *Interaktionsobjekt-Browser*. In Abb. 8 zeigt dieser einen Ausschnitt aus der Struktur des Eigenschaftsfensters. Im Gegensatz zu Browsern, wie sie herkömmlichen Benutzungsoberflächen-Editoren anbieten, werden nicht nur hierarchische Abhängigkeiten zwischen Interaktionsobjekten, sondern beispielsweise auch Verweise auf Layout- und Anwendungsobjekte visualisiert. Verschiedene Objekttypen und -verweise werden unterschiedlich (z. B. durch verschiedene Farben und Symbole) dargestellt. Welche Objekte und Relationen wie visualisiert werden, ist wiederum bei den Interaktionsobjekten selbst repräsentiert. Der Browser kann aus dem Inspektor aufgerufen werden; umgekehrt kann für jedes im Browser dargestellte Objekt der Inspektor aktiviert werden.

2.4   Interaktionsobjekt-Wissensbasis

Die Werkzeuge des Metasystems basieren bezüglich Präsentation und Funktionalität vorwiegend auf Wissen, das bei den Interaktionsobjekten auf Toolkit-Ebene repräsentiert ist und zur Laufzeit abgerufen werden kann. So benötigt der Editor für jedes Interaktionsobjekt Wissen darüber,

- auf welchen anderen Objekten es mittels *Drag&Drop* abgelegt werden kann und welche Aktionen dabei ausgeführt werden;

- wie ein *Prototyp* (eine Instanz mit typischen Attributwerten) oder

- eine *Kopie* (eine Instanz mit gleichen Attributwerten) erzeugt wird;

- wie *Programmcode* zur Reinstantiierung des Objekts generiert wird.

Beim Inspektor und Browser handelt es sich um Wissen,

- durch welche *Attribute* ein Objekt charakterisiert ist und wie diese visualisiert und manipuliert werden;

- welche *Operationen* auf das Objekt angewandt werden können;

- welche *Objekte* und *Relationen* wie dargestellt werden.

Teilweise ist dieses Wissen bereits implizit in den Objekten enthalten. So enthält beispielsweise eine Klassendefinition für jedes ihrer Attribute Angaben über Typ, Initialisierung und Zugriff. Abb. 9 zeigt beispielsweise die Definition einer allgemeinen Klasse für Darstellungsobjekte (`display-object`) und eine ihrer Methoden (`switch-colors`). Aus der Typinformation (`:type`) wird durch Abbildungsregeln ein geeignetes Interaktionsobjekt für die Darstellung im Inspektor generiert. Für die beiden Farbattribute (`background` und `foreground`) sind dies beispielsweise Menüs mit jeweils einem Feld zur freien Eingabe (vgl. Abb. 6).

Anderes Wissen, wie z. B. die Zuordnung von Attributen und Operationen zu Eigenschaftskategorien oder die Behandlung von Attributen beim Kopieren oder bei der Code-Generierung, ist explizit über zusätzliche Methoden der betreffenden Objektklassen repräsentiert. Abb. 9 zeigt einige Beispiele für die Darstellungsobjekt-Klasse: Die Methode `generate-object-code` erweitert die Code-Generierung um Anweisungen zur Initialisierung der Farbattribute. Über `inspector-attributes` und `inspector-operations` werden die für die Klasse definierten Eigenschaften (Attribute und Methoden) dem Inspektor für Präsentationseigenschaften (presentation) zugeordnet.

```lisp
(defclass display-object (interaction-object)
  ((background :type (or (member :none :parent-relative)
                         pixel)
               :accessor background
               :initarg :background
               :documentation "background color")
   (foreground :type pixel
               :accessor foreground
               :initarg :foreground
               :documentation "foreground color"))
  (:documentation "Abstract class for objects displayed
                   in a background and foreground color"))

(defmethod switch-colors ((object display-object))
  "Switch background and foreground color"
  (let ((saved-background (background object)))
     (setf (background object) (foreground object)
           (foreground object) saved-background))
  (display object))


(defmethod generate-object-code APPEND
                                ((object display-object))
  `(:background ,(color-name (background object))
    :foreground ,(color-name (foreground object))))

(defmethod inspector-attributes APPEND
                                ((object display-object)
                                 (category (eql 'presentation)))
  '(background foreground)))

(defmethod inspector-operations APPEND
                                ((object display-object)
                                 (category (eql 'presentation)))
  '(switch-colors)))
```

Abb. 9: Objektdefinitionen und Methoden der Wissenbasis

Bei jeder Klasse wird nur objektspezifisches Wissen abgelegt. Beispielsweise muß ein Textobjekt nur „wissen", daß es über ein Textattribut verfügt, nicht jedoch, daß der Text in einer Vorder- und Hintergrundfarbe dargestellt wird, da dieses Wissen in der hierarchisch übergeordneten Klasse `display-object` repräsentiert ist. Durch *multiple Vererbung, generische Funktionen* und *Methodenkombination* in CLOS wird das für ein konkretes Objekt relevante Wissen zugänglich gemacht. So werden etwa die einzelnen Bestandteile der Code-Generierung aus den über die Klassenhierarchie verteilten Methoden aufgesammelt und (durch APPEND) verkettet. Auf diese Weise steht das gesamte Wissen auch für neue, durch den Benutzer definierte Interaktionsobjekte zur Verfügung. Wird eine neue Objektklasse definiert, so ist das Metasystem automatisch auch auf Instanzen dieser Klasse anwendbar. Falls in der Klasse neue Eigenschaften (Attribute oder Methoden) definiert werden, sind nur wenige Methoden zu ergänzen, um diese auch im Metasystem verfügbar zu machen.

3   Erstellen ergonomischer Benutzungsoberflächen

Das im Interaktionsobjekt-Metasystem eingesetzte Wissen zielt im wesentlichen darauf ab, den Benutzungsschnittstellen-Entwickler bei seiner Arbeit möglichst wenig einzuschränken. Das bedeutet aber auch, daß dieser für die Qualität der erstellten Oberfläche bezüglich funktionaler, ästhetischer und ergonomischer Aspekte weitgehend selbst verantwortlich ist. Herkömmliche Editoren begegnen dieser Anforderung durch Beschränkung auf vorgefertigte Style-Guide-konforme Bausteine. Die konkrete Ausprägung dieser Bausteine und deren Zusammenspiel kann jedoch auch dort zu wenig attraktiven und unergonomischen Benutzungsoberflächen führen.

Der in XIT gewählte wissensbasierte Ansatz eröffnet zahlreiche Möglichkeiten zur Unterstützung des Entwicklers. Durch Erweitern der Wissensbasis um objektspezifische Bedingungen, die beim Bearbeiten von Interaktionsobjekten – automatisch oder auf Anfrage – überprüft werden, kann die *Funktionalität* einer Benutzungsschnittstelle – zumindest auf syntaktischer Ebene – gewährleistet werden. So könnte beispielsweise für jeden Button oder Menüeintrag durch Überprüfen der entsprechenden Attributwerte sichergestellt werden, daß er über eine Beschriftung und eine gültige Aktion verfügt. Auf dieselbe Weise lassen sich auch *ergonomische Gestaltungsrichtlinien* berücksichtigen. Dabei können sowohl objektspezifische Kriterien zur Anwendung kommen, wie z. B. geeignete Fontgrößen oder die Beschränkung der Anzahl von Menüeinträgen, als auch objektübergreifende Aspekte, wie etwa einheitliche geometrische Anordnung von Objekten, einheitliche

Verwendung von Farben und Fonts, sinnvolle Farbkombinationen und konsistenter Einsatz der Maustasten.

Um den Entwickler nicht in seinen Gestaltungsmöglichkeiten einzuschränken, sollte dieses Wissen nur auf Anfrage angewandt werden. Gefundene Mängel können entweder passiv über einer Kritikerkomponente [1] aufgedeckt oder aktiv bei den betreffenden Objekten behoben werden. So verfügt der Interaktionsobjekt-Editor beispielsweise über die Möglichkeit zur automatischen Layoutgenerierung. Die durch direkte Manipulation erfolgte Grobanordnung von Interaktionsobjekten wird dabei auf Regelmäßigkeiten untersucht. Daraus werden entsprechende Layoutobjekte generiert, die Ungenauigkeiten in der Anordnung glätten, auf Veränderungen der Geometrie (z. B. der Größe des umgebenden Fensters) reagieren sowie auch neu eingefügte Objekte entsprechend anordnen.

Richtlinien zur ergonomischen Dialoggestaltung, wie sie etwa in den Normen DIN 66234/8 und ISO 9241/10 festgelegt sind, werden zum Teil bereits durch das XIT zugrundeliegende Modell unterstützt: Die *Individualisierbarkeit* einer Benutzungsschnittstelle kann über das Metasystem erfolgen, das nicht nur dem Entwickler, sondern auch dem Endbenutzer zur Verfügung steht. Die Modellierung der Interaktionsobjekte ermöglicht die *Selbstbeschreibungsfähigkeit* einer Benutzungsoberfläche auf natürliche Weise. So kann beispielsweise jedes Objekt über sein Verhalten und die von ihm repräsentierten Anwendungsfunktionalität Auskunft geben. In XIT wird dies zur Implementierung einfacher generischer Hilfekomponenten eingesetzt. Ein einfaches, aber sehr nützliches Beispiel dafür ist eine ständig sichtbare Benutzerführungszeile (*mouse documentation line* [17]), die Angaben über das Objekt macht, über dem sich der Mauszeiger momentan befindet. Durch Voreinstellung wird dargestellt, welche Benutzereingaben (Events) möglich sind und zu welchen Aktionen diese führen; die benötigte Information wird aus dem Objekt generiert. Für jedes Objekt können auch explizit Hilfetexte angegeben werden.

4   Abschlußbemerkungen und Ausblick

Es wurde gezeigt, wie durch objektorientierte Wissensrepräsentation Werkzeuge zur visuellen Programmierung graphischer Benutzungsschnittstellen generisch erstellt werden können, die bestehende Werkzeuge bezüglich Funktionalität, Flexibilität und Erweiterbarkeit übertreffen. Neben den beschrieben Komponenten zur Visualisierung statischer Strukturen (Inspektor und Browser) umfaßt XIT$_{visual}$ auch *Tracer* zur Visualisierung dynamischer Vorgänge, wie etwa das Empfangen und Verarbeiten von Events.

Der gewählte Ansatz basiert auf einem Toolkit, der es erlaubt, Interaktionsobjekte dynamisch zu erzeugen und zu manipulieren. Er konnte mit den in COMMON LISP und CLOS verfügbaren mächtigen Sprachkonstrukten einfach realisiert werden, ist jedoch nicht auf das XIT gewählte Toolkit-Modell und dessen Implementierung beschränkt. So konnten die graphischen Werkzeuge beispielsweise in wesentlichen Teilen unter SMALLTALK-80 reimplementiert werde. In beiden Fällen war es möglich, die Benutzungsschnittstellen-Entwicklungsumgebung vollständig in die bestehende Programmierumgebung zu integrieren, um so auch von allgemeinen Programmierwerkzeugen wie Objektinspektoren, Klassenbrowsern oder Methodentracern zu profitieren. Dies ist insbesondere dann wichtig, wenn die klare Trennung zwischen Benutzungsschnittstelle und interner Funktionalität nicht mehr möglich ist, wie z. B. bei direkt-manipulativen Anwendungen.

Die fehlende Trennung zwischen Entwicklungs- und Anwendungsumgebung spiegelt sich im Verzicht auf den für herkömmliche Benutzungsoberflächen-Editoren typischen Konstruktions- und Testmodus wider. Daraus ergibt sich allerdings folgendes grundlegendes Problem [15]: Wann wird ein Interaktionsobjekt bearbeitet, und wann wird es im Sinne der Anwendung bedient? Dies wurde durch Einführen von *Metaobjekten* (z. B. Paletten, Metasystem-Button und Property-Sheets) und *Meta-Interaktionen* (z. B. Drag&Drop von/auf Metaobjekte) gelöst. Diese Meta-Benutzungsschnittstelle ist in XIT selbst implementiert und deshalb auf sich selbst anwendbar.

XIT ist eine experimentelle Entwicklungsumgebung, die im wesentlichen für prototypische Anwendungen eingesetzt wird. Das Metasystem eignet sich neben der Möglichkeit zum schnellen Erstellen und Testen von Entwurfsalternativen auch als Lernumgebung, mit der bestehende Benutzungsschnittstellen untersucht und explorativ modifiziert werden können. Ergonomische Aspekte sind bisher nur ansatzweise repräsentiert. Sie sollen jedoch im Zusammenhang mit der Erstellung einer integrierten Programmier- und Lernumgebung weiter untersucht werden.

5   Literatur

[1]   G. Fischer, A. C. Lemke, T. Mastaglio, A. I. Morch. Using Critics to Empower Users. In *Human Factors in Computing Systems, CHI '90 Conference Proceedings*, S. 337–347, 1990.

[2]   J. Foley, C. Gibbs, W. C. Kim, S. Kovacevic. A Knowledge-Based User Interface Management System. In *Human Factors in Computing Systems, CHI '88 Conference Proceedings*, S. 67–72, 1988.

[3]   J. Herczeg. A Design Environment for Graphical User Interfaces. In D. J. Gilmore, R. Winder, F. Détienne (Hrsg.), *User-Centred Requirements for Software Engineering Environments*, Band 123 von *NATO ASI Series F: Computer and Systems Sciences*, S. 213–223. Springer-Verlag, Berlin - Heidelberg - New York, 1994.

[4]   J. Herczeg, H. Hohl, M. Ressel. Progress in Building User Interface Toolkits: The World According to XIT. In *Proceedings of the ACM Symposium on User Interface Software and Technology*, S. 181–190, 1992.

[5]   J. Herczeg, H. Hohl, M. Ressel. A New Approach to Visual Programming in User Interface Design. In *Proceedings of the Fifth International Conference on Human-Computer Interaction, HCI International '93*, S. 74–79, 1993.

[6]   M. Herczeg. Eine objektorientierte Architektur für wissensbasierte Benutzerschnittstellen. Dissertation, Fakultät Mathematik und Informatik der Universität Stuttgart, Dezember 1986.

[7]   M. Herczeg. USIT: A Toolkit For User Interface Toolkits. In *Proceedings of Third International Conference on Human-Computer Interaction*, S. 605–612, 1989.

[8]   K. Kimbrough, L. Oren. *Common Lisp User Interface Environment*. Texas Instruments Incorporated, Dallas, TX, Juli 1990.

[9]   M. A. Linton, P. R. Calder, J. M. Vlissides. InterViews: A C++ Graphical Interface Toolkit. Stanford Technical Report CSL-TR-88-358, Stanford University, Juli 1988.

[10] B. A. Myers, W. Buxton. Creating Highly-Interactive and Graphical User Interfaces by Demonstration. *Computer Graphics, Proceedings of SIGGRAPH '86*, S. 249–258, August 1986.

[11] B. A. Myers, R. G. McDaniel, D. S. Kosbie. Marquise: Creating Complete User Interfaces by Demonstration. In *Human Factors in Computing Systems, INTERCHI '93 Conference Proceedings*, S. 293–302, 1993.

[12] B. A. Myers, M. B. Rossen. Survey on User Interface Programming. In *Human Factors in Computing Systems, CHI '92 Conference Proceedings*, S. 195–202, 1992.

[13] B. A. Myers et al. Garnet: Comprehensive Support for Graphical, Highly Interactive User Interfaces. *IEEE Computer*, 23(11):71–85, November 1990.

[14] G. Singh, C. H. Kok, T. Y. Ngan. Druid: A System for Demonstrational Rapid User Interface Development. In *Proceedings of the ACM SIGGRAPH Symposium on User Interface Software and Technology*, S. 167–177, 1990.

[15] R. B. Smith, D. Ungar, B.-W. Chang. The Use-Mention Perspective on Programming for the Interface. In B. A. Myers (Hrsg.), *Languages for Developing User Interfaces*, Kapitel 5. Jones and Bartlett Publishers, Boston - London, 1992.

[16] SunSoft. *Open Windows Developer's Guide 3.0, User's Guide*. Sun Microsystems, Inc., Mountain View, CA, November 1991.

[17] Symbolics, Inc., Cambridge, MA. *Symbolics Reference Manual: Programming the User Interface – Concepts*, Februar 1988.

[18] P. Szekely, P. Luo, R. Neches. Beyond Interface Builders: Model-Based Interface Tools. In *Human Factors in Computing Systems, INTERCHI '93 Conference Proceedings*, S. 383–390, 1993.

[19] B. F. Webster. *The NeXT Book*. Addison-Wesley Publishing Company, 1989.

Jürgen Herczeg
Forschungsgruppe DRUID
Universität Stuttgart, Institut für Informatik
Breitwiesenstr. 20-22, D-70565 Stuttgart
E-Mail: herczeg@informatik.uni-stuttgart.de

Ergonomische Gestaltung von Software auf Grundlage handlungsorientierter Fehleranalysen[1]

Andreas Kensik, Jochen Prümper, Michael Frese

DATA TRAIN GmbH, Dr. Prümper & Partner, Justus-Liebig-Universität Gießen

Zusammenfassung

Ausgehend von einer handlungsorientierten Fehlertaxonomie werden für einzelne Nutzungsprobleme in der Mensch-Computer Interaktion allgemeine Softwaregestaltungsmaßnahmen zur Fehlervermeidung und zum Fehlermanagement vorgestellt. Anschließend werden diese theoretischen Überlegungen im Rahmen der praktischen Überarbeitung einer Software in konkrete software-ergonomische Designvorschläge umgesetzt. Es zeigt sich, daß bei insgesamt 305 Nutzungsproblemen für 210 (68.9%) Fehlervermeidungs- und für 295 (96.7%) Fehlermanagementvorschläge entwickelt werden konnten. Systematische Fehleranalysen bieten damit nicht nur eine gute Möglichkeit zur Software-Evaluation, sondern in umfangreichem Maße auch Hinweise zur Software-Gestaltung.

1 Einleitung

Fehler und Probleme im Umgang mit der Software gehören zum Alltag jedes Anwenders. Etwa 10 Prozent ihrer Computerarbeitszeit verbringen Benutzer damit, aufgetretene Probleme zu bewältigen (Brodbeck, Zapf, Prümper & Frese, 1993). Damit verursachen Fehler nicht nur Streß (Frese, 1987), sondern auch hohe Kosten für die Unternehmen (Shneiderman, 1993) - sie stellen somit ein psychologisches als auch ein ökonomisches Problem dar.

Aus software-ergonomischer Sicht bedeuten Fehler damit für jeden Software-Entwickler eine große Herausforderung. So legt dann auch die Untersuchung von Booth (1990) nahe, daß systematische Fehleranalysen - umgesetzt zu konkreten Designvorschlägen - zu deutlichen Verbesserungen im Umgang mit Software führen können.

Zur systematischen Analyse von Fehlern wurde von Zapf, Brodbeck und Prümper (1989) eine *handlungstheoretische Fehlertaxonomie* vorgestellt. Zur Entwicklung

[1] Dieser Beitrag entstand im Rahmen des Projektes „VERLAG 2000: Eine benutzerfreundliche integrierte Lösung für die mittelständische Verlags- und Druckereibranche unter Berücksichtigung von zu verbessernden Arbeitsbedingungen für die Beschäftigten" der DATA TRAIN GmbH (einer Firma der SOBA Unternehmensgruppe). Dieses Projekt wird gefördert vom BMFT, Projektträger „Arbeit und Technik" (FKZ: 01 HK 601/8). Die in diesem Aufsatz beschriebene handlungstheoretische Fehlertaxonomie wurde in dem ebenfalls vom BMFT geförderten Arbeit und Technik-Projekt „FAUST: Fehleranalyse zur Untersuchung von Software und Training" entwickelt (FKZ: 01 HK 806/7). Wir danken Michaela Wagner und Dieter Zapf für ihre Anregungen.

konkreter Designvorschläge bietet sich eine Unterscheidung in Möglichkeiten zur *Fehlervermeidung* und Möglichkeiten zum *Fehlermanagement* an (Frese, 1991; Frese, Irmer & Prümper, 1991).

2   Theorie

2.1 Die handlungstheoretische Fehlertaxonomie

In der handlungstheoretischen Fehlertaxonomie werden Fehler als Nutzungsprobleme nach Schritten im *Handlungsprozeß* und nach *Regulationsebenen* aufgeschlüsselt (siehe Abb.1).

Regulationsgrundlage	Wissensfehler		
Regulationsebenen	**Schritte im Handlungsprozeß**		
	Ziele/Planung	**Gedächtnis**	**Rückmeldung**
Intellektuelle Regulationsebene	Denkfehler	Merk-/Vergessensfehler	Urteilsfehler
Ebene der flexiblen Handlungsmuster	Gewohnheitsfehler	Unterlassensfehler	Erkennensfehler
Sensumotorische Regulationsebene	Bewegungsfehler		

Abb. 1:   Handlungsorientierte Taxonomie der Nutzungsprobleme (aus: Zapf, Brodbeck & Prümper, 1989, S. 181)

Der *Handlungsprozeß* läßt sich grob so zusammenfassen, daß am Anfang der Handlung ein Ziel generiert und ein entsprechender Plan aufgestellt wird, der anschließend ausgeführt wird. Dazu werden die zuvor erstellten Pläne eine Zeitlang im Gedächtnis präsent gehalten, um sie dann bei der Handlungsrealisierung abzurufen. Am Schluß der Handlungen oder Teilhandlungen stehen dann jeweils Rückmeldungen, ob das angestrebte Ziel erreicht wurde oder nicht.

Die *Regulationsebenen* lassen sich unterteilen in die sensumotorische Regulationsebene, die Ebene der flexiblen Handlungsmuster und die intellektuelle Regulationsebene. Handlungen auf unterschiedlichen Regulationsebenen unterscheiden sich hinsichtlich ihres Geübtheits- und Automatisierungsgrades (Hacker, 1986). So wie Handlungen treten auch Fehler auf unterschiedlichen Regulationsebenen auf.

Zusätzlich zu den Ebenen der Handlungsregulation existiert die *Regulationsgrundlage*, die konkretes Wissen einer Person über die eigene Arbeitstätigkeit enthält.

2.2 Fehlervermeidung und Fehlermanagement

Grundsätzlich gibt es zwei Strategien, um mit dem Fehlerproblem umzugehen: *Fehlervermeidung* und *Fehlermanagement* (Frese, 1991; Frese, Irmer & Prümper, 1991). Bei der Strategie der Fehlervermeidung bemüht man sich, durch ein gutes Softwaredesign, die Fehleranzahl, bei der Strategie des Fehlermanagements, die negativen Fehlerauswirkungen zu reduzieren. Obwohl die *Strategie der Fehlervermeidung* im Rahmen einer Softwaregestaltung immer im Auge behalten werden muß, sollte sie nicht ausschließlich verwendet werden, da sie gerade aus arbeitswissenschaftlicher Sicht in drei Bereichen zu ungewollten Nebenwirkungen führen kann (vgl. Frese, 1991; Frese, Irmer & Prümper, 1991):

⟨ *Fehlervermeidung durch Arbeitsteilung:* Die Annahme, daß Fehler häufiger bei komplexen Tätigkeiten vorkommen, führt zu der Vorstellung, man könne Fehler durch Verringerung der Qualifikationsanforderungen vermeiden. Dies versucht man häufig durch verstärkte Arbeitsteilung zu erreichen. Möglicherweise führt diese Strategie tatsächlich zu einer Reduzierung von Fehlern, allerdings ist zu befürchten, daß dadurch gleichzeitig eine Verringerung der Arbeitsinhalte, der Handlungsspielräume, der Anforderungen an das eigenständige Denken und der Vollständigkeit und Sinnhaftigkeit der Arbeit bewirkt wird.

⟨ *Ironie der Automatisierung:* In einem Mensch-Maschine-System wird meistens der Mensch als das „fehleranfällige Subsystem" identifiziert. Als Konsequenz wird versucht, durch Automatisierung die Anteile des Menschen am (Produktions-) Prozeß gegenüber der Maschine zu reduzieren. Wenn die Maschinen aber versagen, ist es wieder der „fehlerbehaftete" Mensch, der dann einspringen muß. Nur ist er dann nicht mehr so qualifiziert, da er weitgehend aus dem Prozeß ausgeschaltet wurde, weniger direkte Erfahrungen mit dem System sammeln konnte. Bainbridge (1983) nennt dies „Ironie der Automatisierung".

⟨ *Dilemma der Fehlervermeidung:* Da letztlich immer Menschen die Arbeitsprozesse determinieren, werden auch stets Fehler auftauchen. Da Fehlervermeidung mit einer Verringerung der aktiven Handlungen korrespondiert, steigen die negativen Konsequenzen eines Fehlers mit dem Grad der versuchten Fehlervermeidung. Frese (1991) nennt dies das „Dilemma der Fehlervermeidungsstrategie".

Aufgrund dieser möglichen negativen Konsequenzen einer reinen Fehlervermeidungsstrategie, sollte stets auch die Möglichkeit des *Fehlermanagements* realisiert werden. Fehlermanagement bedeutet ein sinnvolles Herangehen an einen Fehler mit dem Ziel Folgefehler, negative Effekte und den Aufwand für die Fehlerbewältigung zu minimieren (Frese, Irmer & Prümper, 1991). Grundlegend sollte dieses Fehlermanagement im Rahmen des Fehlerprozesses bei der *Fehlerentdeckung*, der *Fehlererklärung* und der *Fehlerbehebung* stattfinden (Frese, 1991; Zapf, Frese, Irmer & Brodbeck, 1991).

Abbildung 2 zeigt Ansatzpunkte, wie sich Fehlermanagement und Fehlervermeidung im Rahmen der handlungsorientierten Fehlertaxonomie realisieren lassen.

Regulationsgrundlage	*Fehlervermeidung* •*Ein gutes konzeptionelles Modell liefern*: wähle Begriffe und Bildschirmsymbole aus dem Alltagsverständnis bzw. aus der Arbeitswelt des Benutzers. •*Dinge sichtbar machen*: mache die für die Aufgabenerledigung relevanten Informationen auf dem Bildschirm verfügbar. Lit.: Norman (1989); Norman & Lindsay (1981)		*Fehlermanagement* • Stelle diejenigen Informationen bereit, die zur Fortsetzung der Handlung benötigt werden. Werkzeuge, die dies unterstützen, sind: *Hilfesysteme, kontextsensitive Hilfefunktionen, Listings, Suchfunktionen, Apropos-Funktionen.* • Sorge für *gute Explorationsmöglichkeiten.* Lit.: Zapf et al. (1991)		
	Schritte im Handlungsprozeß					
Regulationsebenen	**Ziele/Planung**		**Gedächtnis**		**Rückmeldung**	
Intellektuelle Regulationsebene	*Fehlervermeidung* ·*Transparenz*: Mache die Systemlogik durchschaubar. ·*Konsistenz*: Gestalte Systemreaktionen nach dem *Prinzip der geringsten Verwunderung.* · Stelle alle zur Ziel- und Planentwicklung relevanten Informationen dar. Lit.: Frese (1991); Zapf et al. (1991)	*Fehlermanagement* · Ermögliche die *Rekonstrunktion des Handlungsverlaufs* durch Handlungsprotokolle oder History-Funktionen. · Ermögliche eine *Unterbrechung der Arbeitshandlung* und *Sicherung der Ergebnisse.* · Biete einen *Explorationsmodus.* · Erleichtere die *Navigation im System.* Lit.: Zapf et al. (1991); Frese & Brodbeck (1989)	*Fehlervermeidung* · *Erhaltung von Informationen*: Transportiere einmal eingegebene oder im System vorhandene Informationen mit (z.B. durch Zwischenspeicher, Kopier- oder elektronische Notizblockfunktionen). Lit.: Nielsen (1993)	*Fehlermanagement* · Ermögliche das leichte *Einfügen fehlender Teilhandlungen.* · Ermögliche eine einfache *Rückkehr zum Ausgangspunkt.* · Ermögliche eine *Unterbrechung der Arbeitsvorgänge* und *Sicherung der Arbeitsergebnisse.* Lit.: Zapf et al. (1991)	*Fehlervermeidung* · Gestalte *Fehler- und Systemmeldungen* verständlich. · Gib Meldungen als umittelbares Feedback auf die Handlungen der Benutzer. · Verwende eindeutig interpretierbare Piktogramme. Lit.: Shneiderman (1982, 1993); Zapf et al. (1991); Staufer (1987)	*Fehlermanagement* Gib Hinweise auf: · Fehlerursache, · Fehlerort und · das weitere Vorgehen in *System- und Fehlermeldungen.* Lit.: Shneiderman (1982, 1993), Zapf et al. (1991)

Ebene der flexiblen Handlungsmuster	*Fehlervermeidung*	*Fehlermanagement*	*Fehlervermeidung*	*Fehlermanagement*	*Fehlervermeidung*	*Fehlermanagement*
	· *Konsistenz*: Verwende Syntax von Kommandos, Grammatik und Benennungen konsistent und belege Tasten einheitlich. · *Erwartungskonformität*: Gestalte Dialoge entsprechend der Kenntnisse der bisherigen Arbeitsabläufe, der Ausbildung des Benutzers sowie allgemein anerkannter Übereinkünfte. Lit: Frese & Brodbeck (1989); ISO 9241/10 (1994)	· Ermögliche schnelles *Rückgängigmachen von falschen Handlungsschritten*. Nützlich sind Undo- und Redo-Funktionen, sowie Backspace- und Delete-Tasten. Lit.: Zapf et al. (1991); Dzida, Wiethoff & Arnold (1993)	· Lasse notwendige Handlungen soweit wie möglich vom System übernehmen. · Implementiere *Sicherheitsabfragen* an kritischen Stellen. · Blockiere fehlerhafte Handlungen durch „*Zwangsfunktionen*". Lit.: Norman (1989)	· *Fehlertoleranz*: ergänze und korrigiere unvollständige Eingaben durch das System. · Ermögliche eine *Unterbrechung der Arbeitshandlung* und *Sicherung der Ergebnisse*. · Ermögliche das *Zurückgehen in der Arbeitshandlung* und das *Einfügen unterlassener Handlungsschritte*. Lit.: Zapf et al. (1991); ISO 9241/10 (1994)	· Präsentiere *Systemmeldungen* lange genug, deutlich sichtbar und an einer standardisierten Position. Lit.: Herczeg (1994)	· *Monitoring der Handlung*: Erkenne Handlungen, die auf eine Systemmeldung hin unzulässig sind und gib konkrete Hinweise zum weiteren Vorgehen. Lit.: ISO 9241/10 (1994)
Sensumotorische Regulationsebene	*Fehlervermeidung* · Ziele darauf ab, unnötige Bewegungen zu minimieren: Ermögliche Auswahl aus Listen, belege Felder weitgehend vom System vor. · Schaffe durch die offensichtliche Position und Größe der Felder und Steuerlemente einen eindeutigen Ansprungspunkt für den Cursor. Lit.: Prümper, Heinbokel & Küting (1993)			*Fehlermanagement* · Unterstütze Fehlerentdeckung durch Vergleiche der Eingaben mit zugrundeliegenden Daten. Hierzu dienen Rechtschreibprogramme, Straßen-, Orts- und Telefonverzeichnisse, sowie Schlüsseltabellen. · Überprüfe, ob es sinnvoll ist, Korrekturen automatisch durchzuführen. Lit.: Zapf et al. (1991)		

Abb. 2: Fehlervermeidung und Fehlermanagement bei Nutzungsproblemen

3   Empirische Untersuchung

3.1 Methode

Untersucht wurde eine IBM AS/400-Adreßlösung eines mittelständischen Verlages. Dazu wurden vier Mitarbeitern dieses Verlages zehn Standardaufgaben zur Adreßerfassung vorgelegt, die sie mit ihrem vorhandenen Adreßprogramm bearbeiteten. Diese Aufgaben resultierten aus früheren Arbeitsplatzbeobachtungen. So sollten z.B. Firmenadressen mit abweichenden Liefer- und Rechnungsadressen oder Privatadressen mit temporärer Änderung der Adresse aufgrund eines Urlaubsaufenthalts des Kunden angelegt, Bankverbindungen oder Ansprechpartner hinzugefügt werden. Die Untersuchungen fanden an den Arbeitsplätzen der Benutzer statt.

Die vier Benutzer, drei Frauen und ein Mann, waren im Durchschnitt 31.5 Jahre alt, zwei von ihnen arbeiteten in der Vertriebs-, einer in der Anzeigen- und einer in der EDV/Rechnungswesenabteilung. Im Durchschnitt waren sie seit 10.9 Jahren in ihrem Unternehmen beschäftigt. Mit Computern arbeiteten die Anwender im Mittel seit 11.9 Jahren, mit der beurteilten Software durchschnittlich seit 6.0 Jahren. Pro Woche arbeiteten sie mit Computern im Schnitt 25.0 Stunden, mit der beurteilten Software 9.5 Stunden. Bei ihrer Arbeit verwendeten sie durchschnittlich 4.5 Programme, wovon 3.3 Großrechnerprogramme und 1.3 PC-Programme waren. Auf die Frage „Wie gut beherrschen Sie die beurteilte Software?", antworteten die Benutzer auf einer siebenstufigen Skala von „sehr schlecht" (1) bis „sehr gut" (7) im Mittel mit 4.5.

Während der Aufgabenbearbeitung wurde der Bildschirm der Anwender mit einer Videokamera aufgenommen. Anhand dieser Videoprotokolle wurden die aufgetretenen Fehler zunächst von zwei Beurteilern analysiert und gemeinsam den Kategorien der Fehlertaxonomie zugeordnet. Anschließend wurden die einzelnen Fehler daraufhin untersucht, inwieweit für sie konkrete Fehlervermeidungs- und Fehlermanagementstrategien entwickelt werden konnten.

3.2 Anzahl der Fehler

Bei den vier Benutzern wurden insgesamt 305 Nutzungsprobleme registriert. Bei einer Bearbeitungsdauer von 2-3 Stunden pro Benutzer entspricht dies ca. 30 Nutzungsproblemen pro Stunde. Abbildung 3 zeigt die Verteilung der Fehler über die acht Kategorien der Fehlertaxonomie.

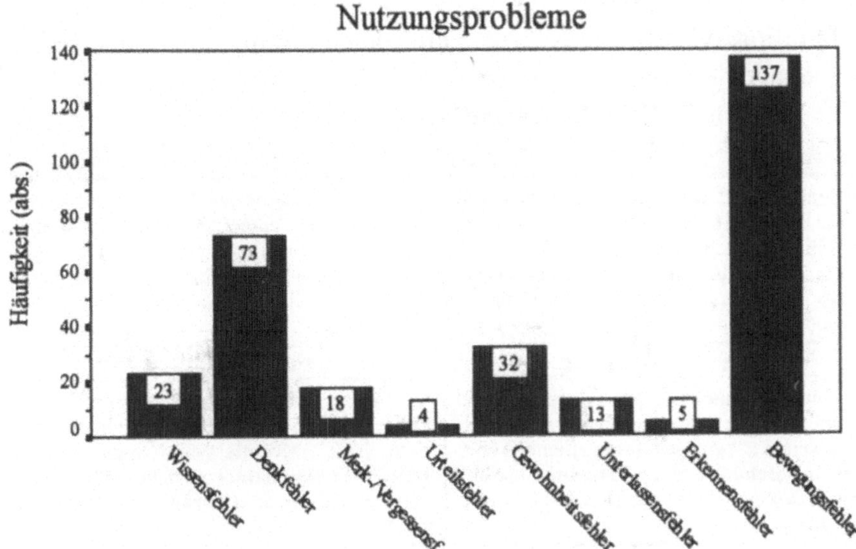

Abb. 3: Anzahl der Fehler für die einzelnen Kategorien

Zur Überprüfung, inwieweit die bei den vier Benutzern aufgetretenen Fehler repräsentativ sind, wurden die in den einzelnen Kategorien beobachteten Fehler mit den aufgetretenen Fehlern bei Zapf, Brodbeck, Frese, Peters und Prümper (1992) verglichen. Dort nahmen 259 Büroangestellte an der Untersuchung teil, 1306 Fehler flossen in die Auswertung ein. Es ergibt sich eine Korrelation von r=.87 (p<0,01, N=8 Kategorien). Trotz der geringen Anzahl der Versuchsteilnehmer entspricht die aufgetretene Fehlerverteilung damit der anderer Untersuchungen (für eine ausführliche Darstellung dieses Projektes vgl. Frese & Zapf, 1991).

4 Ergebnisse

In diesem Abschnitt werden zunächst Beispiele der aufgetretenen Fehler in den einzelnen Kategorien der Fehlertaxonomie dargestellt. Anschließend werden für jedes Fehlerbeispiel konkrete Fehlervermeidungs- und Fehlermanagementmöglichkeiten entwickelt. Da Softwareentwicklung ein konstruktiver Prozeß ist, in dem vielfach neue, kreative Problemlösungen gefragt sind (Keil-Slawik, 1990), handelt es sich bei diesen Vorschlägen zur Softwaregestaltung in erster Linie um Denkanstöße, und nicht etwa um „ausschließliche" Lösungen.

4.1 Qualitative Ergebnisse zur Fehlervermeidung und zum Fehlermanagement

4.1.1   Die Regulationsgrundlage

Wissensfehler
Der Benutzer soll zu einem vorgegebenen Banknamen die zugehörige Bankleitzahl suchen. Obwohl durch die Eingabe eines „?" in dem Feld 'BLZ' eine Maske aufgerufen werden kann, in der durch Eingabe des Banknamens und des Ortes die zugehörige Bankleitzahl ermittelt wird, holt der Benutzer sich ein Bankleitzahlenbuch und schlägt die entsprechende Zahl dort nach. Zu vermerken ist, daß das System keinen Hinweis darauf bietet, daß für das entsprechende Feld eine „?"-Funktion hinterlegt ist.

Fehlervermeidung	Fehlermanagement
Durch eine einheitlich farbliche oder graphische Hervorhebung der Felder, bei denen man durch Eingabe eines „?" Auswahllisten erhalten kann, hätte der Benutzer einen Hinweis darauf erhalten, daß eine Suche nach der Bankleitzahl möglich war.	In diesem Fall wäre ein Hilfesystem von Nutzen, das bei Aufruf ganz spezifisch über das Feld 'BLZ' informiert und angibt, welche weitere Funktionalität sich hinter dem Feld verbirgt und wie man Zugang zu ihr bekommt.

4.1.2   Die intellektuelle Regulationsebene

Denkfehler
Der Benutzer möchte eine Firmenadresse mit Ansprechpartner anlegen. Für die Eingabe der einzelnen Namensbestandteile sind das Feld 'Anredekennzeichen' sowie vier Namenszeilen vorgesehen, die jedoch nicht weiter strukturiert sind. Er trägt in das Feld 'Anredekennzeichen' den Schlüssel für „Herr" ein. In die erste Namenszeile trägt er den Namen der Firma „Deutsche Sterne-Abtlg. Raumfahrt" ein, in die zweite und dritte Namenszeile den Nachnamen und Vornamen des Ansprechpartners. In dem Feld 'Briefanrede' generiert das Programm „Sehr geehrter Herr Deutsche Sterne-Abtlg. Raumfahrt". Für den Benutzer ist nicht klar, welche Zeilen bei welchem Anredekennzeichen zur Konstruktion der Briefanrede herangezogen werden.

Fehlervermeidung	Fehlermanagement
Zur Fehlervermeidung wäre die Unterteilung der Namenszeilen in mehrere Felder sinnvoll, in die klar definierte Namensbestandteile, wie Nachname, Vorname, Titel, Firmenname usw. einzugeben sind. Das System setzt dann je nach gewähltem Anredekennzeichen die im Alltag gebräuchliche Briefanrede aus den erforderlichen Feldern zusammen.	Als Fehlermanagement dient die Rückmeldung der aus den Eingaben automatisch konstruierten Briefanrede. Diese Vorgabe kann manuell geändert werden. Selbstverständlich muß dafür gesorgt werden, daß die Briefanrede auch wahrgenommen wird, damit sie gegebenenfalls korrigiert werden kann (hier war dies häufig nicht der Fall).

<table>
<tr><td colspan="2" align="center">Merk-/Vergessensfehler</td></tr>
<tr><td colspan="2">Der Benutzer sucht eine Bankleitzahl. Als Suchbegriff gibt er den Banknamen ein. Daraufhin erhält er vom System eine Liste mit Filialen dieser Bank in ganz Deutschland. Er geht eine Maske zurück und gibt noch den Ortsnamen als zusätzlichen Suchbegriff ein.</td></tr>
<tr><td align="center">Fehlervermeidung</td><td align="center">Fehlermanagement</td></tr>
<tr><td>Vor der Eingabe der Bankverbindung hatte der Benutzer bereits die vollständige Adresse eingegeben. Das System hätte in diesem Fall den Wohnort in das Feld des Bankortes kopieren können, da in den meisten Fällen beide Orte identisch sind. Der Benutzer wäre in dem Fall nicht gezwungen gewesen, bereits eingegebene Informationen ein weiteres Mal zu erfassen. In den Fällen, in denen Wohn- und Bankort voneinander abweichen, hätte er das vorbelegte Feld einfach überschreiben können.</td><td>Im Sinne des Fehlermanagements konnte die fehlende Information nachgetragen werden. Die bereits eingegebene Information, der Bankname, blieb erhalten, und mußte nicht erneut erfaßt werden. Noch besser wäre die Positionierung des Cursors auf das Feld ‚Ort‘, in dem das System als Vorschlagswert den Ort aus der zuvor erfaßten Adresse einträgt. Der Benutzer hat dann die Möglichkeit mit diesem Vorschlagswert die Suche zu beginnen, oder einen abweichenden Ort einzugeben.</td></tr>
</table>

<table>
<tr><td colspan="2" align="center">Urteilsfehler</td></tr>
<tr><td colspan="2">Der Benutzer trägt in dem Feld ‚A-Sort-Begriff‘ „KÖNIG-VERLAG“ ein. Der Benutzer erhält die Fehlermeldung „Alpha-Sortierbegriff fehlt oder falsch (Keine Sonderzeichen + Umlaute)“. Er löscht alles hinter „KÖNIG“ und bestätigt. Daraufhin erhält er die gleiche Fehlermeldung. Er ändert den Umlaut.</td></tr>
<tr><td align="center">Fehlervermeidung</td><td align="center">Fehlermanagement</td></tr>
<tr><td>Der Benutzer interpretiert die erste Meldung so, daß der Bindestrich ein unerlaubtes Sonderzeichen sei und löscht es mit dem folgenden Wort. Der eigentliche Auslöser für die Meldung ist aber der Umlaut im ersten Wort. Um den Fehler zu vermeiden, wäre es nötig gewesen, daß in einer Software, die für den deutschen Sprachraum entwickelt wurde, auch Umlaute für Suchbegriffe zulässig sind.</td><td>Zur Vereinfachung des Fehlermanagements könnte der unzulässige Buchstabe beim Erscheinen der Fehlermeldung invertiert dargestellt werden, so daß dem Benutzer auf Anhieb die Fehlerursache klar wird. Auch sollte in der Fehlermeldung nicht der Hinweis auf einen möglicherweise fehlenden Sortierbegriff gegeben werden, da dies in diesem Fall nicht zutraf.</td></tr>
</table>

4.1.3　　　Die Ebene der flexiblen Handlungsmuster

Gewohnheitsfehler
Der Benutzer möchte den Cursor in das Feld ‚Herkunft Adresse' bewegen. Er betätigt mehrmals die RETURN-Taste, um von Feld zu Feld zu springen. Beim Überspringen des Feldes ‚Vorwahl' wird der dortige Feldinhalt gelöscht.

Fehlervermeidung	Fehlermanagement
Zur Vermeidung dieses Gewohnheitsfehlers sollte die RETURN-Taste nur eine Funktion erfüllen. Die „Doppelbelegung" - ‚Löschen bis Feldende' und ‚Cursorsprung zum nächsten Feld' - führt bei der Verwendung der Taste, mit dem Ziel nur eine der beiden Funktionen zu benutzen, entweder zu versehentlichem Löschen oder zu falscher Cursorpositionierung. Die Funktionen der RETURN-Taste sollten auf zwei Tasten verteilt werden.	Abhängig von der Fehlerentdeckung lassen sich z w e i u n t e r s c h i e d l i c h e Fehlermanagementstrategien realisieren. Bei Entdeckung des Fehlers durch den Benutzer wird der irrtümlich gelöschte Feldinhalt durch Aktivierung einer UNDO-Funktion wieder eingefügt. Entdeckt der Benutzer den Fehler nicht, so kann die Software den Benutzer bei Verlassen der Adresse durch eine Meldung auf das nicht gefüllte Feld aufmerksam machen.

Unterlassensfehler
Der Benutzer möchte aus dem Hauptmenü heraus eine neue Adresse anlegen. Er trägt in das Feld ‚Option' die Ziffer 1 für „Neuanlage" ein und bestätigt. Daraufhin erhält er die Meldung, daß eine Adresse mit der Kundennummer bereits existiert und nicht noch einmal angelegt werden kann. In dem Feld ‚Adreßnummer' stand noch die Nummer der zuvor bearbeiteten Adresse. Der Benutzer löscht die Nummer durch Betätigung der RETURN-Taste und gelangt so in die Maske zur Adreßerfassung.

Fehlervermeidung	Fehlermanagement
Die Fehlervermeidung kann dadurch erreicht werden, daß das System bei bereits vorhandener Kundennummer und gewählter Option für „Neuanlage" die Vorbelegungen des Feldes ‚Adreßnummer' ignoriert und direkt in die Maske zur Adreßerfassung verzweigt.	Fehlermanagement wird hier unterstützt, indem die Vorbelegung des Feldes ‚Adreßnummer' gelöscht werden kann. Besser wäre in der Meldung die Frage, ob die nächstmögliche „freie" Adreßnummer zur Bearbeitung freigegeben werden soll.

Erkennensfehler
Der Benutzer läßt sich eine Liste mit möglichen Adreßverzweigungen anzeigen. Er versucht in der Liste weiter nach unten zu blättern - es geschieht nichts. Wenn es noch weitere Einträge gibt, dann zeigt das System dies mit einem kleinen '+'-Zeichen in der rechten unteren Ecke der Liste an. In diesem Fall wurde kein Zeichen angezeigt, da die Liste keine weiteren Einträge enthielt.

Fehlervermeidung	**Fehlermanagement**
Zur Fehlervermeidung wäre es günstiger, wenn das System einen eindeutigen Hinweis auf das Ende der Liste geben würde, anstatt dies lediglich durch die Unterdrückung des ‚+‘-Zeichens zu induzieren.	In dem Moment, in dem der Benutzer versucht über das Ende der Liste hinauszublättern, sollte ihm eine Rückmeldung gegeben werden, daß das Ende der Liste bereits erreicht wurde.

4.1.4 Die sensumotorische Regulationsebene

Bewegungsfehler
Der Benutzer möchte in das Feld 'Vorwahl' „030" eintragen. Er gibt die Ziffern mit der numerischen Tastatur ein. Versehentlich drückt er die Komma-Taste, die neben der Ziffer 0 liegt, so daß in dem Feld „,30" steht.

Fehlervermeidung	**Fehlermanagement**
Da der Benutzer vorher bereits die Anschrift eingegeben hatte, hätte das Programm das Feld Vorwahl vorbelegen können, so daß der Benutzer hier überhaupt keine Eingabe mehr hätte tätigen müssen.	Es sollte darauf aufmerksam gemacht werden, daß in dem Feld 'Vorwahl' ein unzulässiges Zeichen eingegeben wurde, welches über ein Telefonverzeichnis im System automatisch korrigiert werden kann.

4.2 Quantitative Ergebnisse zur Fehlervermeidung und zum Fehlermanagement

Im Rahmen des Softwareentwicklungsprojektes „Verlag 2000" entwickelten die Autoren konkrete Fehlervermeidungs- und Fehlermanagementvorschläge zu den in den Felduntersuchungen beobachteten Nutzungsproblemen und meldeten diese an die verantwortlichen Softwareentwickler zurück. So erhielten die entsprechenden software-ergonomischen Empfehlungen Eingang in den iterativen Designprozeß der zu entwickelnden Verlagslösung (siehe auch: Frese, Prümper & Solzbacher, 1994; Prümper, 1993).

Von den 305 Nutzungsproblemen konnten für 210 (68,9%) Fehlervermeidungs- und für 295 (96,7%) Fehlermanagementvorschläge entwickelt werden. Aufgeschlüsselt nach einzelnen Fehlerkategorien ergeben sich folgende Fehlervermeidungs- (vgl. Abb. 4) und Fehlermanagementmöglichkeiten (vgl. Abb. 5).

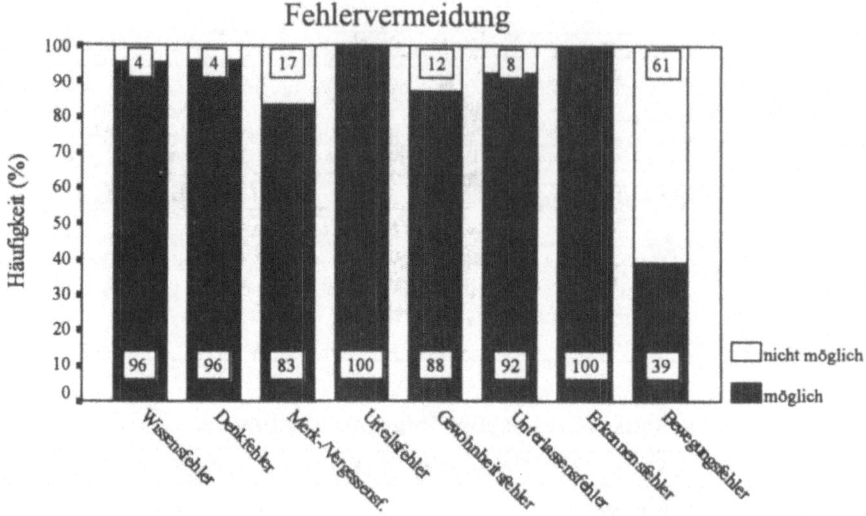

Abb. 4: Möglichkeiten zur Fehlervermeidung

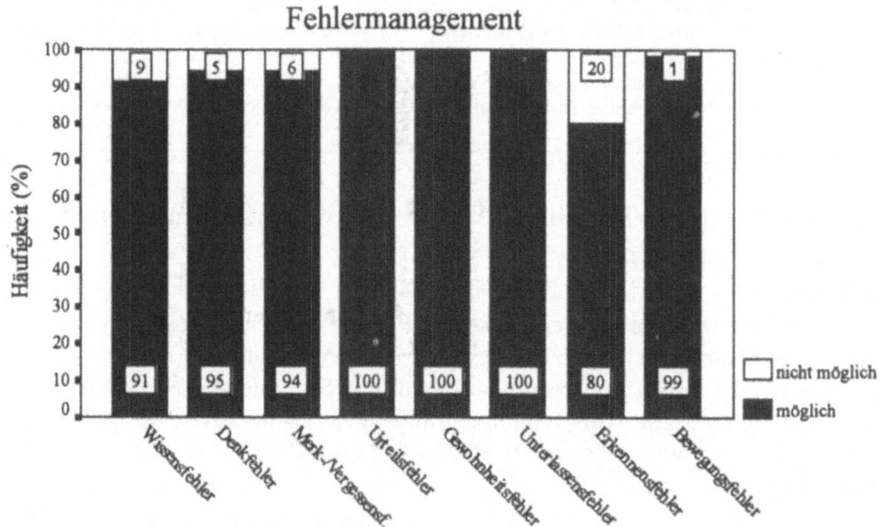

Abb. 5: Möglichkeiten zum Fehlermanagement

Auffällig ist die hohe Anzahl von Bewegungsfehlern, für die keine
Fehlervermeidungsmöglichkeiten generiert werden konnten. Gleichzeitig handelt es
sich in unserer Untersuchung - wie in den Untersuchungen von Zapf et al. (1992) -
bei den Bewegungsfehlern um die am häufigsten vorkommende Fehlerart (vgl. Abb.

3). Deshalb ist es hier besonders wichtig, Fehlermanagementmöglichkeiten zur Verfügung zu stellen (Zapf et al., 1991). Wie aus Abbildung 5 ersichtlich wird, konnte dies auch für 99% der Bewegungsfehler bewerkstelligt werden.

5   Fazit und Ausblick

Ausgehend von der handlungsorientierten Fehlertaxonomie von Zapf, Brodbeck und Prümper (1989) wurden für einzelne Nutzungsprobleme in der Mensch-Computer Interaktion allgemeine Softwaregestaltungsmaßnahmen zur Fehlervermeidung und zum Fehlermanagement vorgestellt. Anschließend wurden diese theoretischen Überlegungen im Rahmen der praktischen Überarbeitung eines Adreßmoduls für das mittelständische Verlagswesen in konkrete software-ergonomische Designvorschläge umgesetzt. Es zeigte sich, daß systematische Fehleranalysen eine fruchtbare, effektive Basis für konkrete Hinweise zur Softwaregestaltung liefern. In Untersuchungen zu weiteren Meßzeitpunkten soll festgestellt werden, wie diese Vorschläge von den Software-Entwicklern umgesetzt werden und ob sie tatsächlich zu gezielten software-ergonomischen Verbesserungen führen.

Deutlich wurde aber auch, daß nicht für alle Fehler ein Fehlervermeidungs- bzw. ein Fehlermanagementvorschlag gemacht werden konnte. Um festzustellen, was den Fehlern gemeinsam ist, für die keine Vorschläge generiert werden konnten - sei es nun auf der Fehlervermeidungs- oder der Fehlermanagementseite - bedarf es weiterer Analysen und Auswertungen.

Kritisch mag vielleicht angemerkt werden, daß in unserer Untersuchung mit lediglich vier Benutzern eine relativ kleine Stichprobe untersucht wurde. Es sollte jedoch aufgezeigt werden, daß systematische Fehleranalysen nicht an ökonomischen Beschränkungen scheitern müssen und ohne weiteres in jedem kommerziellen Softwareentwicklungsprojekt zum Einsatz kommen können. Der Umstand, daß bereits bei einem so kleinen Benutzerkreis eine recht große Anzahl von 305 Fehlern - d.h. ca. 30 Fehler pro Stunde und Benutzer - registriert werden konnte, mag in Anbetracht der Ergebnisse der Studie von Zapf et al. (1992) zunächst verwundern. Dort wurden lediglich ca. 5 Fehler in der Stunde pro Benutzer beobachtet (Brodbeck et al., 1993).

Dies liegt zum einen an der verwendeten Erhebungsmethode: Die Videoprotokollierung erweist sich hinsichtlich der Fehlerentdeckungsrate gegenüber der teilnehmenden Beobachtung als überlegen (Holz auf der Heide, 1993). Vor allem liegen die Gründe aber in der Art der von den jeweiligen Untersuchungspartnern zu bearbeitenden Arbeitsaufgaben. Während Zapf et al. (1992) die Probanden bei der Bewältigung ihrer individuellen, im Arbeitsalltag anfallenden Aufgaben beobachteten, wurden in der vorliegenden Studie Aufgaben vorgegeben. Diese Vorgabe von Aufgaben erleichtert zum einen dem Beobachter die Entdeckung und

Klassifikation der Fehler, da Ziele und Subziele des Benutzers offensichtlicher sind, birgt zum anderen aber prinzipiell die Gefahr in sich, daß möglicherweise für den einzelnen Arbeitsplatz untypische Arbeitsaufgaben gestellt werden, und damit Fehler zur Auswertung kommen, die in der alltäglichen Computernutzung nicht zu beobachten wären. Wie jedoch der korrelative Vergleich der vorliegenden Fehlerverteilung mit derjenigen der Studie von Zapf et al. (1992) zeigte, erweist sich die Fehlertaxonomie als ein robustes Verfahren, womit diese Sorge unbegründet erscheint.

Aufgrund dieser effektiven und ökonomischen Einsatzmöglichkeiten der handlungsorientierten Fehlertaxonomie empfiehlt sich dieses Instrument für den praktischen Einsatz in der Software-Ergonomie. Mit dem „Fehlertrainingsbogen für die Mensch-Computer Interaktion" von Prümper (1994) existiert zudem ein Verfahren, welches für „uneingeweihte Fehlerforscher" als Basisinstrument im Rahmen einer Beurteilerschulung für die handlungsorientierte Fehlertaxonomie dienen kann.

6 Literatur

[1]   Bainbridge, L. (1983). Ironies of automatization. *Automatica, 19*, 775-779.

[2]   Booth, P. A. (1990). ECM: An investigation of its use by designers. *International Journal of Human-Computer Interaction, 2,* 307-322.

[3]   Brodbeck, F. C., Zapf, D., Prümper, J. & Frese, M. (1993). Error handling in office work with computers: A field study. *Journal of occupational and organizational psychology, 66,* 303-317.

[4]   Dzida, W., Wiethoff, M., & Arnold, A. (1993). *ERGOguide. The qualitiy assurance guide to ergonomic software.* Sankt Augustin: Gesellschaft für Mathematik und Datenverarbeitung (GMD).

[5]   Frese, M. (1987). The industrial and organizational psychology of human-computer interaction in the office. In C. L. Cooper & I. T. Robertson (Eds.), *International review of industrial and organizational psychology* (S. 117-165). Chichester: Wiley.

[6]   Frese, M. (1991). Fehlermanagement: Konzeptionelle Überlegungen. In M. Frese & D. Zapf (Hrsg.), *Fehler bei der Arbeit mit dem Computer. Ergebnisse von Beobachtungen und Befragungen im Bürobereich* (S.139-150). Bern: Huber.

[7]   Frese, M. & Brodbeck, F. C. (1989). *Computer in Büro und Verwaltung. Psychologisches Wissen für die Praxis.* Berlin: Springer.

[8]   Frese, M., Irmer, C. & Prümper, J. (1991). Das Konzept Fehlermanagement: Eine Strategie des Umgangs mit Handlungsfehlern in der Mensch-Computer Interaktion. In M. Frese, C. Kasten, C. Skarpelis & B. Zang-Scheucher (Hrsg.), *Software für die Arbeit von morgen* (S. 241-251). Berlin, Heidelberg, New York: Springer.

[9]   Frese, M., Prümper, J. & Solzbacher, F. (1994). Eine Fallstudie zur Benutzerbeteiligung und Prototyping. In F.C. Brodbeck & M. Frese (Hrsg.), *Produktivität und Qualität in Software-Projekten. Psychologische Analyse und Optimierung von Arbeitsprozessen in der Software-Entwicklung* (S. 135-143). München: Oldenbourg.

[10] Frese, M., & Zapf, D. (1991). *Fehler bei der Arbeit mit dem Computer. Ergebnisse von Beobachtungen und Befragungen im Bürobereich.* Bern: Huber.

[11] Hacker, W. (1986). *Arbeitspsychologie.* Bern: Huber.

[12] Herczeg, M. (1994). *Software-Ergonomie. Grundlagen der Mensch-Computer-Kommunikation.* Bonn: Addison-Wesley.

[13] Holz auf der Heide, B. (1993). Welche software-ergonomischen Evaluationsverfahren können was leisten? In K.H. Rödiger (Hrsg.), *Software-Ergonomie '93 - Von der Benutzungsoberfläche zur Arbeitsgestaltung* (S. 157-171). Stuttgart: Teubner.

[14] ISO 9241/10 (1994). *Ergonomic requirements for office work with visual display terminals (VDTs) - Part 10: Dialogue principles.* Revised Draft International Standard.

[15] Keil-Slawik, R. (1990). *Konstruktives Design.* TU Berlin: Unveröff. Habilitationsschrift.

[16] Nielsen, J. (1993). *Usability engeneering.* Boston: AP Professional.

[17] Norman, D. A. (1989). *Dinge des Alltags. Gutes Design und Psychologie für Gebrauchsgegenstände.* Frankfurt, New York: Campus.

[18] Norman, D. A. & Lindsay, P. H. (1981). *Einführung in die Psychologie.* Berlin: Springer.

[19] Prümper, J. (1993). Benutzerorientierte, iterative Software-Entwicklung in der Praxis. In W. Coy, P. Gorny, I. Kopp & C. Skarpelis (Hrsg.), *Menschengerechte Software als Wettbewerbsfaktor* (S. 630-647). Stuttgart: Teubner.

[20] Prümper, J. (1994). *Fehlerbeurteilungen in der Mensch-Computer Interaktion. Reliabilitätsanalysen und Training einer handlungstheoretischen Fehlertaxonomie.* Münster: Waxmann.

[21] Prümper, J., Heinbokel, T., & Küting, H.J. (1993). Virtuelle Prototypen als Werkzeuge zur benutzerzentrierten Produktentwicklung. Anwendung einer handlungstheoretischen Fehlertaxonomie auf reale und virtuelle Oberflächen von Waschmaschinen. *Zeitschrift für Arbeitswissenschaft, 3*, 160-167.

[22] Shneiderman, B. (1982). System message design: Guidelines and experimental results. In: Badre, H. & Shneiderman, B. (Eds.). *Directions in human-computer interaction* (S. 55-78). Norwood, New Jersey: Ablex Publishing Co.

[23] Shneiderman, B. (1993). *Designing the user interface. Strategies for effective human/computer interaction.* Reading, MA: Addison-Wesley.

[24] Staufer, M. (1987). *Piktogramme für Computer.* Berlin: deGruyter.

[25] Zapf, D., Brodbeck, F. C., Frese, M., Peters, H. & Prümper, J. (1992). Errors in working with office computers: A first validation of a taxonomy for observed errors in a field setting. *International Journal of Human-Computer Interaction, 4*, 311-339.

[26] Zapf, D., Brodbeck, F. C., & Prümper, J. (1989). Handlungsorientierte Fehlertaxonomie in der Mensch-Computer Interaktion. Theoretische Überlegungen und eine erste Überprüfung im Rahmen einer Expertenbefragung. *Zeitschrift für Arbeits- und Organisationspsychologie, 33*, 178-187.

[27] Zapf, D., Frese, M., Irmer, C. & Brodbeck, F. C. (1991). Konsequenzen von Fehleranalysen für die Softwaregestaltung. In M. Frese & D. Zapf (Hrsg.), *Fehler bei der Arbeit mit dem Computer. Ergebnisse von Beobachtungen und Befragungen im Bürobereich* (S.177-191). Bern: Huber.

Dipl.-Psych. Andreas
Kensik
SOBA GmbH
Linzer Straße 21
53604 Bad Honnef

Dr. Jochen Prümper
Dr. Prümper & Partner
Holzhofstraße 8
81667 München

Prof. Dr. Michael Frese
Justus-Liebig-Universität
Otto-Behaghel-Straße 10
35394 Gießen

OBSM: Objektorientierte Modellierung von Benutzungsschnittstellen

Birgit Kneer
Universität - GH Paderborn
Heinz Nixdorf Institut
D-33098 Paderborn
`kne@hni.uni-paderborn.de`

Gerd Szwillus
Universität - GH Paderborn
Fachbereich Mathematik/Informatik
D-33098 Paderborn
`szwillus@uni-paderborn.de`

Zusammenfassung

Die Entwicklung von Benutzungsschnittstellen ist dadurch gekennzeichnet, daß mehrere qualitativ verschiedene Aspekte gleichzeitig oder in aufeinanderfolgenden Entwicklungsphasen Gegenstand der Betrachtung sind. Die Aspekte für sich werden jeweils von dafür entwickelten Spezifikationssprachen und -techniken unterstützt - eine gesamtheitliche Darstellung fehlt jedoch, so daß der Entwickler komplexe Transformationen zwischen verschiedenen Sichten durchführen muß. Wir schlagen eine visuelle Spezifikationstechnik vor, die verschiedene Aspekte des Entwurfs von Benutzungsschnittstellen in einer durchgängigen Notation vereinigt. Sie beruht im wesentlichen auf dem Paradigma der Objektorientierung, Constraints, der Darstellung von Interaktionssequenzen und temporalen Relationen.

1 Einleitung

Die Entwicklung von Benutzungsschnittstellen ist durch die Notwendigkeit gekennzeichnet, eine Vielzahl verschiedener Aspekte betrachten zu müssen. Diese sind bereits für sich genommen komplex, müssen aber auch in ihren gegenseitigen Abhängigkeiten stimmig behandelt werden. Diese Sichten betreffen qualitativ verschiedene Aspekte, die gleichzeitig oder aber in aufeinanderfolgenden Entwicklungsphasen zu behandeln sind (vgl. Abbildung 1).

(1)   Zunächst ist hier der komplexe, mehrschichtige Übergang von einer semi-formalen Beschreibung der Benutzeraufgaben (Aufgabenmodell, wie etwa in MAD [19]) zu einer Spezifikation des Verhaltens der Benutzungsschnittstelle (Dialogmodell) zu nennen. Diese Transformation beinhaltet den Übergang von der postulierten oder tatsächlichen Handlungsweise des Benutzers zum Verhalten eines Artefakts, der diese Benutzeraktionen ermöglicht, aufgreift und widerspiegelt.

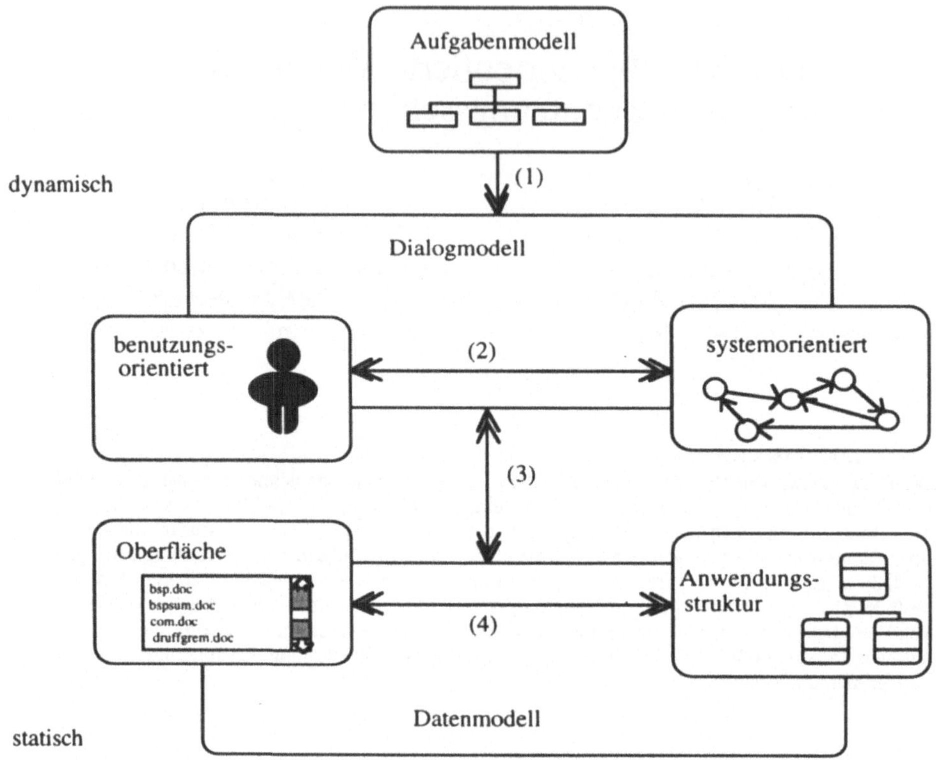

Abb. 1: Entwurfsaspekte von Benutzungsschnittstellen

(2) Innerhalb der Dialogmodellierung kann man eine benutzerorientierte
     Sichtweise (wie sie etwa mit UAN [7] beschreibbar ist) oder eine
     systemorientierte Sichtweise (zum Beispiel beschreibbar durch
     Transitionsnetze [8], DSN [13]) einnehmen. In der Sicht von außen sind die
     Benutzeraktionen, ihre Reihenfolge und die erfahrbaren Rückmeldungen
     Entwurfsgegenstand; in der Systemsicht geht es um die Gestaltung der
     Struktur der Abarbeitung dieser Interaktionen und ihrer Abbildung auf
     Softwarekomponenten (Objekte).

(3) Die Betrachtung der Softwarekomponenten einer Benutzungsschnittstelle
     muß einerseits mit Bezug auf die Datenstrukturen (z.B. Objekte, Klassen,
     Verbünde) geschehen (statische Sicht), andererseits aber auch das
     dynamische Verhalten dieser Komponenten berücksichtigen. Dieses generelle
     Grundproblem von Software-Entwicklung bzw. Programmierung wird bei
     Benutzungsschnittstellen durch den hohen Grad an Dynamik durch die
     Benutzerinteraktion verschärft.

(4)   Schließlich stellt die Kooperation der Benutzungsschnittstelle mit der Applikation ein klassisches Problem des Entwurfs von Benutzungsschnittstellen dar. Die gewünschte Trennung und gleichzeitige enge Kopplung der beiden Anteile motivierte zunächst das Seeheim-Modell [15] und führte mittlerweile zu Ansätzen kooperierender Objektmodelle [3].

Für einzelne dieser Aspekte existieren jeweils Notationen und Spezifikationsmethoden, sowie zum Teil Werkzeuge zum effizienten Arbeiten mit diesen Notationen. Problematisch bleiben jedoch stets die Übergänge zwischen verschiedenen Methoden, da sie ohne Unterstützung etabliert und überprüft werden müssen.

Wir schlagen die visuelle Spezifikationstechnik OBSM (Objektorientierte Benutzungsschnittstellen-Modellierung) vor, die wenige, kompakt notierbare Sprachmittel anbietet, mit denen die angesprochenen Übergänge fließend mit einer durchgängigen Notation behandelt werden können. Sie erlaubt es, Beziehungen herzustellen zwischen

- Benutzeraufgaben (user tasks) und Interaktionsfolgen aus Benutzersicht,

- Interaktionsfolgen und Abarbeitung von Ereignissen,

- Reaktion auf Ereignissen und den Objektstrukturen der Benutzungsschnittstelle sowie

- der Objektwelt der Benutzungsschnittstelle und der Applikation

und diese Beziehungen in einer konsistenten Darstellung zu repräsentieren. Dadurch ermöglicht OBSM die übergreifende Behandlung aller Aspekte in einem gemeinsamen Rahmen, wodurch die gegenseitigen Abhängigkeiten deutlich gemacht und zum expliziten Entwurfsgegenstand werden.

OBSM wurde in einer Diplomarbeit [10] aufbauend auf Ideen der Modellierung objektorientierter Strukturen in Benutzungsschnittstellen [20] entwickelt. Wesentlich beruht es auf dem Konzept der Objektorientierung, angereichert um constraint-basierte Konstrukte, die zur deklarativen Beschreibung von Zusammenhängen verschiedenster Art eingesetzt werden. Zusätzlich werden unter anderem Symbole für Aufgaben, ihre zeitlichen Beziehungen (temporale Relationen) und Interaktion verwendet.

Das Papier behandelt im folgenden Kapitel genauer die oben skizzierten Probleme der Übergänge zwischen verschiedenen Entwurfsaspekten, beschreibt im dritten

Kapitel die Grundzüge der OBSM-Notation und behandelt in einem kurzen Ausblick die Arbeiten an der Werkzeugunterstützung und geplanten Erweiterungen.

2 Entwurfsaspekte von Benutzungsschnittstellen

2.1 Vom Aufgabenmodell zum Dialogmodell

Das in der Aufgabenanalyse entstehende Aufgabenmodell ist eine Beschreibung dessen, wie der Anwender seine Tätigkeiten ausführt bzw. wie die einzelnen Aufgaben gemeinsam von ihm und dem System zu erfüllen sind. Die hierbei zwischen Mensch und Maschine auszutauschenden Informationseinheiten und die Struktur der wechselseitigen Kommunikation wird im Dialogmodell festgehalten.

Beide Modelle werden durch unterschiedliche Methoden unterstützt. Techniken zur Aufgabenmodellierung (wie etwa das GOMS-Modell [2]) sind explizit für diese und nicht im Hinblick auf die Dialogmodellierung entwickelt worden. Vielfach sind die durch eine Benutzeraktion ausgelöste Systemreaktion nicht darstellbar und es wird von fehlerfreien, nicht-unterbrechbaren Abläufen ausgegangen. In der Phase des Dialogentwurfs werden daher zwangsläufig andere Modelle eingesetzt, was einen Wechsel in der Vorgehens- und Darstellungsweise bedingt. Typisch ist hier der Einsatz von Transitionsnetzen [8], [22], ereignisorientierten Spezifikationssprachen [6], Grammatiken [4] oder regelbasierten Notationen [13].

Der Übergang vom Aufgaben- zum Dialogmodell wird methodisch nur geringfügig unterstützt; vielfach beschränken sich die entsprechenden Techniken auf Kriterien, anhand derer, ausgehend vom Aufgabenmodell, erste Hinweise für die Gestaltung der Benutzungsschnittstelle ableitbar sind (MAD, [19]). Auch beim Aufgaben-Kontext-Modell [12] läßt sich trotz spezieller Ausrichtung auf die Dialogerstellung dessen systemorientierte Spezifikation, die mit Produktionsregeln erfolgt, nicht unmittelbar aus dem Aufgabenmodell ableiten.

Objektorientierten Methoden (OOM) mangelt es allgemein an Konzepten zur aufgaben- und benutzungsorientierten Entwicklung. Diese Gesichtspunkte gehen in einigen Ansätzen (OOSE [9], OBA [18]) nur indirekt über die Modellierung von Szenarien mit ein, in denen die Verwendung der einzelnen Objekte dargestellt wird. Durch das Prinzip der Objektorientierung in Analyse und Design wird zwar der Übergang zwischen diesen Phasen vereinfacht, jedoch müssen OOM für eine benutzungsorientierte Entwicklung durch entsprechende Techniken ergänzt werden.

2.2   Dialogmodellierung: Benutzer- und Systemsicht

Da ursprünglich die Benutzungsschnittstellen von den Entwicklern der Anwendungs-Software mit entworfen wurden, sind die etablierten Techniken zur Dialogmodellierung systemorientiert, d.h. sie beschreiben die Interaktionen aus Systemsicht. Zu den klassischen Modellen zählen hier die schon erwähnten kontextfreien Grammatiken, Zustands-Übergangs-Diagramme sowie ereignisbasierte Systeme. Diese Techniken unterstützen das Design und die Implementierung der Benutzungsschnittstellen-Software, nicht aber die adäquate Modellierung von Interaktionsfolgen des Benutzers mit der Schnittstelle, dessen Sichtweise durch die Durchführung einer Aufgabe geprägt ist.

Demgegenüber ist die UAN-Notation [7] eigens für ein aufgaben- und benutzerorientiertes Design entwickelt worden. Eine UAN-Spezifikation enthält jedoch keine Aussagen darüber, welche Komponenten der Software für die Verarbeitung der einzelnen Interaktionen verantwortlich sind. Sie muß durch ein vom Entwickler durchzuführendes systemorientiertes Design ergänzt werden, da dieses nicht direkt ableitbar ist.

2.3   Dynamische und statische Sicht

Bei der Entwicklung der Software von Benutzungsschnittstellen sind neben der Beschreibung des Verhaltens (Dynamik) die Datenstrukturen (statische Sicht) zu definieren, die durch die Interaktionen erzeugt, zerstört bzw. modifiziert werden. Es wird allgemein als ein schwieriges Problem angesehen, dynamische und statische Aspekte in einer einzigen konsistenten Darstellung (z.B. einem Diagramm) unter der Vorgabe zu repräsentieren, daß beide Sichtweisen eindeutig und klar daraus hervorgehen und korrekt in Beziehung gesetzt sind*.

Für die Darstellung der statischen bzw. dynamischen Aspekte werden oft unterschiedliche Notationen benutzt. Zur Beschreibung der Dynamik werden etwa in objektorientierten Modellen systemorientierte Verhaltensbeschreibungen eingesetzt, die in das eigentliche Objektverhalten zu übersetzen sind ([1], [17], [21]). Der Einsatz der Diagramme als Kommunikationsmittel ist erschwert, da die

---

*)   Beispielsweise wurden in SADT (Structured Analysis and Design Technique, [16]) getrennte Modelle zur Beschreibung von Aktivitäten (dynamisches Modell) und Daten (statisches Modell) aufgestellt. Es ist weder den Erfindern noch anderen gelungen, klar zu definieren, wann die beiden SADT-Modelle zu einem Problem konsistent zu nennen sind; letztlich wurde durch die für die Zeit typische Dominanz der funktionalen Sicht das Datenmodell meist weggelassen.

Zusammenhänge zwischen Ereignissen und Datenänderungen oft Diagrammen entnommen werden müssen, die sich keiner einheitlichen Darstellungsart bedienen und daher ineinander transformiert werden müssen. Dieser Übersetzungsprozeß ist von den Benutzern der Notationen zu leisten.

Ziel einer Notation sollte es jedoch sein, den kognitiv-mentalen Aufwand des Benutzers einer Beschreibungstechnik möglichst gering zu halten, damit dieser sich auf die eigentlichen Informationen konzentrieren kann. In diesem Sinne kann auch von der Forderung nach "Direktheit" für Notationen gesprochen werden.

2.4   Objektmodell der Anwendung und der Oberfläche

Graphische Benutzungsschnittstellen (*graphical user interfaces, GUIs*) werden technisch als Hierarchie graphischer Objekte implementiert, die von sog. Toolkits wie OSF/Motif [14], XView [5] und InterViews [11] angeboten werden. Die Applikation (AP) einer Anwendungs-Software weist in der Regel eine Struktur auf, die sich wesentlich von der Struktur des GUI unterscheidet. Die Kopplung beider Komponenten ist somit durch eine i.a. nicht-triviale Abbildung charakterisiert. Vielfach sind nicht nur die Daten oder Objekte der AP selbst, sondern auch deren Beziehungen zueinander durch Objekte des GUI zu repräsentieren. Beispielsweise können die Vorgänger- bzw. Nachfolgerbeziehungen in einer Baumstruktur innerhalb der AP mit Zeigern realisiert und in der Oberfläche als Pfeile dargestellt werden.

3   Modellierung mit der OBSM-Notation

Mit der OBSM-Notation werden die im vorangegangenen Kapitel erläuterten Aspekte einschließlich der Übergänge zwischen diesen unterstützt. OBSM ist eine graphische Notation, die für den praktischen Einsatz werkzeugunterstützt sein muß. Dies erlaubt dem Notations-Benutzer, den Informationsgehalt der Darstellungen durch "Ein- und Ausblenden" selbst zu bestimmen, d.h. er kann sich gezielt Diagrammkomponenten anzeigen lassen. In diesem Sinne werden in den folgenden Beispieldiagrammen immer nur die in einem bestimmten Kontext relevanten Informationen dargestellt.

3.1   Aufgabenmodellierung

Die graphische Darstellung eines Aufgabenmodells umfaßt im allgemeinen die Aufgaben, deren hierarchische Struktur sowie die Reihenfolge ihrer Bearbeitung.

Die OBSM-Notation kann zur Visualisierung dieser Gesichtspunkte eingesetzt werden und bietet darüber hinaus Sprachelemente, mittels derer die zeitlichen Beziehungen der Ausführungssequenzen sowie Ausnahme- und Fehlersituationen ausgedrückt werden können. Daneben besteht die Möglichkeit, Systemaufgaben zu integrieren sowie zwischen benutzer- und systembedingten Verzweigungen zu unterscheiden. Bezüge zum Objektmodell erlauben es, bereits auf dieser Ebene im Aufgabenkontext relevante Objekte einzubinden. Durch die Vielschichtigkeit ihrer Ausdrucksmöglichkeiten, von denen einige im folgenden anhand eines Beispiels vorgestellt werden sollen, kann die OBSM-Notation als Darstellungstechnik für vorhandene Methoden der Aufgabenanalyse eingesetzt werden.

Als Beispiel soll die Verwaltung einer Telefonliste dienen. Abbildung 2 zeigt die Aufgabe **Telefonliste verwalten** und ihre Verfeinerung in **Eintrag verwalten** mit den Unteraufgaben **Eintrag lesen**, **Eintrag ändern** sowie **Eintrag löschen**. Die hierarchische Aufgabengliederung wird durch Ineinanderschachtelung der entsprechenden Rechtecksymbole repräsentiert. Diese Darstellungsweise ist ähnlich der für Klassen- und Objekten, bei denen ebenfalls verfeinernde Komponenten im Klassen- bzw. Objektrechteck enthalten sind und die Spezifikationsgenauigkeit mit der Schachtelungtiefe zunimmt (s. Abschnitt 3.2).

Die zwischen den einzelnen Aufgaben bestehenden zeitlichen Beziehungen werden mittels temporaler Relationen modelliert. In der Telefonliste sollen beliebig oft einzelne Einträge verwaltet werden können (Wiederholungspfeil). Die Aufgabe **Eintrag verwalten** selbst besteht aus den sich zeitlich ausschließenden Unteraufgaben zum Lesen, Ändern oder Löschen eines Eintrags (Auswahlsymbol). Der Benutzer soll also genau eine dieser Teilaufgabe bearbeiten und sich anschließend für die Aufgabe **Eintrag verwalten** erneut entscheiden können. Die Unteraufgabe **Eintrag löschen** ist hier im Gegensatz zu den anderen nicht unterbrechbar (durchgezogene Rechtecklinie des Aufgabensymbols).

Generell werden mit unterbrochenen Linien
die Stellen markiert, an denen andere
Aufgaben eingeschoben werden dürfen. Zur
Bestimmung derjenigen Aufgaben, die dann
als Unterbrechung auftreten dürfen, werden
neben den Unterbrechungssymbolen die Positionen in der Aufgabenhierarchie ausge wer tet
[10]. Auf diesen Beschrei bungsmechanismus
selbst soll hier nicht eingegangen werden.

Objektintegration

Aufgaben stehen in engem
Zusammenhang mit den Objekten, auf
die sie sich beziehen. Insbesondere
beziehen sich die Aufgaben, die in
Kooperation mit einer Anwendung
erledigt werden, auf Objekte des
Softwaresystems. Daher ist es
naheliegend, diesen Zusammenhang
auch im Modell explizit darzustellen.
Dabei bleibt es dem Anwender der
OBSM-Notation überlassen, bis zu
welchem Grad er seinem
Aufgabenmodell Objekte hinzufügen möchte - dies erlaubt einen fließenden
Übergang einer benutzungsorientierten zu einer systemorientierten Sicht.

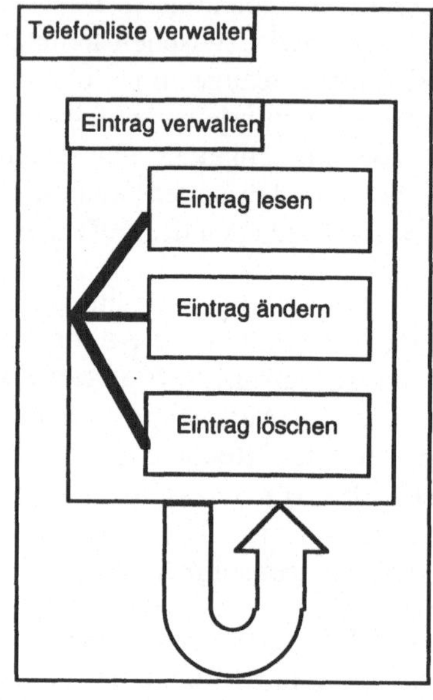

Abb. 2: Aufgabenmodell

Abbildung 3 zeigt die Verfeinerung der Aufgabe **Eintrag löschen**. In ihrer ersten
Unteraufgabe soll zunächst ein Eintrag ausgewählt und anschließend entfernt
werden* . In der Anwendung ist ein Objekt notwendig, das diese Funktionalität
durch eine geeignete Methode unterstützt. Derartige Zusammenhänge, d.h. die
Bearbeitung einer Aufgabe und der damit verbundene Aufruf einer Software-
Prozedur, werden durch einen "Blitz" notiert. Dieser dient als generelles
Gestaltungsmittel zur visuellen Hervorhebung der Benutzeraktivitäten, die quasi
von außen auf das Softwaresystems "treffen". Dementsprechend zeigt im
Diagramm ein Blitz von der Aufgabe **entfernen** auf die durch das Dreieck

---

*)   Zur Festlegung dieser zeitlichen Beziehung zeigt ein Sequenzpfeil von der ersten zur
nachgelagerten Aufgabe.

symbolisiert Methode **l ö s c h e n** des Anwendungsobjektes **Telefonliste** und stellt so deren Aufruf durch eine Benutzeraktion dar. Die Parameter, die hierbei übergeben werden (im Beispiel **Name**), legen fest, welche Eingaben vom Benutzer erwartet werden und somit Bestandteil seiner Aufgaben sind.

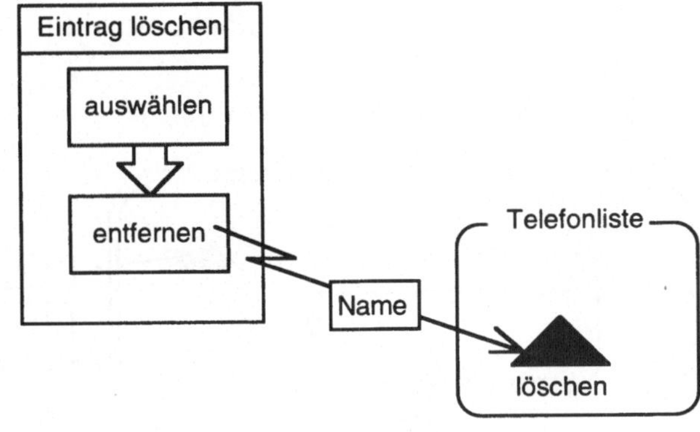

Abb. 3: Objektanbindung

Mit fortlaufender Verfeinerung werden zunehmend Objekte integriert. Es kristallisieren sich auf diese Weise diejenigen heraus, die von Benutzer und Anwendung gemeinsam benutzt werden und es entsteht parallel zum Aufgabenmodell der sogenannte Objektkern ([20]). Dieser enthält die in der Oberfläche sichtbar zu machenden Objekte sowie die Methoden, die dem Benutzer zugänglich sein müssen. Er bildet die Basis für die anschließende Software-Entwicklung, in der die Benutzungsschnittstelle und die Applikation weitgehend unabhängig voneinander erstellt werden können.

Statische und dynamische Sicht

Implizit wird in der obigen Darstellung auf zwei (konzeptionellen) Zeichenflächen agiert: auf einer "statischen" und einer "dynamischen" Zeichenfläche* . Bei der Modellierung der Dynamik steht die Fläche der Symbole für die zeitliche Dimension, d.h. für die Zeit, die eine Aufgabe benötigt oder die durch eine temporale Relation beschrieben wird. So legt der Pfeil für die Sequenz nicht nur die Reihenfolge der Vorgänge fest, sondern repräsentiert auch die Zeit zwischen den Vorgängen. Das Symbol nimmt Platz ein, "verbraucht" also Zeit. Auf der anderen Zeichenfläche steht der Raum, den ein Objekt belegt, für den Speicherplatz, den es zur Laufzeit einnimmt. Die flächige Darstellung von Klassen und Objekten erlaubt es, die jeweils zugehörigen Attribute, Methoden und Constraints graphisch innerhalb der Symbole darzustellen (s. Abschnitt 3.2).

---

*) Zur übersichtlicheren Darstellung werden hier in den Diagrammen das Modell der Dynamik auf der linken und statische Aspekte auf der rechten Seite gezeichnet.

3.2 Anwendungs- und Oberflächenstruktur

3.2.1 Objekte des Kerns und der Oberfläche

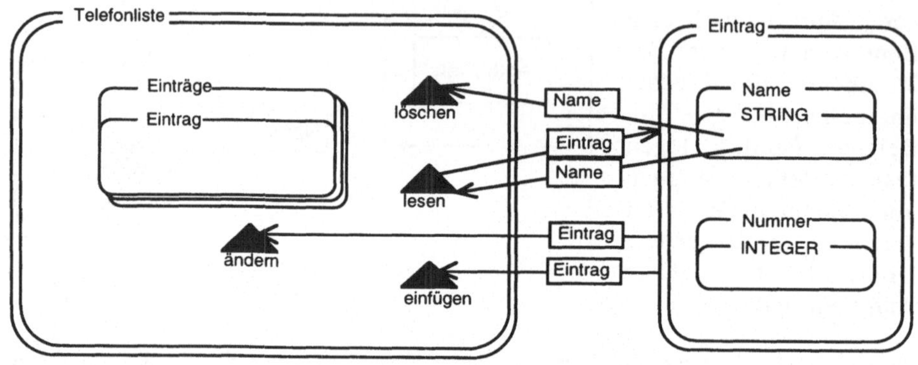

Abb. 4: Klassen des Objektkerns

Den Objektkern für das obige Beispiel zeigt Abbildung 4. Im Diagramm der Klasse Telefonliste - zur Unterscheidung von Objekten wird das Klassensymbol mit einer Doppellinie gezeichnet - werden die einzelnen Listeneinträge in der Objektmenge Einträge zusammengefaßt. Ihre Elemente sind Objekte der Klasse Eintrag und besitzen alle ein Attribut Name vom Typ STRING und ein INTEGER-Attribut Nummer. Zum Zugriff auf Einträge stehen die Methoden lesen, einfügen, ändern und löschen zur Verfügung. Die zugehörigen Parameter (Eintrag, Name) werden durch Pfeile mit den Methodendreiecken verbunden. Hierbei legt die Pfeilrichtung den jeweiligen Inhalt als Ein- oder Ausgabewert fest, die Verlängerung des Pfeiles zum Klassensymbol Eintrag bzw. Name definiert hingegen den Parametertyp.

Mit der Bestimmung der Kernobjekte ist noch nicht festgelegt, wie diese dem Benutzer dargestellt werden sollen. Dies ist Gegenstand der Entwicklung der Benutzungsschnittstelle, d.h. mit ihrer Gestaltung sind u. a. die Objekte zur graphischen Repräsentation der Kernobjekte

Abb. 5: BO-Darstellung

in der Oberfläche festzulegen (BO-Objekte). Im Beispiel soll jeder Eintrag der Telefonliste gemäß Abbildung 5 angezeigt werden. Die notwendigen Vereinbarungen enthält die Klasse BO-Eintrag (Abbildung 6).

Die Klasse enthält Objekte, die den am Bildschirm dargestellten Text (Name und Nummer) und das Telefonsymbol (Telefon) repräsentieren. TEXT ist ein graphisches Element, das - im Gegensatz zum STRING - Textattribute zur graphischen Darstellung festlegt, also zum Beispiel den Font, die Größe und die *Bounding Box* (**BBox**). **ICON** ist ebenfalls vordefiniert und repräsentiert Bitmuster. Das Gleichheitsconstraint ⬡ bestimmt, daß beide Texte übereinander positioniert werden sollen; das **LINKS-VON**-Constraint definiert, daß das Telefonsymbol links von beiden Texten stehen soll. Die Constraints werden in dieser und den folgenden Abbildungen nur in abstrakter Form verwendet. Analog zur Vorgehensweise bei der Beschreibung von Aufgaben und Klassen bzw. Objekten werden sie innerhalb ihres Symbols spezifiziert. Hierbei werden sie mit Durchgriff auf die Elementarkomponenten der verknüpften Objekte durch Verbindung einfacher Constraints hierarchisch aufgebaut. Letztlich werden atomare Constraints durch mathematische Beziehungen zwischen Zahlen, z.B. x- und y-Koordinaten, festgelegt. Auf diesen Mechanismus soll hier ebenfalls nicht näher eingegangen werden.

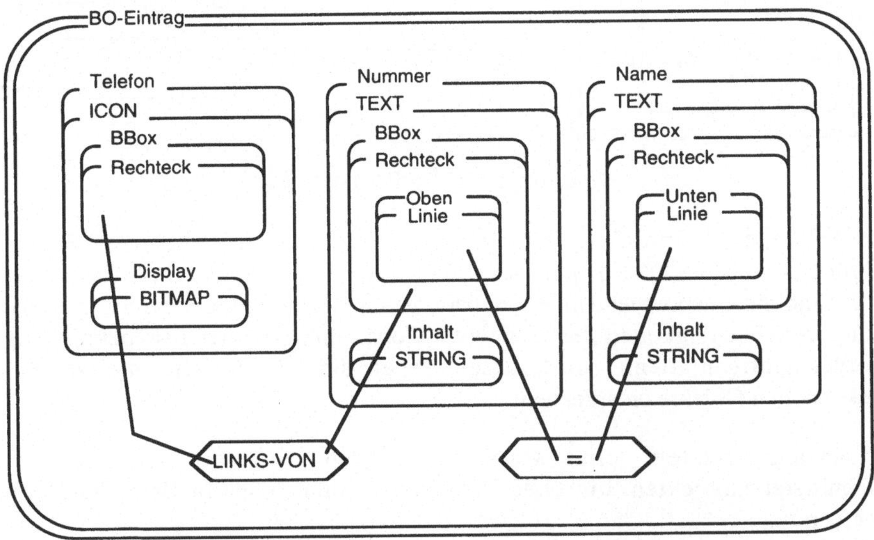

Abb. 6: Objekt zur graphischen Repräsentation des Kernobjektes Eintrag

3.2.2   *Kopplung der Kern- und Oberflächenobjekte*

Objekte des Kerns und der Oberfläche können auf vielfältige Weise miteinander gekoppelt sein. In unserem Beispiel soll zur Laufzeit der Anwendung für jedes Kernobjekt Eintrag ein Oberflächenobjekt **BO-Eintrag** existieren. Diese Forderung wird durch ein sog. Existenz-Constraint (in Abbildung 7 durch **EX** gekennzeichnet) realisiert.

Der Objektrand wird in diesem Zusammenhang als boolescher Wahrheitswert interpretiert (true: Objekt existiert, false: Objekt existiert nicht), was durch die andersartige Pfeilspitze zusätzlich verdeutlicht wird. Das Instanziieren und Löschen der Einträge im Kern und in der BS erfolgt somit in wechselseitiger Abhängigkeit (Instanzkopplung), wobei dann jeweils Name und Nummer in beiden Objektstrukturen übereinstimmen müssen.

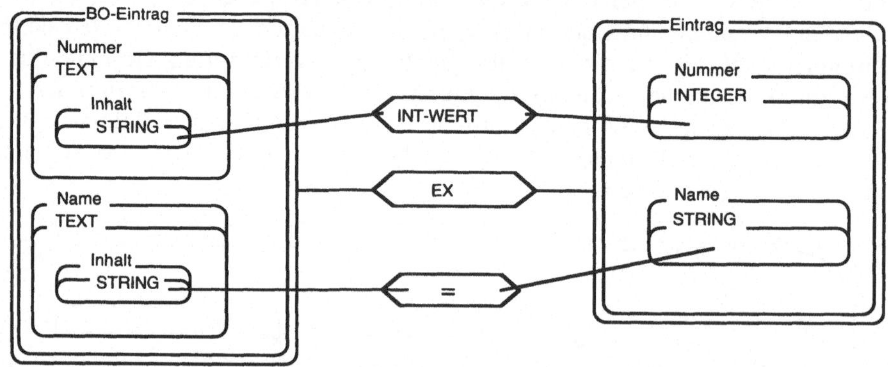

Abb. 7: Instanzkopplung: Eintrag und BO-Eintrag

In Abbildung 7 sind diese Bedingungen für jeweils ein beliebiges Objekt aus Eintrag und BO-Eintrag spezifiziert. Da dies für alle Objekte der Klassen gelten soll, sind die entsprechenden Vereinbarungen auf Klassenebene zu treffen. Hierbei sind die Abhängigkeiten der Gleichheit-Constraints von der Instanzkopplung zu berücksichtigen, denn zur Laufzeit sollen sich immer nur die Werte der gekoppelten Objekte entsprechen.

Insgesamt entsteht eine stark verflechtete Struktur aus Kern- und Schnittstellenobjekten, die über Constraints miteinander in Beziehung gesetzt werden.

3.3   Interaktionssequenzen

Bisher wurden das Aufgabenmodell und der Objektkern mit dessen Repräsentationen in der Oberfläche festgelegt. Die Querbezüge zwischen den Aufgaben und den Kernobjekten markieren hierbei die Stellen, an denen das Softwaresystem die einzelnen Tätigkeiten des Benutzers unterstützen soll. Es werden so die Aufgabenkontexte bestimmt, in denen Kernmethoden aufrufbar sein sollen; offen ist hingegen, wie der Benutzer deren Aktivierung veranlassen kann.

Dies ist Gegenstand der Benutzerinteraktionsfolgen, in denen das Interaktionsverhalten zwischen Benutzer und System modelliert wird.

Das folgende Diagramm (Abbildung 8) zeigt ein Beispiel für die Interaktion innerhalb der Aufgabe **auswählen** (vgl. Abbildung 3): Der Benutzer soll durch das Drücken einer Maustaste (**MDown**) über dem zugehörigen Oberflächenobjekt (**BO-Eintrag**) einen Eintrag selektieren können.

Der Blitz symbolisiert in den Interaktionssequenzen Benutzeraktionen, die außerhalb des Softwaresystems ausgelöste Ereignisse darstellen und immer einem Objekt zugeordnet werden. Dieser Darstellungsweise liegt die Idee zugrunde, daß jedes (gültige) Ereignis von einem oder mehreren

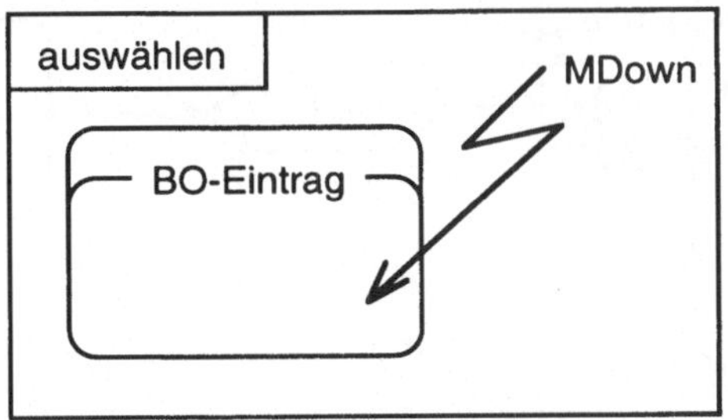

Abb. 8: Verfeinerung der Aufgabe **auswählen**

Objekten verarbeitet wird. Gerade in graphischen Oberflächen bezieht sich eine Benutzeraktion fast ausschließlich auf ein (auf dem Bildschirm sichtbares) Objekt. So können bei der Modellierung des Benutzerverhaltens in vielen Fällen die Aktionen zusammen mit dem jeweils betroffenen BO-Objekt dargestellt werden. Ist dies nicht möglich, wie bei manchen Tastaturkommandos, so ist auf jeden Fall ein Objekt in der Benutzungsschnittstelle, das nicht BO-Objekt ist, für das Ereignis zuständig. Mit zunehmendem Detaillierungsgrad des Modells wird festgehalten, welche Objekte die verschiedenen Ereignisse verarbeiten.

Abbildung 9 zeigt eine mögliche Verfeinerung der kompletten Aufgabe **Eintrag löschen**: Nach der Auswahl des BO-Repräsentanten eines Telefoneintrags soll durch dessen Verschieben in einen Papierkorb die Löschmethode aktiviert werden. Dies wird durch ein wiederholtes **MDMove** (Mausbewegung mit gedrückter Taste) "auf" **BO-Eintrag** modelliert. Die Übereinstimmung zwischen dem zuvor ausgewählten und jetzt zu verschiebenden Eintrags wird durch den Parameter E an der Stelle des Objektnamens hergestellt. Für die Dauer dieser Aktionen wird das Constraint **AN** aktiviert, wodurch das Piktogramm den Mausbewegungen folgt. Da der bisherige Mauszeiger nun nicht mehr angezeigt werden soll, wird seinem Sichtbarkeitsattribut der Wert *false* zugewiesen. Wird unter diesen Vorbedingungen die Maustaste über dem Papierkorb losgelassen, soll der Eintrag aus der Objektmenge **Einträge** von **Telefonliste** entfernt werden.

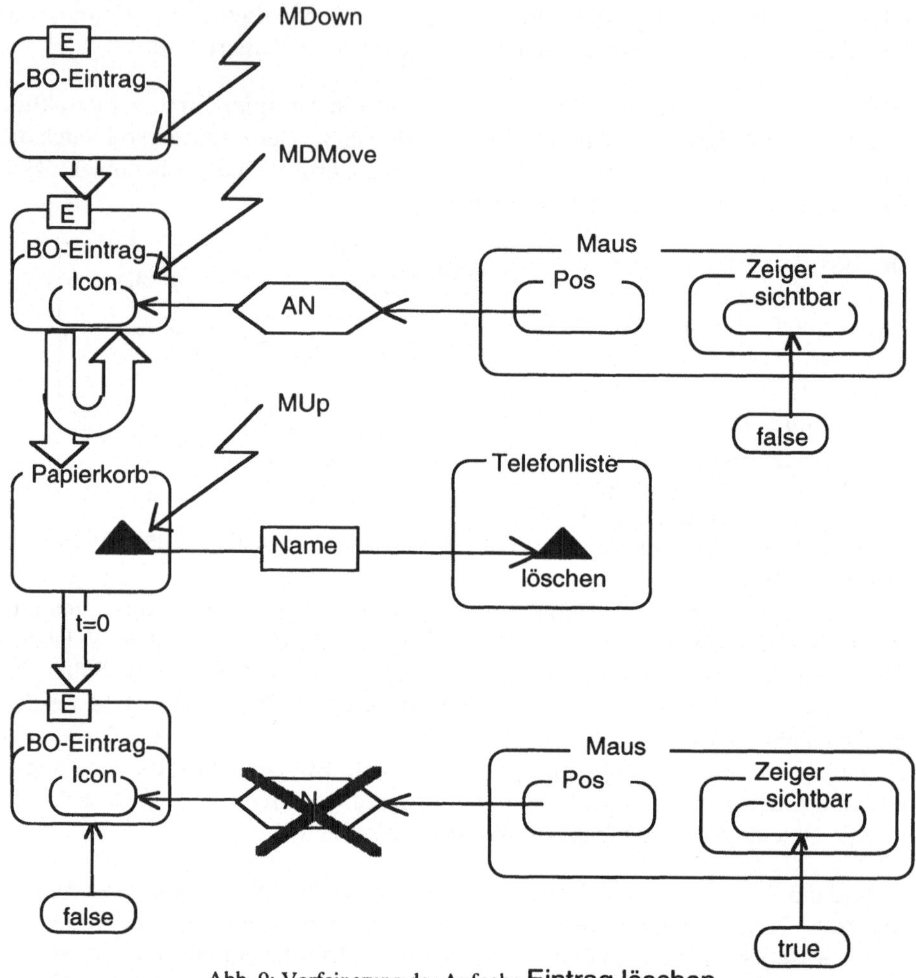

Abb. 9: Verfeinerung der Aufgabe **Eintrag löschen**

An dieser Stelle zeigt sich exemplarisch der Übergang von der abstrakten
Darstellung eines Methodenaufrufs innerhalb einer Aufgabe (Abbildung 3) zur
Spezifikation eines programmtechnischen Ereignisses. Die Aktivierung von
löschen folgt also nicht direkt aus einem Benutzerereignis, sondern durch ein in
einer anderen Methode generiertes (System-) Ereignis. Zur klaren Unterscheidung
der Systemereignisse von den Benutzeraktionen werden erstere mit einem
einfachen Pfeil dargestellt.

Direkt anschließend (Zusatzspezifikation t=0 am Sequenzpfeil) soll das Oberflächenobjekt entfernt werden. Zur Modellierung wird dem hier ebenfalls als Wahrheitswert interpretierten Objektrand der Wert *false* zugewiesen. Die Instanzkopplung bewirkt an dieser Stelle auch das Löschen des zugehörigen Kernobjektes. Die Aufgabe löschen ist mit dem Deaktivieren des Constraints und dem Sichtbarmachen des ursprünglichen Mauszeigers beendet.

Die Modellierung der Aufgaben und somit auch die der Interaktionsfolgen befindet sich immer auf der dynamischen Zeichenfläche, während bezüglich der angebundenen Objekte die statische und die dynamische Sicht möglich sind. Werden sie zur dynamischen Zeichenfläche gezählt, so wird neben deren aktueller Verwendung gezeigt, in welchem Zustand sie sich befinden müssen, d.h. auf welche Werte innerhalb eines Kontextes einzelne Attribute gesetzt sein müssen. Dabei werden jeweils nur die relevanten Attribute gezeichnet. Die in einer Situation nicht dargestellten Attribute dürfen beliebige Werte haben.

Mit der Interpretation als statische Zeichenfläche werden die Auswirkungen der Benutzeraktionen auf das Objektmodell betrachtet. Mit dieser Sichtweise stellen die Querbezüge zwischen Aufgabenmodell und Objektkern gleichzeitig die Wechselwirkungen zwischen dynamischer und statischer Zeichenfläche her.

In den Interaktionsfolgen bedingt der zunehmende Einsatz der Objektsymbole auf der dynamischen Zeichenfläche eine noch stärkere Vermischung mit der statischen Sicht. Auf beiden Zeichenflächen werden für Objekte, Ereignisse, Methoden und Constraints jeweils die selben Symbole verwendet. Hierdurch soll der Übergang zwischen beiden Sichtweisen erleichtert werden, da die Darstellungen nicht zwischen den verschiedenen Zeichenflächen übersetzt werden müssen.

3.4   Aspekte des systemorientierten Modells

Für eine vollständige Modellierung müssen mit den Benutzerinteraktionsfolgen alle möglichen Zweige der Benutzungsschnittstelle erfaßt werden. Aus den verschiedenen Interaktionssequenzen können wesentliche Aussagen zum Objektmodell abgeleitet werden: So legen die Spezifikationen zu einer Objektart an verschiedenen Stellen dar, welche Komponenten die Klasse enthalten und welche Ereignisse sie verarbeiten muß, wobei letzteres Hinweise auf die in ihr zu implementierenden Methoden gibt. Die Gesamtheit aller Benutzerinteraktionsfolgen beschreibt das Verhalten der Benutzungsschnittstelle. Für eine Überprüfung des Modells und insbesondere für eine Simulation der verschiedenen Abläufe ist es erforderlich, nach jedem Dialogschritt den Standpunkt des Benutzers (Zustand des Dialogs) und die möglichen Folgeaktionen vollständig darzustellen.

Für diesen Aspekt bieten Zustands-Überführungs-Diagramme eine vorteilhafte Darstellung. Für jeden Zustand zeigt der ihn repräsentierende Knoten anhand der ausgehenden Kanten alle in ihm möglichen Folgeaktionen, ankommende Kanten stellen die Aktionen dar, mit denen der Zustand erreicht wurde. Hiermit verglichen bilden die verschiedenen Benutzerinteraktionsfolgen einzelne Pfade durch einen derartige Graphen und zeigen meist nur eine Teilmenge der Alternativen.

Zur Gewinnung der Zustandsinformationen aus den Benutzerinteraktionsfolgen wurde ein System dynamischer Kellern entwickelt, auf dessen Grundlage Zustands-Überführungs-Diagramme abgeleitet werden können [10]. Das Kellerverhalten ist direkt aus den Interaktionsfolgen generierbar. Eine wesentliche Aufgabe ist hierbei die Umsetzung der temporalen Relationen, da sie zusammen mit der hierarchischen Aufgabengliederung ausschlaggebend für die zulässigen Folgeaktionen sind.

Jede mögliche Aufgabe bzw. Aktion wird durch ein Kellersymbol repräsentiert. Die Gesamtheit der jeweils obersten Elemente gibt die aktuellen Handlungsalternativen des Benutzers wieder. Abbildung 10 zeigt dies für sechs aufeinanderfolgende Kellerzustände der Telefonlistenverwaltung und eines Hilfesystem (in der Abbildung mit H gekennzeichnet).

Die Entscheidung des Benutzers für die Aufgabe **Telefonliste verwalten** führt zur Ersetzung des zugehörigen Zeichens gemäß ihrer Unterstruktur. Die unterste Kellerzeile steht hier für die Möglichkeit, die Aufgabe erneut auszuführen, wobei der Leerkeller die Alternative keiner erneuten Wiederholung symbolisiert.

Die Verwaltung eines Eintrags besteht aus dem Lesen, Ändern oder Löschen. Die Auswahl einer dieser Aufgaben, im Beispiel **Eintrag löschen**, führt zu einer erneuten Ersetzung und zum Entfernen der Lese- und Ändern-Kellerzeichen, da die zugehörigen Aufgaben nicht mehr als Alternativen zur Verfügung stehen. Ebenso kann der Benutzer an dieser Stelle das Hilfesystem nicht mehr aufrufen, da **Eintrag löschen** in Abbildung 2 als nicht-unterbrechbar vereinbart wurde. Der Keller des Hilfesystems wird daher durch den Querstrich als oberstes Symbol als gesperrt gekennzeichnet. Dies gilt solange, bis die als nicht unterbrechbar gekennzeichnete Aufgabe beendet ist und damit der Keller wieder "geöffnet" wird.

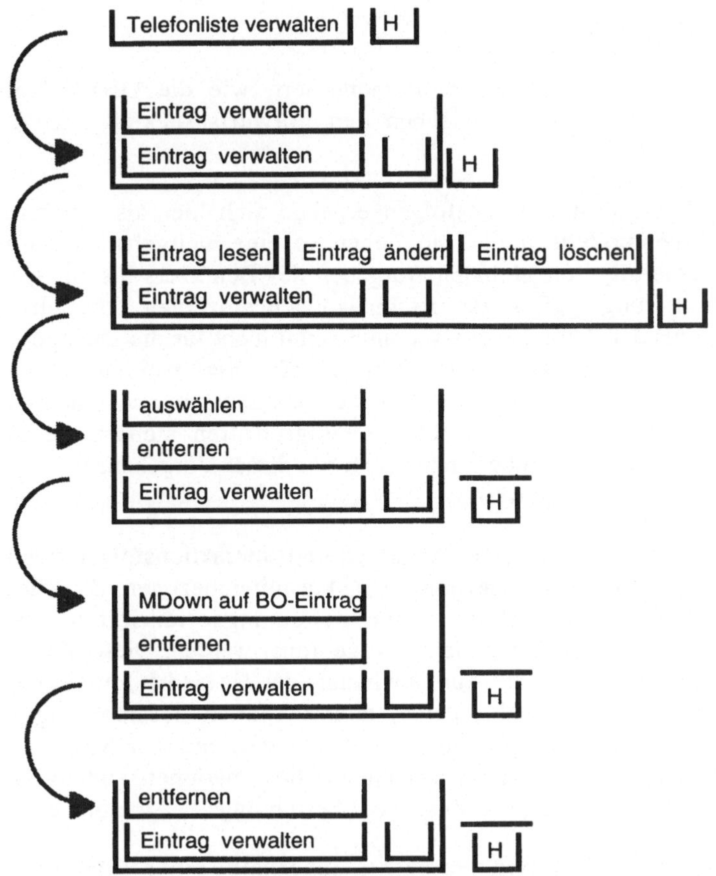

Abb. 10: Beispiel eines Kellersystems

Mit der Bearbeitung der Aufgabe auswählen bzw. der Aktion MDown wird der Keller, da er keine weitere Unteraufgabe enthält, wieder entfernt und als nächstes kann die Löschoperation durchgeführt werden.

Diese am Beispiel gezeigte Darstellung läßt sich aus beliebigen Interaktionssequenzen erzeugen, womit eine OBSM-Spezifikation simulierbar wird.

3.5   Zusammenfassung

Anhand eines Beispiels wurde demonstriert, wie die OBSM-Notation die verschiedenen Übergänge zwischen den Entwurfsaspekten (vgl. Kapitel 2) ermöglicht:

(1)   Die Benutzerinteraktionsfolgen ergeben sich hier als Verfeinerung des Aufgabenmodells und ermöglichen so eine weitgehend kontinuierliche Modellierung von globalen Aufgaben des Benutzers bis zur gewünschten Detaillierung, ggf. sogar bis hin zu den einzelnen Tastendrücken. Das Ergebnis der Aufgabenanalyse muß somit nicht für die Dialogbeschreibung in eine andere Darstellungsform transformiert werden. Dadurch ist es möglich, innerhalb einer Notation die Konsistenz zwischen den modellierten Aufgaben und den einzelnen Dialogschritten sicherzustellen und zu dokumentieren. Darüberhinaus kann mit dieser Vorgehensweise eine leichte Änderbarkeit erzielt werde.

(2)   Bei der Modellierung der Aufgaben und Interaktionsfolgen werden die im jeweiligen Kontext relevanten Objekte miteinbezogen, d.h. die einzelnen Aktionen können bereits im benutzungsorientierten Design den verantwortlichen Komponenten zugeordnet werden. Diese Vorgehensweise und das formale System zur Gewinnung der Zustandsinformationen erlauben es, wesentliche Aspekte der systemorientierten Beschreibung zu folgern, bzw. im Zusammenhang mit der benutzungsorientierten Sicht zu entwerfen. Daneben sind aus den vorhandenen Informationen andere "klassische" Darstellungsweisen (z.B. Zustands-Überführungs-Diagrammme) ableitbar.

(3)   Die OBSM-Notation versucht, innerhalb einer Darstellung durch Unterscheidung zweier "virtueller Zeichenflächen" eine Brücke zwischen dynamischer und statischer Sicht zu bauen. Für die darstellungsrelevanten Komponenten werden auf den sich einander ergänzenden Zeichenflächen jeweils gleiche Symbole mit gleicher Semantik eingesetzt, die in ihrer Gesamtheit ein logisches Diagramm der Benutzungsschnittstelle bilden. Die hierbei bewußt in Kauf genommene Vermischung der ohnehin nicht voneinander unabhängigen statischen und dynamischen Aspekte erlaubt es dem Notations-Benutzer, dieselben Diagramme aus verschiedenen Blickwinkeln zu betrachten. Zusätzlich bewirken die Querbezüge zwischen Aufgabenmodell und Objektkern sowie zwischen Benutzerinteraktionsfolgen und Objekten eine enge Kopplung dieser Perspektiven. Aufgaben und Ereignisse stehen also wie in der realen Welt in direktem Zusammenhang mit den beteiligten Objekten, so daß man von einer frühen, bereits auf Aufgabenebene erfolgenden, Objektanbindung sprechen kann. Gleichzeitig

wird durch die graphischen Bezüge dokumentiert, für welche Aufgaben die jeweiligen Komponenten des Objektkerns entwickelt wurden.

(4)   Das Objektmodell der Anwendung und der Oberfläche ergeben sich durch sukzessive Integration der Objekt bei der Aufgaben- und Interaktionsmodellierung wobei die Kopplung zwischen GUI und AP mittels Constraints definiert wird. Einen speziellen Einsatz von Constraints bildet die Instanzkopplung, die es erlaubt, mittels Constraints auch die Existenz eines Objektes von einem Datum oder der Existenz eines anderen Objektes abhängig zu machen.

Insgesamt hat der Benutzer von OBSM einen großen Freiraum bei der Modellierung, da die Notation von sich aus nahezu keine Abgrenzungen vornimmt. Prinzipiell stehen ihm in allen Arbeitsphasen sämtliche Darstellungstechniken zur Verfügung. So können Informationen zu den Zeitpunkten des dynamischen Entwicklungsablaufs festgehalten werden, an denen sie anfallen und relevant sind, und müssen nicht aufgrund einer Trennung von Phasen oder Modellen in ihrer Erfassung verlagert werden.

4   Ausblick

Für den praktischen Einsatz der OBSM-Notation ist eine Werkzeuglandschaft zur Unterstützung der Modellierungsprinzipien notwendig. Eine wesentliche Aufgabe ist die Bereitstellung eines graphischen Editors, zu dem bereits ein Prototyp existiert. In einer Folgeversion soll es u. a. möglich sein, reale Oberflächenobjekte, d.h. deren graphischen Darstellungen, in die Modellierung miteinzubeziehen.

Weitgehende Hilfe kann zudem bei der Hervorhebung und Generierung verschiedener Sichten geleistet werden. Bei der Beschreibung einer BS entsteht konzeptionell ein großes Diagramm, dessen Informationen je nach Betrachtungsstandpunkt mehr oder weniger relevant sind. Dem Werkzeug-Benutzer müssen geeignete Mechanismen zum Ein- und Ausblenden verschiedener Modellkomponenten geboten werden, so daß er die Bildschirmausgabe weitgehend seinem eigenen Informationsbedürfnis anpassen kann. Hierbei können Abstraktionen wie z.B. Piktogramme und vereinfachte Darstellungen diese Vorgehensweise wesentlich unterstützen.

Neben der Bereitstellung verschiedener Sichten auf das Gesamtdiagramm können Simulationen einen wesentlichen Beitrag zum korrekten Verständnis des Gesamtsystems und der dynamischen Abläufe leisten. Für das zu entwickelnde

Werkzeug bietet sich eine visuelle Simu-lation an, die auf der OBSM-Notation basiert und so die Abläufe anhand der erstellten Diagramme zeigt. Für Beschreibung und Simulation existiert dann ein einziges Modell, wodurch kein Mehraufwand für die Erstellung eines separaten Simulationsmodells zu leisten ist. Zudem wird auf diese Weise erneut ein mentaler Übersetzungsaufwand bei den Benutzern vermieden, da die Diagramme selbst animiert werden.

Die Simulation kann eine bereits während der Modellierung stattfindende Verifikation des zukünftigen Systemverhaltens bieten und zur zusätzlichen Konsistenzprüfung und Absicherung der vorliegenden Ergebnisse führen, so daß sich ein iteratives Vorgehen beim Entwurf bis hin zum Prototypen ergibt.

Daneben soll eine möglichst automatisiert Generierung des systemorientierten Modells aus der benutzerorientierten Darstellung bis hin zum Programmgenerator realisiert werden. Denkbar ist hier, die Komponenten der statischen und der systemorientierten Sicht möglichst weitgehend aus den Benutzerinteraktionsfolgen abzuleiten.

5　　Literatur

[1]　　Booch, G.: *Object Oriented Design with Applications,* Benjamin/Cummings, Redwood City, Ca, 1991

[2]　　Card, S.; Moran, T.; Newell, A.: *The Psychology of Human-Computer Interaction,* Lawrence Erlbaum Associates, Hillsdale, New Jersey, 1983

[3]　　Dance, J.; Granor, T.; Hill, R.; Hudson, S.; Meads, J.; Myers, B.; Schulert, A.: *The Run-time Structure of UIMS-Supported Applications,* Computer Graphics, Vol. 21, No. 2, 1987

[4]　　Green, M.: *A Survey of Three Dialogue Models,* ACM Transaction on Graphics, July 1986

[5]　　Heller, D.: *XView Programming Manual - The Definitive Guide to the X Window System* (Vol. VII), O'Reilly & Associates, Sebastopol (CA), 1989

[6]　　Hill, R.: *Event-Response Systems - A Technique for Specifying Multi-Threaded Dialogues,* Proceedings of CHI'87, Toronto, April 1987

[7]　　Hix, D.; Hartson, R.: *Developing User Interfaces: Ensuring Usability through Product & Process,* John Wiley & Sons, New York, 1993

[8]   Jacob, R.: *A State Transition Diagram Language for Visual Prgramming,*
      IEEE Computer, August 1985

[9]   Jacobson, I.; Christerson, M.; Jonsson, P.; Övergaard, G.: *Object-Oriented
      Software Engineering: A Use Case Driven Approach,* Addison-Wesley, 1992

[10] Kneer, B.: *Visuelle Modellierung von Bnutzungsschnittstellen
      objektorientierter Applikationen,* Diplomarbeit, Universität - GH Paderborn,
      Fachbereich Mathematik/Informatik, März 1994

[11] Linton, M.; Vlissides, J.; Calder, P.: *Composing User Interfaces with
      InterViews,* Stanford University, 1990

[12] Monk, A.; Curry, M.; Maidment, B.: *Task-Based Interface Specification,*
      University of York, Technical Report, 1991

[13] Monk, A.; Curry, M.: *Dialog Specification as a Link between Task Analysis
      and Implementation,* University of York, Technical Report, 1992

[14] Open Software Foundation:, *OSF/Motif Programmers Guide,* Prentice-Hall,
      Englewood Cliffs (NJ), 1990

[15] Pfaff, G.: *User Interface Management Systems,* Springer Verlag, 1985

[16] Ross, D.: *Structured Analysis (SA): A Language for Communication Ideas,*
      IEEE Trans-actions on Software Engineering, SE-3, Januar 1977

[17] Rumbaugh, J.; Blaha, M.; Premerlani, W.; Eddy, F.; Lorensen, W.: *Object-
      Oriented Modeling and Design,* Prentice Hall, Englewood Cliffs, New Jersey,
      1991

[18] Rubin, K.; Goldberg, A.: *Object Behavior Analysis,* Communications of the
      ACM, 35(9), 1992

[19] Sebillotte, S.: *Task Analysis and Formalisation according to MAD:
      Hierarchical Task Analysis, Method of Data Gathering and Examples of Task
      Descriptions,* Technischer Report, Amsterdam, 1992

[20] Szwillus, G.: *Visuelle Spezifikation des Kerns objektorientierter
      Applikationen,* GI Informatik aktuell: Informatik - Wirtschaft - Gesellschaft,
      GI-Jahrestagung '93, Springer-Verlag, Berlin, 1993

[21] Walters, N.: *Using Harel Statecharts to Model Object-Oriented Behavior,*
      ACM SIG-SOFT, Software Engineering Notes, 17(4), 1992

[22] Wasserman, A.: *Extending State Transition Diagram for the Specification of
      Human-Computer Interaction,* IEEE Transactions on Software Engineering,
      SE-11, 8, August 1985

Ein Vorgehensmodell zur integrierten Anforderungsanalyse von Benutzungsschnittstelle und Anwendung

Georg Kösters, Josef Voss
FernUniversität Hagen

Zusammenfassung

Wir stellen in diesem Artikel ein Vorgehensmodell für die frühen Phasen der Softwareentwicklung vor, das in besonderer Weise die Entwicklung direkt-manipulativer graphischer Benutzungsschnittstellen einbezieht. Anforderungen an Anwendung und Benutzungsschnittstelle werden aufeinander abgestimmt und parallel entwickelt. Zentrales Beschreibungsmittel für Anforderungen an die Benutzungsschnittstelle sind abstrakte Sichten, die realisierungsunabhängig formuliert werden. Aus der vollständigen Anforderungsanalyse kann die Softwarearchitektur von Benutzungsschnittstelle und Anwendung systematisch entwickelt werden.

1 Einleitung

Die Entwicklung von direkt-manipulativen graphischen Benutzungsschnittstellen ist bekanntermaßen mit erheblichem Aufwand verbunden. Es ist also dringend geboten, diesen Teil der Software in geeigneter Weise in den Software-entwicklungsprozeß zu integrieren. Vielfach wird in diesem Zusammenhang die Idee des Prototyping propagiert, um durch die frühzeitige Erstellung einer lauffähigen Benutzungsschnittstelle gemeinsam mit dem Auftraggeber die Anforderungen an die Funktionalität und die Benutzungsoberfläche validieren zu können. Bei Systemen mit komplexem Funktionsumfang ergeben sich aber erhebliche Probleme: Wie gelangt man schnell und systematisch zu einem Prototypen und wie gewinnt man aus einem Prototypen präzise und verbindliche Vorgaben für die weitere Entwicklung? Zur Lösung derartiger Probleme schlagen wir ein Vorgehen für die frühen Phasen der Entwicklung vor, in dem die Prototyperstellung als ein nachgeordneter Schritt innerhalb eines differenzierten Vorgehensmodells auftritt.

Wir stellen im folgenden ein Vorgehen vor, das die traditionelle Aufteilung in Anforderungsanalyse und Softwareentwurf mit entsprechenden Dokumenten im wesentlichen beibehält, aber die jeweiligen Tätigkeiten und Ergebnisse neu definiert. Wir plädieren dabei insbesondere für die Formulierung von Anforderungen an die Benutzungsschnittstelle bereits in der Anforderungsanalyse.

Wir formulieren Benutzungsschnittstellenanforderungen realisierungsunabhängig und lassen insbesondere Details von Bildschirmlayout und Dialogablauf noch offen. Die Anforderungen werden allerdings so weit präzisiert, daß die wesentliche Struktur der Benutzungsoberfläche vorgegeben ist, so daß die Softwarearchitektur von Benutzungsschnittstelle und Anwendung daraus systematisch entwickelt werden kann, und zwar unabhängig von und parallel zur weiteren detaillierten Festlegung von Bildschirmlayout und Dialogablauf.

2   Verwandte Arbeiten

Die methodische Integration der Benutzungsschnittstellenentwicklung in den Software Life Cycle ist bislang kaum systematisch verfolgt worden. Die meisten Ansätze beschäftigen sich mit der möglichst effizienten Konstruktion von Benutzungsschnittstellen. Eine Forschungsrichtung ist die modellbasierte Generierung. Im allgemeinen wird dabei zunächst ein bestimmtes Modell der Anwendung erstellt, das dann zur (halb-) automatischen Konstruktion der Benutzungsschnittstelle genutzt wird (z.B. [2], [5], [7]). Auf zwei dieser Ansätze gehen wir kurz ein:

- Bei den Arbeiten von Foley u.a. (z.B. [5]) besteht das Anwendungsmodell aus einer präzisen Definition von Operationen, die mit der Oberfläche realisiert werden, einschließlich deren Semantik in Form von Vor- und Nachbedingungen. Kritik: In der Analyse hat man es zunächst mit unscharfen Anforderungen zu tun. Ein solches Anwendungsmodell kann also nicht gleich zu Beginn der Entwicklung in einem Schritt bestimmt werden.

- Ein differenzierteres Vorgehen wird von Ziegler u.a. verfolgt (z.B. [7]): Ausgehend von einer Problemweltanalyse in Form eines ER-Modells werden Sichten zur Aufgabenbearbeitung definiert und mit Hilfe angereicherter ER-Modelle dokumentiert. Anschließend wird mit Dialognetzen der globale Dialogablauf definiert. Ablauf und Sichten dienen als Ausgangsmaterial für die Generierung der Benutzungsschnittstelle. Kritik: Einzelne Aufgaben werden nicht explizit formuliert. Dialognetze zur Darstellung des globalen Ablaufs sind wenig intuitiv und basieren zum Teil schon auf syntaktischen Details der Oberfläche.

Ein allgemeines Problem dieser Ansätze besteht darin, daß die ersten Schritte der Entwicklung stark auf die jeweiligen Generierungswerkzeuge zugeschnitten sind, wobei insbesondere zu früh Details festgelegt werden. Außerdem ist das Vorgehen jeweils auf ein eng umrissenes Anwendungsgebiet begrenzt.

3   Das Vorgehensmodell im Überblick

Zur Formulierung von Anforderungen an graphisch-interaktive Programme werden, vereinfacht ausgedrückt, Gegenstände und Tätigkeiten untersucht. Wir behandeln beide Aspekte gleichrangig und verfolgen in der Anforderungsanalyse zwei parallele Entwicklungslinien: Eine an den Gegenständen orientierte Problemweltmodellierung sowie eine an den Tätigkeiten orientierte Aufgabenanalyse.

Der zeitliche Ablauf der einzelnen Entwicklungsschritte wird in Abb. 1 idealisiert dargestellt. Geänderte Anforderungen oder erkannte Fehler können natürlich jederzeit Rückkopplungen erfordern. Ein Entwicklungsschritt in einer bestimmten Ebene ist von den Ergebnissen der darüberliegenden Schritte abhängig. Ein erster stabiler Zwischenzustand – angedeutet durch die gestrichelte Linie – ist nach der Aufgabenanalyse und bei der Problemweltmodellierung vor der Bestimmung der Operationen erreicht. Das Ergebnis der gesamten Anforderungsanalyse setzt sich aus dem vollständigen Problemweltmodell, benannten Objekten sowie abstrakten Sichten zusammen. Auf die einzelnen Punkte werden wir im folgenden genauer eingehen.

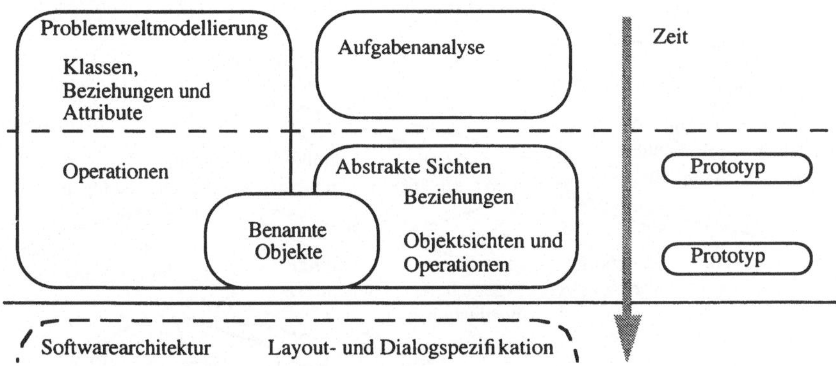

Abb.1: Die Entwicklungsschritte des Vorgehensmodells im Überblick.

4   Aufgabenanalyse und Problemweltmodellierung

In der Aufgabenanalyse gehen wir der Frage nach, was ein Benutzer mit dem Programm bewerkstelligen will. Ausgehend von einer informellen Problemstellung oder einer Grobanalyse, die eine Sammlung potentiell zu realisierender Anforderungen enthält, wird festgelegt, welche Aufgaben tatsächlich mit dem Programm bearbeitet werden, und welche nicht – z.B. weil ihre Realisierung zu aufwendig ist, sie durch existierende Programme oder von Hand erledigt werden.

In der Regel wird ein Softwareprodukt nicht ein völlig neues Aufgabengebiet definieren, sondern existierende Arbeitsvorgänge sollen sicherer und effizienter bearbeitet werden. Zur Bestimmung der Aufgaben orientieren wir uns daher an (existierenden) Arbeitsvorgängen und Arbeitsschritten sowie an den zu bearbeitenden Gegenständen. Die Aufgaben werden umgangssprachlich formuliert ggf. weiter zerlegt oder zusammengefaßt (vgl. [3], [9]). Einzelne Aufgaben werden nicht bis auf eine syntaktische oder lexikalische Ebene heruntergebrochen. Diese Ebenen werden erst in späteren Entwicklungsphasen betrachtet. Ihre Beschreibungen bleiben vielmehr auf einer konzeptuellen Ebene und benutzen das Vokabular der Problemwelt.

Zur Strukturierung werden Aufgaben hierarchisch angeordnet: Komplexere Aufga-

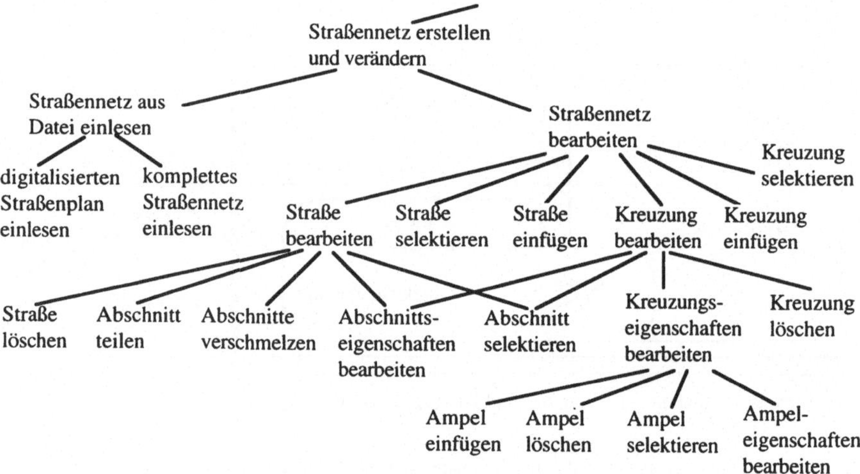

Abb.2: Auszug aus der Aufgabenhierarchie des Verkehrssimulationssystems.

ben setzen sich aus einfacheren Teilaufgaben zusammen. Zusammengefaßte Aufgaben sind durch inhaltliche Nähe und insbesondere durch die Bearbeitung zusammengehöriger Gegenstände gekennzeichnet. Als Beispiel betrachten wir ein Verkehrssimulationssystem, dessen Aufgabenhierarchie Abb. 2 auszugsweise darstellt. Die Beziehungen zwischen den Aufgaben haben hier noch keine präzise

Semantik und werden als „ist Teilaufgabe von" oder auch als expliziter Wechsel zur Bearbeitung einer anderen Aufgabe interpretiert.

Parallel zur Aufgabenanalyse wird in der Problemweltmodellierung die Struktur der vom Programm bearbeiteten Gegenstände untersucht. Die anfängliche Problemstellung und ggf. auch schon Ergebnisse der Aufgabenanalyse können bereits insofern einfließen, als daß sie bestimmen, welche Gegenstände überhaupt relevant sind und bis zu welchem Detaillierungsgrad deren Struktur untersucht wird. Zur Problemweltmodellierung verwenden wir die OOA-Methode und -Notation von Coad/Yourdon [4]. Im ersten Schritt verzichten wir allerdings auf die Einbeziehung von Operationen (Diensten). Abb. 3 zeigt einen Auszug der Problemweltmodellierung des Verkehrssimulationssystems.

Die gegenseitige Beeinflussung von Problemweltmodellierung und Aufgabenanalyse ist hauptsächlich durch die Tatsache begründet, daß zur Formulierung von Aufgaben Objekte, deren Attribute, Beziehungen und Komponenten aus dem OOA-Modell verwendet werden. Gegenstände, die in der Aufgabenanalyse auftreten, aber im OOA-Modell nicht berücksichtigt werden, deuten auf ein unzureichendes OOA-Modell hin. Umgekehrt müssen Objekte des OOA-Modells, die sich keiner Aufgabe zuordnen lassen, hinterfragt werden. Im Idealfall werden beide Schritte

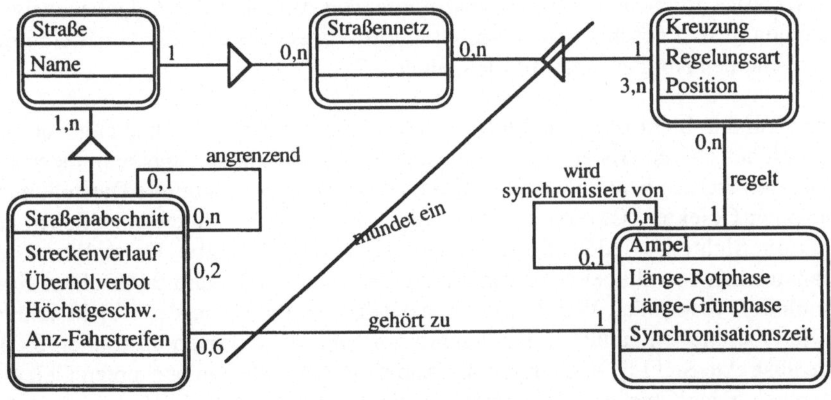

Abb.3: Auszug aus dem OOA-Modell des Verkehrssimulationssystems.

deshalb parallel und mit häufigen Rückkopplungen durchgeführt. Mit Beendigung dieser Schritte erreicht die Anforderungsanalyse einen im Hinblick auf Konsistenz und Vollständigkeit zu validierenden Zwischenstand.

5   Abstrakte Sichten

Als Analyseendergebnis ist der vorliegende Zwischenstand noch unzureichend.
Aus Sicht des Systementwicklers fehlt die Präzisierung der operativen Anforderungen als Vorgabe für die weitere Entwicklung. Aus Sicht des Auftraggebers
fehlen Skizzen oder ein Oberflächenprototyp, um die Vollständigkeit bzw.
Richtigkeit der Anforderungsanalyse beurteilen zu können. Wir ergänzen deshalb
die Anforderungsanalyse und präzisieren die bislang lediglich umgangssprachlich
formulierten Aufgaben und halten dies in geeignet formalisierter Weise fest. Zu
diesem Zweck stellen wir mit abstrakten Sichten und benannten Objekten im
folgenden zwei wichtige Hilfsmittel vor, die den Dialogablauf strukturieren und den
Zusammenhang zu den dabei behandelten Gegenständen bzw. Objekten herstellen.

Unter abstrakten Sichten verstehen wir Fenster bzw. Teilfenster, die dadurch charakterisiert sind, daß sie bestimmte Attribute, Beziehungen und Komponenten von
benannten Objekten darstellen sowie Operationen anbieten, die zu Veränderungen
an abstrakten Sichten oder benannten Objekten führen. Wir sprechen von abstrakten Sichten, weil das Augenmerk auf dem Informationsgehalt und den logischen
Verbindungen der zugehörigen Objekte und nicht auf deren konkretem Aussehen
liegt. Den Ausgangspunkt zur Bestimmung der abstrakten Sichten bilden die zentralen, voraussichtlich am häufigsten auszuführenden Aufgaben der Aufgabenanalyse,
die einen in sich geschlossenen Arbeitsschritt repräsentieren. Wir assoziieren mit
jeder zentralen Aufgabe eine abstrakte Sicht.

Eine abstrakte Sicht ist durch die enthaltenen benannten Objekte und angebotenen
Operationen charakterisiert. Dabei können nicht nur benannte Objekte in mehreren
abstrakten Sichten benutzt werden, auch Operationen und bestimmte Darstellungen
benannter Objekte – Objektsichten genannt – finden sich in unterschiedlichen abstrakten Sichten wieder. So wird beispielsweise dieselbe Darstellung des
Straßennetzes sowohl beim Verändern des Netzes als auch beim Beobachten einer
Simulation verwendet. Wir definieren deshalb benannte Objekte, Objektsichten,
sowie Operationen zunächst unabhängig von einer bestimmten abstrakten Sicht.
Jede abstrakte Sicht ist dann durch die Namen der enthaltenen benannten Objekte,
der zugehörigen Objektsichten und der angebotenen Operationen definiert.

5.1   Benannte Objekte

Eine genauere Betrachtung der Aufgabenbeschreibungen liefert die Erkenntnis, daß
dort im Unterschied zur Problemweltmodellierung von bestimmten Gegenständen
gesprochen wird, und zwar nicht von abstrakten Begriffen oder Objekttypen oder

-klassen, sondern von konkreten Exemplaren, die im Kontext einer Aufgabe eine bestimmte Bedeutung haben. Wir extrahieren diese Gegenstände als sog. benannte Objekte.

Für jedes benannte Objekt wird festgehalten, welchen Typ es hat, in welchen abstrakten Sichten es bekannt ist, bzw. benutzt wird und ggf. von welchen anderen benannten Objekten es abhängig ist (als Komponente, Element oder Teilmenge). Beim Typ der benannten Objekte handelt es sich im allgemeinen um Strukturen, etwa Listen oder Mengen, die Objekte aus den Klassen der Problemweltmodellierung enthalten. Existieren mehrere Ausprägungen einer Sicht, so existieren auch von den in dieser Sicht lokal bekannten Objekten Ausprägungen in entsprechender Anzahl. Zu jeder Ausprägung der abstrakten Sicht „Abschnittsbearbeitung" existiert beispielsweise eine Ausprägung des benannten Objekts „bearbeiteter Abschnitt". Die folgende Aufstellung beschreibt einige benannte Objekte des Verkehrssimulationssystems.

benanntes Objekt: Name	Menge und Typ	bekannt/benutzt in abstrakter Sicht	Teilkomponente von
Straßennetz	ein Straßennetz	global bekannt	
Abschnittselektion	0-2 angrenzende Straßenabschnitte	Netz- und Kreuzungsbearbeitung	Straßennetz
bearbeiteter Abschnitt	ein Straßenabschnitt	Abschnittbearbeitung	Straßennetz

5.2 Objektsichten – Skizzen

In einer abstrakten Sicht werden genau die benannten Objekte dargestellt, die für die assoziierten Aufgaben relevant sind, allerdings nicht mit all ihren Eigenschaften, sondern vielmehr auf Sicht-spezifische Aspekte reduziert. Eine Objektsicht definiert, welche Attribute und Beziehungen des benannten Objekts auf dem Bildschirm dargestellt werden. Bei zusammengesetzten Objekten wird zusätzlich festgelegt, welche Komponenten sichtbar sind, d.h. bis zu welcher Ebene und mit welchen Attributen und Beziehungen auf der jeweiligen Ebene ein Objekt dargestellt wird.

Auch bei der Festlegung der Objektsichten geht es uns lediglich um den Informationsgehalt und nicht um deren Präsentation auf dem Bildschirm. Falls erforderlich ergänzen wir die textuelle Beschreibung einer Objektsicht um eine Skizze, z.B. wenn die Entwicklung eines geeigneten Bildschirmlayouts anwendungsspezifische Kenntnisse erfordert. Eine Skizze beschreibt also Anforderungen hinsichtlich einer bestimmten graphischen Darstellung, die allein durch die textuelle

Beschreibung der Objektsicht nicht ausgedrückt werden können. Die folgende Aufstellung gibt zwei durch Skizzen ergänzte Objektsichten unseres Beispiels wieder.

Name der Objekt-sicht	*dargest. benanntes Objekt*	*dargest. Komponenten, in Beziehung stehende Objekte*	*dargest. Beziehungen*	*dargest. Attribute*	*Skizze, Darstellungs-vorgabe*
Netz-topologie-sicht	Straßen-netz	Straßen und deren Abschnitte, Kreuzungen	angrenzend, mündet-ein	Straße: Name, Abschnitt: Streckenverlauf Kreuzung: Position	Fleyer Str.
Ampel-kreuzungs-sicht	bearbeitete Kreuzung	Straßen und Ampeln an der Kreuzung	wird synchronisiert von	Straße: Name, Ampel: Richtung	Feithstr. Fleyer Str Feithstr.

5.3 Operationen

Bisher beschreiben Aufgaben die Tätigkeiten, die Benutzer mit dem System durchführen. Die in den Aufgaben bisher nur implizit enthaltenen vom System bereitzustellenden Operationen werden jetzt explizit beschrieben. Zur Beschreibung einer Operation werden im wesentlichen nur die beteiligten benannten Objekte angegeben. Beteiligt sind zum einen die Argumente der Operation, zum anderen die als Folge ihrer Ausführung veränderten benannten Objekte.

Die folgende Aufstellung beschreibt einige Operationen der abstrakten Sicht „Netzbearbeitung". Das Löschen von Straßenabschnitten wird von verschiedenen Sichten angeboten, deshalb ist bei der Operation „Abschnitt löschen" die Verwendung eines Parameters vom Typ „Abschnitt" angezeigt.

Operation: Name und Beschreibung	*Argumente/ Parameter*	*veränderte benannte Objekte*	*veränderte abstrakte Sichten*
Abschnitt (de)selektieren		Abschnittselektion	
Wechsel zur Abschnittbearbeitung	Abschnitt-selektion		erzeugt neue Abschnittbearbeitung mit Abschnittselektion

Abschnitt löschen	Straßennetz, ein Abschnitt	Straßennetz, Abschnittselektion, bearbeitete Kreuzungen	zerstört Abschnittbearbeitung, die den gelöschten Abschnitt bearbeitet

6 Beziehungen zwischen abstrakten Sichten

Die einzelnen Fenster einer Oberfläche sind nicht völlig unabhängig voneinander. Wir geben im folgenden die wichtigsten Abhängigkeiten an und modellieren sie als Beziehungen zwischen abstrakten Sichten.

* Wechsel zwischen sichtbaren Fenstern: Im Normalfall können Benutzer zwischen aktiven Sichten bzw. der Bearbeitung der assoziierten Aufgaben beliebig wechseln. Solche Wechsel bezeichnen wir als implizit.

* Explizites Sichtbarmachen von Fenstern: Ein expliziter Wechsel wird durch eine in der Ausgangssicht vorgegebenen Operation ausgelöst. Explizite Wechsel sind durch die hierarchische Anordnung der Aufgabenanalyse vorstrukturiert.

* Variabel viele Ausprägungen von Fenstern: Sichten können dynamisch erzeugt und zerstört werden. Die Erzeugung einer neuen Ausprägung einer abstrakten Sicht ist immer mit einem expliziten Wechsel in die erzeugte Sicht verbunden.

* Fenster können voneinander abhängig sein, etwa in der Form, daß die Zerstörung eines Fensters die Zerstörung von untergeordneten Fenstern nach sich zieht: Eine Kontrollbeziehung strukturiert abstrakte Sichten baumartig. Die Deaktivierung bzw. Zerstörung einer Sicht ist fest mit der Deaktivierung bzw. Zerstörung aller von ihr kontrollierten Nachfolger verbunden. Das Aktivieren einer Sicht erzwingt die Aktivierung aller Vorgänger innerhalb des Kontrollbaumes.

* Die Bearbeitung verschiedener, sich gegenseitig ausschließender Aufgaben in einem Fenster: Ein wechselseitiger Ausschluß verbietet die gleichzeitige Aktivierung von Gruppen abstrakter Sichten. Die Aktivierung einer Sicht einer derzeit inaktiven Gruppe bewirkt die Deaktivierung aller aktiven Sichten der derzeit aktiven Gruppe. Der gegenseitige Ausschluß betrifft genau definierte Gruppen von Sichten und ist auf einen lokalen Bereich begrenzt.

- Modale Dialoge: Ist die Bearbeitung einer Sicht exklusiv, so sprechen wir von einer modalen Sicht. Diese kann nur durch explizite Wechsel erreicht und wieder verlassen werden. Die Aktivierung einer modalen Sicht wirkt sich global aus, da alle impliziten Navigationsmöglichkeiten gesperrt sind, bis die modale Sicht durch einen expliziten Wechsel zerstört bzw. deaktiviert wird.

Explizite Wechsel sind Operationen, die abstrakte Sichten erzeugen, zerstören, aktivieren oder deaktivieren. Aufgrund ihrer Bedeutung führen wir die expliziten Wechsel zwischen abstrakten Sichten als Operationen auf.

Wir fassen die abstrakten Sichten zusammen mit den wichtigsten, Sicht-übergreifenden Informationen in einer mit State Charts [6] verwandten graphischen Darstellung zusammen. Rechtecke repräsentieren abstrakte Sichten, wobei die Zuordnung über ihre Namen erfolgt. Implizite Wechsel zwischen Sichten werden nicht, explizite Wechsel durch gerichtete Kanten dargestellt, wobei die Richtung der Kante die Navigationsrichtung wiedergibt. Die graphische Darstellung eines expliziten Wechsels wird durch einen Kreis ergänzt, falls dieser Wechsel die Erzeugung einer neuen Ausprägung der abstrakten „Ziel"-Sicht beinhaltet. Die Kontrollbeziehung wird durch eine Kante mit einem Dreieck dargestellt, wobei das Dreieck auf die kontrollierende Sicht weist. Der gegenseitige Ausschluß von Sichten wird durch ein in verschiedene Bereiche geteiltes, gestricheltes Rechteck dargestellt. Modale Sichten werden durch eine doppelte Umrandung gekennzeichnet. Abb. 4 zeigt die aus der Aufgabenanalyse abgeleiteten Beziehungen zwischen abstrakten Sichten unseres Beispiels.

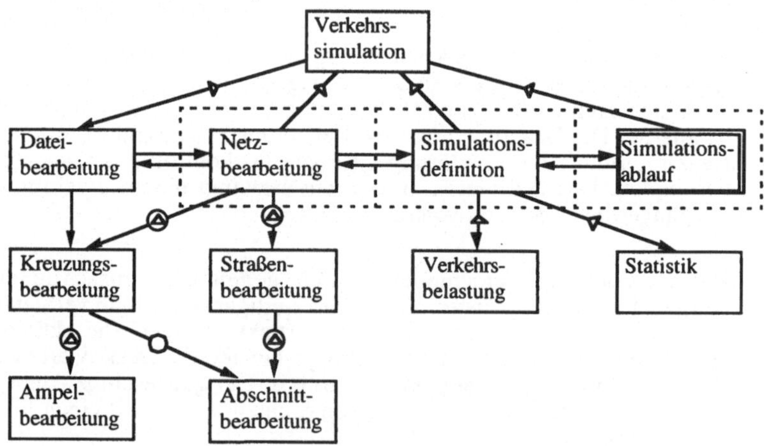

Abb.4: Die graphische Darstellung der abstrakten Sichten des Verkehrssimulationssystems.

7 Diskussion und Ausblick

Das beschriebene Vorgehensmodell erweitert und präzisiert verschiedene Vorversionen. Es stellt ein Gerüst dar, das nicht immer in allen Einzelheiten durchlaufen wird. Je nach Art des Problems sind Vereinfachungen bzw. Spezialisierungen angebracht. Vereinfachungen sind z.B. bei Oberflächen angezeigt, die nur einen geringen Graphikanteil besitzen und sich hauptsächlich aus Eingabemasken zusammensetzen. Spezialisierungen ergeben sich insbesondere für GIS- und CASE-Anwendungen.

Die in Objektsichten enthaltenen Skizzen sind als Mittel zur Veranschaulichung nur begrenzt tauglich. Deshalb werden – wie in Abb. 1 angedeutet – begleitend Prototypen erstellt, wobei es sich in der Anforderungsanalyse um schnell zu erstellende Wegwerf-Prototypen handelt. Wir halten ein abgestuftes Vorgehen für angebracht, wie es z.B. in [1] vorgeschlagen wird. Ein erster Schritt kann die Verwendung eines DTP- oder Zeichenprogramms mit Hypertext-Möglichkeiten sein. Nach der Definition von Objektsichten und Operationen können Prototypen mit detaillierterem Layout und Verhalten erstellt werden. Wir haben an dieser Stellen durchaus gute Erfahrungen mit einer Kombination aus DTP-Programm und User-Interface-Builder gemacht.

Die Anwendbarkeit einer Methode, die die Erstellung umfangreicher und detaillierter Dokumente mit sich bringt, hängt wesentlich von unterstützenden Werkzeugen ab. Die Entwicklung solcher Werkzeuge ist deshalb eines unserer nächsten Ziele. Längerfristig verfolgen wir die Weiterführung der methodischen Überlegungen in Richtung Softwareentwurf. Vergleichende Entwicklungen in verschiedenen Umgebungen zeigen, daß zahlreiche Regeln zur systematischen Umsetzung von Problemweltmodell, benannten Objekten und abstrakten Sichten in einen Softwareentwurf existieren. Der Schwerpunkt bei unseren weiteren Überlegungen wird auf objektorientierte Umgebungen mit integriertem UI-Builder oder UIMS, insbesondere dem von uns entwickelten DIWA-System [8], liegen.

8 Literatur

[1] Bösze J. und Aschacher, D. *Prototyping mit "Hyper-Tools"*. Proceedings Software Ergonomie '91, Teubner, Stuttgart, 1991, 119-129

[2] Balzert, H. *Der Janus-Dialogexperte: Vom Fachkonzept zur Dialogstruktur.* Proceedings Softwaretechnik 93, 62-72

[3] Bass, L. und Coutaz, J. *Developing Software for the User Interface.* Addison Wesley, Reading Massachusetts, 1991

[4] Coad, P. und Yourdon, E. *Object-Oriented Analysis.* Prentice Hall, Englewood Cliffs, NJ, 1990.

[5] Foley, J., Kim, W., Kovacevic, S. und Murray K. *Defining Interfaces at a High Level of Abstraction.* IEEE Software 6,1 (1989), 25-32.

[6] Harel, D. *On Visual Formalisms.* Communcations of the ACM 31,5 (1988), 514-530.

[7] Janssen, C., Weisbecker, A. und Ziegler, J. *Generierung graphischer Benutzerschnittstellen aus Datenmodellen und Dialognetzspezifikationen.* in: Züllighoven, et. al. (Hrsg.) Requirements Engineering '93: Prototyping. Teubner, Stuttgart, 1993, 335-349.

[8] Six, H.W. und Voss, J. *A Software Engineering Perspective to the Design of a User Interface Framework*, in: Proceedings CompSAC 92, IEEE Computer Society Press, Los Alamitos, 1992, 128-134

[9] Ziegler, J. *Aufgabenanalyse und Funktionsentwurf.* in: Balzert, H., et. al. (Hrsg.) Einführung in die Software Ergonomie. de Gruyter, Berlin, 1988, 231-252.

Georg Kösters, Josef Voss
Lehrgebiet Praktische Informatik III
FernUniversität Hagen
58084 Hagen
e-mail: georg.koesters@fernuni-hagen.de
          josef.voss@fernuni-hagen.de

Unterstützung des Benutzungsoberflächen-Designs durch eine Kritikkomponente

Uwe Malinowski, Markus Stolze
Siemens AG, University of Colorado at Boulder

Zusammenfassung

Beim Entwurf grafischer Benutzungsoberflächen muß eine überwältigende Menge von Informationen berücksichtigt werden. Traditionell stehen diese auf Papier als Entwurfsrichtlinien oder Styleguides zur Verfügung und sind damit nur mühsam zugänglich. Eine Kritikkomponente kann durch aktive Präsentation der in einer bestimmten Designsituation benötigten Informationen den Designer unterstützen. Auf der Basis von Erfahrungen aus verschiedenen Projekten werden verschiedene Möglichkeiten der Informationspräsentation diskutiert.

1 Motivation

Der Entwurf einer grafischen Benutzungsoberfläche ist ein aufwendiger Prozeß. In den letzten Jahren sind verschiedene Werkzeuge entwickelt worden, die den Entwurf einer Benutzungsoberfläche vereinfachen. SX/Tools [7] und NextStep [5] sind Beispiele solcher Systeme, die dem Designer erlauben, die Benutzungsoberfläche per direkter Manipulation zu erstellen. Derartige Werkzeuge können zwar den Entwurfsprozeß beschleunigen, eine hohe Qualität der erstellten Benutzungsoberflächen ist damit aber noch nicht garantiert.

Im Unterschied zum Entwurf einer Kommando-orientierten Benutzungsoberfläche hat der Designer beim Entwurf einer grafischen Benutzungsoberfläche wesentlich mehr Gestaltungsmöglichkeiten. Wenn immer alle Freiheitsgrade ausgeschöpft würden, würde dies zu einer Überforderung der Endanwender führen. Um dieser Entwicklung entgegenzuwirken sind unterschiedliche nationale und firmeninterne Entwurfsrichtlinien und Styleguides entwickelt worden (z.B. SNI Grafik-Styleguide 2.0 von 1993). Die Nützlichkeit dieser Informationen ist jedoch dadurch beschränkt, daß sie normalerweise nur in der Form von Handbüchern zur Verfügung stehen, was bei der Menge und Vielfalt der Informationen, nicht angemessen ist.

In letzter Zeit sind verschiedene Ansätze verfolgt worden, die Entwurfsrichtlinien und Fallbeispiele "online" zur Verfügung zu stellen (z.B. [11], [1]). Aber auch die Möglichkeit mittels Datenbankabfragen und Hypertext-Browsing auf Entwurfsrichtlinien zugreifen zu können bringt noch relativ wenig Unterstützung für den Designer. Suchprozesse werden zwar vereinfacht, aber der Designer muß die

Initiative ergreifen, d.h. er muß wissen, daß ihm Informationen fehlen. Weiterhin muß er wissen, welche Informationen ihm fehlen, um danach suchen zu können. Eine weitere Schwäche dieses Ansatzes ist, daß bei der Formulierung der Anfrage nach Informationen wenig Unterstützung gegeben wird, so daß der Designer auch wissen muß, wie auf die Information zugegriffen werden kann.

In Unterschied zu den bisher erwähnten *passiven* Unterstützungstechniken gibt es Methoden, die eine *aktive* Rolle im Designprozeß übernehmen. Constraints beschränken aktiv die Freiheitsgrade in einer bestimmten Designsituation, während eine Kritikkomponente aktiv Informationen zur Verfügung stellt, die in einer bestimmten Situation von Bedeutung sind. Die aktive Rolle geht aber nicht so weit, daß versucht wird, das Design der Benutzungsoberfläche zu automatisieren, was z.B. bei UIDE [13] oder HUMANOID [10] das Ziel ist.

Constraints sind ein wichtiges Element von direkt-manipulativen Systemen. Die grundlegende Idee ist, daß Constraints nur „sinnvolle" Operationen ermöglichen und damit dem Benutzer die Auswahl von Operationen erleichtert. Constraints werden jedoch dann problematisch wenn die Bestimmung der „sinnvollen" Operationen nicht eindeutig ist. In diesen Fällen kann es passieren, daß Designer eine Operation ausführen möchten, welche das System nicht gestattet. Constraints sind deswegen hauptsächlich geeignet für die Kodierung von einfachen und „sicheren" Regeln. Im Benutzungsoberflächen-Design sind mehrdeutige und zum Teil widersprüchliche Designregeln jedoch relativ häufig, so daß sich Constraints als Unterstützungstechnik weniger gut eignen.

Eine Kritikkomponente schränkt die Möglichkeiten von Entwicklern nicht ein, sondern analysiert den momentanen Zustand des Entwurfs, weist Entwickler auf potentielle Probleme und Entwurfsoptionen hin und ermöglichen den Zugriff auf Erklärungen und Hintergrundinformationen [3]. Aufgrund dieser Informationen können Entwickler entscheiden, ob die „Kritik" in der aktuellen Situation gerechtfertigt ist und sie ihren Entwurf anpassen wollen, oder ob sie die Kritik ignorieren wollen, da sie von falschen oder unvollständigen Voraussetzungen ausgeht. Indem der Designer auf mögliche Probleme aufmerksam gemacht wird, aber keine Entscheidung über deren Behandlung getroffen wird bleibt die Kontrolle beim Designer [2]. Aufgrund dieser Eigenschaft sind Kritikregeln besonders geeignet für die Kodierung von komplexeren und „unsicheren" Zusammenhängen.

Im folgenden werden wir verschiedene Möglichkeiten und Probleme der Informationspräsentation durch eine Kritikkomponente diskutieren. Wir wollen jedoch jetzt schon darauf hinweisen, daß im Normalfall eine Integration der Kritikkomponente mit anderen Techniken und Komponenten sinnvoll ist. So können

Constraints -wenn einsetzbar- verhindern, daß eine falsche Designentscheidung überhaupt getroffen wird, was in jedem Fall schneller ist, als eine spätere Korrektur. Für eine optimale Unterstützung sollte dem Designer zudem ein Argumentationswerkzeug und ist eine Katalogkomponente zur Verfügung stehen. Eine Integration dieser Komponenten in einem Designwerkzeug wird als Domain-Oriented Design Environment bezeichnet [3].

2 Eine Kritikkomponente im Benutzungsoberflächen-Design

Idealerweise sollte eine Kritikkomponente die im jeweiligen Designkontext benötigte Styleguide-Information in geeigneter Form zur Verfügung stellen. Grundvoraussetzung zur Erreichung dieses Ziel ist, daß die Styleguide-Informationen erfolgreich in Regeln umgesetzt werden können, eine Spezifikation für die Benutzungsoberfläche in einer für das System verarbeitbaren Form vorliegt, und die Designziele vom Designer definiert werden können. Aber selbst wenn dies gelingt, gibt es noch etliche offene Fragen: Zu welchem Zeitpunkt ist die Präsentation einer Information am effektivsten? Wie soll die Information präsentiert werden? Wo soll die Information präsentiert werden? In verschiedenen Projekten sind verschiedenen Lösungsansätze zu diesen Problemen erarbeitet worden. Diese werden wir in den nächsten Abschnitten vergleichen und in Beziehung zueinander setzen.

2.1 Zeitpunkt der Informationspräsentation

Der frühestmögliche Zeitpunkt der Präsentation von Informationen, die sich auf eine Designaktion beziehen, ist während der eigentlichen Aktion des Benutzers. Im Zusammenhang mit Kritikkomponenten im Bereich Benutzungsoberflächen-Design gibt es hierzu noch keine Erfahrungen. Bei Textverarbeitungsprogrammen und CAD-Systemen ist Feedback durch Veränderung des Cursor oder Signalisieren des Zustands, der erreicht würde, wenn die Aktion in dem aktuellen Moment ausgeführt/abgeschlossen würde, üblich. Ein Werkzeug, welches den Entwurf von Dialogboxen unterstützt, könnte zum Beispiel beim Plazieren eines Textfeldes durch die Veränderung des Cursors anzeigen, wenn dieses zu dicht neben einem anderen Objekt plaziert würde.

Die nächste Möglichkeit ist die Vermittlung einer Information unmittelbar nach der Aktion des Designers. Diese traditionelle Vorgehensweise trägt der Forderung Rechnung, daß die Information so früh wie möglich zur Verfügung gestellt werden soll, um nicht auf einem fehlerhaften Zustand aufzubauen, wenn man es vermeiden kann. In FRAMER [8], einem Werkzeug welches den Design von Symbolics Benutzungsoberflächen unterstützt, wird der Benutzer zum Beispiel direkt nach der

Verschiebung von Fensterbereichen auf nicht erlaubte Überlappungen der bewegten Bereiche hingewiesen.

Es ist offensichtlich, daß einige Analysen nicht unmittelbar nach einem Designschritt durchgeführt werden können, beziehungsweise in der Situation nicht hilfreich sind. Vollständigkeitsüberprüfungen sind nur in bestimmten Designsituationen sinnvoll und würden, nach jedem Schritt durchgeführt, zu einer Flut wenig hilfreicher Meldungen führen. Der Hauptkritikpunkt gegen unmittelbare Präsentation ist, daß der Designer zu sehr von der eigentlichen Aufgabe abgelenkt wird und nicht in der Lage ist, eine Folge von Designschritten ungestört auszuführen.

Um die Störung des Designprozesses möglichst gering zu halten, kann man die Informationen immer nach Abschluß der Arbeit an einer semantischen Einheit präsentieren. Wenn dies gelingt, ist es sicher ein guter Kompromiß zwischen Stören des Designprozesses und frühzeitiger Information. In mehreren Projekten hat sich jedoch gezeigt, daß es sowohl sehr schwierig ist, vorab sinnvolle semantische Einheiten zu definieren, als auch, während des Designprozesses festzustellen, wenn die Arbeit an einer dieser Einheiten abgeschlossen ist.

In VDDE [6], einem System, welches Designer bei der Erstellung von automatischen Telephoninformationssystemen unterstützt, kann eingestellt werden, daß Kritikinformationen nur dann zur Verfügung gestellt werden, wenn der Designer die Arbeit an einem neuen Teildialog aufnimmt. Dies ist jedoch dann problematisch wenn Designer an mehreren Teildialogen gleichzeitig arbeiten und zum Beispiel Unterbereiche eines Dialogs mit einem anderen Dialog austauschen wollen.

Dieses Problem wurde in FRAMER gelöst, indem der Designer durch die Verwendung der „Checklists" dem System explizit mitteilt, wann die Arbeit an einer semantischen Einheit abgeschlossen ist [8]. So sind in FRAMER das Layout der Fensterbereiche und die Zuordnung von Input- und Output-Funktionen zu diesen Bereichen zwei unterschiedliche Designschritte, welche vom Benutzer separat behandelt und „abgehakt" werden können, und dann auch jeweils von der Kritikkomponente separat überprüft werden.

Eine weitere Möglichkeit ist die Analyse des Designs auf Benutzeranfrage. Dieser Ansatz wird im Projekt KRI benutzt [9]. Der Vorteil des Kritik auf explizite Anforderung des Designers ist, daß auch sehr zeitaufwendige Analysen durchgeführt werden können. Dies wird jedoch damit erkauft, daß die eigentliche Stärke des Kritikansatzes, die aktive Präsentation von Informationen wenn sie benötigt werden, verloren geht.

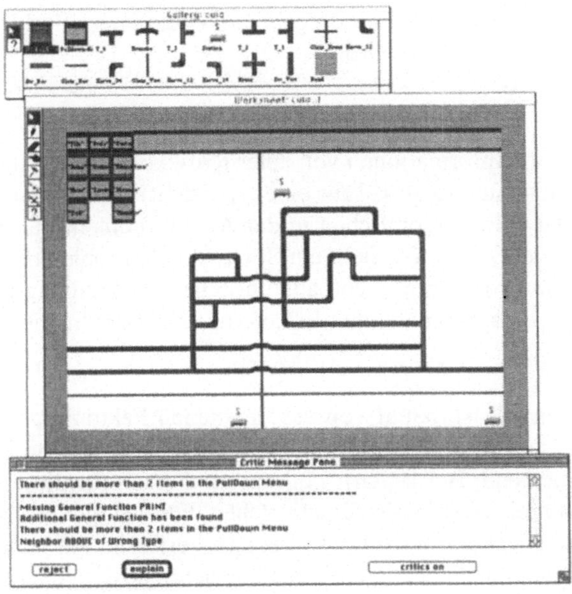

Abbildung 1: Galerie der Designobjekte, Entwurfsfenster und Kritikmeldungen in CUID.

In CUID (Abbildung 1), einem Prototyp einer Kritikkomponente für den Entwurf von grafischen Benutzungsoberflächen für Verkehrsmanagementsysteme, wird dieses Problem umgangen, indem die aktive Informationspräsentation unmittelbar nach der Designaktion mit der Analyse auf Anforderung kombiniert wird. Nach jedem Designschritt werden Regeln überprüft, die sich im wesentlichen auf das Layout der Elemente in der Benutzungsoberfläche beziehen. Das System ermöglicht dem Benutzer aber auch, eine Funktion "Überprüfung des Gesamtentwurfs" auswählen. Mit dieser Funktion wird das gesamte Design unter Verwendung aller verfügbaren Regeln untersucht. Dazu gehören dann z.B. auch Regeln, die überprüfen, ob alle laut Spezifikation benötigten Funktionen im System vorhanden sind und ob zusätzliche, nicht geforderte Funktionen implementiert wurden. Diese Regeln werden während des Designvorgangs nicht überprüft.

2.2 Art und Ort der Informationspräsentation

Da die Art der Informationsdarstellung sehr eng mit dem Ort, an der die Information präsentiert wird, zusammenhängt, werden wir die verschiedenen Ansätze auch unter beiden Aspekten gemeinsam vorstellen. Grundsätzlich unterscheiden wir textuelle

Kritik, in der die Information als Text dargeboten wird, und grafische Kritik, bei der Information graphisch präsentiert wird.

Textuelle Kritik

Traditionell werden die Informationen von einer Kritikkomponente in einem separaten Fenster angeboten (siehe z.B. [2] für eine Übersicht). Die Variation der Art der Darstellung liegt dabei im wesentlichen in der Ausführlichkeit der Informationen. Auf der einen Seite des Spektrums steht ein Stichwort als Problembeschreibung, auf der anderen Seite eine vollständige und ausführliche Beschreibung eines Problems mit allen möglichen Ursachen und einer Erläuterung der Regeln, die dieses Problem identifizieren.

Welcher Ansatz sinnvoll ist, hängt von verschiedenen Faktoren ab. Für erfahrene Designer ist eine Liste von Stichwörtern mit den noch ausstehenden Problemen sinnvoll und oft ausreichend. Bei Bedarf muß natürlich auch für diese Designer eine ausführlichere Beschreibung zur Verfügung stehen. Für weniger erfahrene Designer wie z.B. einen Endanwender der die Benutzungsoberfläche auf seine Bedürfnisse anpaßt, ist die Präsentation einer ausführlicheren Nachricht sinnvoller.

Außerdem spielt die Menge der zu einem bestimmten Zeitpunkt präsentierten Meldungen eine entscheidende Rolle. Wenn viele Meldungen zur gleichen Zeit darzustellen sind, sind kürzere Informationen zu bevorzugen, wenn nur eine Meldung ansteht, kann diese mehr Platz in Anspruch nehmen. In jedem Fall muß der Zugriff auf weitergehende Erläuterungen zur Verfügung stehen.

FRAMER ist ein Beispiel eines Systems in welchem dem die Kritik auf textuelle Weise in einem Fenster dargestellt wird. Durch die Verwendung der Checkliste ist zu jedem Zeitpunkt die Menge der möglichen Kritikinformationen relativ gering und die Texte können daher relativ ausführlich sein. Trotz dieser Ausführlichkeit erlaubt FRAMER aber den Zugriff auf zusätzliche, erklärende Informationen.

Die separat dargestellte textuellen Kritik hat einen entscheidenden Nachteil: Der Bezug zu den von der Meldung betroffenen Objekten im Design ist sehr schwer herzustellen. Dieses Problem kann durch eine Kombination mit grafischen Ansätzen gelöst werden, die wir später diskutieren werden. Ein Vorteil der Trennung des Designs und der Kritikmeldungen ist, daß der Designer durch eine Meldung nicht unmittelbar unterbrochen wird, sondern diese erst wahrnimmt, wenn er seine Aufmerksamkeit bewußt auf das Meldungsfenster lenkt. Andererseits ist gerade dies auch wieder ein Nachteil, da damit trotz aktiver Präsentation durch das System auch ein aktives Nachschauen des Designers notwendig ist.

Eine weitere Möglichkeit der textuellen Kritik, von der uns jedoch bisher keine prototypische Anwendung im Bereich Benutzungsoberflächendesign bekannt ist, präsentiert die Informationen im Stil von "Bubble Help", wie es in Macintosh-Systemen vorhanden ist. Dabei werden die Informationen in der Form einer Sprechblase an dem betroffenen Objekt angezeigt. Dieses Verhalten der Kritikkomponente unterbricht den eigentlichen Designprozeß sehr aggressiv, ein Ignorieren der Meldung ist praktisch nicht möglich. Der Text, der in der Blase angezeigt werden kann, darf nur sehr kurz sein. Außerdem verdeckt die Sprechblase einen Teil des Designs. Trotz dieser Nachteile kann es in einer Kritikkomponente eine Verwendung für diese Art der Präsentation geben, z.B. für sehr dringende Nachrichten, die auf keinen Fall ignoriert werden dürfen.

Ein ähnlicher Ansatz ist die Präsentation der Information in einer modalen Meldungsbox. Dies ist die aggressivste Möglichkeit, den Dialog des Designers mit der Designumgebung auf die Kritikmeldungen zu zwingen.

Eine Möglichkeit der Darstellung, die weder grafische Kritik noch eigentlich textuelle Kritik ist, ist ein Flag, das anzeigt, ob in dem bearbeiteten Design Probleme existieren oder nicht. Ein derartiger Ansatz ist in [14] vorgeschlagen worden, um den Endbenutzer auf Adaptionsvorschläge des Systems hinzuweisen. Dieses Flag könnte im Rahmen des Designfensters untergebracht und damit gut sichtbar sein, wäre auf der anderen Seite leicht zu ignorieren. Auf Wunsch kann der Designer dann die Informationen abrufen, die zur Verfügung stehen. Diese können dann sowohl grafisch als auch textuell präsentiert werden.

Grafische Kritik

Grundsätzlich unterscheiden wir zwei Arten der grafischen Präsentation von Kritikinformationen: Auf der einen Seite kann das Designobjekt selbst grafisch verändert werden, andererseits kann eine Information außerhalb des Designs grafisch dargestellt werden.

Bei der ersten Möglichkeit, der Veränderung des Designobjekts selbst, ist das Problem des Bezugs zwischen einer Information und dem betroffenen Objekt ausgeräumt. Designprobleme oder bestimmte Zwischenzustände, z.B. wenn ein Objekt nicht vollständig spezifiziert ist, können durch Veränderung von Objektattributen angezeigt werden. Besonders geeignet scheinen Farbe, Intensität, Position, Markierungen und, bei Texten, Font zu sein.

Weiterhin können auch Beziehungen zwischen verschiedenen Objekten dargestellt werden, z.B. indem Linien eingezeichnet werden, die eine geometrische Relation offensichtlich machen. Schließlich kann durch verschiedene Cursorformen der zu erwartende Effekt einer Designaktion visualisiert werden.

Es ist offensichtlich, daß durch die beschränkte Anzahl von grafischen Möglichkeiten auch die Menge der darstellbaren Probleme beschränkt ist. Schwerwiegender ist jedoch, daß der Designer die Bedeutung der Präsentation erkennen soll, also die Codierung lernen muß. Zum Ausgleich dafür erlaubt die graphische Darstellung von Informationen erfahrenen Designern ein schnelles und müheloses Verstehen der Kritik. Wird die Kritik im Blickfeld von erfahrenen Designern auf graphische Art dargestellt, dann kann die Information direkt und intuitiv aufgenommen werden – und bei Bedarf ignoriert werden, ohne daß der eigentliche Designprozeß unnötig unterbrochen werden muß.

Werkzeuge für das Design von Benutzungsoberflächen, welche graphische Methoden für die Bereitstellung von Kritikinformationen benutzen, sind uns nicht bekannt. Abbildung 2 vermittelt einen Eindruck, wie Küchen-Designer mittels graphischer Kritik unterstützt werden könnten. Die Spülmaschine (DW) ist in diesem Fall farbig markiert, da in der Spezifikation festgelegt ist, daß diese Küche für einen linkshändigen Benutzer entworfen werden soll. Daher sollte die Spülmaschine auf der anderen Seite des Spülbeckens stehen.

Aufgrund der Möglichkeit, Informationen direkt im Blickfeld des Benutzers anzuzeigen, eignet sich graphische Kritik besonders für die Darstellung von Informationen während und direkt nach der Benutzeraktion. Zudem, sollte für die Anzeige der Kritik während der Benutzeraktion eine Präsentation im Blickfeld des Benutzers gewählt werden damit dem Benutzer unnötige Wechsel der Blickrichtung erspart werden.

Die zweite Variante der grafischen Präsentation von Informationen ist die Darstellung außerhalb des eigentlichen Designs. Neben der Ergänzung von textuellen Erläuterungen durch Bilder und Grafiken kann es auch eine rein grafische Präsentation von Informationen sein, in unserem Beispiel (Abbildung 2) in Form von Balkendiagrammen. Diese Darstellung scheint besonders für die Veranschaulichung von sich ständig ändernden Werten geeignet. Der Designer hat damit die Möglichkeit, während oder direkt nach der Designaktion ohne großen Aufwand festzustellen, ob sich die Aktion den gewünschten Effekt bezüglich der dargestellten Werte hat.

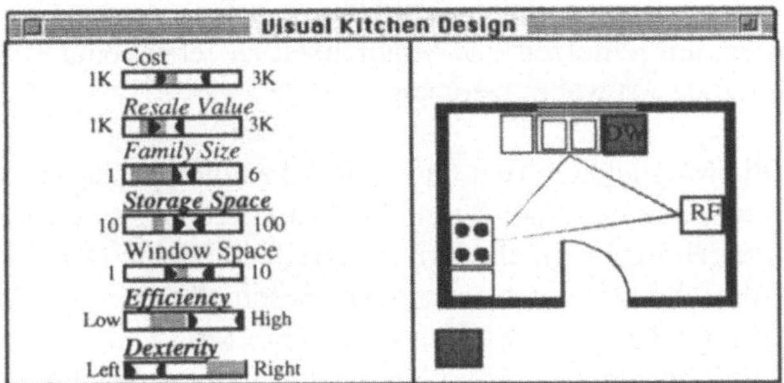

Abbildung 2: Prototyp eines Systems, welches Küchendesigner mittels graphischer Kritik unterstützt.

2.3 Kombination und Integration

Schon aus der Vorstellung der verschiedenen Möglichkeiten der Informationspräsentation wird klar, daß ein sinnvoller Weg nur mit einer Kombination beschritten werden kann. Es ist auch klar, daß der zeitliche Aspekt dabei eine entscheidende Rolle spielt. Wann eine Information präsentiert wird, hängt von zwei völlig verschiedenen und unabhängigen Faktoren ab. Der erste Faktor ist der Aufwand, den die Kritikkomponente betreiben muß, um die Information zu generieren. Wenn es nicht möglich ist, aufwendige Überprüfungen in Realzeit durchzuführen, kann diese Information auch nicht unmittelbar nach einer Designaktion angezeigt werden. Da dieser Aufwand sehr schwer abzuschätzen ist und wir uns hier auch nicht auf ein spezielles Computersystem mit einer speziellen Leistung beziehen, soll dieser Aspekt nicht weiter betrachtet werden.

Wichtiger ist die Frage nach der kognitiven Belastung des Designers: Ist der Gewinn aus der Präsentation der Information zu einem bestimmten Zeitpunkt größer als der Aufwand, den der Designer in die Auffassung der Information investieren muß.

Wenn Informationen präsentiert werden, soll diese Informationsvermittlung so erfolgen, daß der eigentliche Designprozeß möglichst wenig gestört wird. Der Designer soll auch in der Lage sein, die Informationen ggf. zu ignorieren. Auf der anderen Seite sollen die Informationen, je nach Dringlichkeit, unterschiedlich auffällig präsentiert werden. Zusätzlich sollen Informationen auch so dargestellt werden, daß sie dem Designer ohne großen zusätzlichen Aufwand zugänglich sind.

Um dies erreichen zu können, können verschiedene Modelle der Kombination von grafischer und textueller Kritik verwendet werden. Die offensichtlichste Möglichkeit ist, textuelle Informationen durch grafische Mittel, wie Bilder und Grafiken zu ergänzen. Außerdem kann die gleiche Mitteilung als Text und als Graphik redundant dargestellt werden, um Benutzern die Möglichkeit zu geben, eine Mitteilung auf die individuell bevorzugte Art wahrzunehmen.

Zudem gibt es die Möglichkeit Text und Graphik zeitlich gestaffelt einzusetzen. So ist es z.B. möglich, daß Designer durch das Selektieren eines grafisch annotierten Objekts auf die zugehörige textuelle Kritik zugreifen können. Damit können weitere Details der Kritiknachricht oder eine exaktere Beschreibung des Problems gegeben werden. Diese Möglichkeit der Staffelung ist eine Möglichkeit, das Problem des eingeschränkten Codes bei der Annotierung von Designobjekten zu überwinden. Außerdem können textuelle Erklärungen graphisch dargestellte Kritik auch für weniger erfahrene Designer zugänglich machen und ihnen dadurch erlauben den graphischen „Code" zu erlernen.

Aber auch die umgekehrte Folge der Informationsdarstellung kann sinnvoll sein. So können die Objekte, auf die sich eine textuelle Kritik bezieht, graphisch markiert werden. In dem VDDE-System kann der Designer auf eine im Meldungsfenster angezeigte Meldung klicken, um die von der Meldung betroffenen Objekte im Design durch Blinken grafisch hervorzuheben. Darüber hinaus können Kritikinformationen durch grafische Demonstrationen oder Simulationen erläutert werden. Zum Beispiel in PetriNED [12], einer Designumgebung für Petri-Netze, welche graphische Methoden der Kritikdarstellung benutzt, können, ausgehend von einer textuellen Kritik in einem Fenster, Simulationen von Deadlock-Situationen abgerufen werden (siehe Abbildung 3).

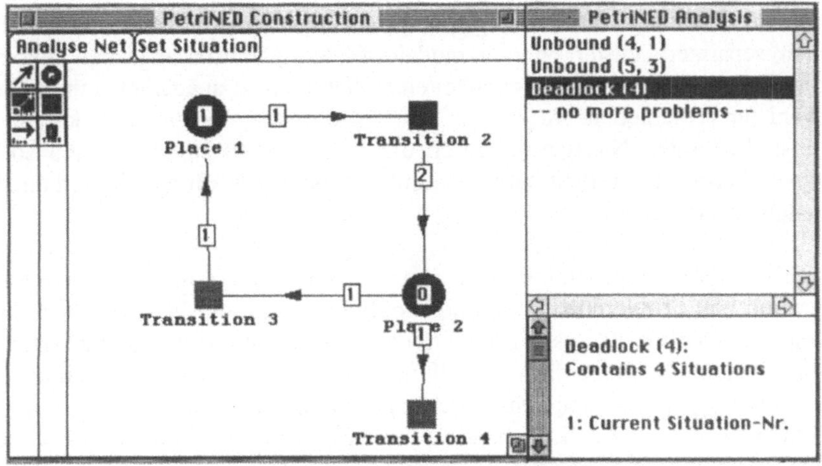

Abbildung 3: Kombination von textueller Kritik (Stichworte oben rechts) und graphischer Erklärung (Simulation im Design, Protokoll unten rechts) in PetriNED.

3 Diskussion und Ausblick

Obwohl es aus den Erfahrungen mit den bisherigen Kritiksystemen offensichtlich ist, daß eine Kombination der verschiedenen Arten der Informationspräsentation notwendig und sinnvoll ist, gibt es hierzu bisher wenig Realisierungen. Im VDDE-System kann der Designer den Bezug zwischen einer Meldung und den betroffenen Objekten durch Klicken auf die Meldung darstellen lassen; in PetriNED sind erste Erfahrungen mit der Präsentation von Simulationen und Demonstrationen von möglichen Problemen gemacht worden. Diese erlauben dem Designer, mehr Informationen über das Systemverhalten zum Ablaufzeitpunkt bereits zum Designzeitpunkt zu berücksichtigen.

Alle anderen dargestellten Aspekte der Kombination der Darstellungsarten und des Zeitpunkts sind bisher nur theoretisch betrachtet worden. In den nächsten Schritten sollen Prototypen gebaut beziehungsweise so verfeinert werden, daß die Aspekte in ausführlichen Benutzertests untersucht werden können. Es ist offensichtlich, daß nicht intuitiv festzulegen ist, wie kompliziert eine Kodierung für die grafische Präsentation von Informationen sein darf, und entsprechend, welche Informationen erst auf Anfrage textuell zur Verfügung gestellt werden sollten.

Neben der Integration von verschiedenen Techniken sollten auch Möglichkeiten für deren Individualisierbarkeit untersucht werden. So sollte der Designer einstellen können, ob er zunächst nur ein Flag, eine grafische Annotation, eine textuelle Kritik in einem separaten Fenster, oder ein modales Meldungsfenster angezeigt bekommen möchte. Diese Einstellungsmöglichkeiten sollen auch so ausgelegt sein, daß diese Auswahl für verschiedene Arten von Meldungen unterschiedlich sein können. Bei der anschließenden Navigation durch die ergänzenden Informationen soll der Designer ebenso die Freiheit haben, die Informationsdarstellung, die ihm am angenehmsten ist, zu wählen.

Ein weitere Forschungsrichtung ist die Untersuchung von Möglichkeiten der Integration von Problemdetektion und Problembehandlung. Wie bereits diskutiert konzentrieren sich Kritiksysteme traditionellerweise auf die Problemdetektion und überlassen dem Designer die Behandlung des Problems. Unserer Meinung nach sollten Kritik-Systeme einen Schritt weiter gehen und nicht nur Probleme aufzeigen, sondern „konstruktive Kritik" geben und den Designer aktiv bei der Behandlung von Problemen unterstützen. Ein relativ einfaches Beispiel der Integration von Problemdetektion und Problembehandlung existiert in PetriNED. PetriNED macht Designer auf mögliche positive Eigenschaften in ihrem Design mittel graphischer Kritik aufmerksam. Wenn zwei Elemente im Design ungefähr ausgerichtet sind, so wird diese positive Eigenschaft dem Designer durch das Zeichnen einer Linie zwischen den beiden Elementen mitgeteilt. Beendet der Designer seine Aktion während eine solche Linie im Design erscheint, dann stellt das System sicher, daß die Elemente genau ausgerichtet werden. Im Benutzungsoberflächen-Design könnte ein Kritik-System welches „konstruktive Kritik" gibt dem Designer zum Beispiel eine Auswahl von sinnvollen Lösungsmöglichkeiten geben. Weitere Möglichkeiten und Techniken der konstruktiven Kritik für die Unterstützung von Benutzungsoberflächen-Designern sollten exploriert werden.

Literatur

[1]	Barber, J., S. Bhatta, A. Goel, M. Jacobson, M. Pearce, L. Penberthy, M. Shankar, R. Simpson & E. Stroulia (1992). AskJef: Integration of Case-Based and Multimedia Technologies for Interface Design Support. In J. S. Gero (Ed.) Artificial Intelligence in Design '92. Amsterdam, NL: Kluwer Academic Publishers. 457-475.

[2]	Fischer, G., A. C. Lemke, T. Mastaglio & A. I. Morch (1991). Critics: An Emerging Approach to Knowledge-Based Human Computer Interaction. *International Journal of Man-Machine Studies* **35**(5): 695-721.

[3]	Fischer, G., R. McCall & A. Morch (1989). Design Environments for Constructive and Argumentative Design. In (Ed.) *Proceedings of ACM CHI'89 Conference on Human Factors in Computing Systems.* 269-275.

[4]	Foley, J., W. C. Kim, S. Kovacevic & K. Murray (1989). Defining User Interfaces at a High Level of Abstraction. *IEEE Software* **6**(1): 25-32.

[5] Garfinkel, S. L. & M. K. Mahoney (1993). *NeXTSTEP Programming. Step One: Object-Oriented Applications.* New York: Springer-Verlag.

[6] Harstad, B., K. Nakakoji & T. Sumner (1994). Perspective-Based Critiquing: Helping Designers Cope with Conflicts Among Design Intentions. In J. S. Gero (Ed.) *Artificial Intelligence in Design '94.* Dordrecht: Kluwer.

[7] Kühme, T. & M. Schneider-Hufschmidt (1992). SX/Tools - An Open Design Environment for Adaptable Multimedia Interfaces. *Proc. Eurographics '92, Cambridge UK, Sept 7-11, 92.*

[8] Lemke, A. C. & G. Fischer (1990). A Cooperative Problem Solving System for User Interface Design. In (Ed.) *Proceedings of AAAI-90, Eighth National Conference on Artificial Intelligence.* Cambridge, MA: AAAI Press/The MIT Press. 479-484.

[9] Löwgren, J. & U. Laurén (1993). Supporting the use of guidelines and styleguides in professional user interface design. *Interacting with Computers* 5(4): 385 - 396.

[10] Luo, P., P. Szekely & R. Neches (1993). Management of Interface Design in HUMANOID. In (Ed.) *Proceedings of ACM INTERCHI'93 Conference on Human Factors in Computing Systems.* 107-114.

[11] Reiterer, H. (1993). Human Factors Based Interface Design. In T. Grechening and M. Tscheligi (Ed.) *Human Computer Interaction, Vienna Conference VCHCI '93.* Wien: Springer-Verlag. 291-302.

[12] Stolze, M. (1994).Visual critiquing in domain oriented design environments: Showing the right thing at the right place, in Gero, J. S. (ed.), *Artificial Intelligence in Design '94,* Kluver, Dordrecht.

[13] Sukaviriya, P., J. D. Foley & T. Griffith (1993). A Second Generation User Interface Design Environment: The Model and the Runtime Architecture. In (Ed.) *Proceedings of ACM INTERCHI'93 Conference on Human Factors in Computing Systems.* 375-382.

[14] Thomas, C. G. & M. Krogsoeter (1993). An Adaptive Environment for the User Interface of Excel. In (Ed.) *Proceedings of the 1993 International Workshop on Intelligent User Interfaces.* 123-130.

Uwe Malinowski
Siemens Zentralabteilung Forschung und Entwicklung (ZFE T SN 5)
D 81730 München
Uwe.Malinowski@zfe.siemens.de

Markus Stolze
Department of Computer Science, University of Colorado at Boulder
Boulder, CO 80309-0430, USA
stolze@cs.colorado.edu

Der Reisesupermarkt: Design und Entwicklung einer Interaktionsphilosophie für die Beratung in Reisebüros

Sabine Musil, Georg Pigel, Manfred Tscheligi
Universität Wien

Zusammenfassung

Der Rationalisierungsdruck schreitet auch im Dienstleistungsgewerbe immer weiter voran. Systeme, die derartige Tätigkeiten erfolgreich automatisieren können, müssen, um von den Kunden auch angenommen zu werden, ganz besonderen Anforderungen genügen: Ihre Funktionsweise muß leicht verständlich sein, sie müssen auch von Ungeübten bedienbar sein und sie müssen dem Kunden attraktiv und für ihn von Vorteil erscheinen. Besondere softwaretechnische Mittel und Methoden müssen diese Eigenschaften schon während der Entwicklung garantieren. Es wurde ein Benutzerschnittstellenprotoyp für ein System, das die dem Verkaufsgespräch vorausgehende einführende Beratungstätigkeit in Reisebüros (Travel Agency Clerk Replacement System, TACRS) automatisiert, realisiert. Der Prototyp realisiert eine auf einer Real World Metapher basierende Interaktionsumgebung - den Reisesupermarkt. Iteratives Prototyping (Papierprototyp, Videoprototyp, Softwareprototyp) war die Hauptsäule innerhalb des Entwicklungsvorganges.

1   Einleitung und Motivation

Mit der zunehmenden Verteuerung von Arbeitskraft in den hochentwickelten Industriestaaten mit dicht gewebtem Sozialnetz der westlichen Welt hat sich in den letzten Jahren herausgestellt, daß auch bisher vom Rationalisierungsdruck mehr oder weniger verschont gebliebene Dienstleistungsbranchen zu effizienterem Einsatz ihres Personals getrieben werden. Ein typisches Beispiel dafür stellen (zumindest in Österreich) die Banken dar: Kontoauszüge, die der Kunde früher beim Schalterbeamten persönlich behoben hat, werden mittlerweile fast ausschließlich vom Kontoauszugsautomaten ausgedruckt, Bargeld wird nur mehr selten vom Kassenschalter geholt, sondern immer häufiger aus dem Bankomaten gezogen. Diese Entwicklung macht auch vor anderen höherwertigen Dienstleistungsbrachen nicht halt, so auch nicht vor Reisebüros. Die dort ausgeübte Beratungstätigkeit ist ausgesprochen personal- und damit kostenintensiv und führt außerdem nur selten direkt zu einem Verkauserfolg, sprich zu einer Buchung. Das hat einerseits zur Überlegung geführt, den Kunden für die Beratung selbst zahlen zu lassen, was natürlich sehr schwer durchsetzbar ist, denn, als Analogon, wer wäre bereit im Schuhgeschäft für das Probieren der Schuhe zu zahlen? Andererseits wird nach Möglichkeiten gesucht, die dem eigentlichen Verkauf vorangehende Beratungstätigkeit zu automatisieren, wozu sich natürlich Weiterentwicklungen von

Tourismusinformationssystemen [14] anbieten. Das persönliche Beratungsgespräch soll also durch ein automatisiertes Beratungssystem, in der Folge mit dem englischen Begriff Travel Agency Clerk Replacement System (TACRS) bezeichnet, ersetzt werden.

2   Anforderungen aus Benutzersicht

TACRS können nur erfolgreich sein, wenn sie vom Kunden akzeptiert werden. Die speziellen Bedingungen der Banken, die den Automatisierungsbestrebungen in diesem Bereich zum Erfolg verholfen haben, weil sie dem Kunden klar erkennbare Vorteile bringen (schnellere Servicezeiten im Falle des Kontoauszugsdruckers bzw. Unabhängigkeit von den überaus eingeschränkten Öffnungszeiten der Banken im Falle des Bankomaten) haben etwaige Widerstände in der Akzeptanz durch die Kunden rasch überwunden. Im Falle der Reisbüros läßt sich ähnliches nicht absehen. Um die potentiellen Kunden eines Reisebüros dazu zu bewegen, die Dienste eines TACRS anstelle eines Beratungsgespräches mit einem Angestellten in Anspruch zu nehmen, muß ein solches System folgende Anforderungen erfüllen:

- Die Funktionsweise des TACRS muß vom potentiellen Kunden problemlos verstanden werden. Auch Menschen, die noch nie zuvor mit Computern in Berührung gekommen sind, müssen die grundsätzlichen Ideen, wie man mit Hilfe des TACRS zu Informationen kommt, verstehen. Die Art und Weise, wie mit dem TACRS interagiert wird, muß auf Wissen aufgebaut sein, von dem angenommen werden kann, daß nahezu jeder darüber verfügt. Die Metaphern [2] sollten deshalb aus Bereichen gewählt werden, in denen möglichst jeder bereits Erfahrungen besitzt.

- Das TACRS muß einfach bedienbar sein. Diese einfache Bedienbarkeit muß während des gesamten Entwicklungsprozesses des TACRS ständig überprüft werden. Nicht nur, daß die zugrundeliegende Metapher sorgfältig ausgewählt werden muß, sondern die Metapher selbst und die Implementierung der Metapher muß wiederholt evaluiert werden. Essentiell dabei ist, daß Anwender in diese Evaluationen eingebunden werden. Dabei spielt iteratives Prototyping eine wesentliche Rolle. Prototyping ist nicht zuletzt auch aufgrund der metaphorischen Umsetzung der Interaktion von großer Bedeutung.

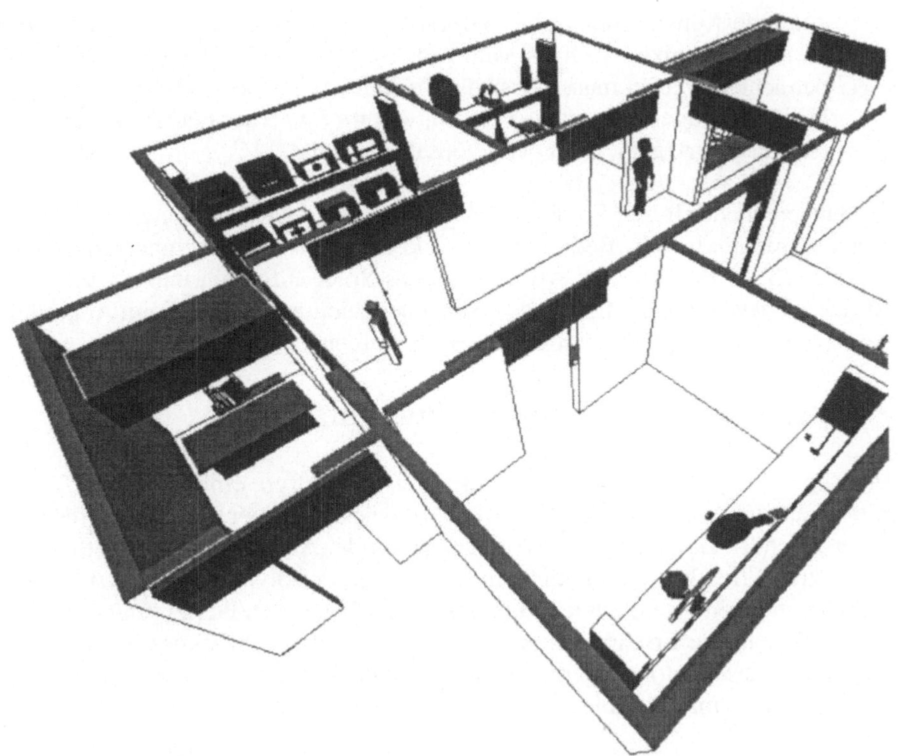

Abb.1   Das TACRS aus der Vogelperspektive

- Das TACRS muß anziehend auf potentielle Kunden wirken. Traditionelle Techniken der Mensch-Computer Interaktion wie das Doppelklicken mit der Maus auf netten kleinen Piktogrammen oder das Zeigen mit dem Finger auf einen Touchscreen sind dazu nicht in dem geforderten Maße instande. Alle diese Interaktionstechniken haben die unangenehme Eigenschaft, Ähnlichkeiten mit Arbeit aufzuweisen, und Arbeit ist wahrscheinlich das Letzte, woran der Kunde eines Reisebüros bei der Planung seines Urlaubs erinnert werden möchte. Urlaub, und alles was damit in Zusammenhang steht, soll aufregend, spannend und sensationell sein. Die aufregendste Entwicklung in der Informatik in jüngster Zeit ist (zumindest wenn man den Berichten und Kommentaren in diversen Medien [1,12,15] Glauben schenkt, aber man darf wohl davon ausgehen, daß das der Wissensstand des Durchschnittsbürgers zum Thema Informatik ist) die Entwicklung von Virtual Realities [10]. Wenn man also außergewöhnliche Reisen verkaufen will, sollte man es mit einem außergewöhnlichen "modernen" (im Sinne von: *en vogue*) System tun, das zumindest einen Hauch von Virtual Reality beinhaltet. Wenn man Virtual Reality Techniken auch als Mittel zur Gestaltung von Benutzerschnittstellen betrachtet, erkennt man, daß sich mit deren Hilfe auch Real World Metaphern [17] überzeugender gestalten lassen, was im Sinne der beiden obigen Punkte ein weiteres Argument für deren Einsatz darstellt.

Ein entsprechend diesen Forderungen gestaltetes TACRS hat gute Chancen vom Kunden nicht als das Bedienen einer komplizierten Maschine, sondern als futuristisches Erlebnis, eine Art aufregendes Spiel aufgefaßt und als solches im Austausch gegen ein einführendes persönliches Beratungsgespräch mit Angestellten des Reisebüros akzeptiert und frequentiert zu werden.

3   Iteratives Prototyping als Mittel zum Zweck

Im Zuge des Projekts wurde ein Software-Prototyp eines solchen TACRS entwickelt, wobei besonderer Wert auf die Entwicklung einer obigen Anforderungen entsprechenden Benutzerschnittstelle gelegt wurde. Als entwicklungsmethodisches Mittel zum Erreichen dieses Ziel wurde iteratives Prototyping eingesetzt, wobei in jeder Iteration ein detaillierterer, mit höherem Aufwand erstellter Prototyp entwickelt wurde, der auf die Erfahrungen und Kritikpunkte, die sich bei der Evaluation des vorhergehenden Prototypen ergeben hatten, Rücksicht nahm. Folgende wesentliche Schritte können im Zuge dieses Prototypings unterschieden werden:

- Das Suchen nach geeigneten Metaphern, auf denen die Interaktion zwischen Kunden und TACRS ausgebaut werden kann.

- Das Entwerfen von Papierprototypen mit deren Hilfe in einer nachfolgenden Evaluation in einer Wizard-of-Oz [16] Technik die geeignet erscheinenden Metaphern auf ihre Verständlichkeit und Interpretierbarkeit [8] in einer Sitzung mit potentiellen Anwendern getestet werden kann.

- Die Erstellung eines Videoprototypen, um während der Evaluation einen genaueren Eindruck von der grundsätzlichen Interaktionsphilosophie des TACRS zu vermitteln, und die Akzeptanz und Verständlichkeit der Interaktionstechniken des Systems zu prüfen.

- Die Entwicklung eines Softwareprotoypen auf Basis eines PC-basierten VR Tools, der als real existierendes Programm dem Einsatz und einer anschließenden Benutzerstudie offensteht.

Jeder einzelne dieser Schritte wurde, wenn nötig, d.h. wenn die Evaluation nicht zufriedenstellend verlaufen ist und es auf Grund der Mängel des existierenden Prototyps zu früh erschien, zum nächstdetaillierten Prototyp überzugehen, wiederholt ausgeführt. So sind beispielsweise mehrere Versionen von Papierprototypen entstanden.

3.1 Finden einer geeigneten Metapher

Zu Beginn des Designs des TACRS galt es, eine Metapher zu finden, deren Funktionsweise als allgemein bekannt vorausgesetzt werden kann, die die Grundlage für die Interaktion mit dem System bildet. In einer ersten Diskussion wurde die Notwendigkeit klar, eine Real World Metapher (eine an einem tatsächlich in der täglichen Realität existierenden System orientierte) und keine abstrakte Metapher als Grundlage der Interaktionsweise zu wählen, da eine solche Metapher einfacher zu verstehen (man muß nicht erst die in der abstrakten Metapher vorhandene Abstraktion eines Sachverhaltes begreifen) und vohandenes Wissen gegenständlicher einsetzbar ist. Bei einem anschließenden Brainstorming der am Projekt beteiligten Mitarbeiter zeichneten sich zwei Metaphern als erfolgversprechend ab:

- Die Metapher eines Supermarktes, in dem der Kunde in separaten Räumen seine Wünsche bezüglich Ort, Klima, Sportmöglichkeiten etc. aus den Regalen auswählt und in einem Einkaufswagen sammelt. An der Kassa erhält er dann an Stelle einer Rechnung Vorschläge für Reiseziele, die seinen Wünschen entsprechen.

- Die Metapher eines Raumes, der Figuren enthält, die verschiedene Aspekte einer Urlaubsreise wie Reiseverkehrsmittel, Klima, etc. verkörpern. Diese Figuren lassen sich Öffnen und erlauben Auswahl und Entnahme der gewünschten Option wie aus einem dreidimensionalen Kartenständer (dreidimensionale Kartenständer erinnern an die Realisierung von Cone Trees in [13]).

Beide Metaphern sind in einer Diskussion mit potentiellen Anwendern auf ihre Plausibilität hin überprüft und für grundsätzlich geeignet befunden worden.

3.2 Papier-und-Bleistift Prototypen

Das Projektteam ging nun daran, Papierprototypimplementierungen dieser beiden Metaphern zu schaffen: Typische Aufgabenstellungen (im Sinne des Task Centered Designs [7]) des Kunden eines TACRS wurden als Folge von Zeichnungen dessen, was der Anwender im tatsächlich implementierten Systems sehen würde, dargestellt. Bei der darauffolgenden Evaluation wurden die Abfolge dieser Bilder der Testperson vorgeführt. Die Ablaufsteuerung im Falle der Alternativen und mögliche Ausgaben oblagen nicht einem Computersystem sondern dem Präsentator des Papierprototypen. Ein solcher Papierprototyp vermittelt also einen ungefähren Eindruck vom Aussehen und der Funktionsweise des zu designenden Systems. Diese Papierprototypen mußten mehrmals erstellt werden, da sich viele Unklarheiten der Metapher oder zusätzlich notwendige Details erst im Laufe der Präsentationen herauskristallisierten. Beispielsweise war im Falle der Kartenständer nicht klar, was mit den ausgewählten Objekten passieren sollte.

Als Konsequenz des Papierprototypings ergab sich, daß die Testpersonen mit der Supermarktmetapher weitaus besser zurecht kamen. Außerdem waren keinerlei Erklärungen darüber, wie diese Metapher aufzufassen sei, nötig, während sich die Kartenständermetapher als weniger intuitiv erwies. Daher wurde beschlossen, die Arbeitskraft des Projetteams zur Gänze der Weiterführung der Supermarktprototypen zu widmen.

Die Interaktionsphilosophie des Reisesupermarktes besteht darin, daß sich der Kunde seinen Wunschurlaub selbst zusammenstellt, indem er die gewünschten Eigenschaften in dinghafter Symbolform den Regalen des Supermarkts entnimmt und in seinem Einkaufswagen sammelt: So geht er etwa in die Sportabteilung und gibt einen Tennisschläger und eine Golfausrüstung in den Einkaufswagen, zum Zeichen dafür, daß er in seinem Urlaub Tennis und Golf spielen möchte. Oder in der Klimaabteilung entnimmt man dem Regal eine Sonne und ein Thermometer, das 25 Grad Celsius anzeigt, als Symbole für das gewünschte Wetter. Es besteht keinerlei Zwang, die Abteilungen des Supermarktes in einer bestimmten Reihenfolge zu durchwandern. Auch das Zurückstellen und Umtauschen von Symbolen steht völlig frei. An der Kasse schließlich bekommt man nicht die Rechnung präsentiert, sondern zeigt an, daß die Zusammenstellung beendet ist, genauso wie in einem wirklichen Supermarkt der Gang zur Kasse das Einkaufen beendet. An Stelle der Rechnung erhält der Kunde eine auf Basis der "eingekauften" Urlaubseigenschaften zusammengestellte Liste von für ihn interessanten Reisezielen. Die Buchung selbst findet noch außerhalb des Systems statt.

3.3 Video als Prototypingwerkzeug

Ein Papierprototyp kann, ungeachtet allen Einsatzes und Geschicks der Zeichner und Präsentatoren, nur einen sehr ungefähren Eindruck darüber, wie das fertige System aussehen soll, vermitteln. Besonders kraß tritt das natürlich zutage, wenn der Papierprototyp ein System darstellen soll, das Techniken der Virtual Reality beinhalten soll. Die softwaremäßige Realisierung eines solchen Systems, und sei es auch nur in der Form des einfachsten Prototypen mit sehr mächtigen Werkzeugen, ist aber zu aufwendig, als daß man sie nur auf Grund einer auf einem Papierprototypen basierenden Evaluation in Angriff nehmen möchte: Die Wahrscheinlichkeit, den ganzen Protoypen neu erstellen zu müssen, ist zu groß.

Ein Mittelding zwischen dem sehr einfach zu erstellenden Papierprototypen, der aber nur einen vagen Eindruck des vollständigen Systems zu vermitteln vermag, und einem aufwendigen Softwareprototypen bietet sich in Form eines mit Video erstellten Prototypen an [4]. Die Kulisse des Ferien-Supermarktes wurde aufgebaut und ein Anwender bei einer typischen Interaktion (Auswahl eines Aktivurlaubs mit speziellen Sportarten an einem südlichen Gewässer) gefilmt. Durch geeignete Schnitte wurde die Illusion eines funktionierenden Systems vorgetäuscht. Da ein Videofilm natürlich keine Interaktion zwischen den Testpersonen und dem System zuläßt, ist die Auswahl eines möglichst typischen und umfassenden Tasks, der im Videofilm abgewickelt wird, von herausragender Bedeutung. Dem Projektteam wurde diese Auswahl aber durch seine Erfahrungen mit der Vorführung der Videoprototypen erheblich erleichtert. Einige technische Unzulänglichkeiten des Videos, die sich mit dem Einsatz besserer technischer Ausrüstung (vor allem bei Einsatz eines Schneidecomputers) hätten vermeiden lassen können, taten der überzeugungskraft in der folgenden Evaluationssitzung keinen Abbruch.

Die Details der Interaktion, die im Papierprototypen nur als Folge für sich statischer Bilder zu sehen war, konnten nun erstmals vorgeführt und schlüssig diskutiert werden. Als Beispiel für die Designunzulänglichkeiten, die durch die Evaluation des Videoprototypen durch die Anwendergruppe gefunden werden konnten, sei das nur einmalige Vorhandensein der Objekte in den Regalen genannt. Als Folge dessen wurde angeregt, aus dem Regal entnommene Objekt (z.B. ein Kletterseil als Symbol dafür, daß jemand in seinem Urlaub Bergsteigen will) vom System sofort durch ein neues zu ersetzen. Das stellt keine Verletzung der Metapher dar, denn auch in einem gutgeführten realen Supermarkt sorgt das Personal dafür, daß der Kunde nicht vor leeren Regalen steht.

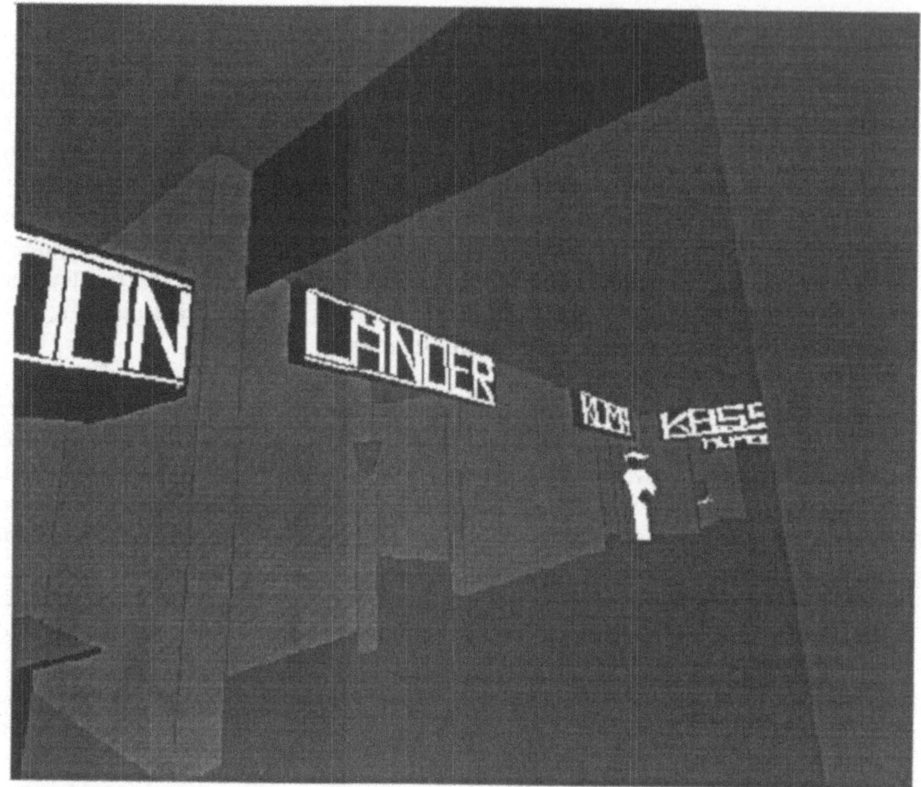

Abb. 2   Ein typischer Gang im TACRS

3.4 Softwareprototyp

Man unterscheidet bei Virtual Reality Systemen zwischen solchen, die den Anwender zwingen, spezielle Hardware wie Helme, HMDs, Datenhandschuhe etc. [9,11,18] zu tragen und solchen die auf diese spezielle Ausrüstung verzichten [5,6]. Das Projektteam bevorzugte die letztgenannten Umgebungen und versuchte, mit einem Bildschirm zur Wiedergabe der dreidimensionalen Darstellung des Systems und einem speziellen Eingabegerät (Spacemaus oder ähnlichem) das Auslangen zu finden. Die Begründung dafür war nicht nur in einem allzu kleinen Projektbudget zu finden:

- Wieviele Kunden würden wohl das TACRS benützen, wenn sie zuerst einen Helm aufsetzen, das HMD adjustieren und in den Datenhandschuh schlüpfen müßten? Pointiert formuliert wäre der Urlaub wahrscheinlich vorbei, bevor der Kunde das TACRS erfolgreich benützt hätte.

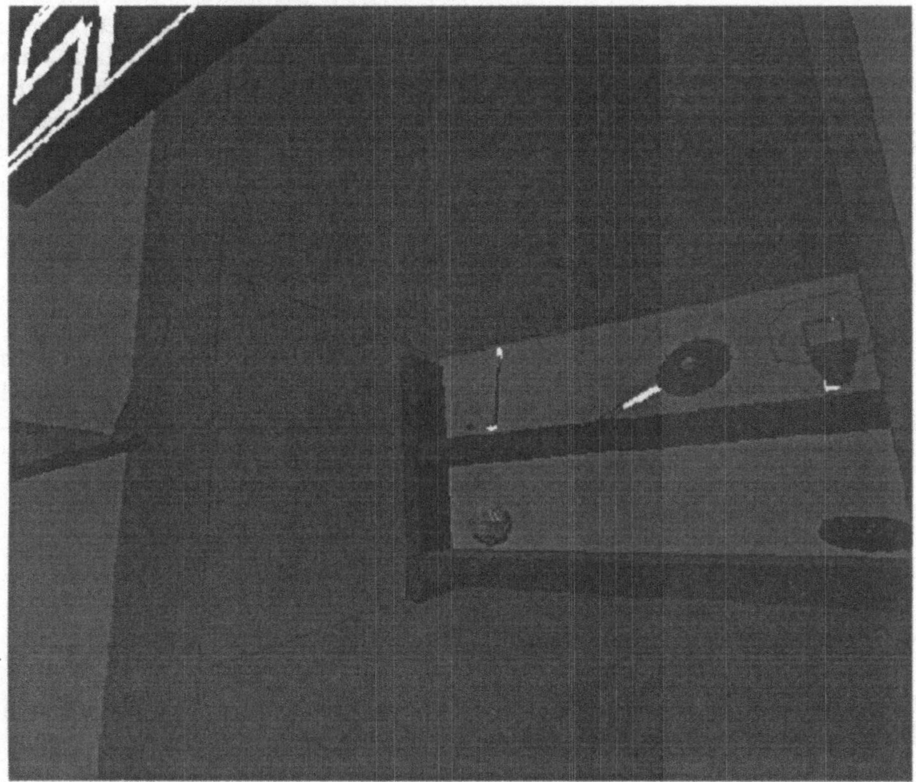

Abb. 3   Der Raum zur Auswahl von Sportarten, die im Urlaub betrieben werden sollen

- Es ist nicht angenehm eine derartige Ausrüstung zu tragen, ganz abgesehen davon, daß die Frage möglicher Gesundheitsschäden durch diese Ausrüstung nicht abgeklärt ist [3].

- Das TACRS soll zur effizienteren Nutzung des Personals dienen. Wenn aber eine spezielle Ausrüstung zu tragen wäre, um das TACRS zu benützen, wäre Personal vorzusehen, um den Kunden beim An- und Ablegen zu helfen. Die wirtschaftliche Sinnhaftigkeit des TACRS wäre nicht gegeben.

Zur Implementierung des Softwareprototyps wurde ein PC-basiertes Werkzeug, VIRTUS Walkthrough, herangezogen. Selbstverständlich ist die Leistungsfähigkeit dieses Werkzeugs nicht mit einem auf High End Graphicworkstations basierenden

Systemen zu vergleichen, für das Ziel dieses Projekts, eine evaluierbare Benutzeroberfläche für ein TACRS zu schaffen, erwies es sich aber als durchaus hinreichend. Wie bei Benutzerschnittstellenprototypen üblich wurde natürlich nicht das gesamte Funktionalitätsspektrum implementiert (die einem fertigen System unterliegende Datenbankfunktionalität war für die Entwicklung der Benutzeroberfläche beispielsweise gänzlich uninteressant) sondern es wurde versucht, vor allem die Teile des TACRS zu implementieren, die für die Evaluation relevante Aspekte beinhalten. Interessant war etwa die Orientierung des Anwenders innerhalb des TACRS (siehe Abb. 2) oder die Entnahme von Objekten aus dem Regal (siehe Abb. 3).

4   Schlußfolgerungen

Die Rationaliserung von Dienstleistungsbranchen verlangt nach Systemen, die auch von Kunden ohne fachspezifisches Vorwissen gerne und mit dem Gefühl, damit gegenüber der Anspruchnahme menschlicher Dienste zu profitieren, bedient werden. Um das zu gewährleisten, muß

- das Funktionsprinzip klar verständlich,

- allgemeine Bedienbarkeit garantiert und

- die Attraktivität des Systems gegeben sein.

Eine Methodik, diese Zielsetzungen zu erreichen, besteht in einer Real World Metapher basierten Benutzeroberfläche, die im Zuge eines iterativen Prototypings entwickelt wird und "latest trends" der Informatik wie Virtual Reality Technology beinhaltet.

In einem beispielhaften Projekt wurde unter Wahrung dieser Methodik ein System für die Rationalisierung von Reisebüros, TACRS, entwickelt.

Die Benutzeroberfläche des resultierenden Protoypen zeigt alle grundlegenden Interaktionstechniken die sich als Folge der vorangegangenen Prototypingzyklen als nötig erwiesen haben. Die zukünftige Arbeit in diesem Projekt wird die Implementierung eines leistungsfähigeren Prototypen umfassen. Auch werden die Interaktionstechniken auf Grund von Tests mit dem derzeitigen System effizienter gestaltet werden können. Das Funktionalitätsspektrum wird sukzessive erweitert. So kann natürlich auch daran gedacht werden, den Buchungsvorgang selbst einzugliedern und Multimediafunktionalitäten in das System zu integrieren.

Ein wesentliches Hauptziel war die Suche nach alternativen Benutzerschnittstellentechniken und alternativen Metaphern für einen Nichtstandardproblembereich. Alternative Benutzerschnittstellenkonzepte können nur bedingt mit den üblichen Werkzeugen gelöst werden. Besonders in den Anfangsphasen gilt es, Techniken zu finden. die das kreative Umsetzen von Alternativen erlauben. Das in diesem Projekt eingesetzte Videoprototyping hat sich als besonders brauchbar erwiesen, Ideen und Designphilosophien entsprechend zu transportieren.

Literaturverzeichnis

[1]   C. Bauer. "Psychotrip im Cyberspace ". OUTPUT + micro, No. 10, pp. 20-23, 1992.

[2]   B. Indurkhya. Metaphor and Cognition. Kluwer Academic Publishers, 1992.

[3]   R.S. Kalawsky. The Science of Virtual Reality and Virtual Environments: A Technical, Scientific and Engineering Reference on Virtual Environments. Addison-Wesley, 1993.

[4]   R. Kolli. "Concept Prototyping: A Designer Based Approach to Envisioning User Interfaces". In: Proceedings of Computergraphics'92, pp. 355-367, 1992.

[5]   M.W. Krueger. "VIDEOPLACE-An Artificial Reality". In: CHI'85 Conference Proceedings, pp. 35-40, ACM, 1985.

[6]   M.W. Krueger. Artificial Reality II. Addison-Wesley, 1991.

[7]   C. Lewis and J. Rieman. Task-Centered User Interface Design - A Practical Introduction. Shareware, P.O. Box 1543, Boulder, CO 80306, 1993. Original files for the book are available via anonymous ftp from ftp.cs.colorado.edu.

[8]   S. Musil and M. Tscheligi. Breaking the Chains: How to Design Non-Standard Interaction Environments. Tutorial Notes CHI'94, Boston, 1994.

[9]   R. Pausch. "Virtual Reality on Five Dollars a Day". In: S.P. Robertson, G.M. Olson, and J.S. Olson, Eds., Reaching Through Technology, CHI'91 Conference Proceedings, pp. 265-270, ACM, Addison-Wesley, 1991.

[10] K. Pimentel and K. Teixeira. Virtual Reality: through the new looking glass. Intel/Windcrest/McGraw-Hill, Inc., 1993.

[11] F. Rabb, E. Blood, R. Steiner, and H. Jones. "Magnetic Position and Orientation Tracking System". IEEE Transaction on Aerospace and Electronic Systems, Vol. 15, No. 5, pp. 709-718, September 1979.

[12] H. Rheingold. Virtual Reality. Summit Books, 1992.

[13] G.G. Robertson, S.K. Card, and J.D. Machinlay. "Information Visualization using 3D Interactive Animation". Communications of the ACM, Vol. 36, No. 4, pp. 57-71, April 1993.

[14] W. Schertler, B. Schmid, A. M. Tjoa, and H. Werthner, Eds. Information and Communications Technologies in Tourism, Springer Verlag Wien New York, Innsbruck Austria, 1994.

[15] Time International Magazine. "Hot Stuff!, From the land of high-tech gizmos, more Techno Toys". p. 51, December 1992.

[16] T. S. Tullis. "Screen Design". In: M. Helander, Ed., Handbook of Human-Computer Interaction, Chap. 18, pp. 377-411, Elsevier Science Publishers B.V. (North Holland), 1988.

[17] K. Väänänen and J. Schmidt. "User Interfaces for Hypermedia: How to Find Good Metaphors?". In: C. Plaisant, Ed., Proceedings of the CHI'94, Conference Companion, pp. 263-264, ACM, ACM ISBN: 0-89791-651-4, 1994.

[18] VPL-Research. DataGlove Model 2 Users Manual. 1987.

Sabine Musil, Georg Pigel, Manfred Tscheligi
    Vienna User Interface Group
    Lenaugasse 2/8, A-1080 Wien
    Universität Wien
    {musil, pigel, tscheligi}@ani.univie.ac.at

PIA - Eine transparente Benutzungsschnittstelle für motorisch behinderte Personen

Christian Piwetz

Universität - Gesamthochschule - Essen

Zusammenfassung

Moderne graphische Benutzungsoberflächen und Anwendungsprogramme sind meist nicht darauf ausgelegt, von Menschen mit Behinderungen bedient zu werden. In diesem Beitrag wird eine prototypische Eingabehilfe für den Apple Macintosh vorgestellt, die motorisch behinderten Menschen die Texteingabe erleichtern soll. Der Anwender interagiert nicht mehr direkt mit dem Anwendungsprogramm, sondern mit Hilfe der Benutzungsschnittstelle, die als eigenständige Schicht über dem Anwendungsprogramm liegt. Die Verbindung zwischen dem Anwendungsprogramm und der Eingabehilfe erfolgt über einen Standardmechanismus des Macintosh-Betriebssystems, so daß prinzipiell alle Programme mit der Eingabehilfe bedient werden können. Zusätzlich werden kontextabhängige Vorhersagen generiert, die der Anwender sehr leicht in seinen eigenen Text übernehmen kann.

1 Einleitung

Die in den letzten Jahren auf den Markt gebrachten Computer, Benutzungsoberflächen und Anwendungsprogramme rühmen sich damit, besonders anwenderfreundlich zu sein. Dies trifft sicherlich zu, wenn ein Anwender dieser Computer und Programme beide Arme vollständig kontrollieren kann. Ist der Anwender jedoch vom Hals abwärts gelähmt, kann er diese Computer geschweige denn eines dieser „modernen" Anwendungsprogramme nicht bedienen. Wie Glinert und York in [8] schreiben: „Most hardware and software has been, and continues to be, designed by and for people who are not disabled.".

Gerade aber der Computer stellt die wichtigste technische Entwicklung der letzten Jahre dar, die in vielen Bereichen des täglichen Lebens zum Einsatz kommt. Durch die fehlende Unterstützung von Menschen mit Behinderungen wird eine große Bevölkerungsgruppe von den Möglichkeiten des Computers ausgeschlossen.

Das hier vorgestellte Programm PIA („Predictive Interface Agent") stellt eine Benutzungsoberfläche für motorisch behinderte Personen dar, die die Standardtastatur nur langsam, mit Mühe oder gar nicht bedienen können. Ausgangspunkt für die Entwicklung von PIA ist das Programm *Reactive Keyboard* [6, 7], das an der Universität Calgary entwickelt wurde. Während beim Reactive Keyboard die Textvorhersage im Vordergrund stand, lag der Schwerpunkt bei der Entwicklung von PIA bei der generellen Unterstützung der Texteingabe für verschiedene

Behinderungen und der vollständigen Loslösung der Eingabehilfe von dem jeweiligen Anwendungsprogramm (Textverarbeitung, Tabellenkalkulation usw.). „Transparent" bedeutet also in diesem Zusammenhang, daß das eigentliche Anwendungsprogramm nichts von der Benutzungsschnittstelle PIA bemerkt.

Die beschriebene Software wurde während einer Kooperation der Universitäten Dortmund und Essen mit dem Forschungsinstitut Technologie-Behindertenhilfe, Wetter-Volmarstein entwickelt und war Teil der Diplomarbeit des Autors [17].

Im folgenden zweiten Abschnitt werden neben einer Erläuterung der möglichen Eingabestrategien bereits existierende Eingabehilfen für Menschen mit Körperbehinderungen besprochen. Der dritte Abschnitt beschreibt die funktionalen Anforderungen an PIA. Die daraus resultierenden Oberflächenelemente werden im vierten Abschnitt erläutert. Im fünften Abschnitt werden einige Aspekte der Implementierung beschrieben. Die Schlußbetrachtung gibt einen Ausblick auf die mögliche weitere Entwicklung von PIA.

2 Texteingabehilfen für körperbehinderte Menschen

In diesem Kapitel werden zuerst die verschiedenen Eingabestrategien für Computerprogramme erläutert. Danach erfolgt ein kurzer Überblick über bereits existierende Eingabehilfen für Menschen mit Körperbehinderungen.

2.1 Eingabestrategien

Bei der Texteingabe mit dem Computer gibt der Anwender Buchstaben und Ziffern ein, die die einzelnen Worte und Sätze des Textes bilden. Zusätzlich macht er Eingaben, die nicht der Texterstellung dienen, d.h. er löscht Zeichen, bewegt die Schreibmarke oder formatiert den Text. Die Zahl der für die Texterstellung möglichen Zeichen beträgt für die deutsche Sprache mindestens 45: die 26 Buchstaben, das Leerzeichen, die Umlaute und das ß sowie die Satzzeichen. Hinzu kommen noch die zehn Ziffern. Wie soll nun ein Computerbenutzer mit Behinderungen, der keine Standardtastatur sondern nur wenige Schalter bedienen kann, diese verschiedenen Zeichen eingeben können? Hierbei werden drei Eingabestrategien [1] unterschieden:

Direkte Auswahl

Es steht für jedes Zeichen jeweils ein Schalter zur Verfügung. Beispiele sind die normale Tastatur, aber auch eine am Bildschirm angezeigte Tastatur, die über ein

Zeigegerät bedient wird. Großbuchstaben und Sonderzeichen werden über zusätzliche Umschalter erzeugt.

Codierung

Jedem Auswahlelement wird ein spezieller Code zugeordnet, den der Benutzer für die Auswahl eingeben muß. So ordnet das Morsealphabet jedem Zeichen eine Kombination aus langen („Strichen") und kurzen („Punkten") Impulsen zu.

Scanning

Die zur Auswahl stehenden Zeichen werden zeilen- und spaltenweise angeordnet. Eine Marke durchläuft in einem bestimmten Takt das Zeichenfeld zuerst zeilen- und dann spaltenweise. Durch Betätigung eines Tasters wählt der Anwender zuerst die Zeile und danach die Spalte aus, in der das gewünschte Zeichen steht.

2.2 Vorhandene Texteingabehilfen

Menschen mit Körperbehinderungen bilden im Gegensatz z.B. zu der Gruppe der blinden Menschen eine sehr heterogene Gruppe, was sich u.a. an der Vielzahl der verfügbaren Eingabehilfen widerspiegelt. Zur Verdeutlichung zwei Beispiele: für einen Computerbenutzer mit einem gelähmten Arm reicht schon Zusatzsoftware aus, die ihm die sequentielle Eingabe von Text ermöglicht. Dagegen ist der Hard- und Softwareaufwand für einen Menschen, der vom Hals abwärts gelähmt ist, sehr viel größer.

Verschiedene Hardwarehilfen (Folientastatur, Fußtastatur, Links- oder Rechtshändertastatur etc.) erweitern bzw. ersetzen die Standardtastatur. Sehr häufig kann aber die Eingabestrategie *direkte Auswahl*, die diesen Eingabegeräten zugrunde liegt, nicht mehr verwendet werden. Für die Kodierung oder das Scanning ist eine Kombination aus Hard- und Software notwendig. Die Auslösung der für das Scanning benötigten Takte ist auf unterschiedlichste Weisen möglich (Shein in [12]). Beispiele sind ein Schlagkissen (ein Impuls) [10], ein Saug-Blas-Rohr (je ein Impuls durch Saugen und Blasen) [23] oder ein durch die Zunge ausgelöster Schalter [5]. Die Umsetzung der Impulse muß über spezielle Software erfolgen.

Alle diese Eingabehilfen machen den Computer für einen behinderten Menschen erst nutzbar. Daneben existiert zur Beschleunigung der Texteingabe, die gerade mit der Eingabestrategie Scanning sehr mühselig und langsam ist, Zusatzsoftware, die Vorhersagen darüber erstellt, wie der gerade eingegebene Text weitergehen könnte. Der Bericht [13] der Universität Edinburgh enthält eine Übersicht über fast alle zum Erscheinungszeitpunkt erhältlichen Vorhersageprogramme für den

Apple Macintosh, MS-DOS-Computer und den in Großbritannien sehr verbreiteten BBC-Computer.

Alle diese Vorhersageprogramme können nur über die Tastatur oder die Maus bedient werden [13], unterstützen also nur die Eingabestrategie *direkte Auswahl*. Bei besonders schweren Behinderungen reicht dies jedoch nicht mehr aus.

3   Funktionelle Anforderungen an PIA

Im folgenden werden die Anforderungen an PIA erläutert, die für das Design der Benutzungsoberfläche und die Implementierung maßgeblich von Bedeutung waren. Dies beinhaltet neben den allgemeinen Anforderungen (transparenter Zugriff und Eingabeunterstützung durch Vorhersagen) die Definition von drei Benutzergruppen.

3.1   Transparenter Zugriff

Unter transparentem Zugriff (engl. transparent access [20, 22]) ist zu verstehen, daß die Eingabehilfe (PIA) für das eigentliche Anwendungsprogramm unsichtbar ist. PIA ist ein eigenständiges Programm, das zusätzlich zum Anwendungsprogramm auf dem Computer abläuft. Die Eingaben des Benutzers werden von PIA entgegengenommen und - zum Teil verändert - an die Zielanwendung weitergesendet. Diese verarbeitet die gesendeten Daten wie Eingaben von der Tastatur oder Maus. Die Zielanwendung „merkt" also gar nicht, daß die Eingabe nicht direkt vom Benutzer, sondern von der Eingabehilfe PIA kommt. Zusätzlich kann der Anwender auch normal mit der Zielanwendung arbeiten. PIA bildet also eine zusätzliche Schicht zwischen der Anwendung und dem Benutzer (Abbildung 1, wobei Pfeile einen Datenfluß kennzeichnen).

Ein Interface Agent ist ein eigenständiges Programm, das für den Anwender selbständig Aufgaben ausführt [15]. PIA arbeitet dabei als Mittler zwischen Benutzer und Anwendungsprogramm [19]. Die Kommunikation zwischen PIA und der Zielanwendung sollte so realisiert werden, daß möglichst alle Anwendungsprogramme für den Apple Macintosh mit PIA zusammenarbeiten können. Dadurch ist der Anwender bei der Wahl der Zielanwendung keinen Beschränkungen unterworfen. Er kann z.B. mit der Textverarbeitung arbeiten, die seinen Ansprüchen am besten gerecht wird.

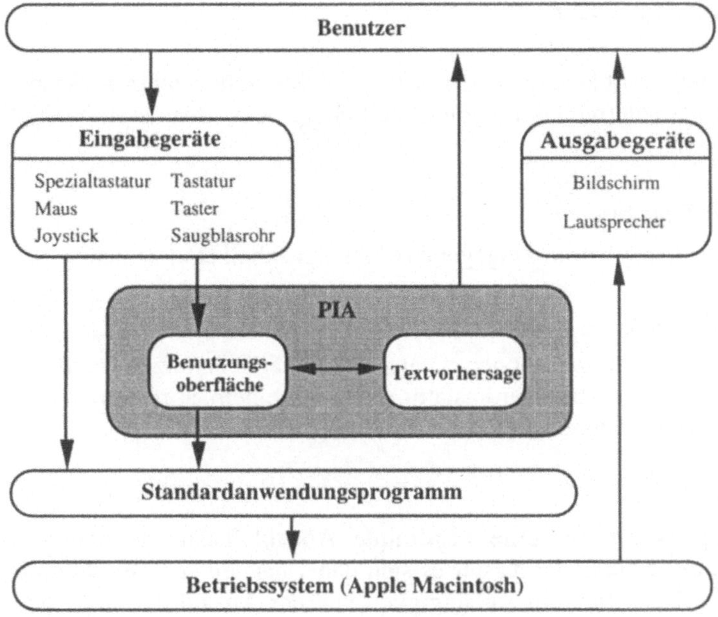

Abb. 1: Schematische Darstellung der Vorgehensweise bei PIA

3.2 Eingabeunterstützung durch Vorhersagen

Venkatagiri [22] hat gezeigt, daß durch zum Eingabekontext passende Vorhersagen die Anzahl der zu betätigenden Tasten / Schalter (Scanning) halbiert werden kann. Higginbotham [9] vergleicht fünf Vorhersageprogramme mit verschiedenen Vorhersagetechniken. Bei den dafür durchgeführten Tests wurden zwischen 36% und 41% der Tastenanschläge eingespart. Aufgrund dieser Ergebnisse ist es zweckmäßig, in PIA einen Vorhersagealgorithmus zu integrieren.

3.3 Die Benutzergruppen

Durch die Vielzahl der möglichen Behinderungen und Kombinationen von Behinderungen kann es kein allgemeines Benutzermodell für behinderte Computeranwender geben. Deshalb wurden für PIA die folgenden Benutzergruppen definiert:

* Benutzer von Zeigegeräten,
* Nur-Tastatur-Benutzer und

- Benutzer von Tastern.

Diese Einteilung stellt keine Klassifizierung von Menschen aufgrund körperlicher Merkmale dar, sondern dient nur zur Erstellung (grober) Benutzermodelle für die Implementierung.

Benutzer von Zeigegeräten

Der Anwender kann nur ein Zeigegerät (Maus, Trackball etc.) bedienen.

Nur-Tastatur-Benutzer

Anwender dieser Gruppe können zwar eine Tastatur benutzen, nicht aber ein Zeigegerät. Neben der Standardtastatur (evtl. mit einem Stirnstab) können hier aber auch Spezialtastaturen zum Einsatz kommen.

Benutzer von Tastern

Diese Gruppe kann nur eine bestimmte Anzahl Taster bedienen. Es ist unerheblich, um welche Art Taster es sich dabei handelt, z.B. ein Schlagkissen oder ein Saug-Blas-Rohr. In Abhängigkeit von der Anzahl der zu bedienenden Taster stehen dem Anwender verschiedene Eingabemöglichkeiten zur Verfügung.

Wenn der Benutzer mindestens fünf Taster bedienen kann, können vier davon für die Richtungssteuerung („Nach oben", „Nach unten", „Nach links" und „Nach rechts") verwendet werden. Der fünfte Taster dient der Bestätigung einer Aktion oder zur Signalisierung von Anfang und Ende einer Auswahl.

Ist der Anwender jedoch nur in der Lage, einen oder zwei Taster zu betätigen, muß die Bedienung der Benutzungsschnittstelle über die Eingabestrategie Scanning erfolgen. Beim automatischen Scanning mit einem Taster wird der Takt, mit dem die zur Auswahl stehenden Elemente durchlaufen werden, vom Computer vorgegeben. Desweiteren hat der Anwender die Möglichkeit, das Zeitintervall zwischen zwei Schritten durch die Betätigung eines zweiten Tasters selber festzulegen. Die Taktrate für den automatischen Durchlauf (z.B. 2 Sekunden) sollte jedoch an die Anforderungen des Anwenders angepaßt werden können.

Aufgrund der definierten Benutzergruppen ergaben sich zwei Eingabemodi für PIA. Zeigegeräte und die Tastatur stellen die Standardeingabegeräte für den Apple Macintosh dar. Wenn diese verwendet werden, verhält sich PIA aus Benutzersicht wie jedes andere Anwendungsprogramm. Der zweite Eingabemodus realisiert das Scanning für die Anwender von Tastern.

4 Die Benutzungsoberfläche von PIA

In PIA werden drei Fenster verwendet, die frei auf dem Bildschirm positioniert werden können. Das Hauptfenster, das als einziges von den drei Fenstern nicht ausgeblendet werden kann, ist das Vorhersagefenster, in dem die Vorhersagen und der aktuelle Eingabekontext angezeigt werden. Daneben kann der Anwender noch wahlweise die Bildschirmtastatur und die sog. Fernbedienung aufrufen. In diesem Abschnitt werden die Fenster und deren Bedienmöglichkeiten für die im vorigen Kapitel definierten Benutzergruppen beschrieben.

4.1 Das Vorhersagefenster

Das Vorhersagefenster (Abb. 2) zeigt neben dem Eingabekontext die aktuellen Vorhersagen. In der Titelleiste des Fensters wird der Name der Datei ausgegeben, in dem das aktuelle Vorhersagemodell gespeichert wird. In diesem Beispiel ist dies die Datei „Phrasen". Die Vorhersagen werden sortiert nach der Wahrscheinlichkeit ihres Auftretens in einem nicht proportionalen Zeichensatz angezeigt, da so die einzelnen Zeichen der Vorhersagezeilen jeweils untereinander stehen. Es können maximal neun Vorhersagen auf einer Bildschirmseite angezeigt werden. Der Rollbalken am rechten Rand dient zum Blättern, wenn mehr als neun Vorhersagen generiert werden.

In dem oberen Textfeld wird der aktuelle Eingabekontext ausgegeben. Das kleine Dreieck zeigt das Ende des Kontextes an. Das Kontextfeld ist ein reines Ausgabefeld, der Anwender kann den Inhalt also nicht verändern. Wird die Returntaste (auf der Standard- oder der Bildschirmtastatur) betätigt, wird der Kontext in das Vorhersagemodell übernommen und das Kontextfeld gelöscht. Würde jedes eingegebene Zeichen oder jede ausgewählte Vorhersage sofort in das Vorhersagemodell eingefügt werden, würden auch alle Tippfehler des Anwenders übernommen werden. So hat der Anwender die Gelegenheit, Tippfehler zu verbessern. Das Vorhersagemodell kann dadurch bessere Vorschläge generieren.

Links neben den einzelnen Vorhersagezeilen sind die Beschriftungen „F1" bis „F8" zu sehen. Diese dienen als Erleichterung für die Auswahl von Vorhersagen über die Funktionstasten (der Standard- oder der Bildschirmtastatur). Es wird das erste Wort der entsprechenden Zeile ausgewählt, z.B. mit F5 das erste Wort der fünften Zeile.

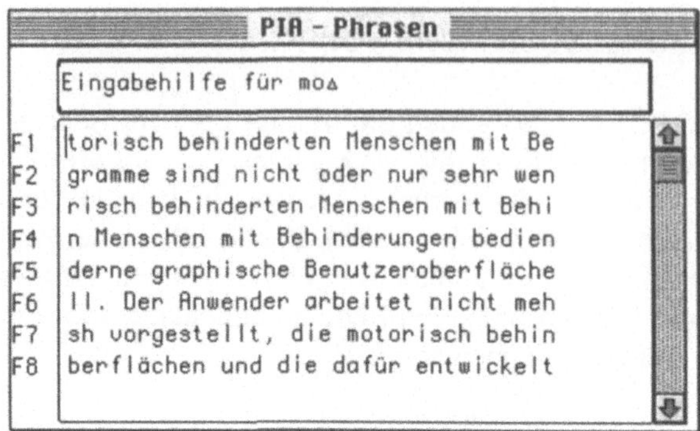

Abb. 2: Das Vorhersagefenster

4.2 Die Auswahl von Vorhersagen

Im folgenden werden für die einzelnen Gruppen die jeweiligen Auswahlmethoden
beschrieben. Zu beachten ist dabei, daß sich die einzelnen Auswahlmethoden
nicht immer eindeutig einer Benutzergruppe zuordnen lassen. So ist die Auswahl
über die Funktionstasten primär für die Nur-Tastatur-Benutzer gedacht, Anwender
eines Zeigegerätes oder des Scan-Modus können diese jedoch über die Bildschirmtastatur verwenden.

Benutzer von Zeigegeräten

Es werden zwar alle Begriffe für die Maus verwendet, wie z.B. „Mausknopf“,
„Mauszeiger“ usw., gemeint sind damit jedoch alle Zeigegeräte. Um eine Vor-
hersage mit einem Zeigegerät auszuwählen, bewegt der Anwender den
Mauszeiger an den Anfang der entsprechenden Zeichenkette und drückt den
Mausknopf. Durch Bewegung der Maus bei gleichzeitig festgehaltenem Maus-
knopf wird die Auswahl aufgezogen, d.h. der Text wird markiert. Wenn der
Mausknopf gelöst wird, ist die Auswahl abgeschlossen, und der Text wird an die
Zielanwendung gesendet. Durch einen Doppelklick auf die Beschriftungen „F1“
bis „F8“ am linken Rand der Vorhersagen wird das erste Wort der entsprechenden
Zeile analog zum Verhalten der Funktionstasten ausgewählt.

Nur-Tastatur-Benutzer

Wie bereits erwähnt, ermöglichen die Funktionstasten *F1* bis *F8* die schnelle
Auswahl des ersten Wortes einer Vorhersagezeile auf der aktuell angezeigten
Seite. Eine individuelle Selektion innerhalb des Vorhersagefeldes kann der An-
wender über die Pfeiltasten durchführen. Zur Markierung von Anfang und Ende

der Selektion ist eine weitere Taste notwendig. Diese Funktion wird bei der aktuellen Implementierung von der Taste *Einf.* (oberhalb der Pfeiltasten) übernommen.

Benutzer von Tastern

Kann der Anwender fünf Taster bedienen, erfolgt die Auswahl von Vorhersagen analog zu den Nur-Tastatur-Benutzern. Für Anwender von ein oder zwei Tastern werden die Vorhersagen von einer Marke zuerst zeilen- und dann spaltenweise durchlaufen. Zur Auswahl einer Zeile oder Spalte muß der Anwender den Auswahltaster betätigen. Die aktuelle PIA-Implementierung arbeitet mit der Returntaste als Auswahl- und mit der Leertaste als Takttaster für den Scan-Modus mit einer benutzerdefinierten Taktrate. Sinnvoller ist hierbei jedoch die Unterstützung externer Taster.

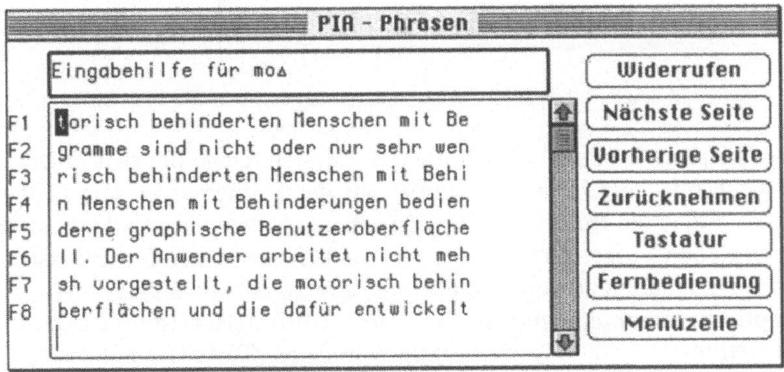

Abb. 3: Das Vorhersagefenster für den Scan-Modus

Das Vorhersagefenster wurde für den Scan-Modus rechts neben dem Rollbalken um sieben Knöpfe (Abb. 3) erweitert, die in den Scan-Durchlauf integriert sind. Diese Knöpfe ermöglichen das Widerrufen einer Auswahl, das Blättern innerhalb der Vorhersagen und den Wechsel zu den anderen PIA-Fenstern. Um Fehleingaben zu verhindern, haben alle anderen Tasten im Scan-Modus keine Funktion und werden ignoriert.

4.3   Die Bildschirmtastatur

Über die Bildschirmtastatur (Abb. 4) können die Benutzer von Zeigegeräten oder des Scan-Modus fast alle Zeichen und Tastenkombinationen der normalen Tastatur eingeben. Die Knöpfe werden entweder mit der Maus angeklickt oder im

Scan-Modus zuerst zeilen- und dann spaltenweise durchlaufen. Das Verhalten aller Knöpfe der Bildschirmtastatur entspricht dem der realen Tasten.

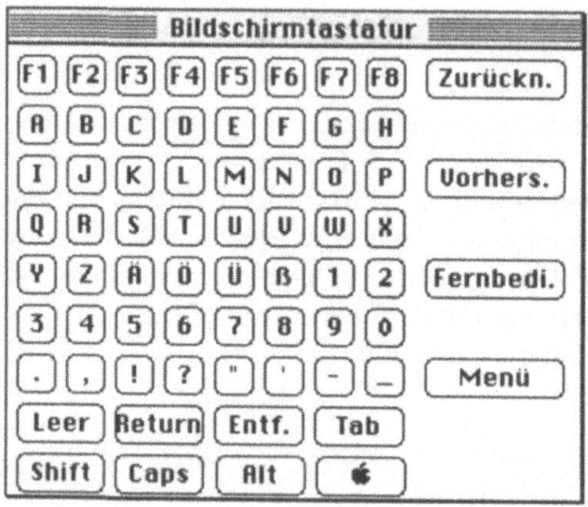

Abb. 4: Die Bildschirmtastatur von PIA

Die oberste Knopfleiste bilden die acht Funktionstasten. Ohne in den Vorhersagedialog zu wechseln, besteht so ein Zugriff auf die Vorhersagen. Die nächsten sieben Reihen bilden die Buchstaben, die Ziffern, die Satzzeichen und die Sondertasten „Leerzeichen", „Return", „Entfernen" und „Tabulator". Die Buchstabentasten sind bei der momentanen Implementierung alphabetisch angeordnet. Die letzte Knopfreihe beinhaltet die Sondertasten „Shift" (Hochstelltaste), „Caps Lock" (Feststelltaste), „Alt" (Wahltaste) und die Apple-Taste (Befehlstaste). Das Verhalten dieser Tasten ist so implementiert, daß der Anwender den Text sequentiell eingeben kann, d.h. die Sondertaste, die der Anwender ansonsten selber festhalten muß, wird von PIA „festgehalten". Die vier Knöpfe am rechten Rand der Tastatur haben dieselben Funktionen wie im Vorhersagedialog.

4.4 Die Fernbedienung

Die bisherigen Funktionen von PIA unterstützen nur die reine Texteingabe, nicht jedoch die Bearbeitung von Dateien (Anlegen, Öffnen, Speichern und Drucken). Ein Mausbenutzer kann mit einem Mausklick in die Zielanwendung wechseln und die gewünschte Funktion aufrufen. Ein Wechsel in die Zielanwendung ist jedoch gerade für Anwender des Scannings nicht möglich oder sehr problematisch, da nach dem Wechseln alle Eingabehilfen von PIA nicht mehr zur Verfügung stehen.

Abb. 5: Die Fernbedienung von PIA

Diese Funktionen der Zielanwendung können über das dritte PIA-Fenster - die sog. Fernbedienung (Abb. 5) - ausgeführt werden. Wie bei einer „echten" Fernbedienung, z.B. für einen Fernseher, braucht sich der Anwender nicht zum eigentlichen Anwendungsprogramm (Gerät) zu begeben, sondern kann die gewünschte Funktion aus der „Ferne" aufrufen. Die Fernbedienung ist wie das Vorhersagefenster und die Bildschirmtastatur in das Scanning eingebunden. Die einzelnen Knöpfe werden dabei von unten nach oben durchlaufen. Die Knöpfe können natürlich auch mit der Maus angeklickt werden. PIA wandelt die Benutzereingaben in entsprechende Anweisungen um und sendet diese an die Zielanwendung.

Während die Funktionen *Neu*, *Drucken* und *Beenden* ohne weitere Abfragen von der Zielanwendung ausgeführt werden können, benötigen die drei anderen Funktionen der Fernbedienung jeweils einen Dateinamen. Die Standarddialoge des Apple Macintosh für die Auswahl von Dateinamen können wegen der starken Fixierung auf die Mausbedienung nicht verwendet werden. Statt dessen wird oberhalb des Vorhersagefensters ein weiteres Dialogfenster eingeblendet. Nun kann der Anwender über die Standardtastatur, die Bildschirmtastatur oder durch die Auswahl von Vorhersagen den gewünschten Dateinamen eingeben. Durch eine erneute Betätigung des entsprechenden Knopfes in der Fernbedienung wird die Eingabe abgeschlossen und der Befehl an die Zielanwendung gesendet. Die übrigen Knöpfe der Fernbedienung haben dieselben Funktionen wie im Vorhersagedialog und in der Bildschirmtastatur.

5 Die Implementierung

In diesem Kapitel werden die Aspekte der Implementierung beschrieben, die für die Umsetzung der Anforderungen aus Kapitel 3 von besonderer Bedeutung sind.

5.1 Die Kommunikation mit der Zielanwendung

PIA sollte als eigenständige Anwendung mit möglichst allen Anwendungsprogrammen für den Apple Macintosh zusammenarbeiten, d.h. eine Anwendung (PIA) soll mit einer anderen Anwendung (der Zielanwendung) kommunizieren. Diese Art von Kommunikation wird unter dem Begriff *Inter-Application-Communication* (IAC) zusammengefaßt [2, 16].

Inter-Application-Communication

Umgangssprachlich bedeutet IAC nichts anderes, als daß zwei Anwendungsprogramme miteinander kommunizieren. Dabei ist es unerheblich, ob diese zwei Anwendungen auf demselben Rechner oder auf zwei durch ein Netzwerk miteinander verbundenen Rechnern laufen. Dies ist jedoch klar von Inter-Prozeß-Kommunikation (Inter-Process-Communication, IPC) und der Netzwerkschicht zu unterscheiden.

Während die Netzwerkschicht die physikalische Verbindung (z.B. Ethernet) und die Übertragungsprotokolle (z.B. TCP/IP) zur Verbindung der Computer behandelt, bezeichnet die Inter-Prozeß-Kommunikation die Sicht des Betriebssystems auf die einzelnen Prozesse und deren Behandlung. Dagegen werden beim IAC die Anwendungen nicht aus der Sicht des Programmierers („Wie sind die einzelnen Bits und Bytes angeordnet?") betrachtet. Entscheidend ist hierbei, was sich in der Anwendung aus Benutzersicht befindet. Ein Anwendungsprogramm stellt ein sog. Kontextmodell (engl. context model) dar. Dieses besteht aus einzelnen Bestandteilen (Worte, Zeilen, Absätze usw.) und den Operationen, die darauf angewendet werden können, z.B. das Anlegen, Bewegen, Kopieren und Löschen von einzelnen Bestandteilen. IAC wird nun als die Kommunikation zwischen zwei Anwendungen mit verschiedenen Kontextmodellen definiert [16]. Im Gegensatz dazu arbeiten bei Netzwerkprotokollen beide Seiten der Kommunikation mit demselben Kontextmodell.

Apple Events und AppleScript

Das Betriebssystem des Apple Macintosh arbeitet ereignisgesteuert. Das Betriebssystem und alle aktiven Anwendungsprogramme befinden sich in Endlosschleifen und warten darauf, daß ein Ereignis (engl. event) eintritt. Ereignisse sind z.B. die

Betätigung einer Taste oder eine Bewegung der Maus. Das Betriebssytem gibt dieses Ereignis nach einer ersten Überprüfung an die betroffene Anwendung weiter, die das Ereignis dann bezogen auf den Anwendungskontext bearbeitet.

Eine spezielle Art von Ereignissen sind auf dem Apple Macintosh die sog. *Apple Events*, die zwischen zwei Anwendungsprogrammen gesendet werden können [2]. Apple hat diese Events als Standardweg für das IAC auf dem Apple Macintosh vorgesehen. Neben einer Datenübertragung können der Zielanwendung auch Anweisungen gegeben werden, z.B. „Drucke das Dokument mit dem Namen "Brief vom 1.12.1994" aus.". Die Zielanwendung sollte solche Anweisungen genauso behandeln wie entsprechende Benutzereingaben über die Tastatur oder die Maus.

Der Mechanismus der Apple Events garantiert alleine noch keine allgemein gültige Anbindung von PIA an andere Anwendungsprogramme, da nicht davon ausgegangen werden kann, daß diese die benötigten Apple Events unterstützen. Auf Basis der Apple Events hat Apple die universelle Makrosprache AppleScript [3] entwickelt. AppleScript stellt für den Anwender eine Möglichkeit dar, einfache Programme für wiederkehrende Aufgaben (z.B. die Datensicherung) zu entwickeln und bei Bedarf auszuführen. Da AppleScript und somit Apple Events von Apple als Schlüsseltechnologien für Macintosh-Computer bezeichnet werden [18], kann man davon ausgehen, daß bald alle wichtigen Anwendungsprogramme AppleScript unterstützen und somit als Zielanwendungen für PIA in Frage kommen.

Die Verwendung von Apple Events für die Realisierung von PIA

Alle Datenübertragungen zwischen PIA und der Zielanwendung sind über entsprechende Apple Events realisiert worden. Dies schließt neben der Textübertragung von Tastatureingaben und Vorhersagen auch das Widerrufen von Eingaben sowie alle Funktionen der Fernbedienung zur Dateibehandlung ein. Eine tiefergehende Beschreibung findet man in [17].

5.2 Der Vorhersagealgorithmus

Grundlage und Motivation für die Entwicklung der Eingabehilfe PIA stellt das Programm *Reactive Keyboard* (RK) dar, beschrieben in [6] und [7]. Dieses wurde an der Universität Calgary, Kanada, von John J. Darragh und Ian H. Witten entwickelt. Die ersten Versionen des RKs für MS-DOS- und UNIX-Computer arbeiten als Eingabeunterstützungen für die Kommandozeileneingabe. Neben den Kommandozeilenversionen des RKs wurde von Mark L. James eine Version für den Apple Macintosh mit dem Namen *RK-Pointer* entwickelt. Der unveränderte Vorhersagemechanismus ist in einen rudimentären Texteditor eingebettet. Die

Vorschläge werden in einem zusätzlichen Fenster angezeigt, während bei den anderen RK-Versionen nur jeweils eine Vorhersage ausgegeben wird.

Der Algorithmus des Reactive Keyboards

Der Vorhersagealgorithmus des RKs unterscheidet sich grundlegend von denen anderer Vorhersageprogramme (z. B. *Magic Typist* [4], *Co:Writer* [11] und *PAL* [14]), die fast alle auf der Basis von Wörterbüchern arbeiten. Bei diesen entnimmt der Algorithmus alle Wörter aus dem Wörterbuch, die zur aktuellen Eingabe des Benutzers passen. Wenn der Benutzer z.B. ein „h" eingibt, sind dies alle Worte, die mit einem „h" beginnen. Um die Vielzahl von möglichen Wörtern in eine sinnvolle Reihenfolge zu bekommen, ist jedem Wort eine Wahrscheinlichkeit zugeordnet. Nach dieser sortiert werden die einzelnen Worte angezeigt. Der Benutzer kann dann einen dieser Vorschläge annehmen oder aber ein weiteres Zeichen eingeben, z.B. ein „a". Der Vorhersagealgorithmus sucht dann alle Worte mit dem Wortbeginn „ha".

Das RK arbeitet hingegen nicht mit einem Wörterbuch, sondern sammelt alle Benutzereingaben in einem Vorhersagemodell. Dieses Vorhersagemodell wird durch einen Baum realisiert, in dem Zeichentupel der Länge k abgespeichert werden. Die sog. Modellgröße k gibt die Zeichenanzahl an, die zwischen dem aktuellen Eingabekontext und dem Inhalt des Vorhersagemodells verglichen werden. Wenn die Modellgröße z.B. gleich zehn ist, werden die letzten zehn Zeichen der Eingabe zur Überprüfung mit dem Vorhersagemodell herangezogen.

Das RK nimmt zur Erstellung von Vorhersagen einen reinen Zeichenvergleich vor. So kann der Fall eintreten, daß dadurch unsinnige Worte entstehen, die zwar auf Grund ihrer Zeichenfolge (syntaktisch) sehr gut in den aktuellen Kontext passen, aber keinerlei Bedeutung (semantisch) haben. Bei Vorhersageprogrammen auf Wörterbuchbasis kann dieser Fall nicht eintreten, da das Wörterbuch nur gültige Worte enthält. Auf der anderen Seite erkennt der Algorithmus des RK aufgrund des reinen Zeichenvergleichs nicht das Ende eines Wortes. So können nicht nur einzelne Worte, sondern ganze Phrasen und Sätze als Vorhersagen erstellt werden.

Das RK erstellt in einem Durchlauf 128 verschiedene Vorschläge, für jedes Zeichen der ASCII-Tabelle einen. Diese Vorschläge haben alle unterschiedliche Anfangszeichen, d.h. für ein mögliches Zeichen, das der aktuellen Eingabezeile folgen könnte, wird immer nur genau ein Vorschlag erstellt. Wenn der Benutzer beispielsweise ein „R" eingibt, kann der erste Vorschlag des RKs „eactive Keyboard" lauten. Es wird nun kein weiterer Vorschlag mit einem „e" als ersten Buchstaben erzeugt, z.B. das Wort „Regenwald". Das RK arbeitet adaptiv, d.h.

alle Vorhersagen werden aufgrund der vorangegangenen Benutzereingaben erstellt. Für eine genauere Beschreibung des Vorhersagealgorithmus wird auf die Originalliteratur [6, 7] verwiesen.

Gründe für die Wahl des RK-Algorithmus

Im Abschnitt 3.2 wurde bereits auf die Wichtigkeit von Textvorhersagen für die Eingabeunterstützung hingewiesen. Bei der Arbeit an PIA fiel die Wahl auf den RK-Algorithmus, da sich dieser durch seine adaptive Arbeitsweise von den anderen Vorhersagealgorithmen unterscheidet. Desweiteren ist der RK-Algorithmus im Quellcode verfügbar und konnte so leicht übernommen und angepaßt werden. Nur die Speicherverwaltung mußte wegen den geänderten Bedingungen unter Macintosh System 7 (32 Bit-Adressierung und Einbau eines Garbage Collectors) erweitert werden. Dadurch konnte der Schwerpunkt bei der Implementierung auf die Benutzungsschnittstelle, die Unterstützung verschiedener Eingabegeräte und die universelle Anbindung an Anwendungsprogramme gelegt werden.

Durch den Einsatz von Apple Events konnte eine sehr flexible und leistungsfähige Verbindung zwischen der Eingabehilfe PIA und der Zielanwendung realisiert werden. Die Unterstützung möglichst vieler Anwendungsprogramme für den Apple Macintosh kann als gegeben angesehen werden. Auch bieten die Apple Events noch viele Ansatzpunkte für Erweiterungsmöglichkeiten. Der Algorithmus des Reactive Keyboards stellt durch sein adaptives Vorhersagemodell eine interessante Alternative zu den anderen Vorhersagemethoden dar.

6   Schlußbetrachtung und Ausblick

PIA liegt als stabiler Prototyp in einer deutschen und einer englischen Version sowohl für den Apple Macintosh als auch für den Power Macintosh vor. Die Arbeiten haben die prinzipielle Realisierbarkeit einer separierten und vielseitig einsetzbaren Benutzungsschnittstelle für körperbehinderte Menschen nachgewiesen. Dennoch enthält PIA noch einige Teile, bei denen Verbesserungen notwendig sind.

Die Vorhersagen, die der Algorithmus des Reactive Keyboards erstellt hat, könnten durch einen zweiten Algorithmus ergänzt werden. Dieser fügt die gebräuchlichsten Wörter passend zum aktuellen Kontext in die Liste der Vorhersagen ein. Weitergehend könnten unsinnige Vorhersagen durch eine Rechtschreibprüfung bereits vor der Anzeige auf dem Bildschirm herausgefiltert werden. Das Layout der Bildschirmtastatur sollte über ein Zusatzprogramm an die Bedürfnisse und Wünsche des Anwenders angepaßt werden können. Die Menüzeile des Apple Macintosh kann nicht über die Tastatur (mit Ausnahme von Tastaturkürzeln) bedient werden. PIA sollte die Nur-Tastatur-Anwender und die Benutzer des Scan-Modus in die Lage versetzen, jeden Eintrag der Menüzeile zu erreichen. Daneben sollte auch die Redigierung von Text stärker berücksichtigt werden.

Literatur

[1]   N. Alm, J. L. Arnott, A. F. Newell. Predictional Conversational Momentum in an Augmentative Communication System. In Communications of the ACM, 35(5): 46-57, 1992.

[2]   Apple Computer, Inc. Inside Macintosh: Interapplication Communication. Addison-Wesley, 1993.

[3]   Apple Computer, Inc. AppleScript-Getting Started Guide. Cupertino, California, USA, 1993.

[4]   L. A. Bardi. Magic Typist 2.0. OLDUVAI Corporation, Miami, Florida, USA, 1990.

[5]   C. Clayton, R. G. S. Platts, M. Steinberg, A. M. Trudgeon. Palatal Tongue Controller. In Computers for Handicapped Persons. R. Oldenbourg, Wien, 1990.

[6]   J. J. Darragh, I. H. Witten. The Reactive Keyboard. Cambridge University Press, 1992.

[7]   J. J. Darragh, I. H. Witten, M. L. James. The Reactive Keyboard. In IEEE Computer, 23(11): 41-49, 1990.

[8]   E. P. Glinert, B. W. York. Computers and People with Disabilities. In Communications of the ACM, 35(5): 33-35, 1992.

[9]   D. J. Higginbotham. Evaluation of Keystroke Savings across Five Assistive Communication Technologies. In AAC Augmentative and Alternative Communication of ISAAC, 8(12):258-272, 1992.

[10] Don Johnston Developmental Equipment, Inc. Catalog 1993: Assistive Technology for Computer Access, Communication & Literacy. Wauconda, Illinois, USA, 1993.

[11] Don Johnston Developmental Equipment, Inc. Co:Writer. Wauconda, Illinois, USA, 1993.

[12] M. Mantei, P. Orbeton. Human Factors in Computing Systems. Conference Proceedings of "Human Factors in Computing Systems", Zürich, Schweiz, 1986.

[13] S. V. Millar, P. D. Nisbet. Accelerated Writing for People with Disabilities. CALL Centre, University of Edinburgh, 1993.

[14] A. N. Newell, J. L. Arnott, L. Booth, W. Beattle, B. Brophy, I. W. Ricketts. Effect of the "PAL" Word prediction System on the Quality and Quantity of Text Generation. In AAC Augmentative and Alternative Communication of ISAAC, 8(12): 304-311, 1992.

[15] J. Nielsen. Noncommand User Interfaces. In Communications of the ACM, 36(4): 83-99, 1993.

[16] K. Piersol. Inter-Application-Communication. Videokassette. University Video Communication, Stanford, California, USA, 1992.

[17] C. Piwetz. Entwicklung einer separierten Benutzungsschnittstelle für Textsysteme auf Basis des IAC-Mechanismus von Macintosh-System 7: Eine adaptive Eingabehilfe für motorisch behinderte Personen. Interner Bericht Nr. 7, Fachbereich Mathematik und Informatik / Systemmodellierung, Universität - Gesamthochschule - Essen, 1994.

[18] R. Priem. Amber schmückt den Macintosh. In MACup, MACup Verlag, Hamburg, 7/1993: 12-13, 1993.

[19] M. Streit. Architekturalternativen für Agentenbasierte Multimodale Interfaces. In J. Kunze, H. Stoyan (Hrsg.): KI-94 Workshop, Extended Abstracts. Springer-Verlag, 5-6, 1994.

[20] U. Thakkar. Ethics in the Design of Human-Computer Interfaces for the Disabled. In SIGCAPH Newsletter, ACM Special Interest Group on Computers and the Physically Handicapped, 42: 1-7, 1990.

[21] G. C. Vanderheiden. Making Software More Accessible for People with Disabilities. In SIGCAPH Newsletter, ACM Special Interest Group on Computers and the Physically Handicapped, 47, 1993.

[22] H. S. Venkatagiri. Efficiency of Lexical Prediction as a Communication Acceleration Technique. In Augmentative and Alternative Communication, ISAAC, 9/1993:161-167, 1993.

[23] VERTIKAL-Informatik. Kommunikationshilfen für Schwerbehinderte (Produktübersicht 1992). Weinsberg, Deutschland, 1992.

Christian Piwetz
Universität - Gesamthochschule - Essen
Fachbereich Mathematik und Informatik
45117 Essen
Tel.: 02 01 / 183 - 2168
e-mail: piwetz@informatik.uni-essen.de

Kooperative Interaktionsunterstützung in Groupware

Matthias Ressel
Universität Stuttgart

Zusammenfassung

Sogenannte Groupware unterstützt die Kooperation, Kommunikation und Koordination von Personen, die gemeinsam eine Aufgabe bearbeiten. Bei zahlreichen Anwendungen ist es wichtig, die Interaktion der Kooperationspartner mit Hilfe des Computers koordinierend zu unterstützen. Die vorliegende Arbeit zeigt, daß die Methode der Operationstransformationen zur Implementierung von Gruppeneditoren unter softwareergonomischen Gesichtspunkten besonders geeignet ist. Sie zeigt, welche Eigenschaften Transformationsregeln erfüllen müssen, um als Grundlage einer benutzerorientierten, kooperativen Koordinationsmethode zu dienen. Solche Transformationen ermöglichen die Erstellung eines einheitlichen Interaktionsmodells, das als Grundlage weiterer unterstützender Module, wie etwa für Gruppen-Undo, dient.

1. Einleitung

Rechnerprogramme, die vereinfachend als Groupware [6] bezeichnet werden, sollen *Kooperation*, also die Zusammenarbeit einer Gruppe von Personen, die gemeinsam an einer Aufgabe arbeiten, unterstützen. Voraussetzung für eine solche Kooperation sind gemeinsame Ziele, gemeinsame Handlungspläne, ein gemeinsamer Kontext, flexible Regelbarkeit und Kontrollierbarkeit [9]. *Koordination* dient dazu, die einzelnen Handlungen zeitlich und inhaltlich aufeinander abzustimmen. Sie soll vor allem die Interaktion innerhalb der Gruppe unterstützen. *Kommunikation* der Kooperationspartner untereinander dient übergreifend zur Einigung über Ziele, zur Herstellung eines gemeinsamen Kontextes und allgemein zur Unterstützung der Interaktion.

Interaktionsunterstützung (IU) läßt sich charakterisieren durch die Eigenschaften vorausschauend, begleitend und rückblickend. *Vorausschauende IU* gibt feste Handlungspläne vor, an die sich die Kooperationspartner zu halten haben. Sie zielt vor allem auf präventive Konfliktvermeidung. *Begleitende IU* erlaubt die Anpassung der Koordinationsstrategie an die tatsächlichen Gegebenheiten. Insbesondere kann sie auf Ausnahmesituationen reagieren. Ihr Schwerpunkt liegt auf der Konflikterkennung und Konfliktbehebung. *Rückblickende IU* schließlich erlaubt es, Einblick auf bereits ausgeführte Handlungen zu nehmen, um daraus zukünftige Koordinationsstrategien abzuleiten.

Die Koordination kann von den Partnern selbst (*intern*) durchgeführt werden. Sie kann aber auch durch eine koordinierende Instanz von außen (*extern*) unterstützt werden. Dieser Koordinator kann ein Mensch, aber auch ein Computer sein.

Im folgenden wird ein Ansatz vorgestellt, der vor allem zur begleitenden und rückblickenden IU in Groupware mit direkt manipulativer Benutzungsoberfläche geeignet ist und der sowohl interne als auch externe Koordination unterstützt. Zuvor definieren wir die wichtigsten Anforderungen und stellen weitere bekannte Ansätze vor. Unser eigener Ansatz verwendet die Methode der Operationstransformationen und erweitert sie um ein einheitliches Interaktionsmodell. Dieses Modell dient als Grundlage weiterer interaktionsunterstützender Module, die u. a. Undo für Gruppen und das nachträgliche Navigieren im Interaktionsraum erlauben. Die prototypische Realisierung eines Texteditors für Gruppen und seine Benutzungsoberfläche werden vorgestellt; ein Vergleich mit anderen vorliegenden Arbeiten und ein Ausblick runden diesen Beitrag ab.

2. Problembeschreibung und Anforderungen

Als Szenario dient die Dokumenterstellung durch mehrere Autoren (*co-authoring*). Dies kann z. B. die Fertigstellung eines Projektantrages, das Verfassen eines Abschlußberichtes, die Erstellung von Bedienungsanleitungen oder einfach das Schreiben eines gemeinsamen Briefes sein; auch die Software-Entwicklung im Team gehört i. w. S. zu diesem Anwendungsbereich.

Solche Anwendungen weisen einige gemeinsame Merkmale auf:

1. Mehrere Personen arbeiten an einer Aufgabe zusammen.
2. Die Personen befinden sich nicht notwendig alle an einem Ort.
3. Die einzelnen Teiltätigkeiten (Schreiben, Zeichnen, Modifizieren, Layout) werden bereits mit einem Computer durchgeführt.
4. Die Tätigkeit wird meist über einen längeren Zeitraum ausgeführt, muß aber zu einem bestimmten Zeitpunkt beendet sein.

Eine wichtige Voraussetzung für eine solche Zusammenarbeit ist das gegenseitige Gewahrsein (*awareness*). Solches Gewahrsein entsteht einerseits durch *explizite Kommunikation* miteinander, anderseits aber auch durch *implizite Kommunikation*: Hierzu gehört die bewußte oder unbewußte Wahrnehmung einer Person und ihrer Tätigkeiten, aber auch das Wahrnehmen von *Spuren*, die andere Personen „in der Welt" hinterlassen haben. Gewahrsein liefert also die Antwort auf Fragen wie: *Wer* wird was machen? *Wer* macht gerade was? *Wer* hat was gemacht? Dieses Wissen ist wichtig für die interne Koordination, d. h. es erlaubt den beteiligten Personen, ihre Handlungen gegenseitig abzustimmen.

Vielfach reicht Gewahrsein aber nicht aus, um Konflikte bei der Interaktion auszuschließen. Durch gleichzeitige, parallele Zugriffe kann es zu inkonsistenten Zuständen von Anwendungsobjekten kommen. Beispiele für Konflikte sind das gleichzeitige Löschen von Objekten, oder widersprüchliches Ändern von Objekteigenschaften. Groupware sollte auf solche Inkonsistenzen aufmerksam machen bzw. helfen, sie zu vermeiden. Groupware sollte verschiedene Möglichkeiten des *Undo* bieten: Es muß möglich sein, auf frühere Zustände zurückzusetzen oder Teile der Interaktion zurückzunehmen. Es stellen sich hierbei folgende Fragen: *Welche* Auswirkung hat es, wenn mehrere Operationen gleichzeitig ausgeführt werden? *Was* passiert bei Konflikten? *Wann* kann eine Operation wieder zurückgenommen werden?

Für Groupware, die Anwendungen mit den oben beschriebenen Merkmalen unterstützt, ergeben sich folgende spezifischen Anforderungen:

1. Den Autoren soll der aktuelle Stand der Aufgabenbearbeitung zur Verfügung stehen, damit inhaltliche Inkonsistenz und unnötige Mehrarbeit vermieden werden können.

2. Die gewohnte Direktheit der Benutzungsschnittstelle darf auch bei räumlich verteiltem Betrieb nicht verloren gehen. Insbesondere müssen solche Aktionen wie Einfügen und Löschen von Zeichen unmittelbar auf dem eigenen Bildschirm zu sehen sein.

3. Das kooperative System sollte nicht allzusehr von den gewohnten Werkzeugen (Editoren) abweichen, die für die Teiltätigkeiten benutzt werden. Auch sollte ein nahtloser Übergang zwischen Ein- und Mehrbenutzerbetrieb ermöglicht werden.

4. Der Zeitdruck, der regelmäßig gegen Ende einer solchen Tätigkeit entsteht, macht eine engere Zusammenarbeit nötig. Parallele Interaktionen – wie gleichzeitiges Editieren einer Datei – sollten daher möglich sein, ohne zu undefinierten Ergebnissen zu führen. Notwendige manuelle Eingriffe sollten sich dabei auf ein Minimum beschränken, damit die Personen nicht von ihrer eigentlichen Tätigkeit des Schreibens (Zeichnens, etc.) abgehalten werden.

Der Schwerpunkt im folgenden wird auf dem letzten Punkt, der parallelen Interaktion, liegen. Es wird sich jedoch zeigen, daß der vorgeschlagene Ansatz zur Unterstützung paralleler Interaktion auch die anderen Anforderungen weitgehend erfüllt.

3. Ansätze zur Interaktionsunterstützung

Zur Koordinierung mehrerer Eingabeströme (*concurrency control*), wie sie bei
paralleler Interaktion mehrerer Akteure auftreten, bieten sich verschiedene Ansätze an
(vgl. hierzu [6], [1]). Sie unterscheiden sich dadurch, ob

- Konflikte von vornherein vermieden werden oder
- Konflikte bei ihrem Auftreten erkannt und geeignet weiterbehandelt werden.

3.1. Methoden zur Konfliktvermeidung

Konflikte durch konkurrierende Operationen können vermieden werden, indem der
Zustandsraum der Anwendung *partitioniert* wird. Verschiedene Personen können
dabei gleichzeitig an verschiedenen Objekten einer Anwendung arbeiten. Ein Beispiel
hierfür ist *Hypertext*, der Informationen in Netzknoten aufteilt. Ein weiteres Beispiel
sind graphische Zeichenprogramme, die für mehrere Benutzer unterschiedliche
übereinandergeschichtete, transparente Zeichenebenen (*Layers*) bereitstellen.
Gleichzeitiges Editieren desselben Objektes ist hierbei ebenso unmöglich wie das
Editieren eines fremden Objekts. Das Ausmaß der Interaktion wird dadurch sehr
eingeschränkt.

Konflikte können durch *Kopieren* der Anwendungsobjekte, wenn nicht vermieden,
so zumindest hinausgezögert werden. Dieses Verfahren wird beispielsweise in
Versionsverwaltungssystemen eingesetzt. Es gilt dabei, aus mehreren Kopien
parallel entwickelter Anwendungsobjekte eine einheitliche, konsistente Version
herzustellen. Da nicht sichergestellt ist, daß dies überhaupt möglich ist, ist meist eine
manuelle Nachbearbeitung erforderlich.

Weitere Verfahren beruhen darauf, den Zugriff auf Anwendungsobjekte nur
sequentiell zuzulassen; *feste Dialogabläufe* werden vorgegeben. Dies wird häufig bei
Spielprogrammen verwirklicht, bei denen die Spielpartner abwechselnd am Zug
sind.

Sperrmechanismen (*Locking*) reservieren Teile des Anwendungsbereichs zeitweise
für einen exklusiven Zugriff; sie stellen daher eine dynamische Art der
Partitionierung dar. Die Sperren können entweder explizit angefordert werden oder
sie werden durch die Ausführung bestimmter Aktionen implizit eingerichtet. Um zu
vermeiden, daß bestimmte Benutzer Sperren unnötig lange halten – etwa bei
Abwesenheit – können solche Sperren nach einer gewissen Zeit, in der der

Sperrenhalter inaktiv war, automatisch (*tickle locks*) oder auf Anfrage durch andere Benutzer (*soft locks*) entfernt werden. Sperren haben allerdings den Nachteil, daß sie – zumindest zeitweise – den freien Zugriff auf Anwendungsobjekte einschränken und damit unsichere modale Zustände erzeugen. Eine visuelle Hervorhebung gesperrter Objekte kann dazu beitragen, diese Probleme zu reduzieren.

Eine andere Möglichkeit ist die explizite Zuteilung eines Schreib- oder sonstigen Zugriffsrechtes (*floor control*). Soziale Protokolle mittels expliziter Absprachen haben den Nachteil, daß sie umgangen werden können, was in diesem Fall zu undefinierten Zuständen des Anwendungssystems führen kann. Sicherer sind technische Protokolle mittels sogenannter *Tokens*, die das Schreibrecht repräsentieren und jeweils bei Bedarf weitergereicht oder zurückgegeben werden können. Tokens stellen eine spezielle Art von Sperren dar, da genau ein Token für einen bestimmten festen Systembereich zuständig ist. Daher können auch bei Floor-Control-Verfahren die von Sperren her bekannten Methoden angewandt werden, um Tokens bei Passivität des Token-Halters wieder freigeben.

Die Sequentialisierung über eine zentralen Koordinator stellt einen weiteren Ansatz dar. Bei parallel abgegebenen Benutzeranforderungen ist hierbei allerdings nicht gewährleistet, daß sie immer im richtigen Zustand ausgeführt werden. Um solche parallele Eingaben zu vermeiden, ist die Benutzungsoberfläche solange zu sperren, bis eine Anforderung ausgeführt werden konnte.

Allgemein läßt sich feststellen, daß Konfliktvermeidung durch Partitionieren, Kopieren oder Sequentialisieren die Vorteile parallelen Arbeitens nicht nutzen kann. Die Aktionsfreiheit des einzelnen geht verloren: „The computer won't let me" [3].

3.2. Methoden zur Konflikterkennung und Konfliktbehebung

Transaktionsmechanismen, wie sie bei Datenbanken eingesetzt werden, versuchen parallele Anforderungen zu sequentialisieren. Treten dabei Konflikte auf, werden solche Transaktionen zurückgesetzt. Ihr Ergebnis wird erst dann öffentlich zugänglich, wenn die Sequentialisierung erfolgreich verlaufen ist. Zwischenzustände sind dadurch für andere Benutzer nie einsehbar. Die Forderung nach Gewahrsein ist nicht erfüllt.

Weitere Verfahren zeichnen sich dadurch aus, daß Benutzeranforderungen lokal immer sofort ausgeführt werden, ohne vorher zu überprüfen, ob die Anforderung mit anderen kollidiert. Sie werden daher als *optimistische* Verfahren bezeichnet.

Ein einfaches Verfahren dieser Art unterbricht die Verarbeitung, wenn es einen Konflikt erkannt hat. Er muß dann von den Benutzern manuell behoben werden. Dies hat den weiteren Nachteil, daß das System keinen konsistenten Zustand garantieren kann.

Eine Verbesserung stellen solche Verfahren dar, die durch eine Rücknahmemöglichkeit (*undo*) von Operationen und durch eine automatische Umsortierung der Interaktionshistorie für eine Konfliktbehebung sorgen.

Die Methode der *Operationstransformation* schließlich versucht, Konflikte durch parallele Eingabeanforderungen dadurch zu lösen, daß sie bei deren Interpretation und Ausführung den Anwendungszustand berücksichtigt, der zum Zeitpunkt der Anforderung gültig war. Bei Bedarf werden Operationen derart transformiert, daß das Resultat für jede mögliche Ausführungsreihenfolge dasselbe ist.

4. Operationstransformationen

Operationstransformationen wurden zum ersten Mal im Gruppen-Outline-Editor GROVE [5] verwendet. Bei diesem Verfahren arbeitet jeder Benutzer mit einem individuellen Editor an Kopien der Anwendungsobjekte. Die Benutzeranforderungen werden im lokalen Editor unmittelbar ausgeführt und dann an die Editoren der anderen Teilnehmer geleitet. Jeder Anforderung wird Information beigefügt, aus der die Empfänger schließen können, in welchem Zustand sich der Editor vor der Eingabe der Anforderung befand. Diese Zustandsinformation spezifiziert die Anzahl der Operationen jedes Teilnehmers, die bereits ausgeführt worden sind. Die verschiedenen Editoren führen die eingehenden Anforderungen unter Umständen in anderen Reihenfolgen aus. Eine Umsortierung wird dadurch möglich gemacht, daß bei der Ausführung fremder Anforderungen berücksichtigt wird, welche zwischenzeitliche Veränderungen des Anwendungszustands durch zusätzlich ausgeführte Operationen entstanden sind. Unter bestimmten Bedingungen ist es hierdurch möglich, in allen Editoren konsistente Zustände zu erzielen, sobald alle Anforderungen abgearbeitet sind.

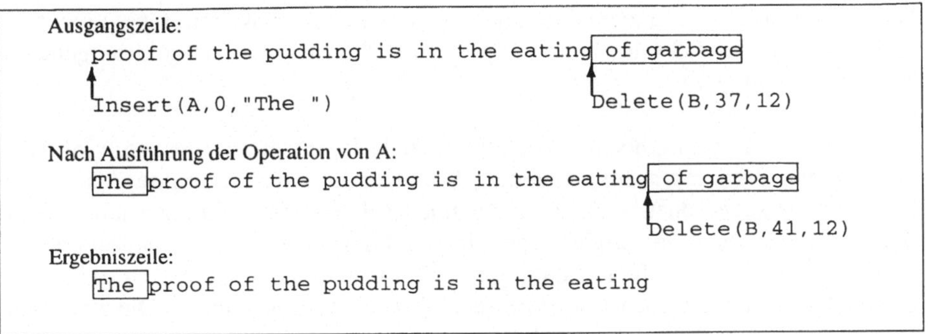

Abb. 1: Synchrones Editieren: Die Einfügeoperation von A muß berücksichtigt werden, wenn die Löschoperation von B danach ausgeführt wird.

Beispiel 1: Zwei Personen, A und B, benutzen gleichzeitig einen Gruppen-Editor. A fügt am Anfang einer Zeile ein Wort ein, gleichzeitig löscht B am Ende derselben Zeile einige Zeichen. Beide Editoren führen zunächst die lokalen Operationen aus. Der Editor von B kann danach das von A eingefügte Wort an derselben absoluten Position einfügen wie A. Der Editor von A hingegen muß beim Löschen der Zeichen berücksichtigen, daß deren Position sich durch die ausgeführte Einfügeoperation verschoben hat (s. Abb. 1).

Wenn, allgemein betrachtet, zwei Operationen o_1 und o_2 im selben Zustand ausgeführt wurden, müssen Operationen o_1' und o_2' so gewählt werden, daß gilt:

$$o_1 o_2' = o_2 o_1', \tag{1}$$

d. h. o_1 gefolgt von o_2' führt zum selben Resultat wie o_2 gefolgt von o_1'.

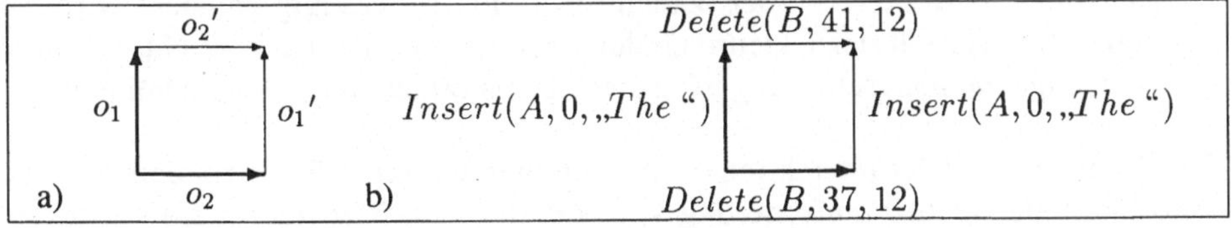

Abb. 2: a) elementare L-Transformation: $o_1 o_2' = o_2 o_1'$, b) Beispieltransformation

Eine Funktion T, die zu zwei Operationen o_1 und o_2 zwei weitere Operationen o_1' und o_1' liefert, so daß Gleichung (1) gilt, wird Transformationsfunktion genannt: $T(o_1,o_2)=(o_1',o_2')$.

Abb. 2a zeigt eine graphische Darstellung. Jede Benutzeranforderung wird dabei durch einen Pfeil repräsentiert. Jedem Benutzer ist eine bestimmte Achse oder Pfeilrichtung zugeordnet. Schließlich erhält jeder dieser Pfeile eine Bezeichnung, die die eigentliche Operation angibt. Operationen, die originale Benutzeranforderungen darstellen, werden i. d. R. fett dargestellt. Auf Grund der Lage der Originalpfeile dieser Operationstransformation, wird sie *L-Transformation* genannt. Die graphische Darstellung des Transformationsschrittes von Beispiel 1 zeigt Abb. 2b.

Gleichung (1) reicht nicht aus, eine L-Transformationsfunktion eindeutig zu definieren. Insbesondere darf der Endzustand einer Transformation nicht von der Reihenfolge der Ausgangsoperationen abhängen.

Beispiel 2 liefert die Definition einer allgemeinen anwendungsunabhängigen L-Transformationsfunktion:

$$T(o_1, o_2) = \begin{cases} (I, \overline{o_1}o_2) & \text{für} \quad p(o_1) > p(o_2), \\ (\overline{o_2}o_1, I) & \text{sonst.} \end{cases} \tag{2}$$

I ist dabei die neutrale Identitätsabbildung, $\overline{o}$ die inverse Operation zu o, deren Existenz vorausgesetzt werden muß, und $p(o)$ die Priorität einer Operation o. Die Priorität berechnet sich aus dem Akteur, der die Operation angefordert hat. Die Akteure lassen sich hierzu auf eindeutige Weise ordnen.

Die Transformation T kann so interpretiert werden, daß eine der parallel ausgeführten Operationen wieder rückgängig gemacht und dafür die andere ausgeführt wird. Welche der beiden Operation rückgängig gemacht wird, ist durch die Priorität der Akteure eindeutig festgelegt. Durch Verzicht auf diese eindeutige Fallunterscheidung wäre diese Transformation nicht wohldefiniert.

Die Abfolge paralleler Interaktionen läßt sich in graphischer Form gut darstellen (s. Abb. 3): Jede Benutzeranforderung wird durch einen Pfeil repräsentiert, dessen Ausgangskoordinaten angeben, wieviele Operationen der einzelnen Benutzer bereits abgearbeitet worden sind.

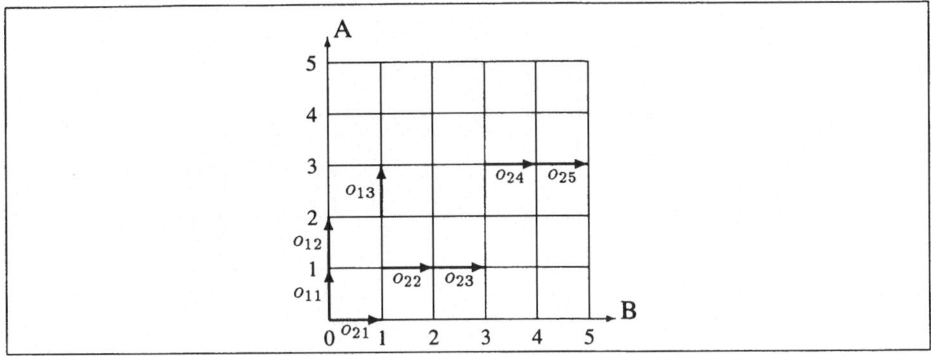

Abb. 3: 2-dimensionales *Interaktionsmodell* mit drei Operationen von Person A und fünf von
Person B.

Zwischen den einzelnen Systemzuständen ist eine partielle (zeitliche) Ordnung definiert: Für einen Zustand s_1, dessen Koordinaten alle kleiner oder gleich den jeweiligen Koordinaten eines anderen Zustands s_2 sind, soll gelten: $s_1 \leq s_2$.

Ein *gültige Ausführungsfolge* ist dadurch definiert, daß darin jede Operation im Originalzustand oder einem späteren ausgeführt werden muß. Die Originaloperationen sind dazu bei Bedarf zu transformieren. Abb. 4a zeigt eine mögliche Ausführungsfolge, für die sechs Elementar-Transformationen nötig sind (mit *T* markierte Quadrate). Abb. 4b zeigt den *Ausführungsraum*, der sich durch Vereinigung aller Ausführungsfolgen ergibt. Damit selbst ausgelöste Operationen möglichst direkt verarbeitet und die Ergebnisse sichtbar werden, werden Operationen i. d. R. lokal sofort ausgeführt. Die obere bzw. untere Begrenzungslinie des Ausführungsraums in Abb. 4b stellt daher typische Ausführungsfolgen für A bzw. B dar.

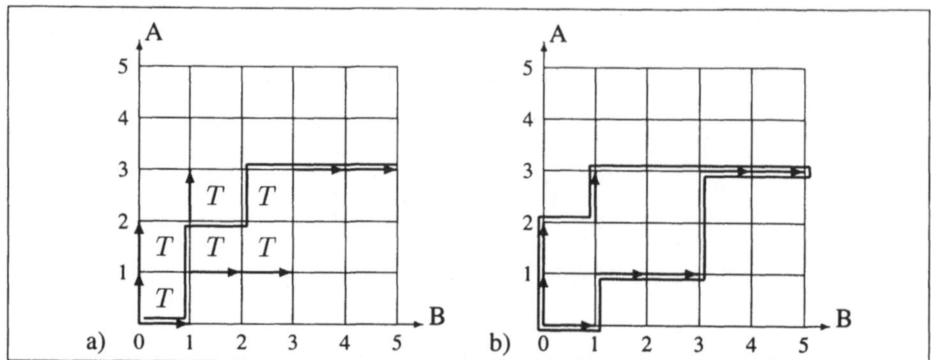

Abb. 4: a) Beispiel einer Ausführungsfolge, b) Ausführungsraum

Im 2-dimensionalen Fall reicht die Transformationsbedingung (1) aus, um zu gewährleisten, daß die Anwendung jeder gültigen Ausführungsfolge auf den Ausgangszustand zum selben Endzustand führt. Der Beweis hierzu läuft induktiv über die Länge der Ausführungsfolge. Bei höheren Dimensionen muß zusätzlich gefordert werden, daß die Transformation einer Operation entlang jeder Ausführungsfolge zur selben resultierenden Operation führt. Dadurch ist gewährleistet, daß jeder Teilschritt einer Ausführungsfolge, der keiner Originaloperation entspricht, eindeutig definiert ist.

Abschätzungen für die Komplexität des Verfahrens ergeben, daß sie im ungünstigsten Fall proportional zum Quadrat der Anzahl der Beteiligten ist. Dies setzt allerdings maximale Parallelität voraus, d. h. auseinanderlaufende Interaktionen, die erst am Ende zusammengefügt werden. In der Praxis wird dies nur dann der Fall sein, wenn die Netzverbindung zeitweise ausfällt.

Aus Benutzersicht ist die Transformationsbedingung (1) nicht hinreichend, da sie auch sinnlose oder triviale Transformationen zuläßt.

Beispiel 3: Eine Transformation T sei definiert durch

$$T(o_1, o_2) = (\overline{o_2}, \overline{o_1}) \tag{3}$$

Diese Transformation bestimmt, daß im Falle zweier gleichzeitig ausgeführter Operationen beide rückgängig gemacht werden. Das Pathologische dieser Transformation zeigt sich z. B. dann, wenn einer der Partner seine Operation unmittelbar rückgängig macht. Als resultierende Operation ergibt sich o_1 und nicht etwa o_2, was intuitiv eher zu erwarten wäre (s. Abb. 5a).

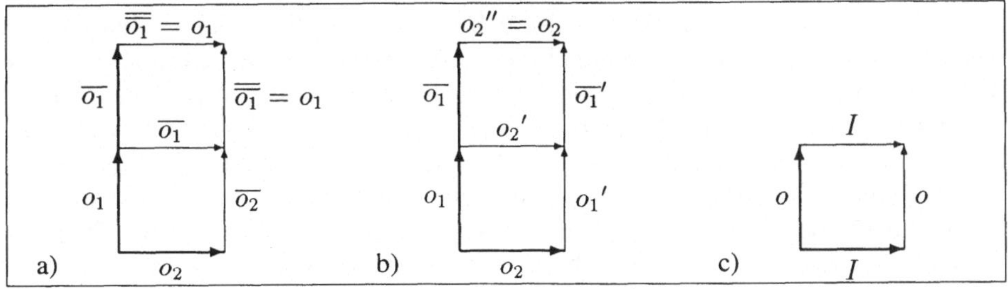

Abb. 5: a) Zu Beispiel 3: $T(o_1, o_2) = (\overline{o_2}, \overline{o_1})$, b) Undo-Eigenschaft, c) Neutralitätseigenschaft

Aus Benutzersicht ist das Verhalten gegenüber zurückgenommenen Operation sehr bedeutend. Die unmittelbare Rücknahme von Operationen soll häufig gerade vermeiden, daß Operationen von anderen dadurch beeinflußt werden.

Eigenschaft 1 (Undo-Eigenschaft): Sei $\overline{o_1}$ die inverse Operation zu o_2, dann soll für alle Operationen o_1 gelten (Abb. 5b):

$$\text{Wenn } T(o_1, o_2) = (o_1', o_2') \text{ und } T(\overline{o_1}, o_2') = (\overline{o_1}', o_2''), \text{ dann } o_2 = o_2'' \qquad (4)$$

Eine einfachere Bedingung, die an Transformationen gestellt werden kann, besteht darin, daß sich die Identitätsabbildung neutral bezüglich Transformationen verhält.

Eigenschaft 2 (Neutralitätseigenschaft): Sei I die neutrale Identitätsabbildung, dann soll für alle Operationen o gelten (Abb. 5c):

$$T(o, I) = (o, I) \qquad (5)$$

Für jede Anwendung müssen spezielle Transformationsregeln gefunden werden, die diese Eigenschaften erfüllen. Falls solche Regeln überhaupt existieren, müssen sie nicht eindeutig sein. Stößt das Computersystem bei den Transformationen auf solche, nicht eindeutig zu klärende Interaktionskonflikte, kann es zunächst eine der Alternativen auswählen und somit den Benutzern das Weiterarbeiten gestatten. Zusätzlich kann das System in solchen Ausnahmefällen die Benutzer warnen und alternative Interaktionen vorschlagen. Durch diese Vorgehensweise werden die

Benutzer möglichst wenig bei ihrer Arbeit unterbrochen. Klärende Dialoge können nachträglich auf Benutzerwunsch angestoßen werden.

Die Attraktivität der Operationstransformationen aus Benutzersicht liegt in der Unterstützung von sowohl interner als auch externer Koordination der Interaktion. Das Computersystem sorgt zunächst für syntaktische Konsistenz. Dazu erhalten die Benutzer durch die Visualisierung fremder Interaktionen jederzeit ausreichend Informationen, um selbst auf inhaltliche Konsistenz achten zu können. Die explizite Repräsentierung des Interaktionsmodell ermöglicht den Benutzern, auch nachträglich den Verlauf vergangener Interaktionen zu inspizieren. Außerdem dient das Interaktionsmodell als Grundlage umfangreicher Rücknahmemöglichkeiten. Damit ist eine kooperative Zusammenarbeit von Mensch und Computer bei der Konfliktbewältigung paralleler Interaktion in Groupware weitgehend verwirklicht.

5. Transformationsregeln für Texteditoren

Die grundlegenden Operationen eines Texteditors sind das Einfügen und das Löschen von Einzelzeichen. Komplexere Operationen lassen sich durch deren Zusammensetzung erzeugen. Wir definieren drei Operationen: *Ins*, *Del* und *Seq*. Eigentlich wären nun $3^2 = 9$ Transformationsregeln zu definieren, für jede Kombination der drei Operationen eine. Dies ist jedoch nicht nötig. Zum einen wurde bereits oben bemerkt, daß Transformationen nicht von der Reihenfolge ihrer Argumente abhängen dürfen. Es reicht also, für ungleiche Paare von Operationen nur die eine Richtung zu definieren. Wird die andere Richtung benötigt, werden die Operation vor und nach der Transformation einfach vertauscht.

Wir definieren (für Personen A und B):

$$T(Ins[A, p_1, c_1], Ins[B, p_2, c_2]) = \begin{cases} (Ins[A, p_1, c_1], Ins[B, p_2 + 1, c_2]), \\ \quad \text{falls } p_1 < p_2 \text{ oder} \\ \quad\quad p_1 = p_2 \text{ und } prio(A) < prio(B), \\ (Ins[A, p_1 + 1, c_1], Ins[B, p_2, c_2]), \\ \quad \text{sonst.} \end{cases} \quad (6)$$

$$T(Del[A, p_1, c_1], Del[B, p_2, c_2]) = \begin{cases} (I, I), \\ \quad \text{falls } p_1 = p_2 \text{ und damit } c_1 = c_2, \\ (Del[A, p_1, c_1], Del[B, p_2 - 1, c_2]), \\ \quad \text{falls } p_1 < p_2, \\ (Del[A, p_1 - 1, c_1], Del[B, p_2, c_2]), \\ \quad \text{sonst.} \end{cases} \quad (7)$$

$$T(Seq[o_1, o_2], o) = (Seq[o_1{}', o_2{}'], o''),$$
$$\text{wobei } T(o_1, o) = (o_1{}', o') \text{ und } T(o_2, o') = (o_2{}', o''). \tag{8}$$

Die Definition einer Regel für das Paar *Ins* und *Del* erfolgt entsprechend.

Diese Definitionen unterscheiden sich an einigen Punkten von jenen in [5]. Zwei Einfügeoperationen desselben Zeichens an derselben Stelle ergeben hier zwei Zeichen, deren Reihenfolge durch die Priorität der beteiligten Partner eindeutig bestimmt ist. Die Originaldefinition unterdrückt eines der Zeichen und erfüllt damit nicht die Undo-Eigenschaft. Tatsächlich unterscheiden sich die beiden eingefügten Zeichen, da sie von unterschiedlichen Autoren stammen. Gruppeneditoren, die – wie unsere Implementierung – die Herkunft von Textteilen farblich markieren, lassen dies auch visuell erkennen. Sollte der Fall eintreten, daß auf diese Weise versehentlich ein Text doppelt eingetragen wird, ist dies auf einen Blick zu erkennen. Es zeigt sich außerdem, daß auf die komplizierte Prioritätsfunktion, wie sie in [5] für den Fall der gleichzeitigen Einfügung an derselben Stelle durch mehr als zwei Benutzer definiert wird, verzichtet werden kann.

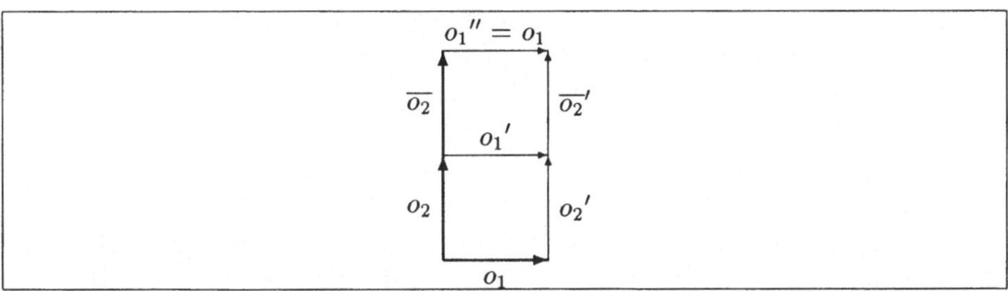

Abb. 6: Kompositionsregel

Die obige Definition (8) einer Transformation für zusammengesetzte Operationen ist neu. Tatsächlich steht sie für fünf einzelne Transformationsregeln. Die Gültigkeit läßt sich leicht anhand des zugehörigen Graphen (s. Abb. 6) ersehen. Diese Regel kann dazu dienen, Einfüge- oder Löschoperationen einer ganzer Zeichenkette als eine elementare Operation aufzufassen. Die Definition (8) garantiert zudem, daß der Effekt derselbe ist, wie wenn unmittelbar aufeinanderfolgende Einzeloperationen ausgeführt würden. Dies führt unter anderem zu der erfreulichen Eigenschaft, daß Einfügungen von mehreren aufeinanderfolgenden Zeichen, die zwei Personen an derselben Position starten, nicht vermischt werden.

Bei der Definition der Transformation eines *Del*-Paares zeigt sich, daß sie einen pathologischen Fall beinhaltet: das Löschen desselben Zeichens. Dieser Fall erfüllt i. d. R. nicht die Undo-Eigenschaft. Wird von einer Partie eine Löschoperation rückgängig gemacht, so wird dies vom System ausgeführt, obwohl eigentlich noch eine weitere, aber bisher unterdrückte Löschoperation vorliegt. Wenn dieser Fall bei einem Transformationsschritt auftritt, sollten die Benutzer informiert werden. Zu beachten ist, daß diese Anomalie am konsistenten Endzustand der einzelnen Editoren nach Ausführung der Löschoperation nichts ändert. Bisher sind keine Transformationsregeln bekannt, die dieses Löschproblem eindeutig lösen.

Allerdings existiert ein anderer Weg, um dieses Problem zu vermeiden. Löschoperationen können auch so interpretiert werden, daß die gelöschten Objekte lediglich als gelöscht markiert und als solche am Bildschirm dargestellt bleiben, etwa durchgestrichen. Ein solches Objekt kann von jeder Person einmal als gelöscht markiert werden. Eine Löschung wird erst dann rückgängig gemacht, wenn alle ihre Löschung zurückgenommen haben. Löschungen sind dann natürlich neutral bezüglich positionsabhängiger Operationen. Offen bleibt hier die Frage nach der inversen Funktion der Einfügeoperation, die Löschoperation scheidet hierzu aus. Die Einführung einer sogenannten Destroy-Funktion könnte helfen, falls diese ausdrücklich als nicht rücknehmbar definiert wird.

6. Implementierung

Der Gruppeneditor „Joint Editor" wurde in CLOS implementiert. Die Benutzungsoberfläche (s. Abb. 7) basiert auf dem Toolkit XIT [8]. Der Gruppeneditor ähnelt von der Funktionalität im wesentlichen einem Einbenutzereditor (als Vorbild diente EMACS in der Version 18 [7]), um den Übergang von einem Ein- zu einem Mehrbenutzereditor zu erleichtern. Textoperationen besitzen einen zusätzlichen Parameter, der den Benutzer angibt, der die Operation angestoßen hat. Der Editor wurde derart erweitert, daß Texte verschiedener Autoren in unterschiedlichen Farben – oder Schriftarten – dargestellt werden. Dies trägt zum gegenseitigen Gewahrsein bei und unterstützt somit die interne Koordination. Pastellartige Farben als Hintergrund zu schwarzer Schrift haben sich als zweckmäßig erwiesen. Farben stellen eine bekannte Metapher dar, um verschiedene Personen zu unterscheiden (vgl. Spielfiguren). Obwohl sie prinzipiell frei wählbar sind, sollte eine feste Zuordnung bevorzugt werden. Bei Bedarf ist die farbliche Markierung auch abschaltbar.

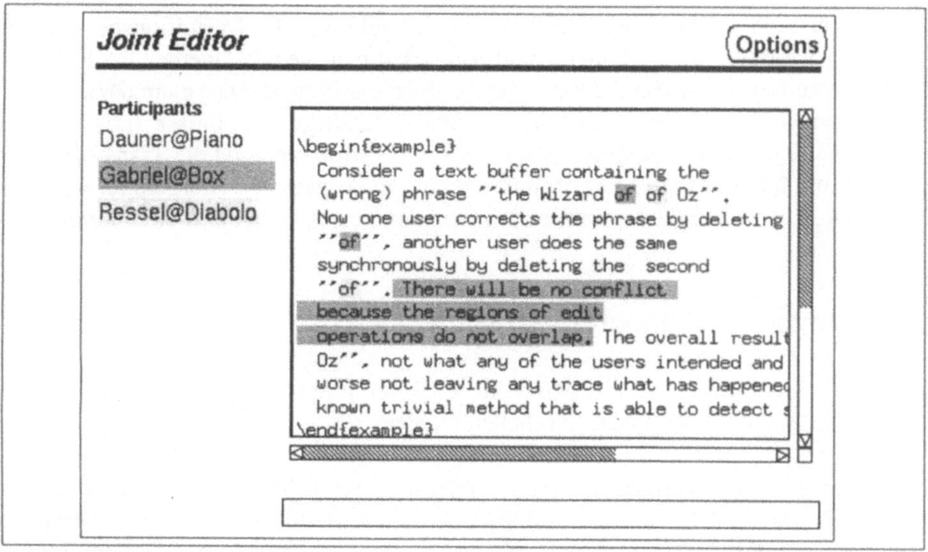

Abb. 7: Oberfläche des Texteditors für Gruppen mit Textbeiträgen von drei Teilnehmern

Die objektorientierte Repräsentation von Operationen erlaubt es, diese in eine Vererbungshierarchie einzubetten, die die Komplexität der Transformationsregeln vereinfacht und die Zahl der Regeln verringert.

Das mehrdimensionale Interaktionsmodell, das während der Verarbeitung aufgebaut wird, kann zusammen mit dem Dokument abgespeichert und wieder geladen werden. Es reicht hierbei aus, die Originaloperationen abzuspeichern, da alle anderen Informationen daraus wieder eindeutig abgeleitet werden können. Eine zusätzliche chronologische, lineare Historie, die natürlich bei jedem Benutzer unterschiedlich sein kann, kann ebenfalls abgespeichert und wieder geladen werden. Sie dient dazu, mit Hilfe einer einfachen Bedienoberfläche, ähnlich der eines Videogeräts, vorhergehende Änderungen und sogar vergangene Sitzungen animationsartig abzurufen.

Ein Menü mit Scroll-Balken stellt die chronologische Historie zusätzlich dar. Einträge in der Historie können ausgewählt werden, um sie selektiv rückgängig zu machen. Hierzu wird lokal eine Undo-Operation erzeugt und zwar im Systemzustand, wie er nach Ausführung der ausgewählten Operation bestand. Mit Hilfe der Transformationsregeln wird diese Operation in den aktuellen Zustand transformiert; das Resultat wird dann wie eine Benutzereingabe behandelt, also lokal ausgeführt und an die anderen Editoren weitergeleitet. Der Definitionsbereich der Transformationsregeln muß hierbei auf solche Fälle ausgedehnt werden, in denen die Akteure zu transformierender Operationen übereinstimmen. Dies kann dazu führen, daß gewisse Transformationen nicht mehr eindeutige Ergebnisse liefern. Dies beeinträchtigt aber die Ermittlung der Undo-Operation nicht, da die Transformation nur einmal lokal ausgeführt wird und erst das Endergebnis weitergeleitet wird. Es kann somit nicht zu Inkonsistenzen zwischen den verschiedenen Editoren kommen.

7. Vergleich mit anderen Arbeiten

Es gibt eine Reihe von Gruppeneditoren oder allgemeinen Ansätzen, die im folgenden mit unserem Ansatz verglichen werden.

GROVE [5] verwendete als erster Editor Operationstransformationen. Der Originalalgorithmus konnte allerdings für Sitzungen mit mehr als zwei Personen noch keine Konsistenz garantieren. Wir erweiterten das Verfahren außerdem um das Interaktionsmodell, das die komplexen zeitlichen Beziehungen repräsentiert und um eine Undo-Funktion. Im Gegensatz zu unserem Editor bietet GROVE die Möglichkeit, Dokumentteile in verschiedene Bearbeitungsmodi (*private, public, shared*) zu versetzen.

ShrEdit [4] verwendet Sperrmechanismen auf der Ebene von Textselektionen. Zwei Einfügemarken an derselben Textposition sind verboten, das Auftreten dieses Konflikts wird akustisch angezeigt. Allerdings verhindert dies z. B. das parallele Editieren eines leeren Textpuffers. Zudem gelten die Einschränkungen, die für Sperrmechanismen gelten.

Der Gruppeneditor DistEdit [10] benützt ebenfalls Sperrmechanismen, um Konflikte zu vermeiden. Sowohl temporäre, implizite als auch explizit anzufordernde Sperren werden unterstützt. Eine den L-Transformationen verwandte Transformationsart (Γ-*Transformationen*) wird dort zum Vertauschen benachbarter Operationen in einer Historie verwendet, um Gruppen-Undo und selektives Undo im allgemeinen zu implementieren. Zusätzlich werden sogenannte Konfliktfunktionen definiert, die testen, ob zwei benachbarte Operationen in einer Historie miteinander vertauschbar sind. Konflikte mit Operationen, die bereits rückgängig gemacht wurden, werden extra behandelt und können so eliminiert werden. Interessant ist in diesem Zusammenhang, daß sich die hier vorgestellten L-Transformationen und die Γ-Transformationen eindeutig aufeinander abbilden lassen.

Abowd und Dix [1] schlagen als Alternative zu Operationstransformationen die Verwendung dynamischer Zeiger (*dynamic pointers*) vor, die Einfügepositionen repräsentieren und bei Modifikationen automatisch verschoben werden. Dieses Vorgehen setzt voraus, daß bei der Ausführung einer Aktion bereits alle davon abhängigen Aktionen bekannt sind. Das Verfahren hilft auch nicht, Konflikte bei identischen Bezugspunkten zu lösen. Für Benutzer ist unklar, ob und welche Zeiger am Rande eines Bereichs mitberücksichtigt werden sollen, wenn darauf Move- oder Copy-Operationen ausgelöst werden. Außerdem wird eine lineare Historie vorausgesetzt. Unser Editor verwaltet zwar auch eine lineare Historie, um für jeden Benutzer chronologisches Navigieren im Interaktionsraum zu ermöglichen; wir halten aber das für alle Benutzer identische umfassende Interaktionsmodell für ebenso wichtig, um den Benutzern die Orientierung im Interaktionsraum zu ermöglichen.

GINA [2] verwaltet einen History-Baum, dessen Äste verschiedene Versionen darstellen. Diese Äste können mittels der Technik des *selektiven Redo* kombiniert werden. Hierzu werden keine Transformationen benutzt, sondern es wird durch Testfunktionen ermittelt, ob der neue Zustand mit dem alten derart kompatibel ist, daß die Operation unverändert ausgeführt werden kann. Dieser Mischprozeß muß manuell vorgenommen werden. Das Ergebnis ist davon abhängig, welcher Ast zu welchem „dazugemischt" wird. Dadurch, daß GINA bei der Ausführung nur Ausgangs- und Zielzustand betrachtet, ist es nicht möglich, bestimmte Konfliktarten zu erkennen. Ein Beispiel ist das Färben und gleichzeitige Kopieren eines Objektes. Das Färben bezieht sich nur auf das Originalobjekt, obwohl es Sinn macht, das Einfärben auch auf die Objektkopie anzuwenden. Unser Ansatz erlaubt dies. Hierzu wird eine der beiden Alternativen standardmäßig bei der Transformation durchgeführt. Gleichzeitig kann eine Meldung bzw. Abfrage erfolgen, ob die andere Alternative durchgeführt werden soll. Es ist denkbar, die Voreinstellung, welche Alternative standardmäßig ausgewählt wird, interaktiv und kooperativ mit Hilfe desselben Verfahrens im laufenden Betrieb zu ändern.

8. Ergebnisse und Ausblick

Operationstransformationen erfüllen die weiter oben gestellten Anforderungen und erweisen sich somit als ein geeignetes Verfahren zur kooperativen Interaktions-unterstützung in Groupware:

1. Ein „feinkörniger" Informationsaustausch und die Visualisierung fremder Aktionen gewährleisten hohes gegenseitiges Gewahrsein.

2. Sofortige lokale Ausführung einer Eingabe garantiert kurze Antwortzeiten, eine wesentliche Voraussetzung für direkte Manipulation.

3. Die replizierte Architektur erleichtert die Erweiterung bestehender Anwendungssysteme für Einzelbenutzer.

4. Die formale Grundlage erlaubt es, Vorhersagen über Konsistenzverhalten zu machen. Der Computer kann jederzeit einen syntaktisch konsistenten Zustand herstellen. Benutzereingriffe können so weitgehend vermieden werden. Über das Interaktionsmodell haben die Benutzer darüber hinaus jederzeit Zugriff auf vergangene Interaktionen.

Damit sie Benutzererwartungen entsprechen, sollten Transformationen zwei Eigenschaften, die Undo- und die Neutralitätseigenschaft, haben. Für die Implementierung eines einfachen Texteditors konnten Transformationen bestimmt werden, die bis auf eine Ausnahme den genannten Eigenschaften entsprechen. Insbesondere ist es mit Hilfe einer zusammengesetzten Operation möglich, auch komplexere Editorfunktionen als elementare Operationen anzusehen. Durch Ausnutzung spezieller Eigenschaften ist es zudem möglich, mit wesentlich weniger Regeldefinitionen auszukommen, als zunächst theoretisch zu erwarten waren. Wir suchen weitere Anwendungsbereiche für Operationstransformationen. Davon können auch Anwendungen profitieren, die selektives Undo mittels Γ-Transformationen implementieren.

Wegen der Toleranz gegenüber Netzausfällen eignet sich das Verfahren auch für asynchrone Anwendungen, bei denen die Teilnehmer nicht gleichzeitig am Computer anwesend sein müssen. Ein Beispiel wäre verteiltes „Brainstorming“: Jeder Beteiligte erweitert lokal eine Brainstorming-Datei; die jeweiligen Änderungen werden per elektronischer Post an alle anderen verteilt, wo sie mittels Operationstransformationen mit den aktuellen Zuständen verrechnet werden.

Das Interaktionsmodell kann schließlich als Grundlage weiterer unterstützender Module dienen. Ein Interaktions-Browser kann beispielsweise das Navigieren durch den Ausführungsraum erlauben und so vergangene Zustände auf einfache Weise zugänglich machen. Darüber hinaus können Planerkennungs- oder Analysekomponenten auf dem Interaktionsmodell aufsetzen. Damit können Interaktionsmuster gesucht und z. B. der Parallelitätsgrad der Zusammenarbeit oder die Häufigkeit von Konfliktfällen analysiert werden.

Literatur

[1]   G. D. Abowd, A. J. Dix. Giving Undo Attention. *Interacting with Computers*, 4(3):317–342, 1992.

[2]   T. Berlage, A. Genau. A Framework for Shared Applications with a Replicated Architecture. In *Proceedings of the UIST '93*. ACM Press, 1993.

[3]   C. Condon. The Computer Won't Let Me: Cooperation, Conflict and the Ownership of Information. In Steve Easterbrook (Hrsg.), *CSCW: Cooperation or conflict?*, Computer Supported Cooperative Work, Kapitel 8, S. 171–185. Springer-Verlag, London, 1993.

[4]   P. Dourish, V. Bellotti. Awareness and Coordination in Shared Workspaces. In Jon Turner, Robert Kraut (Hrsg.), *Proceedings of the CSCW '92*, S. 107–114, Toronto, Canada, 31.Okt.–4.Nov 1992. ACM SIGCHI & SIGOIS, ACM Press.

[5]   C. A. Ellis, S. J. Gibbs. Concurrency control in groupware systems. In *Proceedings of the ACM SIGMOD '89 Conference on the Management of Data*, S. 399–407, Seattle, Washington, 1989. ACM, New York.

[6]   C. A. Ellis, S. J. Gibbs, G. L. Rein. Groupware: Some issues and experiences. *Communications of the ACM*, 34(1):38–58, Januar 1991.

[7]   J. Gosling. *Unix Emacs*. Carnegie-Mellon University, Pittsburgh, 1982.

[8]   J. Herczeg, H. Hohl, M. Ressel. Progress in Building User Interface Toolkits: The World According to XIT. In *Proceedings of the ACM Symposium on User Interface Software and Technology*, S. 181–190, November 1992.

[9]   U. Piepenburg. Ein Konzept von Kooperation und die technische Unterstützung kooperativer Prozesse in Bürobereichen. In J. Friedrich, K.-H. Rödiger (Hrsg.), *Computergestützte Gruppenarbeit (CSCW)*. B. G. Teubner, 1991.

[10]   A. Prakash, M. J. Knister. Undoing Actions in Collaborative Work. In Jon Turner, Robert Kraut (Hrsg.), *Proceedings of the CSCW '92*, Consistency in Collaborative Systems, S. 273–280, Toronto, Canada, 31.Okt.–4.Nov 1992. ACM SIGCHI & SIGOIS, ACM Press.

Danksagung

Prof. Rul Gunzenhäuser danke ich für die Unterstützung und die hilfreichen Kommentare. Jürgen Herczeg und Hubertus Hohl steuerten wichtige Grundlagen der Implementierung bei.

Matthias Ressel
Institut für Informatik
Universität Stuttgart
Breitwiesenstr. 20-22
70565 Stuttgart

Modellierung von graphischen Benutzungsoberflächen im Rahmen des TADEUS-Ansatzes

E. Schlungbaum, T. Elwert
Universität Rostock

Zusammenfassung

Der TADEUS-Ansatz (*ta*sk-based *de*velopment of *us*er interface software) hat das Ziel, Voraussetzungen für eine aufgabenorientierte und benutzergerechte Gestaltung von Interaktiven Graphischen Systemen (IGS) zu schaffen. Dazu werden ein methodischer Rahmen, die notwendigen Modelle zur Beschreibung von IGS und die entsprechenden Werkzeuge zur Unterstützung der Entwickler von IGS erarbeitet. In diesem Artikel wird das TADEUS-Interaktionsmodell, welches die Modellierung moderner graphischer Benutzungsoberflächen gewährleistet, detailliert beschrieben. Das Konzept der Dialoggraphen stellt eine Erweiterung zu den von Janssen [9] vorgestellten Dialognetzen dar.

1 Einleitung

Die aufgabenorientierte und benutzergerechte Gestaltung von Interaktiven Graphischen Systemen (IGS) ist eine wesentliche Voraussetzung für die spätere Akzeptanz durch den Endbenutzer und stellt einen Beitrag für eine anzustrebende, ganzheitliche Arbeitsgestaltung dar (vgl. [11]).

Die Dialogmodellierung und die Daten- und/oder Funktionsmodellierung für ein Dialogsystem erfolgte in der Vergangenheit der Software-Entwicklung mit meist unabhängigen und wenig aufeinander abgestimmten Methoden. Die Konsistenz und Vollständigkeit eines Designs methodenübergreifend sicherzustellen ist eine nichttriviale, fehlerträchtige und aufwendige Aufgabe für die Software-Entwickler (vgl. [7]). Die von der UIMS-Technologie geförderte Separierung von Dialog- und Applikationskomponente, welche aus der Sicht der Entwicklung und Implementation von Dialogsystemen enorme Vorteile bietet, trägt zu einer Verschärfung des zuvor genannten Problems bei.
Mit dem TADEUS-Projekt (*ta*sk-based *de*velopment of *us*er interface software) wird versucht, einen methodischen Rahmen, die notwendigen Modelle zur Beschreibung von IGS und die entsprechenden Werkzeuge zur Unterstützung der

Entwickler von IGS (Dialog- und Applikationsentwickler) zu erarbeiten. Dabei werden folgende Ziele im Mittelpunkt stehen:

- Die Software-Entwicklung für das IGS erfolgt in kontinuierlichen, aufeinander aufbauenden Schritten von der Requirementsanalyse bis zur automatischen Erzeugung eines Prototypen des Software-Produkts.
- Für das IGS erfolgt eine aufgabenorientierte und benutzergerechte Gestaltung der Benutzungsoberfläche. Dazu werden vier Bereichsmodelle, zu welchen neben dem Datenmodell des zu entwickelnden IGS (Problembereichsmodell) auch ein Aufgaben-, ein Benutzer- und ein Interaktionsmodell zählen, erarbeitet.
- Das Interaktionsmodell, welches zunächst ausgehend von Aufgaben-, Benutzer- und Problembereichsmodell generiert und dann durch den Dialogentwickler erweitert wird, muß die Beschreibung moderner graphischer Benutzungsoberflächen gewährleisten.
- Der Prototyp der Benutzungsschnittstelle wird auf der Basis des Interaktionsmodells automatisch erzeugt.

In diesem Artikel wird ein Überblick zum TADEUS-Ansatz für die ganzheitliche Entwicklung von Interaktiven Graphischen Systemen dargestellt und das TADEUS-Interaktionsmodell ausführlich erläutert.

2 Überblick zum methodischen Rahmen in TADEUS

2.1 Der Modellierungsansatz

Der TADEUS-Ansatz für die Entwicklung von Interaktiven Graphischen Systemen unterscheidet vier Bereichsmodelle und ein Vorgehensmodell (vgl. [2]). Der TADEUS-Ansatz empfiehlt für die Modellierung von IGS zwei Abschnitte:

- Die Ergebnisse der Projektanalyse werden durch die Erstellung von Aufgaben-, Benutzer- und Problembereichsmodell dokumentiert.
- Die Dialoggestaltung erfolgt durch die schrittweise Weiterentwicklung des Interaktionsmodells, welches auf der Grundlage von Aufgaben-, Benutzer- und Problembereichsmodell erzeugt wurde.

Falls der Endbenutzer in die Entwicklung des IGS miteinbezogen wird, sind die genannten Modelle für eine Bewertung des aktuellen Entwicklungsstandes durch den Endbenutzer nicht geeignet. Deshalb ist es notwendig, in einem 3. Schritt einen Prototypen der Benutzungsschnittstelle zu erzeugen. Dabei sollte dieser Schritt mit minimalem Aufwand durch den Dialogdesigner verbunden und möglichst flexibel bzgl. einer Modifikation der bisher erzeugten Modelle sein. Hier wurde der Weg einer weitgehenden automatischen Erzeugung der Benutzungsschnittstelle gewählt (vgl. [15]). In Abbildung 1 sind die Beziehungen zwischen den Bereichsmodellen dargestellt.

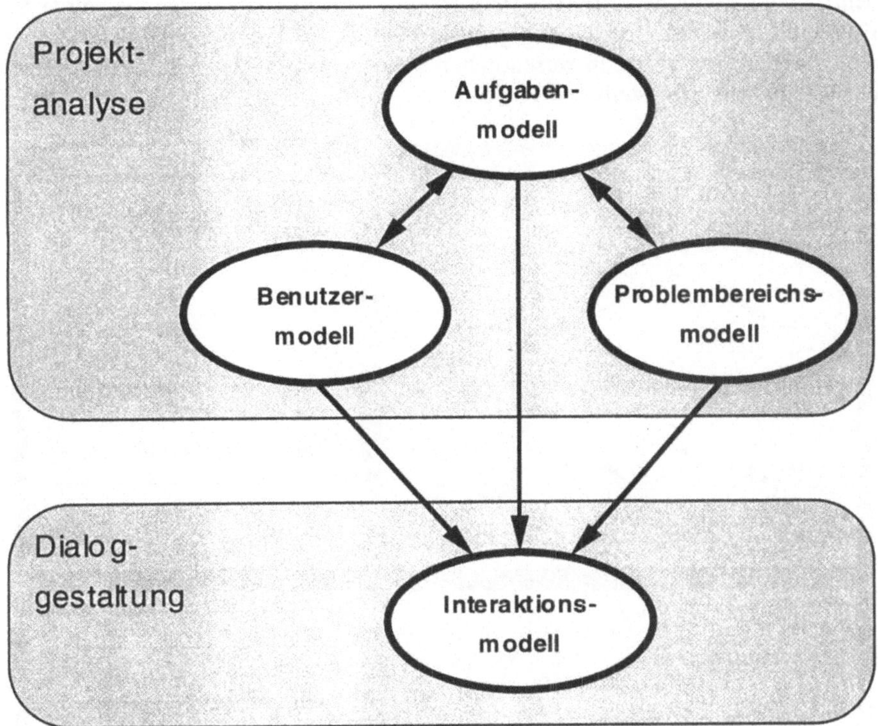

Abb. 1 Beziehungen zwischen den Bereichsmodellen im TADEUS-Ansatz

Neben der eigentlichen Modellierung des Interaktiven Graphischen Systems wird in einem Forschungsprojekt untersucht, in welcher Form der Entwickler bei der Modellerstellung beraten werden kann (vgl. [4],[5]).

2.2 Ein kleines Beispiel

Die Modellierung eines IGS mit dem TADEUS-Ansatz soll an einem Beispiel aus dem zu entwickelnden TADEUS-System selbst demonstriert werden. Es wurde die Komponente Benutzermodellierung ausgewählt. Dazu sind in Abbildung 2 die für die weiteren Ausführungen wesentlichen Teile des Aufgaben- und des Problembereichsmodells dargestellt.

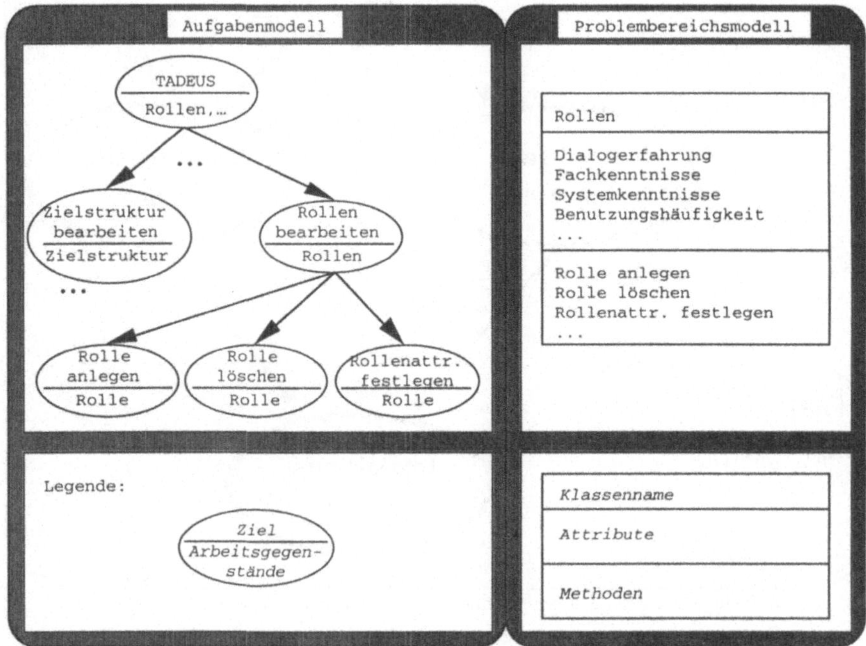

Abb. 2 Aufgaben- und Problembereichsmodell für die Benutzermodellierung (Ausschnitte)

Um das Interaktionsmodell zu generieren (s. Abschnitt 3.2), identifiziert der Dialogdesigner im Aufgabenmodell Knoten der Zielstruktur, welche als Sichten an der Benutzungsoberfläche repräsentiert werden sollen. In diesem Beispiel werden drei Sichten gekennzeichnet: "TADEUS", "Zielstruktur bearbeiten" und "Rollen bearbeiten". Die Spezifik des Arbeitsgegenstandes Rollen, der Dialogdesigner kann mehrere Rollenbeschreibungen (Instanzen der Klasse Rollen) anlegen, erfordert die Generierung einer Sicht zur Bearbeitung aller Rollenbeschreibungen und einer Sicht zur Bearbeitung einer konkreten Rollenbeschreibung.

3   Das TADEUS-Interaktionsmodell

3.1 Beschreibungstechniken für Dialogabläufe

Für die Beschreibung von Dialogabläufen finden sich in der Literatur verschiedene Ansätze (z.B. [6]). Als eine wesentliche Eigenschaft von Dialogspezifikationsmitteln bezeichnet Bay die Möglichkeit zur Visualisierung der Beschreibungskonstrukte, als Idealfall die interaktive graphische Bearbeitung der Dialogspezifikation (vgl. [1]).

Eine bekannte und häufig verwendete Form der graphischen Spezifikation von Dialogabläufen sind Zustandsübergangsdiagramme und ihre Erweiterungen. Allerdings sind mit ihnen in der Grundform nur sequentielle Dialogabläufe beschreibbar, also Formen des alphanumerischen Dialogs. Die Erweiterung von Jacob führte dazu, daß ein Dialogsystem mit mehreren, parallel ablaufbaren Zustandsübergangsdiagrammen beschrieben wurde (vgl. [10]). Hier fehlt eine Möglichkeit der graphischen Darstellung des Zusammenspiels der einzelnen Diagramme.

Eine andere Form der graphischen Spezifikation sind die auf Petrinetzen beruhenden Techniken der Dialogbeschreibung. Durch die Verwendung der Petrinetze, welche in der Grundform bereits Konzepte zur Beschreibung von Nebenläufigkeit beinhalten, wird es möglich, Dialogabläufe, wie sie an modernen graphischen Benutzungsoberflächen auftreten, zu beschreiben. Beispiele dafür sind die Ereignisgraphen von Roudaud et.al. [14] oder die Dialognetze von Janssen [9], wobei die Dialognetze eine Weiterentwicklung der Ereignisgraphen darstellen. Das TADEUS-Interaktionsmodell basiert auf den Dialognetzen und erweitert sie.

Eine weitere Möglichkeit der Beschreibung von graphischen Benutzungsschnittstellen sind objektorientierte Ansätze wie z.B. das' objektorientierte Interaktionsmodell von Hübner [8] oder der Peridot-Ansatz von Myers [12]. Bei diesen Methoden wird ein bottom-up-Weg der Benutzungsoberflächen-Gestaltung verfolgt, welcher die Konstruktion von komplexen Interaktionen (Dialogabläufen) aus elementaren Eingabeaktionen ermöglicht. Der TADEUS-Ansatz dagegen favorisiert ein top-down-Vorgehen. Eine ähnliche, top-down-orientierte Methode zur Entwicklung von IGS beschreibt Sukaviriya et.al. [16]. Die hierarchischstrukturierte Modellierung mit Applikationsaktionen, Interface-Aktionen und Interaktionstechniken ermöglicht eine detaillierte und anwendungsorientierte Dialogbeschreibung. Die fehlende Abstraktion für den Navigationsdialog führt bei der Entwicklung komplexer IGS zu einem umfangreichen Interaktionsmodell, wodurch die Forderung nach einer anschaulichen Visualisierung der Beschreibungskonstrukte verletzt wird.

Dialognetze basieren auf Bedingungs-/Ereignisnetzen mit Erweiterungen für die Darstellung modaler Teildialoge und für die hierarchische Gliederung. Dadurch erlauben sie eine übersichtliche Darstellung von Dialogabläufen, insbesondere für die Ebene der Navigationsdialoge. Aus der Verwendung der B/E-Netze, welche mit einfach markierten Stellen arbeiten, resultiert der Nachteil, daß die mehrfache Inkarnation eines Fensters nicht modelliert werden kann.

Im TADEUS-Interaktionsmodell werden *Dialoggraphen* zur Modellierung von Dialogabläufen verwendet. Es werden die vorteilhaften Eigenschaften der Dialognetze übernommen und für die Modellierung von mehreren Exemplaren eines Fensters erweitert. Durch festgelegte Transformationsregeln kann jeder Dialoggraph in ein Petrinetz mit individuellen Marken überführt werden. Neben der Beschreibung der Dialogabläufe wird damit ihre Simulation vor dem eigentlichen Erzeugen der Benutzungsschnittstelle ermöglicht. Mit den Dialoggraphen können Navigations- und Bearbeitungsdialoge beschrieben werden. Im folgenden soll die Verwendung von Dialoggraphen zur Modellierung von Dialogabläufen im Navigationsdialog erläutert werden. Auf die Darstellung der Bearbeitungsdialoge muß hier aus Platzgründen verzichtet werden (s. [3]).

3.2 Dialoggraphen zur Beschreibung von Dialogabläufen

Das TADEUS-Interaktionsmodell beschreibt die Zustände der Benutzungsoberfläche des Interaktiven Graphischen Systems, mögliche Dialogabläufe und die Interaktionsmöglichkeiten des Benutzers. Der Ausgangspunkt für die Generierung des Interaktionsmodells bildet die Definition von Sichten im Aufgaben- und/oder Problembereichsmodell. Sichten repräsentieren die Möglichkeiten zur Aufgabenbearbeitung, die in einem logischen Zusammenhang stehen und daher gleichzeitig an der Benutzungsoberfläche darzustellen sind. Für die Beschreibung einer graphischen Benutzungsoberfläche sind i.a. mehrere Sichten erforderlich. Wir gehen davon aus, daß jede Dialogsicht durch ein Fenster der zu entwickelnden graphischen Benutzungsoberfläche repräsentiert wird. Die Beschreibung der Interaktionen zwischen den Sichten wird als *Navigationsdialog*, die Beschreibung der Interaktionen innerhalb einer Sicht als *Bearbeitungsdialog* bezeichnet (vgl. [17]).

Das Interaktionsmodell unseres Modellierungsansatzes hat unterschiedliche Ausprägungsformen und reicht vom abstrakten Dialoggraphen bis zu einem mit einem

UIMS erzeugten Prototypen der Benutzungsschnittstelle des IGS. Die Grundlage der Dialoggraphen bilden Petrinetze mit individuellen Marken [13].

Die Dialoggraphen des TADEUS-Interaktionsmodells existieren in verschiedenen Ausprägungsformen. Der *abstrakte Dialoggraph* dient der ausschließlichen Beschreibung von Navigationsdialogen. Der *erweiterte Dialoggraph* ergänzt den abstrakten um Teile des Bearbeitungsdialoges, wobei allerdings von der genauen Repräsentation einzelner Interaktionen des Endbenutzers abstrahiert wird. Der *konkrete Dialoggraph* beschreibt den vollständigen Bearbeitungsdialog und die dazu möglichen Interaktionen des Endbenutzers. Für die Erstellung des Interaktionsmodells ist es erforderlich, daß der Dialogdesigner den abstrakten und den konkreten Dialoggraph entwickelt. Bei geringer Komplexität des zu modellierenden Dialogsystems kann die Entwicklung eines erweiterten Dialoggraphens zur Dialogmodellierung ausreichend sein. Dieser Artikel beschränkt sich auf die Erläuterung des abstrakten Dialoggraphens.

Ein abstrakter Dialoggraph besteht aus Knoten, welche die Dialogsichten repräsentieren und Übergängen zwischen diesen Knoten. Die Knoten eines abstrakten Dialoggraphens werden durch Kreise oder Ovale repräsentiert und mit einem Namen beschriftet. Bei der Darstellung des Navigationsdialoges kann von der Interaktion des Endbenutzers, welche den Übergang zwischen zwei Knoten initiiert, abstrahiert werden. Diese Interaktion wird mit einem schwarzen, unbeschrifteten Punkt auf dem Übergang angedeutet. Es werden nichtmodale, modale und komplexe Knoten unterschieden. In Analogie zu den Dialognetzen von Janssen wird ein nichtmodaler Knoten mit einem einfachen Rand, ein modaler mit einem fetten Rand und ein komplexer mit doppelter Umrandung dargestellt. Zwischen zwei nichtmodalen Dialogsichten können nebenläufige, sequentielle oder objektbezogene-nebenläufige Übergänge existieren. Weiterhin kann aus einer nichtmodalen Dialogsicht in eine modale Dialogsicht verzweigt werden. In Abbildung 3 sind zunächst die graphischen Symbole für die Übergangstypen zusammengefaßt. Ihre Bedeutung soll an einem einfachen Beispiel in Abbildung 4 erläutert werden. Die Definition der Übergangstypen wird im folgenden Abschnitt angegeben.

Um den Zustand der Benutzungsoberfläche des modellierten Interaktiven Graphischen Systems zu erfassen, werden die Knoten des abstrakten Dialoggraphen markiert. Die Markenbelegung eines Knoten beschreibt den Zustand der entsprechenden Dialogsicht (genauer: den Zustand des mit der Dialogsicht assoziierten Fensters an der Benutzungsoberfläche). Die Gesamtheit aller Knoten des abstrakten Dialoggraphen mit ihren Markierungen beschreibt den Gesamtzustand der Benutzungsoberfläche des modellierten IGS.

Die Markierungen der Knoten repräsentieren die Eigenschaften einer Dialogsicht. Es sollen folgende Eigenschaften unterschieden und als Marke repräsentiert werden:

- **s**: Sichtbarkeit - eine entsprechend markierte Dialogsicht ist sichtbar,

- **m**: Manipulierbarkeit - eine entsprechend markierte Dialogsicht ist sichtbar und kann aktiviert werden,

- **a**: Aktivität - die entsprechend markierte Dialogsicht ist sichtbar und wurde aktiviert, besitzt also den Eingabefokus.

Weiterhin existieren variable Markierungen, welche erst zur Laufzeit des IGS bestimmt werden können. Sie sind von den Interaktionen des Endbenutzers abhängig.

Bezeichnung	Graphisches Symbol
nebenläufiger Übergang	 $S_1 \triangleleft\!\!-\!\!\bullet\!\!\blacktriangleright S_2$
sequentieller Übergang	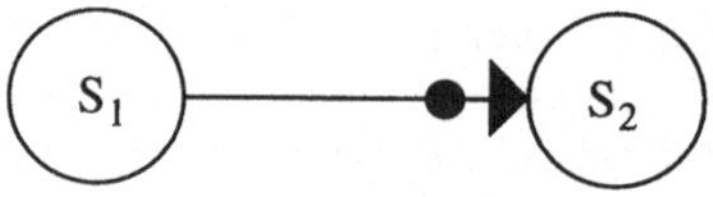 $S_1 -\!\!\bullet\!\!\blacktriangleright S_2$
objektbezogener-nebenläufiger Übergang	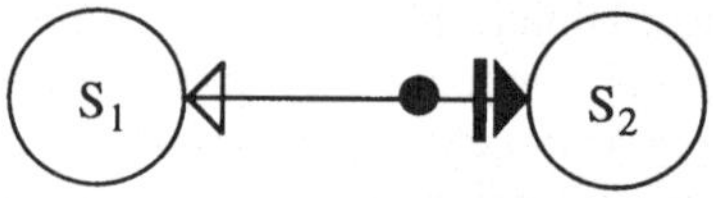 $S_1 \triangleleft\!\!-\!\!\bullet\!\!\parallel\!\!\blacktriangleright S_2$
modaler Teildialog	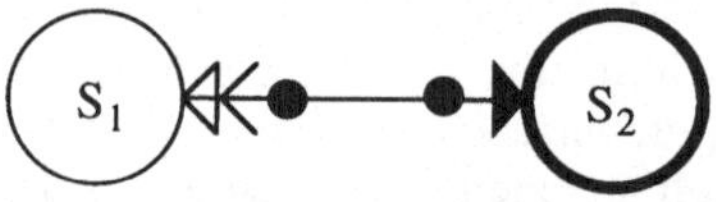 $S_1 \triangleleft\!\!\triangleleft\!\!\bullet\!\!-\!\!\bullet\!\!\blacktriangleright S_2$

Abb. 3 Graphische Symbole der Übergangstypen eines abstrakten Dialoggraphen

Zwischen diesen Eigenschaften einer Dialogsicht bestehen Abhängigkeiten, welche für die Dialogmodellierung ausgenutzt werden. Daraus resultieren folgende Regeln für die Existenz der Marken:

1) In einer graphischen Benutzungsoberfläche kann zu einem beliebigen Zeitpunkt immer nur ein Fenster den Eingabefokus besitzen. D.h., daß die Marke **a** im abstrakten Dialoggraphen nur in einem Exemplar auftreten darf.

2) Falls eine Dialogsicht aktiv ist, muß sie sichtbar und manipulierbar sein.

3)   Falls eine Dialogsicht sichtbar und manipulierbar ist, kann sie durch eine
     Interaktion des Benutzers den Eingabefokus übernehmen. Sie wird
     aktiviert, z.B. ein bereits geöffnetes Fenster wird in den Vordergrund
     geholt.

4)   Falls eine Dialogsicht sichtbar und nicht manipulierbar ist, kann sie nicht
     aktiviert werden. Das bedeutet, daß im modellierten System eine modale
     Dialogsicht aktiv ist.

5)   Eine nicht markierte Stelle bedeutet, daß die entsprechende Dialogsicht
     nicht sichtbar ist.

In der folgenden Abbildung wird der abstrakte Dialoggraph mit einer
Anfangsmarkierung zu dem oben eingeführten Beispiel dargestellt. Die Dialogsicht
"TADEUS Hauptfenster" ist mit (**a**, **s**, **m**) markiert, alle weiteren sind nicht
markiert. Das bedeutet, daß nur das Hauptfenster des modellierten Dialogsystems
geöffnet ist und den Eingabefokus besitzt.

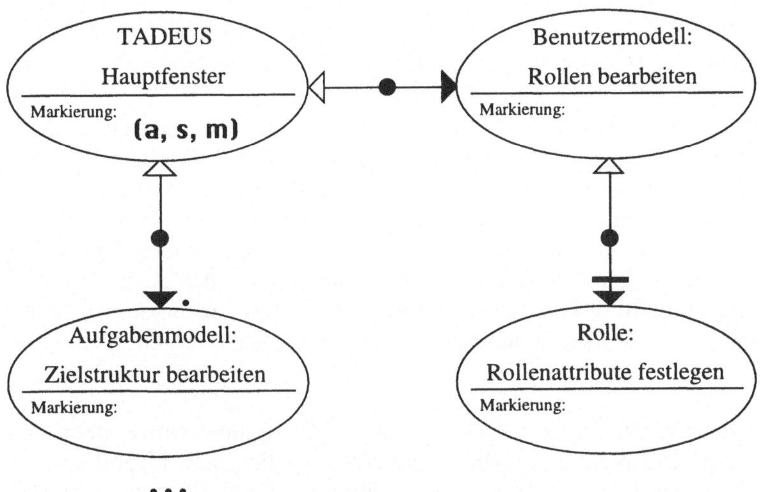

Abb. 4 Ausschnitt aus dem abstrakten Dialoggraphen des TADEUS-Systems

Ausgehend vom Hauptfenster des TADEUS-Systems kann der Benutzer weitere
Fenster zur Bearbeitung von Teilaufgaben bei der Modellierung des zu entwickelnden Interaktiven Graphischen Systems öffnen, wobei das Hauptfenster geöffnet
bleibt. Diese Möglichkeiten wurden mit nebenläufigen Übergängen modelliert.

Eine Besonderheit enthält der rechte Teil des Dialoggraphen. Zwischen den Dialogsichten "Benutzermodell: Rollen bearbeiten" und "Rolle: Rollenattribute festlegen" wurde ein objektbezogener-nebenläufiger Übergang modelliert. Die Sicht "Benutzermodell: ..." ist eine *Containersicht* (vgl. [17]) und faßt die Möglichkeiten zur Manipulation aller definierter Rollen zusammen. Die Sicht "Rolle: ..." ist eine *Einzelsicht* (vgl. [17]) und erlaubt die Modifikation einer konkreten Rollenbeschreibung. In den Abbildungen 5 und 6 werden mögliche Layoutgestaltungen für beide Dialogsichten angegeben. Der objektbezogene-nebenläufige Übergang erlaubt, daß ausgehend von der Containersicht mehrere Fenster der Einzelsicht geöffnet werden können. Das Fenster der Containersicht bleibt ebenfalls erhalten.

Beim sequentiellen Übergang zwischen zwei nichtmodalen Dialogsichten (in Abbildung 4 nicht enthalten) wird der Eingabefokus an die Dialogsicht S_2 übergeben und die Dialogsicht S_1 wird geschlossen. Die bisher genannten Konstrukte des abstrakten Dialoggraphen werden im folgenden Abschnitt durch ihre Abbildung auf Netze mit individuellen Marken definiert.

3.3 Definition der Elemente des abstrakten Dialoggraphen

Netze mit individuellen Marken (vgl. [13]) bestehen wie alle Petrinetze aus Stellen, Transitionen und Kanten zwischen Stellen und Transitionen und umgekehrt. Die Kanten können als Adressen konstante oder variable Kantenbeschriftungen aufweisen. In Petrinetzen mit individuellen Marken besteht für jede Transition die Möglichkeit, Aktionen, welche beim Schalten der entsprechenden Transition ausgeführt werden, zu spezifizieren. Diese Aktionen können den Aufruf von im Problembereichsmodell definierten Methoden der Applikationsobjekte oder spezielle Funktionen für die Steuerung der Dynamik des Interaktionsmodells beinhalten.

In den folgenden Abbildungen zur Definition der Übergänge wurde oben das Symbol für den abstrakten Dialoggraphen und unten das Petrinetz abgebildet. Die Knoten des abstrakten Dialoggraphen werden durch die Stelle S_i modelliert. Die Stellen B_i stellen zusätzliche Bedingungen zum Schalten der zugehörigen Transition T_i dar, welche von den Interaktionen des Endbenutzers abhängig sind. Führt der Endbenutzer eine zur Transition T_i definierte Interaktion I_i aus, so wird die Stelle B_i markiert. Ist gleichzeitig die Transition T_i aktiviert, dann kann die Transition T_i schalten und eine eventuell spezifizierte Aktion wird ausgeführt. Mit $\_x$ und $\_y$ werden in den Petrinetzen die variablen Kantenbeschriftungen gekennzeichnet.

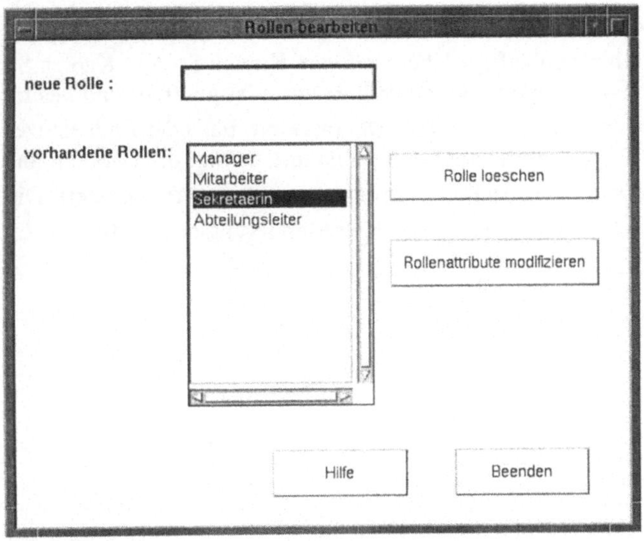

Abb. 5 Layout zur Dialogsicht "Rollen bearbeiten"

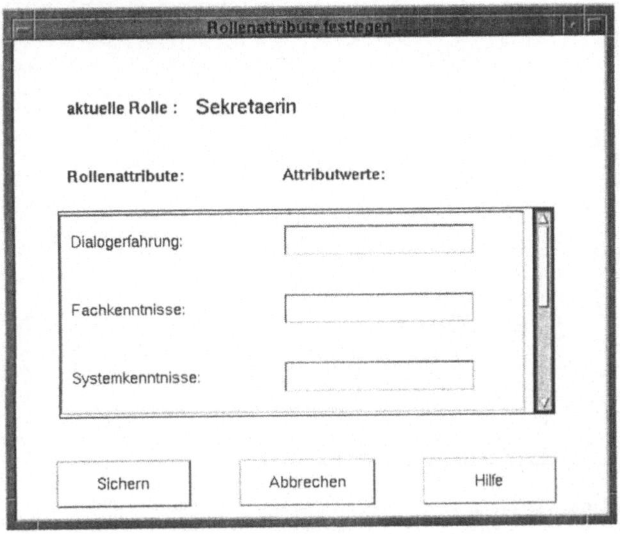

Abb. 6 Layout zur Dialogsicht "Rollenattribute festlegen"

Nebenläufiger Übergang

In Abbildung 7 wurde ein nebenläufiger Übergang vom Knoten S_1 zum Knoten S_2 dargestellt. Das Fenster, welches mit S_1 modelliert wird, ist geöffnet und besitzt den Eingabefokus. Die Stelle S_1 ist mit (**a**, **s**, **m**) markiert. Falls der Endbenutzer eine für die Transition T_1 gültige Interaktion I_1 ausführt, erfolgt die Öffnung und Aktivierung des Fensters S_2. Die Stelle S_2 ist dann mit (**a**, **s**, **m**) markiert. Die Darstellung des Fensters S_1 bleibt erhalten, aktuelle Markierung mit (**s**, **m**).

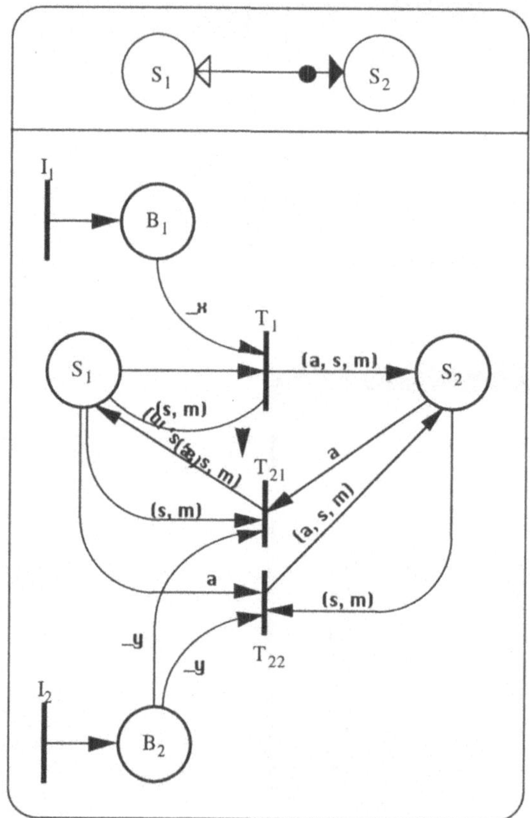

Abb. 7 Definition eines nebenläufigen Überganges

Mit der Interaktion I_2, z.B. einfacher Mausklick, kann der Endbenutzer den Eingabefokus zwischen den geöffneten und manipulierbaren Fenstern S_1 und S_2

wechseln, d.h. von der aktiven Dialogsicht wird die Marke a abgezogen und an die andere Dialogsicht übergeben.

Sequentieller Übergang

Der sequentielle Übergang vom Knoten S_1 zum Knoten S_2 erfolgt durch die vom Endbenutzer ausgelöste Interaktion I_1 und führt zur Öffnung und Aktivierung von Fenster S_2. Gleichzeitig wird Fenster S_1 geschlossen (s. Abbildung 8).

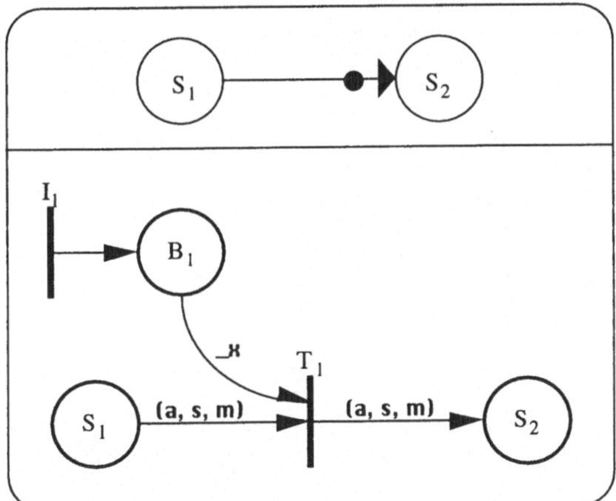

Abb. 8 Definition eines sequentiellen Überganges

Objektbezogener-nebenläufiger Übergang

Die Erläuterung der Funktionsweise des in Abbildung 9 definierten objekt-bezogenen-nebenläufigen Übergangs bezieht sich auf den Ausschnitt des abstrakten Dialoggraphen des TADEUS-Systems (s. Abbildung 4).

Ausgehend vom aktiven Fenster S_1 ("Benutzermodell: Rollen bearbeiten") hat der Endbenutzer die Rolle "Sekretärin" für die weitere Bearbeitung gewählt. Die Transition T_1 schaltet und die Stelle S_2 wird mit (**a, s, m**, "Sekretärin") markiert. Das entsprechende Fenster "Rolle: Rollenattribute festlegen" wird für die Rolle "Sekretärin" geöffnet und aktiviert. Danach kann der Endbenutzer wieder zum Fenster S_1 wechseln und die Rolle "Mitarbeiter" für die weitere Bearbeitung auswählen. Die Transition T_1 schaltet und die Stelle S_2 wird zusätzlich mit (**a, s, m**, "Mitarbeiter") markiert und das entsprechende Fenster geöffnet und aktiviert.

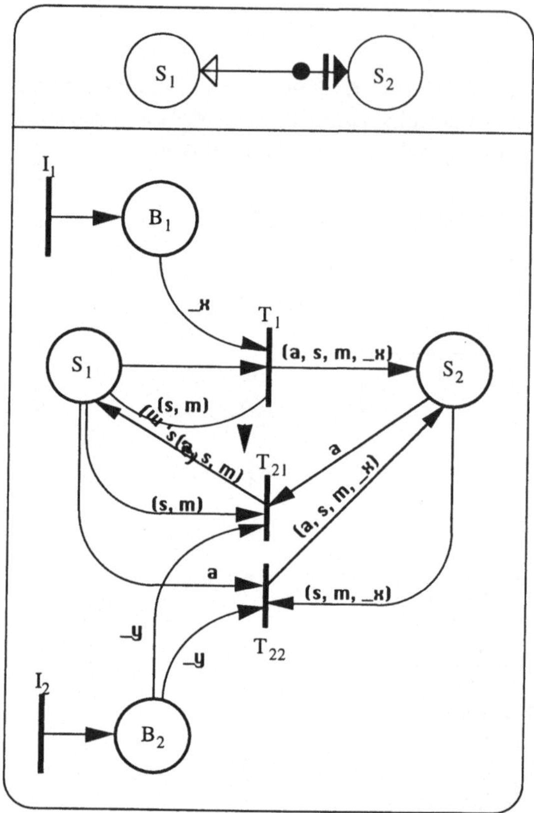

Abb. 9 Definition eines objektbezogenen-nebenläufigen Überganges

Nach diesen Interaktionen ergibt sich folgender Gesamtzustand für den dargestellten Ausschnitt der Benutzungsschnittstelle:

- Das Fenster "Benutzermodell: ..." ist sichtbar und manipulierbar.

  DieStelle S_1 ist mit (**s, m**) markiert.

- Das Fenster "Rolle: ..." ist für die Rolle "Sekretärin sichtbar und manipulierbar und für die Rolle "Mitarbeiter" aktiv.

  Die Stelle S_2 ist mit (**s, m**, "Sekretärin") und (**a, s, m**, "Mitarbeiter") markiert.

Modale Teildialoge

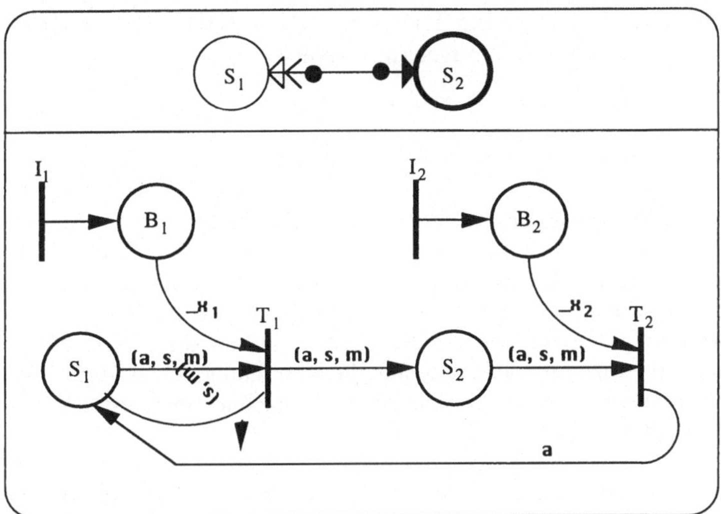

Abb. 10 Definition für modale Teildialoge

Die Dialogsicht S_2 ist eine modale Sicht und wird durch den nebenläufigen Übergang von der Dialogsicht S_1 erreicht. Für die Grundform der modalen Teildialoge wurde hier nur ein, der zum Ursprungsfenster zurückführende Ausgang dargestellt, z.B. Bestätigung einer Meldung in einer Dialogbox. Nach der Bestätigung durch den Endbenutzer (Interaktion I_2) wird die modale Sicht geschlossen und wieder die Dialogsicht S_1 aktiviert. In [3] werden Beispiele für modale Teildialoge mit weiteren Ausgängen demonstriert.

Um für die Dialogsicht S_2 den Zustand einer modalen Dialogsicht zu gewährleisten, ist es erforderlich, für die Transitionen T_1 und T_2 jeweils eine Aktion zur Beeinflussung der Dynamik des Interaktionsmodells zu definieren:

- Transition T_1: Die Funktion entfernt von den Stellen S_i, $i \neq$ aktive modale Stelle, welche mit (**s**, **m**) markiert sind, das Individuum **m** (vgl. Regel 4).

- Transition T_2: Die Funktion markiert alle Stellen S_i, i ≠ aktive modale
  Stelle, welche bereits mit **s** markiert sind, zusätzlich mit
  dem Individuum **m** (vgl. Regel 3).

Die für die Transition T_2 definierte Funktion muß jeweils allen weiteren
Ausgängen einer modalen Dialogsicht zugeordnet werden.

4   Zusammenfassung

Das Konzept der Dialoggraphen, welches im TADEUS-Interaktionsmodell
verwirklicht wird, erweitert die von Janssen eingeführten Dialognetze für die
Modellierung von mehrfachen Exemplaren eines Fensters. Die geringe Anzahl von
Elementen zur Konstruktion des abstrakten Dialoggraphen gewährleistet ein für
den Dialogentwickler gut handhabbares Hilfsmittel. Durch die beschriebene
Definition des abstrakten Dialoggraphen konnte die Vermutung von Janssen (vgl.
[9]), daß beim Übergang zu höheren Petrinetzen die Einfachheit des Formalismus
leiden würde, widerlegt werden.

Neben den beschriebenen Elementen wurden für das Konzept der abstrakten
Dialoggraphen weiterhin Konstrukte für das Zusammenfassen von Teildialogen zu
komplexen Dialogsichten und für das Einbinden des Bearbeitungsdialoges in den
Navigationsdialog definiert.

Mit den aufgezeigten Abbildungen der Elemente auf Petrinetze und mit weiteren
Regeln für die Komposition der einzelnen Elemente zu einem abstrakten
Dialoggraphen kann zu jedem korrekt entwickelten Dialoggraphen ein äquivalentes
Petrinetz mit individuellen Marken erzeugt werden (vgl. [3]). Damit ist die
Simulation und das Testen der Benutzungsoberfläche vor dem eigentlichen
Erzeugen der Benutzungsschnittstelle formal durchführbar. Ein weiterer Vorteil
ergibt sich aus der Nutzung vorhandener Hilfsmittel zur Analyse und Simulation
von Petrinetzen für die Entwicklung eines Werkzeuges zur Verwaltung von
Dialoggraphen.

Die Erarbeitung des Interaktionsmodells für ein Interaktives Graphisches System
ist integraler Bestandteil das TADEUS-Ansatzes für das aufgaben- und
benutzerorientierte Design. Ausgehend von Aufgaben- und
Problembereichsmodell, in welchen der Entwickler Sichten identifiziert hat, wird
ein abstrakter Dialoggraph generiert und dann durch den Dialogentwickler
vervollständigt und zum konkreten Dialoggraph weiterentwickelt. Auf der Basis
des Interaktionsmodells wird dann die Benutzungsschnittstelle weitgehend

automatisch erzeugt. Eine prototypische Werkzeugumgebung zur Unterstützung des TADEUS-Ansatzes befindet sich in der Entwicklung.

Danksagung

Die Idee zu dieser Arbeit entstand in den Diskussionen zum Projekt EXPOSE, welches vom Bundesministerium im Rahmen des Programms "Arbeit und Technik" gefördert wurde. (FKZ 01 HK 291)

Wir danken den anonymen Gutachtern aus dem Programmkommittee, deren konstruktive Bemerkungen zur Verbesserung des vorliegenden Artikels beigetragen haben.

Literaturverzeichnis

[1]   C. Bay: Analyse von Spezifikationsmitteln für graphische Dialoge zur Weiterentwicklung des THESEUS-Dialogmodells. Studienarbeit, Darmstadt: TH Darmstadt, FB Informatik, 1988.

[2]   T. Elwert, P. Forbrig, E. Schlungbaum: Ein Beschreibungsmodell für aufgabenorientiertes Design. Teil 1: Dokumentation der Projektanalyse. Preprint CS-07-94, Rostock: Universität Rostock, FB Informatik, 1994.

[3]   T. Elwert, P. Forbrig, E. Schlungbaum: Ein Beschreibungsmodell für aufgabenorientiertes Design. Teil 2: Entwicklung des Interaktionsmodells. Preprint im Druck, Rostock: Universität Rostock, FB Informatik, 1994.

[4]   P. Forbrig, P. Gorny, (Autorenkollektiv): Konzepte für EXPOSE. 1. Zwischenbericht zum Projekt EXPOSE, Oldenburg, Rostock, 1993.

[5]   P. Forbrig, P. Gorny, (Autorenkollektiv): Der EXPOSE-Prototyp. 2. Zwischenbericht zum Projekt EXPOSE, Oldenburg, Rostock, 1994.

[6]   M. Green: A Survey of Three Dialogue Models. In: ACM Transaction on Graphics, 5(1986)3, S. 245-275.

[7]   T. Greutmann: Datenmodellierung und aufgabengerechte Dialoge: ein Synchronisations-problem. In: K.-H. Rödiger (Hrsg.): Software-Ergonomie '93 Von der Benutzungsoberfläche zur Arbeitsgestaltung. Stuttgart: Teubner, 1993, S. 99-109.

[8]   W. Hübner: Entwurf Graphischer Benutzerschnittstellen. Berlin, Heidelberg: Springer 1990.

[9]   C. Janssen: Dialognetze zur Beschreibung von Dialogabläufen in graphisch-interaktiven Systemen. In: K.-H. Rödiger (Hrsg.): Software-Ergonomie '93 Von der Benutzungsoberfläche zur Arbeitsgestaltung. Stuttgart: Teubner, 1993, S. 67-76.

[10]   R. J. K. Jacob: A Specification Language for Direct Manipulation User Interfaces. In: ACM Transaction on Graphics, 5(1986)4, S. 283-317.

[11]   S. Maaß: Software-Ergonomie Benutzer- und aufgabenorientierte Systemgestaltung. In: Informatik-Spektrum, Berlin: Springer, 16(1993)4, S. 191-205.

[12]   B.A. Myers: Creating Interaction Techniques by Demonstration. In: IEEE Computer Graphics & Application, 7(1987)9, S. 51-60.

[13]   W. Reisig: Systementwurf mit Netzen. Berlin: Springer, 1985.

[14]   B. Roudaud, V. Lavigne, O. Lavigneau, E. Minor: SCENARIO: A new Generation UIMS. In: D. Diaper et.al. (Eds.): Human-Computer Interaction - INTERACT '90. Amsterdam: North-Holland, 1990, S. 607-612.

[15]   E. Schlungbaum, M. Schmidt: Automatische Erzeugung von Beschreibungen für Benutzungsoberflächen. In: Rostocker Informatik Berichte, Rostock, (1994)15, S. 97-106.

[16]   P. Sukaviriya, J.D. Foley, T. Griffith: A Second Genereation User Interface Design Environment: The Model and The Runtime Architecture. In: S. Ashlund et.al. (Eds.): Proceedings INTERCHI '93. Reading: Addison-Wesley, 1993, S. 375-382.

[17]   J. Ziegler: Entwurf graphischer Benutzungsschnittstellen. In: J. Ziegler, R. Ilg (Hrsg): Benutzergerechte Software-Gestaltung: Standards, Methoden, und Werkzeuge. München: Oldenbourg, 1993, S. 145-169.

Dr. Egbert Schlungbaum & Thomas Elwert
Universität Rostock, Fachbereich Informatik
18051 Rostock
{Egbert.Schlungbaum, Thomas.Elwert}@informatik.uni-rostock.de

Integrierte Erstellung von Aufgabenmodell und Dialogspezifikation interaktiver Benutzungsschnittstellen

Khatoun Shahrbabaki
Universität - GH Paderborn
Fachbereich Mathematik/Informatik
D-33098 Paderborn
Tel.: 05251 60-3073
cat@uni-paderborn.de

Gerd Szwillus
Universität - GH Paderborn
Fachbereich Mathematik/Informatik
D-33098 Paderborn
Tel.: 05251 60-2077
szwillus@uni-paderborn.de

Zusammenfassung

Benutzerorientierter Entwurf interaktiver Systeme besteht aus der Iteration einzelner in sich abgeschlossener Phasen, wie Aufgabenanalyse, -modellierung und -spezifikation, die durch Beschreibungsnotationen unterstützt werden, die auf die jeweiligen Ansprüche zugeschnitten sind. Da die Aussagekraft dieser Notationen meistens auf die einzelnen Phasen beschränkt bleibt, gehen vielfach Informationen an den Übergängen zwischen den einzelnen Phasen verloren. Entwurfsentscheidungen können aus diesem Grund von anderen Entwicklungsphasen nicht nachvollzogen werden. In diesem Artikel stellen wir eine Kombination von zwei Notationen (MAD und DSN) vor, die durch ihre Struktur nicht nur Entwicklungsentscheidungen zurückverfolgen lassen, sondern dank ihrer Verflechtung eine konsistente und korrekte Übertragung der Entwurfsentscheidungen gewährleisten.

1 Einleitung

Viele Methoden für den Entwurf von Benutzungsschnittstellen enthalten die zyklischen Phasen Aufgabenanalyse, Aufgabenmodellierung, Dialogspezifikation und Prototyping. Um eine Problemstellung formal spezifizieren zu können und die Erkenntnisse des jeweiligen Entwurfs festzuhalten, sind verschiedene Notationen mit Betonung unterschiedlicher Aspekte entwickelt worden. Notationen wie **UAN** (Hartson et al 1990 [6]), **GOMS** (Card et al 1983 [3]), setzen den Schwerpunkt mehr auf Beschreibung des Benutzerverhaltens, während **PPS** (Olsen 1990 [8]) und **DSN** (Curry und Monk 1991 [2]) das dynamische Verhalten einer Applikation beschreiben. Andere Notationen wie **MAD** (Sebillotte 1992 [10]), und Green's Notation (Green 1981 [4]) modellieren nicht die Interaktionen des Benutzers, sondern formalisieren die Aufgabenstruktur. Da diese Notationen für spezielle Aspekte einzelner Phasen entworfen sind, bereiten die Übergänge zwischen den verschiedenen Entwicklungsphasen und ihren dezidierten Methoden die meisten Probleme im Hinblick auf Korrektheit und Konsistenz.

In einem Beispielprojekt (Shahrbabaki 1993 [11]) haben wir untersucht, wie zwei, originär nicht "füreinander" entwickelte Methoden (nämlich MAD bei der Aufgabenmodellierung und DSN bei der Dialogspezifikation), elegant und konzeptionell klar miteinander verknüpft werden können, wodurch ein fließender Übergang von einer benutzerorientierten Sicht auf die Aufgaben zu einer systemorientierten Sicht auf Kontrollaspekte der Benutzungsschnittstelle ermöglicht wird. Als algorithmisch einfaches Problem mit hinreichend komplexer Benutzungsschnittstelle wurde ein **elektronischer Terminkalender** gewählt.

In dem vorliegenden Papier betrachten wir zunächst die Aufgabenmodellierung mit MAD und ihre Anwendung in unserem Beispielprojekt. Anschließend führen wir kurz in die Dialogspezifikation mit DSN ein und zeigen an unserem Beispielprojekt auf, wie diese Methode bei der Erstellung einer Spezifikation durch eine vorangegangene Aufgabenmodellierung mit MAD effizient unterstützt wird. Zum Schluß diskutieren wir die aufgetretenen Synergieeffekte der beiden Notationen, indem wir von unserem konkreten Anwendungsbeispiel abstrahieren.

2   Aufgabenmodell

In vielen Fällen ist die Erstellung eines Aufgabenmodells eine naheliegende Vorstufe der Entwicklung einer interaktiven Applikation. Typischerweise beschreibt sie in einer semi-formalen Notation die Aufgaben, die der zukünftige Benutzer an oder mit einem System durchführen möchte. Benutzeranforderungen und andere applikationsrelevante Informationen finden dementsprechend dort ihren Niederschlag. Zusätzlich beschreiben einige Modelle die Aufgabe in Szenarien (Curry und Monk 1991 [2]), (Robertson 1995 [9]). Die Verwendung dieser Szenarien ist sinnvoll, da die Entwickler eine realistische Vorstellung des späteren Einsatzumfelds gewinnen können. Dies und die Aufgabenanalyse bilden die Basis eines benutzerzentrierten Entwurfs. Als alleinige Beschreibungsnotation kommen sie wegen ihrer inhärenten Ungenauigkeit nicht in Frage, so daß zusätzliche Notationen für eine formale Spezifikation notwendig werden. In unserem Projekt wurden Szenarien nur als Vorstufe der Modellierung verwendet.

Andere Ansätzen wie **UAN** (Hartson et al 1990 [6]), Command Language Grammar **CLG** (Moran 1981 [7]) integrieren die Benutzeraktionen bereits in das Aufgabenmodell. Damit zieht eine Änderung des Modells meistens auch eine Änderung der Interaktionen nach sich, was die Bearbeitung des Aufgabenmodells erschwert, da beide Aspekte zu eng miteinander verknüpft behandelt werden. Außerdem kann mit diesen Notationen nicht festgestellt werden, in welchem Zustand das System sich gerade befindet. Wir schlagen hier die Verwendung der **MAD**-Notation (Sebillotte 1992 [10]) vor, die eine hierarchische Modellierung der Benutzeraufgaben erlaubt. In unserem Ansatz wurde diese Notation teilweise erweitert und ergänzt.

Im folgenden werden wir zunächst die Grundzüge von MAD erläutern, danach führen wir als Anwendungsbeispiel einen **elektronischen Terminkalender** ein und stellen beispielhaft einige wesentliche Aspekte des Entwicklungsprozesses dar.

2.1 MAD-Methode

MAD ist eine graphische Notation, die ähnlich wie Green's Methode (Green 1981 [4]) die Reihenfolge der Aufgaben mit Vor- und Nachbedingungen festlegt, jedoch zusätzlich mit einer graphischen Darstellung der Aufgabenhierarchie unterstützt wird. Durch ihre graphische Repräsentation und Einführung einer Struktur ist MAD anschaulich und leicht nachvollziehbar. Den Kern des Ansatzes bildet ein Hierarchieplan von zu erledigenden Aufgaben. Jede Aufgabe wird solange in Teilaufgaben zerlegt, bis keine weitere Zerlegung mehr möglich ist. Dadurch entsteht ein hierarchischer Baum mit Knoten als Teilaufgaben, Blättern als atomaren Zielen und Kanten, die die Zusammensetzung einzelner Teilaufgaben *(Task Structure)* widerspiegeln.

Für jeden Baumknoten der MAD-Hierarchie wird ein Schema mit Werten gefüllt, welches Felder besitzt, in denen verschiedene relevante Daten vermerkt werden (siehe Tabelle 1).

- TASK: Bezeichnung der jeweiligen Teilaufgabe

- Lfd. Nr.: Hierarchische Numerierung der Felder

- GOAL: Ziele, die mit dieser Teilaufgabe verfolgt werden

- PRECONDITIONS: Vorbedingungen, die zur Ausführung der Aufgabe erfüllt sein müssen.

TASK:	Lfd. Nr.:
INITIAL STATE:	FINAL STATE:
GOAL:	
PRECONDITIONS:	POSTCONDITIONS:
UPPER LEVEL:	COMPOSITE TASK & STRUCTURE:
ELEMENTARY TASK:	

Tabelle 1: Datenfelder eines Aufgabenobjekts im MAD-Formalismus

- POSTCONDITIONS: Nachbedingungen, welche erst nach dem Erledigen der Aufgabe zutreffen.

- TASK STRUCTURE: Es werden sechs unterschiedliche Strukturierungsarten von MAD zur Verfügung gestellt (LOOP, OPT, OR, PAR, SEQ, SIM). Sie geben an in welcher Reihenfolge die in diesem Knoten involvierten Teilaufgaben ausgeführt werden müssen, dabei haben die im Baum höher angesiedelten Knoten bei der Ausführung eine höhere Priorität. Dies stellt eine unserer Erweiterungen dar. In der ursprünglichen MAD-Version kann eine Aufgabe nur in Teilaufgaben zerlegt werden, die in gleicher Beziehung zueinander stehen. Deshalb ist auch keine Festlegung einer Ausführungsordnung (mit Ausnahme von sequentiellen Aufgaben) notwendig. Wir erlauben in unserer Erweiterung die Entstehung inhomogener Aufgabenstrukturen, die eine definierte Ordnung erforderlich machen.

  - Wiederholende Ausführung (**LOOP**)
  - Optionale Ausführung (**OPT**)
  - Gegenseitiger Ausschluß (**OR**)
  - Beliebige Reihenfolge (**PAR**)
  - Sequentielle Reihenfolge (**SEQ**)
  - Gleichzeitige Ausführung (**SIM**)

- UPPER LEVEL: In diesem Feld wird auf die nächsthöhere Baumebene verwiesen.

- COMPOSITE TASK: Komponenten der zusammengesetzten Teilaufgaben.

- ELEMENTARY TASK: Wenn dieser Knoten ein Blatt darstellt, wird hier 'Ja', sonst 'Nein' eingetragen.

2.2   Das Aufgabenmodell des Terminkalenders

Das gesamte Aufgabenmodell unserer Beispielanwendung umfaßt eine Vielzahl von MAD-Diagrammen. Als repräsentativen Ausschnitt des Modells stellen wir in diesem Papier die Termineingabe vor. Dabei verwenden wir zur optischen Präsentation und zum leichteren Verständnis einen Layoutvorschlag (Abbildung 1), der nach vollständiger Aufgabenmodellierung des Terminkalenders (Shahrbabaki 1993 [11]) und entsprechender Integration der Teilspezifikationen resultierte.

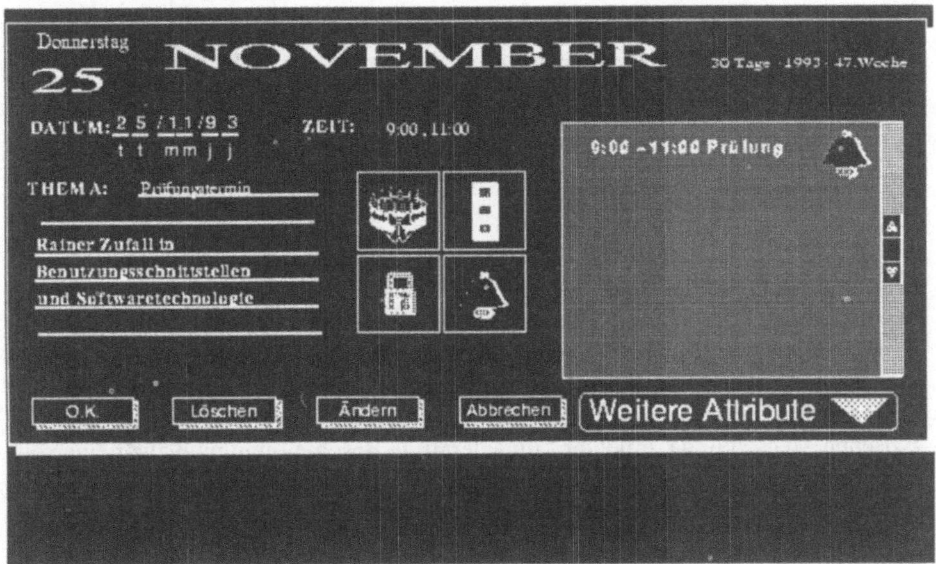

Abbildung 1: Ein mögliches Layout für Termineintragung nach dem Aufgabenmodell

Zu einem vollständigen Termin gehören die Eingaben Datum, Zeit und Thema (Beschreibung des Termins). Ein Aufgabenmodell für die Termineingabe repräsentiert die Abbildung 2 .

Wie man anhand des Modells (Abbildung 2) erkennt, können Datum, Zeit und Thema in beliebiger Reihenfolge angegeben werden (PAR-Konstrukt). Dagegen ist die Eingabe von Attributen (wie z.B. das Einstellen eines Alarms) nicht zwingend

(OPT). Das Quittieren erfolgt als letzte Aktion und steht deshalb als sequentiell (SEQ) zu allen vorherigen Aktionen; die Reihenfolge der Teilaufgaben von oben nach unten ist also, anders als im ursprünglichen MAD-Modell, signifikant. Wir werden später in der Dialogmodellierung sehen, daß eine im Sinne dieser Vorgaben "falsche" Benutzereingabe (z.B. Eintragung eines Termins ohne Datum) zu Fehlermeldungen des Systems führt.

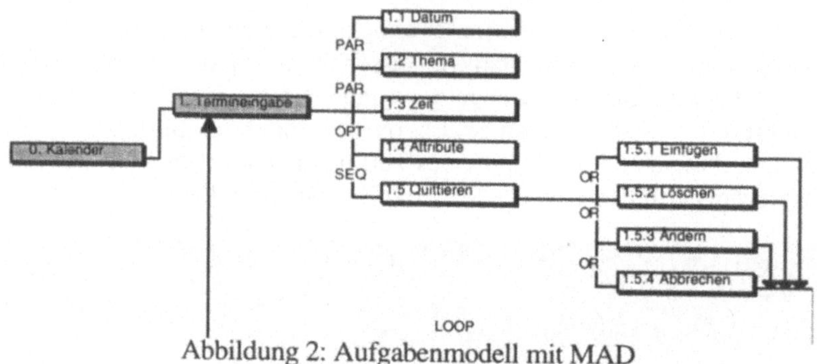

Abbildung 2: Aufgabenmodell mit MAD

3   Dialogspezifikation

Angesichts der Komplexität der Interaktionen zwischen Mensch und Computer benötigen die Entwickler eine Beschreibungsmethode zur Festlegung der Benutzereingabe- und Verzweigungsmöglichkeiten die im Laufe eines Dialogs mit dem System entstehen. In diesem Sinne versteht sich eine formale Dialogspezifikation als eine präzise, eindeutige Beschreibung der Interaktionen zwischen Mensch und Computer. Eine derartige Beschreibung muß zwar einerseits systemorientiert sein, andererseits sollte sie den Entwickler nicht zwingen, alle oder zu viele Details der Interaktionen festlegen bzw. überhaupt betrachten zu müssen.

Es gibt Dialogspezifikationen in verschiedenen Detaillierungsgraden, denen allerdings durchweg eine Betonung der Beschreibung aus der Sicht des Systems - nicht des Benutzers - gemeinsam ist. Hat man vorher, bei der Aufgabenmodellierung die Denk- und Entscheidungsabläufe des Benutzers beschrieben, so behandelt die Dialogspezifikation nun die Eingabemöglichkeiten und die resultierenden Zustandsänderungen des Systems.

Diese Informationen lassen sich nicht mehr mit der MAD-Notation erfassen, da sie nicht Teil eines Aufgabenmodells sind. Wir setzen nun die DSN-Methode ein, um die aus der Aufgabenmodellierung resultierenden Eigenschaften des Dialogs zu beschreiben. DSN, welches unabhängig von MAD entwickelt wurde, eignet sich sehr gut als Anschluß an eine MAD-Analyse.

Mit Hilfe von DSN spezifizieren wir den Dialog, der in unserer Beispielmodellierung bei der Eingabe eines Termins entsteht.

3.1 Dialogspezifikation mit DSN

Die **Dialogue Specification Notation** (DSN) wurde von Curry und Monk entwickelt (Curry und Monk 1991, [2]). DSN modelliert einen Dialog durch ein Produktionssystem, dessen Hauptkomponenten aus dem **Propositional Production System** (**PPS**), einer von Olsen (Olsen 1990 [8]) entwickelten Notation, übernommen sind.

In diesem Produktionssystem werden die gewünschten "Ziele" erreicht, wenn sich durch Benutzeraktionen in einem bestimmten "Kontext" Produktionsregeln ableiten lassen. **Zustandsvektor**, **Ereignisse** und **Regeln** bilden die Hauptkomponenten der DSN, die im folgenden erklärt werden:

- Zustandsvektor

  Der Kontext einer Regel ist mit Bedingungen gleichzusetzen, die bei der Auslösung einer Produktionsregel erfüllt sein müssen. Diese Bedingungen spiegeln sich in den Systemzuständen wider.

  Sämtliche Zustände eines Systems sind mit XOR* in verschiedenen Mengen (*State-Space Conditions*) gruppiert. Der Zustandsvektor setzt sich aus genau einem Zustand aus jedem dieser Mengen zusammen; mathematisch wird also ein Kreuzprodukt gebildet. Elemente des Zustandsvektors werden als "aktive" Zustände bezeichnet. Somit bilden die aktiven Zustände des Systems (oder der Zustandsvektor) den zum "Feuern" einer Regel erforderlichen Kontext. Alle Zustände werden mit dem Präfix # von anderen Komponenten der Notation unterschieden.

---

*) Diese Relation drückt hierbei aus, daß sich das System in genau einem Zustand aus dem Vektor befindet, der mit aktivem Zustand bezeichnet wird. Beispielsweise werden die Zustände "Mauseingabe" und "Tastatur" in einem Vektor zusammengefaßt, wenn das System nicht gleichzeitig von Maus und Tastatur eine Eingabe erhalten sollte.

- Ereignisse

  In DSN wird zwischen den vom Benutzer und den vom System ausgelösten Ereignissen differenziert. Erstere stellen die Benutzeraktionen dar und werden mit dem Präfix "$\downarrow$" eingeleitet. Ähnlich wie die Systemzustände werden sie in einem Vektor *User Events* zusammengefaßt. Letztere repräsentieren die Rückmeldungen des Systems und werden mit einem nach oben gerichteten Pfeil ($\uparrow$) gekennzeichnet. Sie werden in dem Vektor *Application Events* zusammengefaßt.

- Regeln

  Man kann die Regeln abstrakt als eine Abbildung sehen, die das zu bearbeitende Problem auf eine mögliche Lösung abbildet.

  Auf der linken Seite einer Produktion werden die Parameter aufgeführt, die als Vorbedingung einer möglichen Lösung des Problems gelten sollen. Darunter fallen aktive Zustände, Benutzeraktionen und Systemmeldungen. Auf der rechten Seite einer Regel werden Konsequenzen des Systems in Form von Zustandsänderungen eingetragen.

Zu DSN gehört ein Simulationstool **DDT** (Dialogue Design Tool), das zu jedem gegebenen Zeitpunkt die Regel, die "gefeuert" werden muß, und die mögliche Benutzerangaben anzeigt. Außerdem kann der Entwickler Skizzen anfertigen, die er einzelnen Regeln zuordnet, welche die in der Regel beschriebene Situation optisch repräsentiert. Damit kann der Entwickler sehr grob einzelne graphische Aspekte der entworfenen Applikation schon im Vorfeld einer Weiterentwicklung testen. Die gebotenen Möglichkeiten des existierenden Werkzeugs sind jedoch noch weit von einem echten Prototyping entfernt.

3.2   Die Dialogspezifikation des Terminkalenders

Die Spezifikation der Teilaufgaben beginnt mit der Beschreibung der Benutzeraktionen. Da zu einem Zeitpunkt immer nur ein Ereignis vom Anwender ausgelöst werden kann, ist ein gleichzeitiges Eintreten verschiedener Benutzereingaben ausgeschlossen. Aus diesem Grund werden alle Eingaben in einem Vektor *User-Event* (F1) zusammengefaßt. Der Vektor F1 kann im Laufe der Entwicklung immer wieder um neue Eingabemöglichkeiten erweitert werden. Des weiteren können Benutzereingaben erst "grob" und später feiner granuliert werden. Beispielsweise kann die Eingabe eines Datums zunächst abstrakt als ($\downarrow$*Datum*) bezeichnet werden, um später auf eine speziellere Eingabe, wie z.B. Tag-,

Monat- und Jahreingabe verfeinert zu werden. Wir bleiben vorerst bei folgenden abstrakten Benutzereingaben:

USER EVENTS:
    F1 (↓datum, ↓zeit, ↓thema, ↓ok, ↓abbrechen, ↓löschen, ↓ändern)

Die möglichen Reaktionen der Applikation werden in einem Vektor *System-Feedback-Event* (F2) zusammengefaßt. Auch diese können zunächst abstrakt und später detailliert formuliert werden.

SYSTEM FEEDBACK EVENTS:
    F2 (↑OK, ↑ERROR)

Betrachten wir nun das in Abschnitt 2.2 entworfene Aufgabenmodell. Die Dialogspezifikation in DSN läßt sich direkt aus den in MAD modellierten Teilaufgaben ableiten. Dort wurde die Teilaufgabe **Termineingabe** zunächst in (**Datum, Thema, Zeit, Attribute** und **Quittieren**) zerlegt. Die ersten drei Eingaben wurden mit der Option (*PAR*) zusammengefaßt, die eine Ausführung dieser Teilaufgaben in beliebiger Reihenfolge erlaubt (siehe Abbildung 3).

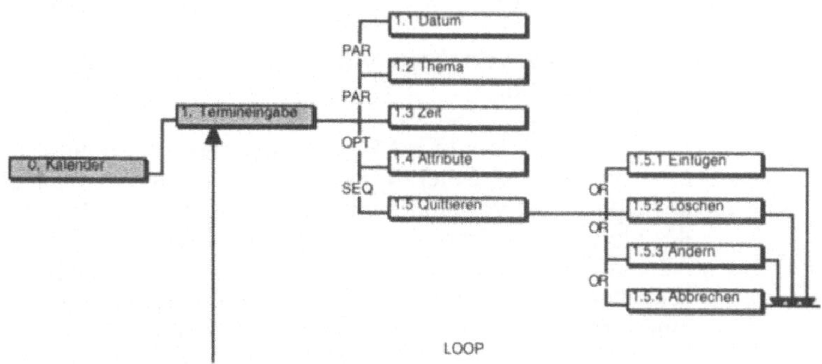

Abbildung 3: Teilaufgaben der Termineingabe

Wir zeigen nun am Beispiel auf, wie eine derartige Teilstruktur in entsprechende DSN-Konstrukte umgesetzt werden kann.

Als erstes werden für jede einzelne Teilaufgabe, die eine "*PAR*-Struktur" besitzt, zwei komplementäre Zustände erzeugt. Im Falle der Teilaufgabe *Datum* sind dies die

Zustände #Datum und #KeinDatum, die aufgrund gegenseitigen Ausschlußes in einem Zustandsvektor (F4) zusammengefaßt werden können.

F4 (#Datum, #KeinDatum)

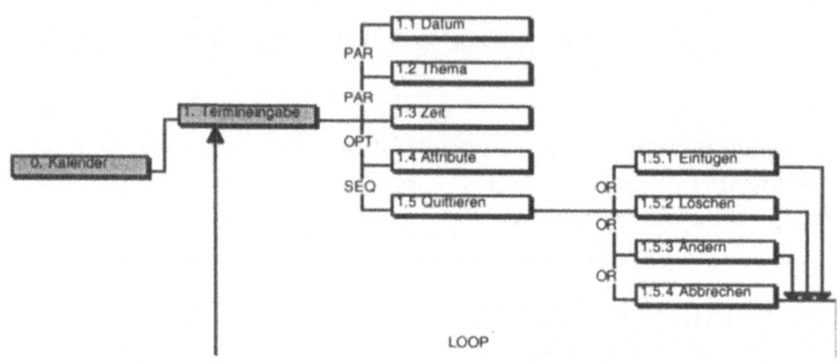

Abbildung 4: Teilaufgaben des Quittierens

Entsprechend verfahren wir mit weiteren Zuständen, die mit *SIM* oder *OPT* strukturiert sind.

Die Teilaufgaben (**Einfügen, Löschen, Ändern** und **Abbrechen**) schließen eine gleichzeitige Selektion vom Benutzer aus, weshalb sie im Aufgabenmodell mit der Option *OR* verbunden sind (siehe Abbildung 4). Deshalb können sie nicht auf der gleichen Art und Weise in den Dialogspezifikation einfließen, wie die vorher behandelten Teilaufgaben. Da solche Zustände sich per Definition ausschließen, brauchen keine komplementären Zustände eingeführt werden. Sie können direkt in einem Zustandsvektor zusammengefaßt werden.

F8 (#Einfügen, #Löschen, #Ändern, #Abbrechen)

Es resultieren nun folgende Zustandsvektoren:

SYSTEM STATES

F3 (#Eingabe, #KeineEingabe)

F4 (#Datum, #KeinDatum)

F5 (#Thema, #KeinThema)

F6 (#Zeit, #KeineZeit)

F7 (#Attr, #KeinAttr)

F8 (#Einfügen, #Löschen, #Ändern, #Abbrechen)

Die Produktionsregeln beschreiben nun die bei der Anwendung erzielten Reaktionen bzw. das dynamische Verhalten der Applikation. Jede Regel ist mit einer Nummer gekennzeichnet, die kanonisch aus der Numerierung des entsprechenden Aufgabenobjektes hervorgeht. Wir wollen nun anhand der folgenden Abbildung den Dialogverlauf nachvollziehen.

Start (#KeineEingabe, #KeinDatum, #KeineZeit, #KeinThema, #Abbrechen)

1.1. #KeinDatum ↓datum Æ #Datum

1.2. #KeinThema ↓thema Æ #Thema

1.3. #KeineZeit ↓zeit Æ #Zeit

1.4. #KeinAttr ↓att Æ #Attr

1.5.0 #KeineEingabe #Zeit #Datum #Thema ↑OK Æ #Eingabe

1.5.1 #Eingabe ↓ok Æ

    #Einfügen #KeineEingabe #KeinDatum #KeinThema #KeineZeit #KeinAttr

1.5.2 #Eingabe ↓löschen Æ

    #KeineEingabe #Löschen #KeinDatum #KeinThema #KeineZeit #KeinAttr

1.5.3 #Eingabe ↓ändern Æ

    #KeineEingabe #Ändern #KeinDatum #KeinThema #KeineZeit #KeinAttr

1.5.4 #Abbrechen ↓abbrechen Æ

    #KeineEingabe #KeinDatum #KeinThema #KeineZeit #KeinAttr

Die Regeln 1.1 bis 1.4 zeigen, daß der Benutzer in beliebiger Reihenfolge Zeit, Thema, Datum angeben kann. Erst wenn alle drei Angaben eingetragen sind, erlaubt das System die Eintragung (Regel 1.5.0). An dieser Stelle kann man durch entsprechende Regeln die Systemreaktionen um sinnvolle Fehlermeldungen erweitern. Beispielsweise können wir die Regeln 1.5.0.1, 1.5.0.2, 1.5.0.3 einfügen, die auf unvollständige Eingaben hinweisen und keine Eintragung zulassen.

1.5.0.1 #KeineEingabe #KeineZeit #Datum #Thema ↑ERROR Æ

    #KeineEingabe {Zeit fehlt}

1.5.0.2 #KeineEingabe #Zeit #KeinDatum #Thema ↑ERROR Æ

    #KeineEingabe {Datum fehlt}

1.5.0.3 #KeineEingabe #Zeit #Datum #KeinThema ↑ERROR Æ

    #KeineEingabe {Thema fehlt}

Wir können in diesem Papier nicht auf alle Einzelheiten von DSN und des Anwendungsbeispiels eingehen und beschränken uns daher auf die Aussage, daß wir die oben aufgeführten Regeln mit DDT animiert und damit ihr korrektes Verhalten verifizieren konnten. Zur einer tiefergehenden Betrachtung verweisen wir auf die Arbeiten von Monk und Curry (Curry und Monk 1991 [2]).

3.3 Allgemeine Transformationsregel zwischen MAD und DSN

Beispielhaft haben wir eine Übertragung von MAD nach DSN im vorigen Abschnitt vollzogen. An dieser Stelle geben wir einen kurzen Einblick in die formalen Transformationsregeln, die wir bei der Übertragung von MAD nach DSN eingesetzt haben. Auf eine ausführlichere Diskussion der Regeln verzichten wir aus Platzgründen, da dies eine detaillierte Erklärung der DSN erforderlich machen würde. Der Beitrag dieses Abschnitts zielt damit nur auf die Vermittlung der Grundideen der vereinfachten Transformationsregeln und erhebt keinen Anspruch auf Vollständigkeit. Er soll es dem Leser ermöglichen, eine bessere Vorstellung von der schematisch vorgenommenen Übertragung zwischen den Notationen zu gewinnen. Wir behandeln nacheinander die Aspekte Ereignisse (*User Events & Application Events*), Zustände und Produktionsregeln.

- **Ereignisse**, also Eingaben des Benutzers und Reaktionen des Systems, werden in einer MAD-Spezifikation syntaktisch nicht betrachtet. Hier kann eine Transformation nicht schematisch vorgehen - vielmehr besteht hier Entwurfsspielraum bzw. -bedarf bei der endgültigen Dialogmodellierung. Dies sind Sachentscheidungen, die in jedem Transformationsfall individuell und aus der Semantik heraus festgelegt werden müssen.

- **Zustände** werden in den Transformationsregeln schematisch eingeführt, um die Semantik der in MAD festgelegten Zeitabläufe im DSN-Modell nachzuvollziehen. Wir ordnen Aufgaben DSN-Zustände zu, die den Beginn bzw. den Abschluß der Durchführung einer Aufgabe repräsentieren; abhängig vom gewählten MAD-Konstruktor gehen wir verschieden vor:

  - *OR* oder *SEQ* -verknüpfte Aufgaben $a_1,...,a_n$ werden durch einen Vektor von sich gegenseitig ausschließenden Zuständen $(\#a_1,...,\#a_n)$ repräsentiert.

  - *OPT*, *SIM* oder *PAR* -verknüpfte Aufgaben $a_1,...,a_n$ werden durch n Vektoren sich jeweils paarweise ausschließender Zustände $(\#a_1, \#\overline{a_1}),...,$ $(\#a_n, \#\overline{a_n})$ dargestellt.

- *LOOP* -verknüpfte Aufgaben wirken sich lediglich in den Produktionsregeln aus, indem der Kontext wiederhergestellt wird, der vor Ausführung der Schleife gültig war.

- **Produktionsregeln** stellen im DSN-Modell die Beschreibung der Übergänge zwischen Zuständen sicher. Dabei lassen sich die beteiligten Zustandsmengen schematisch ableiten, um dem MAD-Modell zu genügen; die Festlegung von Benutzereingaben und Systemreaktionen, die in den Produktionsregeln ausgewertet werden, ist eine Entwurfsentscheidung bei der Dialogmodellierung. Die allgemeinen Transformationsregeln, die wir hier behandeln wollen, können daher nur die Zustandstransformationen betreffen und nicht die auslösenden Ereignisse behandeln.

Bei der Übertragung des Aufgabenmodells zu Produktionsregeln spielt die Ordnung der MAD-Knoten (graphisch von oben nach unten), die wir in das Modell eingeführt haben, eine entscheidende Rolle. Dadurch werden die inhomogenen Teilstrukturen möglich, die das ursprüngliche MAD-Konzept nicht enthielt. Gemäß dieser Ordnung wird zuerst die Beziehung zwischen den zwei jeweils ersten Knoten ausgewertet; dann werden diese mit dem jeweils nächsten Knoten verknüpft.

In den folgenden Schemata haben wir aus Gründen der einfacheren Darstellung Benutzeraktionen als das Signal gewählt, welches jeweils das Ende einer Sequenz markiert. Im konkreten DSN-Modell kann dies jedoch sowohl eine vom System ausgelöste Reaktion als auch eine Aktivierung eines Zustands sein. Außerdem betrachten wir der Einfachheit halber lediglich Paare von Aufgaben.

Seien im folgenden also T, a_1 und a_2 und Knoten eines MAD-Modells. Wir wollen einige der MAD-Konstruktoren anführen, bei denen die formale Transformation besonders gut nachzuvollziehen ist. Es gelte also die Beziehung $T = a_1 \oplus a_2$, wobei $\oplus \in \{OR, PAR, SIM\}$

($\oplus$ = OR): Wir erzeugen die Zustandsvektoren F0 und F1 und die Produktionsregeln R1 und R2.

$$F0 = (\#T, \#\overline{T})$$
$$F1 = (\#a_1, \#a_2)$$

$$\text{R1)} \quad \#T\#a_1 \quad \downarrow a_1 \quad \rightarrow \quad \#T\#a_1$$
$$\text{R2)} \quad \#T\#a_2 \quad \downarrow a_2 \quad \rightarrow \quad \#T\#a_2$$

$\#\overline{T}$ repräsentiert den Zustand, "T ist durchzuführen".

$\#T$ entspricht dem Zustand "T ist durchgeführt".

Das Eintreffen der Benutzereingabe a_1 repräsentiert das Ereignis, daß die Aufgabe a_1 abgeschlossen ist; analog ist a_2 zu verstehen.

($\oplus$ = PAR oder SIM): Hier werden F0, F1 und F2 erzeugt, um zu symbolisieren, inwieweit die Gesamtaufgabe T, und die Teilaufgaben a_1 und a_2 initiiert bzw. abgeschlossen sind. Es ergeben sich folgende Zustände und Regeln:

$$F0 = (\#T, \#\overline{T})$$
$$F1 = (\#a_1, \#\overline{a_1})$$
$$F2 = (\#a_2, \#\overline{a_2})$$

$$\text{R1)} \quad \#\overline{T}\,\#\overline{a_1}\,\#\overline{a_2} \quad \downarrow a_1 \quad \rightarrow \quad \#T\#a_1\,\#\overline{a_2}$$
$$\text{R2)} \quad \#\overline{T}\,\#\overline{a_1}\,\#\overline{a_2} \quad \downarrow a_2 \quad \rightarrow \quad \#T\#\overline{a_1}\,\#a_2$$
$$\text{R3)} \quad \#\overline{T}\,\#\overline{a_1}\,\#a_2 \quad \downarrow a_1 \quad \rightarrow \quad \#T\#a_1\,\#a_2$$
$$\text{R4)} \quad \#\overline{T}\#a_1\,\#\overline{a_2} \quad \downarrow a_2 \quad \rightarrow \quad \#T\#a_1\,\#a_2$$

3.4 Integration von Aufgabenmodell und Dialogspezifikation

In praktischen allen Bereichen der Softwareentwicklung hat sich die hierarchische Dekomposition als wesentliches Strukturierungsprinzip sowohl für die erzeugten Artefakte, als auch für die Aufgliederung der Arbeitsphasen etabliert. Aus der Theorie komplexer Systeme (wie z.B. beschrieben in (Booch 1991 [1])) ist bekannt, daß dieses Prinzip nicht künstlich aufgesetzt ist, sondern vielmehr als den verschiedenen Problemen inhärent angesehen werden kann. Daher reflektieren die Notationen zur Beschreibung von Systemen typischerweise diese hierarchischen Strukturen, was ein Vorgehen nach dem *Divide-&-Conquer*-Prinzip erlaubt, bzw. sogar nahelegt. Im Bereich der Entwicklung von interaktiven Systemen zeigt sich das z.B. an der Baumstrukturierung von GOMS-Spezifikationen, welche sich nach Auffassung der Erfinder von GOMS an den postulierten hierarchischen Speicher- und Lernstrukturen des menschlichen Handelns orientieren.

Auch die MAD-Methode setzt daher naheliegend Hierarchien ein, um komplexe Aufgabenstrukturen in einfachere Teile zu zerlegen. Die nachfolgende Phase der Dialogspezifikation, also der Modellierung von Benutzereingaben, Ausgaben an den Benutzer, Unterbrechungen und Zustandsübergängen, manchmal auch bezeichnet als das sogenannte *Kontrollmodell*, bricht jedoch im Kern mit der hierarchischen Betrachtungsweise: Das kanonische Beschreibungsmittel, der endliche Automat, entspricht in seinen Strukturierungsmöglichkeiten gerichteten Graphen und nur in Teilaspekten Baumstrukturen. Man erkennt dies unmittelbar an der wohlbekannten Veranschaulichung endlicher Automaten durch Graphen mit Kreisen als Zustandsrepräsentationen und Pfeilen als gerichteten Übergängen dazwischen*.

Ein wesentlicher Anteil der Probleme beim Übergang von Aufgabenmodellierung zu Dialogspezifikation besteht in diesem Wechsel des Beschreibungskalküls: Im hierarchischen Aufgabenmodell finden sich nicht explizit Zustände und Übergänge dazwischen - im automatenorientierten Dialogmodell sind die Zustände im wesentlichen flach nebeneinander angeordnet und besitzen keine hierarchische Struktur. Die Kombination der Methoden MAD und DSN erlaubte es nun, durch leichte Modifikationen diesen Übergang systematisch durchzuführen und die Sicherstellung der Konsistenz zwischen diesen beiden Phasen deutlich zu erleichtern.

Eine wichtige Voraussetzung für die Durchführung der systematischen Transformation war eine Modifikation der MAD-Methode zur reicheren Gestaltung der Aufgabenstrukturen. MAD, wie in (Sebillotte 1992 [10]) definiert, erlaubt nur eine homogene Strukturierung der Teilaufgaben: alle Unteraufgaben einer Aufgabe stehen in einer einzigen der angebbaren Beziehungen (PAR, SEQ, ...). In der praktischen Durchführung zeigt sich schnell, daß diese Festlegung sehr unflexibel ist und dazu führt, daß aus rein technischen Gründen in der Aufgabenmodellierung

---

*)   Statecharts (Harel 1988 [5]) kombinieren diese Grapheigenschaft in eleganter Weise mit Hierarchien - das beherrschende Stilelement bei der Spezifikation von Kontrolle in Benutzungsschnittstelle ist jedoch der unstrukturierte Übergang von einem Zustand zum nächsten.

Zwischenknoten eingeführt werden müssen, um Teilaufgaben homogen zu strukturieren. Resultat sind unangemessene Strukturierungen, die zunehmend unlesbar werden, und damit ihren Anspruch der Semiformalisierung umgangssprachlich formulierter Aufgabenstrukturen nicht mehr erfüllen.

In der modifizierten Fassung kann man Teilaufgaben erstens zeitlich anordnen, zweitens jede einzelne Teilaufgabe mit dem angemessenen Konstrukt versehen, und drittens neben den von MAD gegebenen Konstruktoren auch optionale Teilaufgaben mit OPT spezifizieren. In der graphischen Darstellung repräsentiert die Reihenfolge von "oben nach unten" eine entsprechende zeitliche Reihenfolge und die Konstruktoren an den Kanten werden an die Eingänge in die Teilaufgaben, statt an den Ausgang bei der übergeordneten Aufgabe notiert. Diese moderaten Änderungen erlauben ein wesentlich "flüssigeres" Verwenden der Notation - viel weniger als bei der urprünglichen Methode muß man über die Notwendigkeiten der Notation nachdenken. Gleichzeitig erreicht man mit den neuen Mitteln eine Anreicherung des Aufgabenmodells um Anfänge einer Verhaltensbeschreibung, die die Dialogspezifikation vorbereitet, ohne von der in dieser Phase durchzuführenden Aufgabenbeschreibung abzulenken. Es ist sogar sinnvoll denkbar, mit einer Modellierung ohne zeitliche Reihenfolgen und Konstruktoren auszugehen und diese Informationen schrittweise anzureichern.

Beginnt man nun mit der Dialogmodellierung, muß man in die Abläufe aus Benutzersicht die Eingaben an das System und die Ausgaben des Systems integrieren. Wie bereits gesagt, kommt hier zur Spezifikation das Zustandskonzept hinzu, das sich im Aufgabenmodell nur sehr indirekt bzw. überhaupt nicht widerspiegelt. Das Dialogmodell strukturiert seine Darstellung analog der Beschreibung von Zustandsübergängen und Ausgaben abhängig von den Benutzereingaben in endlichen Automaten. Diese Beschreibung muß möglichst in die Darstellung des Anwendungs eingebettet sein und es andererseits erlauben, die große Fülle von möglichen Interaktionstechniken zu beschreiben und einzusetzen. Nach unseren Erfahrungen ist DSN hierfür hervorragend geeignet, da es - ohne auf bestimmte Techniken oder Abstraktionsebenen festgelegt zu sein - erlaubt, komplexe Zustandsübergänge und Systemreaktionen zu beschreiben. Anders als die endlichen Automaten selbst, sind die Beschreibungen bei DSN strukturierter und damit kompakter, da die Kreuzproduktbildung als wesentliches Konstruktionsmerkmal verwendet wird.

Wie im vorangegangenen Abschnitt exemplarisch beschrieben, kann das gemäß unserer MAD-Variante fomulierte Aufgabenmodell sehr naheliegend und nachvollziehbar in eine DSN-Spezifikation transformiert werden. Damit hat man den Übergang von der hierarchischen zur automatenorientierten Beschreibung durchgeführt und kann anschließend auf DSN-Ebene selbst weiter verfeinern, bis man zu den verfeinerten Interaktionstechniken kommt. Zentral hierbei ist die Einbettung der einen Beschreibungsform in die andere; überläßt man den Transformationsaufwand alleine dem Anwender von Notationen ohne eine entsprechende Einbettung anzubieten, entstehen eine Reihe von Problemen: Vergessen von Ideen aus einer Phase in der nächsten, Fehler beim Übertragen von Abläufen der einen Phase in die nächste, Verpflichtung zur Sicherstellung der Konsistenz zwischen den Phasen und mangelhafte Dokumentation von Entwurfs aus einer Phase in der darauf folgenden. Bei der Kombination von MAD und DSN kommt hinzu, daß das Werkzeug DDT eine sehr frühe Animation der DSN-Spezifikation zuläßt - damit läßt sich, bei der engen Kopplung von Aufgabenmodell und Dialogmodell, auch das Aufgabenmodell indirekt überprüfen.

Ein anderer Ansatz zur Lösung der Probleme der Übergänge zwischen Entwurfsphasen könnte in einer einheitlichen Notation für die Modellierung verschiedener Aspekte bestehen. Allerdings liegt die Betonung in verschiedenen Phasen auf derart unterschiedlichen Aspekten, daß kaum eine Notation in der Lage ist, alle diese zu umfassen. Die singulären Notationen sind ohne diesen Anspruch entwickelt wurden, was vielleicht auch darauf zurückzuführen ist, daß ein globaler Überblick auf das gesamte interaktive System durchaus nicht immer im Vordergrund stehen muß. Diese globale Erkenntnis von Notationen notiert Whitehead in seinem fundamentalen Werk über Mathematik

> *"By relieving the brain of all unnecessary work, a good notation sets it free to concentrate on more advanced problems"* (Whitehead 1958 [12])

und bringt damit zum Ausdruck, daß man Notationen aufgabengerecht ausstatten soll, damit sie den Anwender von unnötigen Details befreien und ihm erlauben, auf der "richtigen" Abstraktionsebene zu denken und Entscheidungen zu treffen. Die für die Phasen Aufgabenmodellierung und Dialogspezifikation entwickelten Notationen MAD bzw. DSN sind gut geeignet, um in ihrer jeweiligen "Welt" eingesetzt zu werden. Da man verschiedene Probleme angeht, sollen und müssen solche Notationen verschieden sein, was zu den erwähnten Problemen bei den Übergängen zwischen zwei derartigen Spezifikationen führt. Die Erfahrung, die wir in diesem Papier vermitteln wollen, besteht darin, daß gerade diese beiden Methoden auf Grund ihrer Struktur sehr gut zueinander "passen" und einen kanonischen Übergang von der Aufgabenmodellierung zur Dialogspezifikation erlauben.

4 Ausblick

Das vorliegende Papier stellt einen Versuch dar, existierende moderne Methoden der Entwicklung von Benutzungsschnittstellen miteinander zu kombinieren, um die notwendigen Phasenübergänge innerhalb des Entwicklungsprozesses zu erleichtern. Dies kann nur ein Anfang sein, da einerseits diese Methodik keineswegs in allen Fällen anwendbar sein kann, andererseits aber auch die umliegenden Phasen, insbesondere die vorangehende Aufgabenanalyse und das sich anschließende Prototyping (Simulation) in die Betrachtungen einbezogen werden müssen. Es sollte aber hier einmal der Versuch gemacht und dokumentiert werden, wie man die inhärenten Mängel von "Insellösungen" angehen kann.

Literatur

[1] Booch, G.: *Object Oriented Design with Applications*, Benjamin Cummings, Redwood City, CA 1991

[2] Curry, M. B.; Monk, A. F.: *Dialogue modelling of graphical user interfaces with a production system*, Technical Report, Department of Psychology, University of York, 1991

[3] Card, S. K.; Moran, T. P.; Newell, A: *The Psychology of Human Computer Interaction*, Erlbaum, Hillsdale, NJ, 1983

[4] Green, M.: *A methodology for the specification of graphical user interfaces*, Computer Graphics, Vol. 15, No.3, 1981

[5] Harel, D.: *On Visual Formalisms*, Communications of the ACM, Vol 31 No.5, pp 514-540, 1988

[6] Hartson, H.R.; Siochi, A.C.; Hix, D.: *The UAN: A User- Oriented Representation for Direct Manipulation Interface Designs*, ACM Transactions on Information Systems, Vol. 8, pp 181-203, 1990

[7] Moran, T.P.: *The Command language Grammar: A representation for the user interface of interactive computer systems*, International Journal of Man-Machine Studies, pp 15-51, 1981

[8]   Olsen, D.R.: *Propositional Production Systems for dialogue description*, Human Factors in Computer Systems, CHI '90 Conference Proceedings, pp 57-63, 1990

[9]   Robertson, S.P.: *Generating object-oriented design representations via scenario queries*, In: J. Caroll (Hrsg.) *Scenario-based design: Envisioning Work in technology and system design*, (im Druck)

[10]   Sebillotte, S.: *Task Analysis and Formalization according to MAD: Hierarchical Task Analysis Method of Data Gathering and Examples of Task Descriptions,* Technical Report, 1992

[11]   Shahrbabaki, K.: *Beiträge zu einer partizipativen Entwurfsmethodik von Benutzungsschnittstellen*, Diplomarbeit, Universität - GH Paderborn, Fachbereich Mathematik/Informatik, 1993

[12]   Whitehead, A.: *An Introduction to the Mathematics,* New York, Oxford University Press, 1958

Ein modellbasierter Ansatz zur Dialogsteuerung in Benutzungsoberflächen

Friedrich Strauß, Stefan Hügel, Kirsten Winter, Britta Schinzel
Universität Freiburg

Zusammenfassung

User Interface Management Systeme (UIMS) ermöglichen eine getrennte Entwicklung von Benutzungsoberfläche und Applikation. Für die Konstruktion von direkt manipulativen Oberflächen ist jedoch die Integration von Anwendungswissen in die Dialogsteuerung notwendig.

Dieser Beitrag zeigt auf, wie anwendungsabhängiges Wissen durch eine modellbasierte Beschreibung der Dialogsteuerung in das UIMS aufgenommen werden kann. Dabei wird unsere situationsorientierte Sichtweise der Benutzungsschnittstelle vorgestellt, die Grundlage für eine modulare Dialogbeschreibung ist. Die Zerlegung der Oberfläche in Teilschnittstellen wird durch die situationsorientierte Konstruktion der Dialogsteuerung erreicht. Die Beschreibung der Anwendungsabhängigkeiten innerhalb bzw. zwischen den Teilschnittstellen ist so in einfacher Weise möglich. Die realisierte Dialogsteuerung erlaubt außerdem die Steuerung von Präsentationsoberflächen mit variabler Anzahl von Objekten und Fenstern und bietet so eine weitergehende Unterstützung für direkt manipulative Oberflächen als bisherige Ansätze.

Ein derartiges Modell der Anwendung wird zusätzlich zur Erzeugung kontextsensitiver Hilfe genutzt. Darüberhinaus kann die Dialogbeschreibung die Analyse der Dialogsteuerung vereinfachen und ihre Adaptierbarkeit verbessern.

1   Einleitung

Graphisch interaktive Benutzungsoberflächen (GUIs) mit direkter Manipulation als primärer Interaktionsform bilden die geeignete Grundlage für werkzeugartige Anwendungen. Die Programmierung dieser Oberflächen ist häufig genauso komplex und umfangreich, wie die Programmierung der eigentlichen Anwendung. Nach verschiedenen empirischen Untersuchungen werden bis zu 88% des Codes für die Gestaltung der Oberfläche aufgewendet (vgl. [10]). Oberflächeneditoren (Interface Builder), die auf objektorientierte Fenstersysteme aufbauen, vereinfachen die Programmierung der Präsentationskomponente erheblich, indem standardisierte Oberflächenelemente (Widgets) graphisch interaktiv zu einem Rumpf für die Benutzungsoberfläche zusammengesetzt werden [11]. Problematisch ist unter softwaretechnischen wie softwareergonomischen Aspekten die Programmierung der eng mit der Oberfläche verzahnten Dialogsteuerung innerhalb der Anwendung [11].

Mit dem Seeheim-Modell wurde eine Architektur für User Interface Management Systeme (UIMS) vorgeschlagen, die eine konzeptuelle Trennung von Oberfläche (inkl. der Dialogsteuerung) und Anwendung vorsieht. Diese Aufteilung bietet eine Reihe von Vorteilen: die Oberflächenprogrammierung kann durch spezielle Sprachen und Werkzeuge unterstützt werden, die Konsistenz der Präsentation und der Dialogstruktur kann erhöht werden, kontextsensitive Hilfen können einfacher zur Verfügung gestellt werden etc.

Die direkte Manipulation [8], die sich in allen modernen graphisch interaktiven Anwendungen wiederfindet, widerspricht allerdings der sequentiellen Struktur der frühen UIMS-Ansätze, die eine getrennte lexikalische, syntaktische und semantische Bearbeitung vorsehen. Das Prinzip der direkten Manipulation beruht u.a. wesentlich auf direkter *syntaktischer und semantischer Behandlung* von Interaktionen. Bei direkt manipulativen Oberflächen bietet sich häufig die Darstellung beliebig vieler Objekte in einem Fenster an, die bearbeitet und z.B. auch zwischen Fenstern verschoben werden sollen. Ein graphisch interaktiver Dateimanager ist eine typische Anwendung mit solchen Anforderungen, die im folgenden auch als Illustrationsbeispiel verwendet wird.

Die frühen auf dem Seeheim-Modell basierenden Dialogsteuerungen können solche Oberflächen nicht unterstützen. Eine ausführliche Diskussion früher und aktueller (amerikanischer) Ansätze findet sich in [13]. Neuere Ansätze zur Dialogsteuerung erlauben direkt manipulative Interaktionen, stellen aber häufig keine Mittel zur Verfügung, eine beliebige Anzahl von Objekten oder Fenster zu verwalten (z.B. [9]). Andere Ansätze stellen eher eine abgespeckte objektorientierte Programmiersprache zur Verfügung (vgl. [19]), deren Repräsentation dann allerdings u.a. den Einsatz von Hilfemodulen und Analysewerkzeugen erschwert, oder nutzen Constraints zur Beschreibung graphischer Abhängigkeiten.

Modellbasierte UIMS-Ansätze basieren auf einem funktionalen Modell der Anwendung, das zur automatischen Konstruktion der graphischen Darstellung und/oder zur Steuerung der Oberfläche genutzt wird. Wesentlich ist hier die Ausdrucksmächtigkeit und Adäquatheit der Modellbeschreibungssprache. Die meisten Ansätze nutzten eine objektorientierte Sichtweise zur Beschreibung des Anwendungsmodells, die allerdings durch ihre uniforme Beschreibung von Interaktionen Beschränkungen bzgl. der Ausdrucksmächtigkeit von semantischen Abhängigkeiten implizieren (vgl. Abschnitt 2.4). Wir schlagen deshalb neben einer objektorientierten Realisierung der Präsentationskomponente zur graphischen Umsetzung der direkt manipulativen Interaktionen eine *logisch orientierte* Beschreibung der Interaktionen bzw. Funktionen einer Anwendung (der Anwendungslogik) vor, da so eine größere Ausdrucksmächtigkeit für semantische Eigenschaften zur Verfügung gestellt und eine explizite Dialogbeschreibung realisiert werden kann, die für weitere Werkzeuge zur Verfügung steht. Dieser Ansatz ist auf einer höheren Abstraktionsebene angesiedelt als graphische Constraints, die im wesentlichen zur direkten Umsetzung graphischer Abhängigkeiten dienen.

Im folgenden Abschnitt stellen wir den SUSI-Ansatz vor, der diese logisch- und situations-orientierte Steuerung von Oberflächen erlaubt. Als erste realisierte Erweiterung skizzieren wir Abschnitt 3 die kontextsensitiven Hilfefunktionen, die mit dem zur Dialogsteuerung realisierten Hilfemodul zur Verfügung gestellt werden und einer von mehreren Aspekten darstellt, die erst durch eine explizite Dialogbeschreibung ermöglicht werden. Abschließend werden mögliche Erweiterungen der SUSI-Dialogsteuerung diskutiert. Neben Evaluation und Analyse der Dialogsteuerung erscheint uns die dynamische Anpassung der Steuerung durch das Hinzufügen weiterer Aktionsregeln zur Dialogbeschreibung besonders hilfreich.

2   Die SUSI-Dialogsteuerung

In diesem Abschnitt wird SUSI (Situation Oriented User Interface Model) vorgestellt, ein auf Aktionsregeln und logischen Beschreibungen basierendes Konzept zur Dialogsteuerung in Benutzungsschnittstellen (vgl. [6, 7, 16]).

In SUSI wird der Dialog in strukturierter Weise durch *Aktionsregeln* beschrieben, deren Ausführung vom Eintreten von Ereignissen (repräsentiert durch *Trigger*) und der Gültigkeit von Vorbedingungen abhängig ist. Bei der Verarbeitung einer Aktionsregel können Operationen in der Anwendung ausgelöst, die Darstellungsschicht verändert, und das Wissen über den aktuellen Zustand beeinflußt werden.

2.1   Modularisierung der Beschreibung durch Situationen

Graphische Benutzungsschnittstellen sind meist aus mehreren Teilschnittstellen zusammengesetzt, die _ beispielsweise in einer Datenbankanwendung _ verschiedenen Benutzerprozessen oder unterschiedlichen Sichten auf den Datenbestand zugeordnet sind. In der Regel werden diese Teilschnittstellen durch verschiedene *Fenster* visualisiert, Bereichen, in denen zum Teil unabhängig voneinander gearbeitet wird (z.B. nebenläufige Prozesse) und die zum Teil miteinander kommunizieren (z.B. Verzeichnisfenster, zwischen denen Dateien verschoben werden).
Bei der Beschreibung einer solchen Schnittstelle ist es damit einerseits sinnvoll, unabhängige Teilschnittstellen unabhängig voneinander zu beschreiben; gleichzeitig muß jedoch eine Kommunikation zwischen ihnen möglich bleiben.

In SUSI werden die einzelnen Sichten durch *Situationstypen* repräsentiert. Jedem Fenster ist eine *Situation* als Ausprägung eines Situationstyps zugeordnet; gleichartige Fenster, wie z.B. Verzeichnisfenster eines Dateimanagers, haben den gleichen Situationstyp, d.h. eine festgelegte Mengen von *Aktionsregeln*, wie z.B.

> *trig (drag (X, Y)), fileviewer(Y), Y:cwd(Z), movable (X, Z), move (X, Z)*
-->
> *not (fileIcon (X)), trig (window2:fileviewer display (X));*

die das dynamische Verhalten bzw. die Reaktion auf Benutzereingaben festlegen und eine festgelegte Menge von *Nebenbedingungen* (constraints), wie z.B.

file (A), folder (B), user (C), owns (A, C), owns (B, C) --> movable (A, B);

die die Definition von anwendungsbezogenen Konzepten ermöglichen, die in Vorbedingungen von Aktionsregeln genutzt werden können. Der momentane Zustand einer Situation wird durch eine Menge von *Fakten* (Relationen im Sinne der Prädikatenlogik erster Stufe) beschrieben, die sich mit der Zeit verändert. Die Kommunikation zwischen den Situationen erfolgt durch das Versenden von Triggern; dadurch kann in einer anderen Situation die Ausführung einer Aktionsregel angestoßen werden. Das Zusammenwirken ist in Abb. 1 dargestellt.

2.2   Der Dialogmanager

Eine Benutzungsschnittstellen kann in eine Darstellungsschicht (Präsentation und lexikalische Behandlung von Interaktionen), Dialogsteuerungsschicht (syntaktische und semantische Behandlung von Interaktionen) und Anwendungsschnittstelle (Ausführung von Anwendungsfunktionen) unterteilt werden. Im folgenden wird beschrieben, wie diese Schichten in SUSI realisiert sind. Zusätzlich wird kurz der Aspekt der Darstellung der Anwendungsdaten angesprochen.

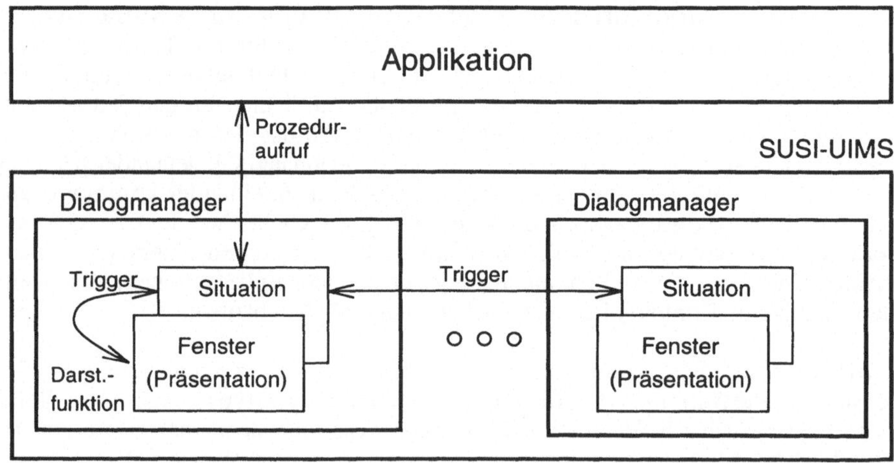

Abb. 1: Präsentationsumgebung und Situationen in der Benutzungsoberfläche zur Laufzeit.

Durch die lexikalische Verarbeitung werden Benutzeraktionen an der Oberfläche registriert und _ entsprechend codiert _ an die Dialogsteuerung weitergeleitet. Dies erfolgt in SUSI durch die Übersetzung in Trigger: Relationen (Literale), die Ereignisse repräsentieren. Aktionsregeln werden aktiviert, indem der übergebene Trigger mit einem ausgezeichneten Literal der Regel unifiziert wird; im Erfolgsfall wird diese Regel weiter bearbeitet. Bei der Unifikation können gleichzeitig Variablen instantiiert werden; auf diese Weise erfolgt eine Parameterübergabe.

Ist ein von der Darstellungsschicht erzeugter Trigger mit dem entsprechenden Literal einer Aktionsregel unifizierbar, so wird diese Regel weiter verarbeitet. Zunächst wird die *Vorbedingung* abgeleitet, eine Konjunktion von Relationen, die entweder innerhalb der aktuellen Situation definiert sind, d.h. Aspekte des Zustands dieser Situation beschreiben, die in einer anderen Situation definiert sind, d.h. sich auf deren Zustand beziehen und von *prozeduralen* Fakten, die durch den Aufruf einer Funktion abgeleitet werden und ihren Wert abhängig vom Erfolg dieses Funktionsaufrufs erhalten.

Konnte die Vorbedingung erfolgreich abgeleitet werden, d.h., wurde die Interaktion ausgeführt, so wird die *Nachbedingung* behandelt. Sie ist ebenfalls eine Konjunktion von Literalen und beschreibt den Zustand nach der Ausführung der Aktionsregel. Dazu werden die in ihr enthaltenen positiven Literale etabliert, d.h. der Faktenmenge werden entsprechende Fakten hinzugefügt, und es werden in der Nachbedingung enthaltene negative Literale aus der Faktenmenge gelöscht. Gleichzeitig werden ggf. Seiteneffekte erzeugt (s.u.). Außerdem können durch die Ausführung der Aktionsregel neue Trigger erzeugt werden, die die Auswahl weiterer Regeln bestimmen.

An der Anwendungsschnittstelle werden die in der Dialogsteuerung erzeugten Befehle in Operationen der Anwendung umgesetzt. Dies erfolgt durch prozedurale Fakten in den Vorbedingungen von Regeln, die als Seiteneffekte die entsprechenden Anwendungsaktionen aktivieren. Das Kopieren einer Datei beispielsweise wird durch eine solche Funktion ausgeführt; die Relation ist *wahr*, wenn das entsprechende Betriebssystemkommando erfolgreich abgeschlossen wurde. Ansonsten ist die Relation *falsch*; durch das Prüfen dieses Wertes kann so entschieden werden, ob die Aktionsregel erfolgreich ausgeführt werden kann oder ob eine (andere) Regel zur Fehlerbehandlung aktiviert werden muß.

Rückmeldungen von Zustandsänderungen in der Darstellungsebene erfolgen durch eine Kopplung einzelner Fakten an Objekte der Oberfläche. Werden bestimmte Fakten bei der Verarbeitung der Nachbedingung einer Aktionsregel etabliert oder aus der Faktenmenge entfernt, werden ihnen zugeordnete Funktionen aufgerufen, die die gewünschten Effekte auf die Darstellungsebene realisieren. Ist beispielsweise eine Datei in der Dialogsteuerung durch ein Fakt und auf der Präsentationsebene durch ein Sinnbild (Icon) repräsentiert, so wird durch solche Funktionsaufrufe bewirkt, daß das Sinnbild dargestellt wird, wenn das Fakt gültig wird und wieder verschwindet, wenn das Fakt ungültig wird.

2.3   Ein Anwendungsbeispiel

Anhand der zuvor dargestellten Aktionsregel bzw. Nebenbedingung soll die Arbeitsweise verdeutlicht werden: Wird eine Datei *file1* aus dem Fenster *window1* in das Fenster *window2* (beide vom Typ *fileviewer*) gezogen, ergibt dies einen Trigger *drag (file1, window2)*, der die Aktionsregel (siehe 2.1) aktiviert. Dabei werden X und Y entsprechend instantiiert.

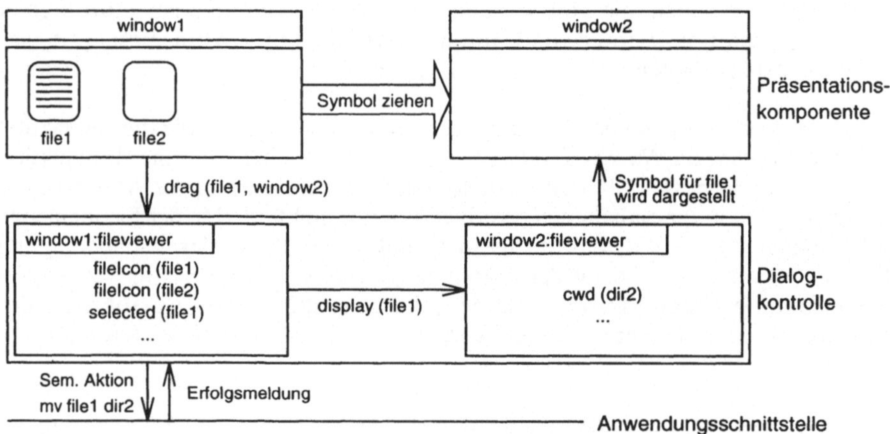

Abb. 2: Der Kontrollfluß in SUSI: Aus *window1* soll eine Datei durch Ziehen mit der Maus auf *window2* verschoben werden.

Nachdem das aktuelle Verzeichnis *dir2* aus der Situation *window2* ausgelesen wurde, wird mit Hilfe der Nebenbedingung das Literal *movable (file1, window2)* abgeleitet; dies geschieht durch Resolution. Die Auswertung des prozeduralen Fakts *move (file1, dir2)* löst die Dateiverschiebung im Betriebssystem aus. Dann wird durch die erste Nachbedingung das Fakt *fileIcon (file1)* gelöscht, das das Sinnbild in *window1* repräsentiert. Als Seiteneffekt wird gleichzeitig dieses Sinnbild von der Oberfläche entfernt. Zuletzt wird ein Trigger erzeugt, der in der Situation *window2* eine Regel zur Darstellung aktiviert. Den Kontrollfluß bei dieser Operation verdeutlicht Abb. 2.

2.4 Die Dialogbeschreibung als funktionales Modell der Anwendung

Ein Ziel bei der Entwicklung der Dialogbeschreibungssprache ist, mit der Beschreibung der Dialogsteuerung auch ein funktionales Modell der Anwendung (auf Interaktionsebene) aufzustellen. Wir sehen im Gegensatz zu anderen Ansätzen, wie HUMANOID, GENIUS oder UIDE davon ab, die Oberfläche automatisch aus einem abstrakten (aufgaben- bzw. funktionsorientierten) Modell der Anwendung zu generieren (vgl. [18]). Ein wesentliches Problem ist bei solchen Ansätzen die adäquate Erzeugung der Präsentationskomponente zu der Wissen über den Anwendungsbereich und fundierte Kenntnisse über die (nicht formalisierbaren) Regeln zur softwareergonomischen Gestaltung notwendig ist [20].

Das funktionale Modell der Anwendung _ die Anwendungslogik _ kann durch die Situationstypen und die dazu durch das Triggerkonzept realisierte Kommunikation einfach modularisiert werden und ermöglicht eine explizite Repräsentation durch Aktionsregeln und Nebenbedingungen. Da durch Aktionsregeln beliebige Fakten innerhalb einer Situation etabliert werden können und zusätzlich Konzepte durch Hornformeln definiert werden können, sind auch komplexe Zusammenhänge zwischen verschiedenen Anwendungsfunktionen in den Vorbedingungen der Aktionsregeln darstellbar. Ob bestimmte Informationen als Fakten innerhalb der Situation dauerhaft vorliegen oder bei Bedarf durch prozedurale Fakten bestimmt werden, kann anwendungsabhängig umgesetzt werden. So werden z.B. im Dateimanager die Informationen über die Zugriffsrechte und Besitzrechte von Dateien aus Effizienzgründen als Fakten in der Situation repräsentiert.

Die meisten modellbasierten Ansätze stellen eine objektorientierte Darstellung von Anwendungsfunktionen und Interaktionen zur Verfügung, die allerdings nur eine eingeschränkte Formulierung von semantischen Bedingungen erlaubt. Zum Beispiel können in UIDE, analog zu unseren Vorbedingungen, semantische Bedingungen an die Argumente einer Interaktionen bzw. einer Funktion geknüpft werden [18]. Als Bedingungen lassen sich Eigenschaften an das Argument formulieren (Typ des Objektes, Ungleichheit zu einem anderen Argument). In UIDE ist es jedoch nicht möglich, Eigenschaften an Attribute eines Argumentes zu formulieren, wie es in unserem Dateimanager mit dem Konzept *movable* als Vorbedingung für eine Verschiebe-Aktionsregel realisiert wurde (pers. Kommunikation N. Sukaviriya).

3 Kontextsensitive Hilfe

Handhabbarkeit und Benutzerfreundlichkeit von Anwendungssystemen wird zum großen Teil durch die Selbstbeschreibungsfähigkeit der Benutzungsschnittstelle bestimmt. Unter diesem Begriff fordert die DIN-Norm, da "*[...] jeder einzelne Dialogschritt unmittelbar verständlich ist oder der Benutzer auf Verlangen zu dem*

jeweiligen Dialogschritt entsprechende Erläuterungen erhalten kann. [...] Beschrei-
bungen sollen situationsabhängig gegeben werden." [4]. Die Verständlichkeit einer
graphischen Oberfläche wird häufig durch ein einsichtiges Design suggeriert. Die
gestalterische Ausdrucksmöglichkeit ist dabei durch die statischen Gegebenheiten des
Interaktionsbereichs beschränkt, hingegen kann die situationsgebundene Dynamik
einer Interaktion auf diese Weise nur unzureichend beschrieben werden. Diese
(dynamische) Selbstbeschreibungsfähigkeit läßt sich jedoch durch ein Hilfesystem
verbessern. Die Aufgabe eines Hilfesystems ist, auf Anfrage Einsatzzweck, Lei-
stungsumfang, Funktionsweise und Handhabung des Systems wiederzugeben (vgl.
[12]).

Im SUSI-Konzept werden diese Anforderungen an ein Hilfesystem durch folgende
Eigenschaften erfüllt: Vermittelt wird ein *benutzerorientiertes Modell* des Gesamtsystems, das durch die Repräsentation der Anwendungslogik gestützt wird. Zusätzlich wird das Hilfesystem mit kontextsensitiven Eigenschaften versehen. *Kontextsensitivität* bedeutet, daß Fragestellungen in Abhängigkeit vom aktuellen Zustand
des Systems beantwortet werden und damit auch die vorhandenen Informa-
tionsmenge auf den Ausschnitt begrenzt wird, der im aktuellen Zustand für Benutzer
und Benutzerin Relevanz hat. Dargestellt werden die aktuellen Möglichkeiten und
deren Auswirkungen. Auf diese Weise werden Benutzer und Benutzerin beim
Aufbau eines mentalen Modells vom System und seiner Funktionsweise unterstützt.

3.1 Funktionalität eines Hilfemoduls

Für den Entwurf eines Hilfesystems müssen Designentscheidungen bzgl. verschiedener Fragen getroffen werden: Wie bzw. wann können Fragen an das Hilfesystem
gerichtet werden? Was soll das Hilfesystem beschreiben? In welcher Weise wird das
Vorwissen der Fragestellenden berücksichtigt? Wie werden Erklärungen erzeugt?
Hinsichtlich der Problemstellungen, die hier nur ausschnittweise aufgelistet sind, legt
das SUSI-Konzept folgende Gestaltung nahe (siehe auch [7, 21]):

Entsprechend den Anforderungen der DIN-Norm [4] und der im vorigen Abschnitt
explizierten Forderung nach Kontextsensitivität kann das Hilfesystem Funktionsweise und Handhabung des Anwendungssystems sowohl unabhängig als auch
abhängig vom aktuellen Systemzustand beschreiben. Kontextsensitive Erklärungen
können u.a. in Bezug auf ein bestimmtes Widget gestellt werden und beschreiben die
im aktuellen Zustand möglichen oder unmöglichen Aktionen.
Der Vorgang des Erklärens wird zunächst als ein Kommunikationsprozeß aufgefaßt,
der auf einer Sequenz von Fragen basiert (siehe Abbildung 3). Darin haben Benutzer
und Benutzerin die Möglichkeit weiterführende bzw. vertiefende Fragen zu einem
Hilfetext zu stellen, sofern ihr Erklärungsbedürfnis nicht befriedigt werden konnte.
Auf diese Weise wird der Wissensstand der Fragenden berücksichtigt. Diese Sichtweise auf den Vorgang legt eine Hypertext-ähnliche Struktur der Hilfetexte nahe.

Kontextsensitivität kann nur erreicht werden, falls Wissen über die Anwendungslogik adäquat repräsentiert ist. Im SUSI-Konzept wird die Dialogbeschreibung als Wissensbasis genutzt. Die Erzeugung von Hilfetexten ist daher vom Aufbau der Dialogbeschreibungssprache geleitet. Fragestellungen _ insbesondere weiterführende Fragestellungen _ beziehen sich implizit auf die Konzepte der Dialogbeschreibung. Durch diese enge Beziehung zur Dialogbeschreibung wird eine automatische Generierung von Hilfetexten möglich. Wie das Hilfemodul zur Erzeugung von kontextsensitiver Hilfe konzipiert ist, stellt der nächste Abschnitt dar.

3.2   Konzeption des Hilfemoduls

Das Hilfesystem adaptiert ein Modell des Anwendungssystems, indem die Struktur der Dialogbeschreibung aufgenommen wird: Die Frage nach möglichen Aktionen läßt sich anhand ableitbarer Aktionsregeln beantworten. Die Nichtausführbarkeit einer bestimmten Aktion läßt sich mit Hilfe der nichterfüllten Fakten oder Nebenbedingungen in der betreffenden Aktionsregel begründen. Das Sprachkonzept der Nebenbedingung in SUSI bietet für die Hilfe den Vorteil, daß für die Anwendung wesentliche Eigenschaften von Objekten zusammengefaßt und erklärt werden können. Damit bilden die Sprachkonzepte der Dialogbeschreibung elementare Erklärungseinheiten für die zu generierende Hilfe. Zu diesem Zweck wird jedem Sprachelement vor Laufzeit des Systems ein Erklärungstext zugeordnet, der den Begriff zunächst unabhängig vom Systemzustand beschreibt.
Die Kontextsensitivität des Hilfesystems, die Bezugnahme zum aktuellen Systemzustand ist dadurch gewährleistet, daß das Hilfesystem Anfragen an den Dialoginterpreter richten kann. Der Dialogmanager stellt zwei Funktionen zu Verfügung, um auf die Anfragen zu reagieren:

- *eine Routine zur "Simulation" einzelner Aktionen*, die die Ausführbarkeit einer Aktion im aktuellen Kontext testet, ohne die Aktion auszuführen;
- *eine Routine zur Evaluation von Nebenbedingung*, die überprüft, ob eine Nebenbedingung im aktuellen Systemzustand erfüllt ist.

Bei der automatischen Generierung von Hilfetexten verknüpfen Generierungsmethoden zur Laufzeit die passenden Erklärungseinheiten entsprechend der Fragestellung zu einem Hilfetext und nehmen die Aktualisierung der zugehörigen Erklärungstexte vor. Der Bezug zum aktuellen Systemzustand wird durch die Anfragen an den Dialogmanager hergestellt.

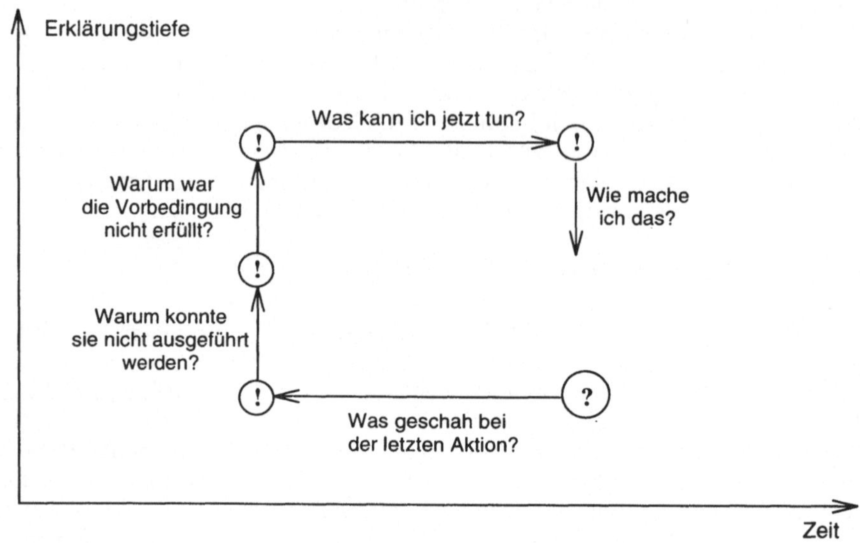

Abb. 3: Sequenz von Fragestellungen

Diese Hilfe unterstützt u.a. den Benutzer beim Aufbau eines mentalen Modells von der Funktionsweise der Anwendung, indem Abhängigkeiten der Interaktionen vom aktuellen Systemzustand erläutert werden und setzt zum Beispiel auch die Hemmschwelle zum Explorieren herab, indem die Auswirkungen einzelner Funktionen skizziert werden (vgl. *Sichten* und *Neutral-Modus* in [14]).

4   Realisierung

Zur SUSI Dialogbeschreibungssprache wurde ein Dialogmanager realisiert, der die hier vorgestellten Funktionen realisiert und als graphische Komponente in der aktuellen Version den Common Lisp Interface Manager (CLIM) nutzt [7]. Der Dateimanager ist als Anwendungsbeispiel mit diesem System realisiert worden; eine Adreßdatenbank, die komplexere semantische Beziehungen zwischen den einzelnen Anwendungsobjekten erlaubt, ist als nächste Anwendung vorgesehen. Geplant ist außerdem eine C++ Version des Dialogmanagers, der dann in Verbindung mit den üblichen auf C oder C++ aufsetzenden Interface Managern als UIMS-System genutzt werden kann. Die Konstruktion einer Dialogsteuerung, die Eingabe der als Hypertextschablone organisierten Hilfetexte oder auch die Umsetzung einer aufgabenorientierten Beschreibung in eine funktionale Beschreibung durch Situationstypen (vgl. Sichten in [1]) wird momentan noch nicht durch spezielle Werkzeuge unterstützt.

5   Mögliche Erweiterungen

Die situationsorientierte Dialogbeschreibung stellt ein funktionales Modell zur Anwendung zur Verfügung. Im folgenden wollen wir zwei Aspekte skizzieren, die durch diese expliziten Dialogbeschreibung ermöglicht werden. Die Dialogbeschreibung kann auf Mängel hin analysiert werden und kann Daten zur Evaluation der Anwendung bereitstellen. Zudem kann die dynamische Anpassung der Anwendung durch eine Ergänzung der Dialogbeschreibung vorgenommen werden.

5.1   Evaluation und Analyse

Ein Vorteil expliziter Dialogsteuerungen innerhalb eines UIMS ist die Möglichkeit, die Dialogstruktur zu evaluieren und zu analysieren. Für die Evaluation einer Anwendung und einer Oberfläche werden die Interaktionen des Benutzers mit der Anwendung abgespeichert. Da neben den primitiven Ereignisse der einzelnen Widgets auch die angestoßenen Anwendungsfunktionen (Aktionsregeln) von der Dialogsteuerung abgespeichert werden können, können effizienzorientierte Keystroke-Level- und GOMS-Analysen (siehe [3]) einfach durchgeführt werden. Die Zusammenfassung mehrerer Interaktionen zu einzelnen (kognitiven) Handlungen kann auch hier nicht automatisch geleistet werden, da innerhalb SUSI kein aufgabenorientiertes Modell vorliegt (vgl. [2]).

Eine Analyse der Dialogsteuerung kann mehrere Zielsetzungen haben. Fragestellungen nach der Erreichbarkeit von bestimmten Systemzuständen oder der Interaktionslänge für bestimmte Aufgaben können durch eine Analyse der Aktionsregeln beantwortet werden. Softwareergonomisch motivierte Fragestellungen, wie die der max. Anzahl der Schritte um bestimmten Aktionen ausführen zu können (Undo, Hilfe, Ende etc.) oder die Länge von ununterbrechbaren Sequenzen, können ebenfalls untersucht werden. Durch Rückwärtsverkettung von Aktionsregeln zur Untersuchung von Situationsveränderungen kann die kombinatorische Explosion für die Analysen eingeschränkt werden.

5.2   Adaptierbare Oberflächen

Adaptierbare Steuerungen werden z.B. bei Dateimanagern eingesetzt. Hier können weitere Dateitypen und diesen Typen zugeordnete Aktionen definiert werden. Tabellenkalkulationen sind eine weitere Klasse von Anwendungen, die von den Benutzern erfolgreich selber programmiert bzw. adaptiert werden. Da in SUSI eine explizite und strukturierte Darstellung der Dialogsteuerung vorliegt, die nicht kompiliert, sondern interpretiert wird, ist auch mit SUSI eine dynamische Adaptierung der Steuerung in weitreichendem Maße möglich. Durch Veränderung oder Erweiterung der Aktionsregeln können weitere oder komplexere Interaktionen durch den Benutzer definiert werden. Da die Interaktionen zu einem Widget zur Interpretation an die Dialogsteuerung weitergegeben werden, kann die Funktionalität

der Anwendung auf einfache Weise erweitert werden, indem Aktionsregeln
eingeführt oder verändert werden, die diese Interaktion bearbeiten. Solange die zur
Verfügung gestellte Oberfläche (die Präsentationskomponente) und die Anwendungsfunktionen für die Erweiterungen ausreichend sind, ist dies im Gegensatz zu
klassisch realisierten Anwendungen durch die Veränderung der Dialogbeschreibung
einfacher möglich. Ein gutes Verständnis der Funktionsweise der genutzten Anwendungsfunktionen sowie der Dialogsteuerung ist allerdings nötig, damit Benutzer
selbständig die Oberfläche erweitern können. Durch eine Markierung sollten
wesentliche Aktionsregeln und kritische Anwendungsfunktionen vor Benutzerver-
änderungen geschützt, und so der Anpassungsbereich eingeschränkt werden.

6   Zusammenfassung

User Interface Management Systeme stellen einen wesentlichen Fortschritt für die
Entwicklung von Oberflächen dar, indem sie eine adäquate Auftrennung zwischen
Oberflächenentwicklung und Anwendungsentwicklung ermöglichen und so den
unterschiedlichen Anforderungen an Oberflächenprogrammierung und Anwendungs-
programmierung Rechnung getragen werden kann. Die hier vorgestellte Dialogbeschreibung ermöglicht die Integration von Anwendungswissen (insbesondere der
Anwendungslogik) in die Dialogsteuerung und ist in der Lage, auch komplexe *direkt
manipulative* Interaktionen adäquat in der Oberfläche zu bearbeiten.

Durch die Faktenbasis und die Nebenbedingungen wird die objektorientierte Präsentationskomponente um eine logische Sicht zur Beschreibung von Oberfläche und
Anwendungszustand erweitert. Die Interpretation von Interaktionen erfolgt durch
Aktionsregeln, die durch die Nachbedingungen die Zustandsveränderungen von
Anwendung und Oberfläche beschreiben. Dadurch stellt die Dialogbeschreibung ein
funktionales Modell der Anwendung zur Verfügung. Diese logische Sicht ist eng
verwandt mit den stärker objektorientierten Modellbeschreibungen, wie in UIDE
oder HUMANOID, erlaubt jedoch eine einfachere bzw. ausdrucksstärkere Formulierung von semantische Bedingungen und der zugrundeliegenden Umgebung (Situation) zur Interpretation von Interaktionen. Allerdings verfolgen wir mit unserem
Ansatz bewußt keine automatische Konstruktion der Präsentationskomponente aus
einem Modell der Anwendung. Wir haben außerdem gezeigt, wie diese Dialogbeschreibung für ein kontextsensitives Hilfemodul genutzt wird und skizziert, wie
Analyseverfahren und dynamische Anpassungen der Applikation auf solch eine
Dialogsteuerung aufsetzen können. Interessante Erweiterungsmöglichkeiten sind das
automatische Sperren und Entsperren von Interaktionen anhand der Ausführbarkeit
der Aktionsregeln und die Einbindung von graphischen Constraints, um u.a. eine
vereinfachte Realisierung von semantischen Rückkopplungen zu ermöglichen.
Die modulare Beschreibung der Oberfläche durch Situationen, die über Ereignisse
kommunizieren können, sind u.E. ein gutes Beschreibungsmittel für den Übergang
von einer aufgabenorientierten Beschreibung der Schnittstelle zu einer rechnerorien-
tierten bzw. schnittstellenorientierten Beschreibung ihrer Funktionalität. Im Rahmen

einer größeren Anwendung ist die weitere Evaluation der praktischen Nutzbarkeit und Beschreibungsfähigkeit der Dialogbeschreibung sowie des Hilfesystems durchzuführen und SUSI um Entwicklungswerkzeuge zu erweitern, die das Design und die Konstruktion einer Schnittstelle unterstützen.

7   Literaturverzeichnis

[1]   Astrid Beck, Christian Janssen: Vorgehen und Methoden für Aufgaben- und Benutzerangemessene Gestaltung von graphischen Benutzungsschnittstellen, in: Wolfgang Coy, Peter Gorny, Ilona Kopp, Constantin Skarpelis (Hrsg.): Menschengerechte Software als Wettbewerbsfaktor. Forschungsansätze und Anwenderergebnisse aus dem Programm "Arbeit und Technik", Teubner, Stuttgart, 1993, S. 200-221.

[2]   Michael Byrne, Scott Wooe, Piyawadee Sukaviriya, James Foley, David Kieras: Automating Interface Evaluation, in: Proccedings of CHI '94, Addison Wesley, Ney York, 1994, S. 232-237.

[3]   Stuart K. Card, Thomas P. Moran, Alan Newell: The Psychology of Human-Computer Interaction, Hillsdale, NJ, Lawrence Erlbaum Associates, 1983.

[4]   DIN-Norm 66234 Teil 8, 1991.

[5]   James D. Foley, Won Chul Kim, Srdjan Kovacevic, Kevin Murray: UIDE - An Intelligent User Interface Design Environment, in: J. Sullivan, S. Tyler (Hrsg.): Architectures for Intelligent Interfaces: Elements and Prototypes, Addison Wesley, MA, 1991.

[6]   Stefan Hügel: Logische Modellierung von Benutzungsschnittstellen - Realisierung einer Sprache zur Beschreibung von Kontexten und Aktionen, Diplomarbeit, Institut für Informatik und Gesellschaft, Universität Freiburg und Institut für Logik, Komplexität und Deduktionssysteme, Universität Karlsruhe, 1994.

[7]   Stefan Hügel, Kirsten Winter, Friedrich Strauß: SUSI - Situationsorientierte Modellierung von Benutzungsschnittstellen mit integrierter kontextsensitiver Hilfe, IIG-Bericht 9/94, Institut für Informatik und Gesellschaft, Universität Freiburg, 1994

[8]   Rolf Ilg, Jürgen Ziegler: Direkte Manipulation, in: Helmut Balzert, Heinz Hoppe, Reinhard Oppermann, Helmut Peschke, Gabriele Rohr, Norbert Streitz (Hrsg.): Einführung in die Software-Ergonomie, de Gruyter, 1988.

[9]   Christian Janssen: Dialognetze zur Beschreibung von Dialogabläufen in graphisch-interaktiven Systemen, in: [15], S. 67-76.

[10] Brad A. Myers: Creating User Interfaces by Demonstration, Boston, MA, 1988.

[11] Brad A. Myers: State of the Art in User Interface Software Tools, in: H. Rex Hartson, Deborah Hix (Hrsg.): Advances in Human-Computer Interaction, Vol. 4, Ablex Publishing Corporation, Norwood, New Jersey, 1993.

[12] Reinhard Oppermann, Bernd Murchner, Harald Reiterer, Manfred Koch: Softwareergonomische Evaluation: Der Leitfaden EVADIS II, de Gruyter, Berlin, 1992.

[13] Dan R. Olsen Jr.: User Interface Management Systems: Models and Algorithms, Morgan Kaufmann Publishers, San Mateo, California, 1992.

[14] Hansjürgen Paul: Das Explorative Modell als konzeptioneller Ansatz zur Gestaltung interaktiver Systeme, in: [15], S. 77-86.

[15] Karl-Heinz Rödiger (Hrsg.): Software-Ergonomie '93 - Von der Benutzungsoberfläche zur Arbeitsgestaltung, Teubner, Stuttgart, 1993.

[16] Friedrich Strauß: Situation Oriented Description of User Interfaces, in: Patrick Brezillon (Hrsg.): Proceedings of the IJCAI'93 (13th International Joint Conference on Artificial Intelligence) Workshop on "Using knowledge in its Context", Rapport Interne du LAFORIA 93/13, Institut Blaise Pascal, Paris, 1993.

[17] Friedrich Strauß: Contextsensitive Help-facilities in GUIs through Situations, in: T. Grechenig, M. Tscheligi (Hrsg.): Proceedings of the VCHCI'93 (Vienna Conference on Human Computer Interaction), Lecture Notes in Computer Science 733, Springer-Verlag, 1993, S. 79 - 90.

[18] Piyawadee "Noi" Sukaviriya, James Foley, Todd Griffith: A Second Generation User Interface Design Environment: The Model and The Runtime Architecture, in: Proceedings of Interchi '93, 1993, S. 375-382.

[19] Bernhard Trefz: Diamant - A User Interface Management System for Object Oriented Interfaces, in: Margaret Galer, Susan Harker, Jürgen Ziegler: Methods and Tools in User-Centered Design for Information Technology, North Holland, Amsterdam, 1992, S. 319-343.

[20] Jean Vanderdonckt, Missiri Ouedraogo, Banta Ygueitengar: A Comparison of Placement Strategies for Effective Visual Design. In: G. Cockton, S.W. Draper, G.R.S. Weir: Proceedings of HCI '94, Cambridge University Press, 1994, S. 125 - 144.

[21] Kirsten Winter: Kontextsensitive Benutzerunterstützung in graphischen Benutzerschnittstellen, Diplomarbeit, Institut für mathematische Maschinen und Datenverarbeitung, Universität Erlangen-Nürnberg, 1994.

Friedrich Strauß, Stefan Hügel, Kirsten Winter, Britta Schinzel
Universität Freiburg
Institut für Informatik und Gesellschaft, Abt. Modellbildung und soziale Folgen
Friedrichstraße 50, D-79098 Freiburg i. Brsg., Telefon 0761/203-4954
{frieder, stefan, kirsten, britta}@modell.iig.uni-freiburg.de

Unterstützung bei der Gestaltung von Benutzungs-schnittstellen durch die Bereitstellung von Software-Ergonomie-Wissen in einem Informations- und Beratungssystem

H. Wandke & J. Hüttner
Humboldt-Universität zu Berlin

Zusammenfassung

Die Entwicklung einer ergonomischen Benutzungsschnittstelle ist eine vielschichtige und komplexe Tätigkeit, die durch wissensorientierte Unterstützungsmittel erleichtert werden kann. Der Unterstützungsbedarf von Programmierern bei der Gestaltung benutzungsfreundlicher Software ist enorm. Hilfsmittel sind kaum vorhanden und nicht für die Praxis entwickelt. Mit dem Projekt *inra* (Teil vom Verbundprojekt WEDA, gefördert vom BMFT [Förderkennzeichen 01 HK 790-8] im AuT-Programm) wird ein Unterstützungsmittel für Software-Entwickler erstellt. Ein Hypertextsystem (für Windows) und ein Begleitbuch bieten vielfältige Informationen aus dem Bereich der Software-Ergonomie. Ein Schwerpunkt im Hypertextsystem ist die Bereitstellung von Wissen in Beispielen, mit diesen kann in einem speziellen Modus gearbeitet werden. Damit sollen Entwickler zu einem Perspektivenwechsel animiert werden, sie sollen in die Rolle eines Benutzers mit festgelegten Aufgaben schlüpfen. Das Unterstützungsmittel ist in der Software-Entwicklungs-praxis getestet und anschließend verbessert worden. Dieser Prozeß ist noch nicht abgeschlossen. Das System kann sowohl in der Ausbildung (z.B. Universitäten, Weiterbildung) als auch zum selbstgesteuerten Lernen am Arbeitsplatz genutzt werden.

Einleitung

Software-Entwickler stehen vor neuen Aufgaben. Sie sollen - im Gegensatz zu den Ansprüchen noch vor wenigen Jahren - nicht nur Programme entwickeln, die korrekt und zuverlässig laufen, sondern die von späteren Benutzern einfach zu erlernen und leicht zu benutzen sind. Usability ist ein bestimmender Marktfaktor geworden, dessen Wert mit einer Übernahme der EU-Bildschirmrichtlinie (für die BRD wahrscheinlich 1995) noch steigen wird (vgl. Cakir, 1991).
Um diese neuen Anforderungen erfüllen zu können, müssen Software-Entwickler sehr viel über Benutzer und ihre Aufgaben, über Arbeitsorganisation und über Arbeitsgestaltung wissen. Eine Arbeitsgruppe der Gesellschaft für Informatik unter der Leitung von Susanne Maaß hat 1993 für ein Software-Ergonomie-Curriculum insgesamt 29 verschiedene Lernziele in neun Qualifikationsgebieten der Software-Ergonomie zusammengetragen, die im Rahmen der Informatikausbildung erreicht werden sollten (Maaß u.a. 1993). Diese Wissensinhalte sind vielen Entwicklern in der Praxis weitgehend unbekannt (Hüttner & Wandke, 1993).

Ergonomie-Wissen für die Benutzungsschnittstellengestaltung

Die erforderlichen Kenntnisse auf psychologischem und ergonomischem Gebiet werden auch zunehmend bedeutsamer, weil in der jüngsten Zeit die Gestaltung der Benutzungsschnittstelle immer weiter in den Mittelpunkt der Entwicklung von Softwaresystemen rückt . Während Smith und Mosier (1984) noch feststellten, daß ca. 30 - 35 % aller Software-Entwicklungskapazität in die Benutzungsschnittstelle gesteckt wird, ergab eine Umfrage bei Software-Herstellern in den USA von Myers und Rosson (1992), daß durchschnittlich 48 % des Programmcodes der Benutzungsschnittstelle gewidmet ist und daß der Arbeitszeitaufwand für die Gestaltung der Benutzungsschnittstelle sogar deutlich über 50 % der Gesamtentwicklungszeit liegt.

Diese und ähnliche Umfrageergebnise basieren auf einem Schnittstellen-verständnis, wie es in der Praxis der Software-Entwicklung üblich ist. Legt man das IFIP-Benutzungsschnittstellenmodell (Dzida, 1984) zugrunde, so verstehen Praktiker häufig unter Benutzungsschnittstelle (der tatsächliche Sprachgebrauch im „Alltag" ist *Benutzerschnittstelle*) nur die Ein-/Ausgabeschnittstelle und die Dialogschnittstelle, jedoch selten die Werkzeugschnittstelle und fast nie die Organisationsschnittstelle. Berücksichtigt man, daß aus software-ergonomischer Sicht alle Hard- und Softwarekomponenten zur Benutzungsschnittstelle gehören, mit denen Benutzer nach der "klassischen" Definition von Moran (1981) in Kontakt kommen, so ist der Arbeitsaufwand, der in die Entwicklung der Benutzungsschnittstelle investiert wird, wahrscheinlich noch größer. Unterstützt wird diese eingeschränkte Problemsicht durch die Annahme, daß schon allein das Befolgen von Style Guide Regeln eine ergonomische Benutzungsschnittstelle entstehen läßt.

Es ist ein Ziel von *inra*, diese Problemsicht zu erweitern und auf Gestaltungsprobleme aufmerksam zu machen, die über das "look and feel" einer Benutzungsschnittstelle hinausgehen. Dies ist schwierig, da es in der Software-Ergonomie auf den unteren Ebenen einer Benutzungsschnittstelle (z.B. Informationsdarbietung, visuelle Gestaltung) sehr viel mehr Gestaltungswissen gib, das außerdem sehr viel stärker formalisierbar ist, als auf den höheren Ebenen (Aufgabenebene, semantische Ebene). *inra* bietet deshalb auf drei verschiedenen Wegen Gestaltungswissen für eben diese höheren Ebenen der Benutzungsschnittstelle an:

1. Die Gestaltung der Funktionalität wird in die Beispiele des *inra* aufgenommen. Da die meisten Gestaltungsprobleme auf höheren Ebenen nicht durch bloßes Betrachten einer Oberfläche erkannt werden können, sind im *inra*-System Aufgaben integriert, die im Rahmen der Beispiele bearbeitet werden sollen.

2.	Durch zahlreiche Hinweise in der Textkomponente des *inra* vom Typ "Dieses Problem können Sie nur lösen, wenn Sie die Aufgaben der Benutzer kennen, ...wenn Sie Benutzer fragen, ...wenn Sie Benutzertests durchführen, ...".

3.	Durch die Aufnahme einer Zusammenstellung von empirischen Evaluationsmethoden der Software-Ergonomie in das *inra*-Begleitbuch. Hier wird erläutert, wie Benutzerbefragungen, -beobachtungen u.ä. durchgeführt werden, welche Vor- und Nachteile diese Methoden besitzen und welche Voraussetzungen erfüllt sein müssen, um sie anwenden zu können.

Unterstützung der Entwickler bei der Gestaltung von Benutzungsschnittstellen

Die Gestaltung von Benutzungsschnittstellen ist ein Prozeß der drei unterschiedliche Vorgehensweisen beinhalten kann, die mit verschiedenen Mitteln unterstützt werden können. Zum einen kann er als Technologie, d.h. als eine hochgradig formalisierte und durch Regeln bestimmte Vorgehensweise aufgefaßt werden, insbesondere dann, wenn Bauelemente, Verfahren oder Formeln benutzt werden können. Das ist beispielsweise beim Zusammenstellen einer grafischen Benutzungsoberfläche aus einer Menge von widgets der Fall. Hier können Baukastensysteme mit integrierten „Ergonomiekontrollen" den Entwickler entlasten, indem sie beispielsweise Teile seiner Aufgaben automatisch realisieren und als Lösung anbieten. Da diese technologisch orientierte Gestaltung durch andere Unterstützungsmittel (vgl. z.B. Projekt JANUS; Balzert, 1993 oder Projekt IDA; Reiterer, 1992) bereits gut abgedeckt ist, wird im *inra* darauf nicht eingegangen.

Zugleich kann Gestaltung aber auch erfahrungsgeleitet erfolgen. Entwickler, die bereits viele Benutzungsschnittstellen gestaltet haben, zu diesen Rückmeldungen von Benutzern erhalten haben, können auf bestimmte Musterlösungen zurückgreifen und diese neuen Anforderungen anpassen. *inra* vermittelt zu großen Teilen solches Erfahrungswissen, das nicht formalisierbar ist und das - im Gegensatz zum technologischen Vorgehen - nicht direkt angewendet werden kann, sondern vom Entwickler aufgenommen und "umgesetzt", d.h. auf seine konkreten Gestaltungsprobleme übertragen werden muß. Dies wird unterstützt durch die Beispiele und Kommentare, die eine Verallgemeinerung des Erfahrungswissens ermöglichen sollen.

Neben technologischen und erfahrungsgeleiteten Aspekten schließt Gestaltung auch kreative Aspekte ein. Kreatives Gestalten kann nur sehr begrenzt unterstützt werden. *inra* bietet zwei notwendige, aber keineswegs hinreichende Rahmenbedingungen für kreatives Gestalten.

1.	Es stellt (in Textform) psychologisches Hintergrundwissen bereit, das helfen kann, neue Gestaltungslösungen zu finden.

2. Es weist mit den Kommentaren für die Beispiele und mit den Texten auf die Vorläufigkeit und Offenheit der bisherigen Gestaltungslösungen - u.a. abhängig von der zur Verfügung stehenden Hard- und Software - hin. Die Gestaltungsbeispiele des *inra* sind keine perfekten, kopierbaren Gestaltungslösungen, sondern sollen anregen, wie man es anders und (noch) besser machen kann.

Informations- und Beratungssystem - *inra*

In einer Untersuchung wurden Entwickler mit einem Fragebogen nach ihren derzeitig verfügbaren Hilfsmitteln und nach ihren Vorstellungen von einem praxistauglichen Unterstützungsmittel befragt (Beimel, Hüttner & Wandke, 1993). Sie entschieden sich in fast allen Aspekten der Benutzungsschnittstellengestaltung für eine Unterstützung durch ein klassisches Buch mit leicht verständlichen Texten *und* ein rechnerbasiertes Unterstützungsmittel. Die eher klassischen Unterstützungsmittel, wie Schulungen, Lehrbücher und externe Experten wurden weniger gefordert. Zusätzlich zu den technischen Vorteilen, die ein Computersystem als Unterstützungsmittel bietet, wird es auch aus motivationalen Gründen nahezu unmöglich sein, Personen, die ständig am Rechner arbeiten, zum Wissenserwerb ohne einen Computer zu motivieren.
Ein weiterer Gesichtspunkt ist die Freude von Software-Entwicklern am Explorieren. Das Erkunden und Ausprobieren einer Software ermöglicht einen individuellen Zugang zum jeweiligen Wissenskörper - eine neue Anwendungs-Software oder ein wissensorientiertes Unterstützungsmittel (vgl. den weitreichenden Vorschlag von Paul [1993, 1994] interaktive Systeme explorationsfreundlich zu gestalten).
Mit dem Informations- und Beratungssystem *inra* soll Software-Entwicklern ein modernes Unterstützungssystem zur Verfügung gestellt werden, das zwar nicht alle, aber doch wesentliche Anforderungen und Erwartungen der befragten Entwickler erfüllen kann. *inra* ist ein Unterstützungsmittel, das zusammen mit anderen (vgl. z.B. die Beiträge von Bachmann u.a., Heintzen u.a., Malinowsi und Stolze, in diesem Band) für verschiedene Teilaufgaben und Phasen der Benutzungsschnittstellengestaltung herangezogen werden kann. Dieses System soll den Programmierern helfen, Kompetenz für einen Teil ihrer Arbeitsaufgabe zu erwerben. Spezifisches Fachwissen aus dem Forschungsgebiet Software-Ergonomie muß so aufbereitet werden, daß es für Personen außerhalb der kleinen scientific community verständlich, nachvollziehbar und damit erst nutzbar wird. Das Informations- und Beratungssystem besteht aus einem Hypertextsystem für IBM-kompatible PC's (entwickelt mit ToolBook) und einem Begleitbuch (Hüttner, Wandke & Rätz, 1994). Die folgenden Darstellungen beziehen sich vor allem auf das Hypertextsystem.
Das Gegenstandsgebiet von *inra* - die ergonomische Gestaltung von Benutzungsschnittstellen - verändert sich in hohem Tempo. Heute Gültiges kann schon in wenigen Jahren überholt sein, neue Techiken erfordern wahrscheinlich

andere ergonomische Lösungen. Ständig verbesserte Rahmenbedingungen in der Hard- und Softwaretechnik bieten die Voraussetzungen für die Entwicklung neuer Interaktionstechniken. „Das aktuell Gelernte kann also höchstens exemplarisch sein und sollte zu Innovationen befähigen und ermuntern." (Oberquelle, 1994, S. 24). Das Wissen, das durch *inra* vermittelt und bereitgestellt werden soll, ist deshalb weniger technologiebezogen, sondern stärker auf grundsätzliche Eigenschaften menschlicher Informationsaufnahme und -verarbeitung sowie auf die Steuerung von Handlungen ausgerichtet. Es wurde bei der Entwicklung des Systems auch nicht versucht, das Ergonomie-Wissen vollständig abzubilden.

Das Ziel von *inra* ist es also nicht, für jede Gestaltungsfrage eine konkrete Antwort zu bieten, sondern den Software-Entwickler in die Lage zu versetzen, selbst eine Antwort zu finden.

Bevor Entwickler diese Antworten finden können, müssen sie allerdings in der Lage sein, die richtigen Fragen stellen zu können. *inra* soll auch dazu einen Beitrag leisten. Durch die Arbeit mit dem System sollen Entwickler von Software auf Schwierigkeiten bei der Benutzungsschnittstellengestaltung aufmerksam gemacht werden. Dies wird vorrangig durch einen Perspektivenwechsel erreicht: die Entwickler sollen sich in die Rolle eines Benutzers mit festgelegten Aufgaben begeben und versuchen, mit einer Beispiel-Software diese Aufgaben zu lösen. Sie sollen anhand von komplexen Beispielen für Benutzungsschnittstellen erkennen können, an welchen Stellen und durch welche Gestaltungsdetails Benutzer Schwierigkeiten beim Umgang mit Systemen haben können. Es soll eine Sensibilisierung der Entwickler für die Belange der Benutzer erreicht werden. "The first step toward improving computer-human interfaces *involves changing the attitude of the software professional.* Software professionals need to be convinced that a high-quality user interface is infact important." (Mayew, 1992, S.3, kursiv im Original). Zugleich sollen die Beispiele aber auch Anregungen bieten, wie man eine Benutzungsschnittstelle verbessern kann.

Inra dient sowohl der Wissensvermittlung (längerfristige Nutzung) als auch der Bereitstellung von Wissen (kurzfristige Nutzung). Primär ist *inra* dafür gedacht, daß Entwickler durch das Bearbeiten von Beispielen in einer Art "Selbsterfahrung", durch das explorative, von Kommentaren unterstützte Durchprüfen von Beispielen und durch das Lesen von Empfehlungen und Hinweisen Wissen erwerben, das sie später bei der Gestaltung von Benutzungsschnittstellen einsetzen können, ohne *inra* wiederholt zu benutzen. Die Empfehlungen und Hinweise werden in einer kurzen und knappen Form am Bildschirm dargeboten und sind ergänzend, in einer ausführlichen, auch stärker auf Begründungen und Hintergrundinformation eingehenden Form in dem Begleitbuch zusammengestellt.

Inra ist aber auch dazu gedacht, als externer Wissensspeicher benutzt zu werden. Software-Entwickler, die eine spezifische Information benötigen, können *inra* heranziehen, um diese Information in den Texten, Beispielen und Kommentaren zu finden. Dazu gibt es in dem Hypertextsystem verschiedene Such- und Navigationsfunktionen, die es dem Entwickler einfach machen sollen, schnell die entsprechende *inra*-Komponente zu finden (siehe ein Beispiel in Abb. 1). Darüber hinaus bietet ein Glossar Informationshilfe durch die Erläuterung psychologischer und software-ergonomischer Begriffe. Dies ist allerdings eine eher wissensvermittelnde Komponente als eine, die Wissen zur direkten Nutzung bereitstellt. Ein kommentiertes Literaturverzeichnis liefert Hinweise auf weitere Informationsquellen. Diese Komponente ist geeignet, die Unabgegrenztheit und Offenheit von *inra* zu verdeutlichen, verweisen doch die Literaturangaben auf weitere, außerhalb von *inra* liegende externe Wissensspeicher.

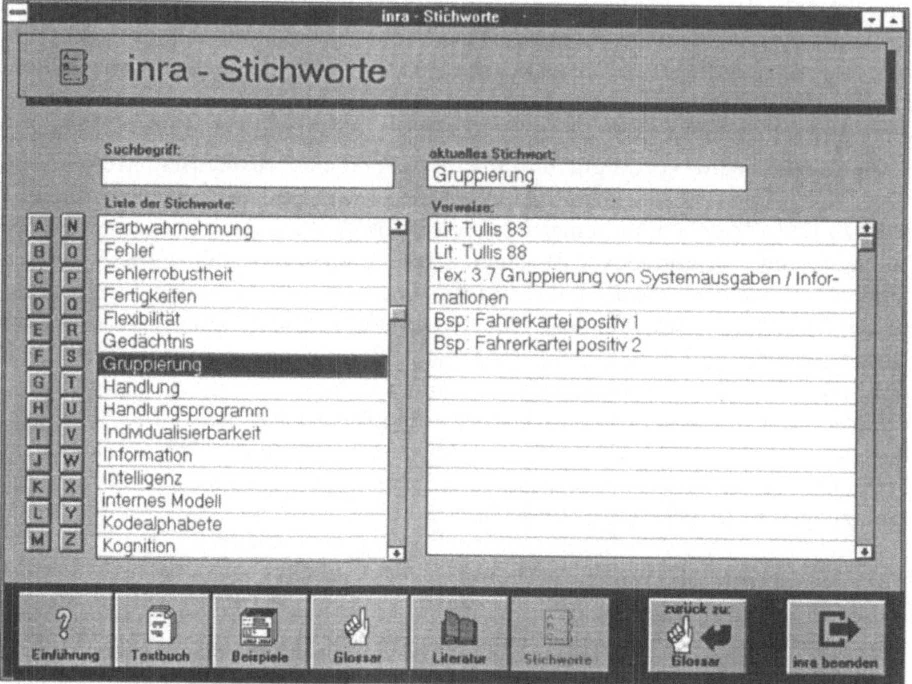

Abb.1: Das *inra*-Stichwortverzeichnis ist ein Beispiel für einen schnellen Zugang zu verschiedenen Informationen zu einem Stichwort (hier „Gruppierung").

Alle sechs *inra*-Komponenten (Textteil, Glossar, Einführung, Literaturverzeichnis, Beispielsammlung und Stichwortverzeichnis) sind nach einem einheitlichen Schema aufgebaut. Ein Überschriftsbereich enthält den Namen sowie ein Symbol für die jeweilige Komponente. Am unteren Rand jeder Komponente gibt eine Leiste mit Funktionsknöpfen einen Überblick über die anderen erreichbaren Komponenten und die gerade aktive Komponente - hier ist der Knopf eingedrückt und grau dargestellt. Alle Komponenten - ausgenommen das Stichwortverzeichnis - können wie Bücher seitenweise durchgeblättert werden. Auf der ersten Seite wird eine Übersicht über den jeweiligen Komponenten-Inhalt gegeben. Die weiteren Seiten enthalten die Informationen in unterschiedlicher Form. Die Inhalte der einzelnen Komponenten sind in einer Hypertextstruktur über Führungsworte miteinander verbunden.

Komponente	Inhalte	Unterstützungsfunktionen
Einführung	Beschreibung der *inra*-Komponenten	Einführung bei Erstbenutzung / Erläutert den Umgang mit *inra*
Stichworte	Suchworte vorr. zu Gestaltungsprobl.	schnelle Navigation / Unterstützt den Zugriff auf das Wissen in den Komponenten
Textbuch	Software-Ergonomie-Wissen in kurzen Texten	Vermittlung von Grundwissen in Form von Empfehlungen u.ä. /Wissenserwerb für die langfristige Nutzung / Querverweise
Beispiele	Beispiele mit Erklärungen zum Ausprobieren	Sensibilisierung, Exploration, Entdeckung von Problemen, Fragen / Wissenserwerb durch selbstgesteuertes Lernen
Glossar	Begriffe aus der Psychologie / Arbeitswissensch.	Erklärung der Bedeutung der Begriffe / Unterstützt den Wissenserwerb / Schaffung einer deklarativen Wissensstruktur
Literatur	Quellenverweise und erklärende Hinweise	Weiterführende Informationen / „Weiterlernen" / Erweiterung des Suchraums für Gestaltungsfragen

Tab. 1: Die Inhalte der Komponenten und deren Funktionen im *inra*-Hypertextsystem.

Die Entwicklung der Beispiele im Hypertextsystem

Die beiden wesentlichen Vorteile des entwickelten Hypertextsystems sind der flexible, individuelle Zugang zur Information und die Realitätsnähe der Beispiele auf dem Bildschirm. Die Beispiele wurden in einen inhaltlichen Rahmen gefaßt, um damit einzelne Wissensinhalte für eine fiktive, relativ komplexe Software-Anwendung umzusetzen. Nur dadurch ist es möglich, Wechselwirkungen zwischen Einzel-Gestaltungsentscheidungen zu illustrieren. Eine hinreichende Komplexität ist auch notwendig, wenn nicht nur Wissen zur Ein-/Ausgabeschnittstelle, sondern auch für die höheren Ebenen der Benutzungsschnittstelle vermittelt werden soll. Die fiktive Software-Anwendung darf allerdings kein Spezialwissen erfordern, zum Verständnis sollte Alltagswissen ausreichen. Es wurde deshalb eine stark vereinfachte und ausschnitthaft simulierte Tätigkeit eines Angestellten in einer Spedition als Anwendung gewählt. Dadurch, daß alle Einzelbeispiele in diesem Arbeitskontext angesiedelt sind, fügt sich jedes Einzelbeispiel in einen umfassenden Rahmen und die Wissensvermittlung geht über Einzelelemente hinaus. Um eine gute

Kontrastwirkung zu ermöglichen und damit den Lerneffekt zu erhöhen, wurde angestrebt von jedem komplexen Beispiel eine positive und eine negative Lösung zu entwickeln.

Die Entwicklung der Beispiele erfolgt in einem iterativen Prototyping-Prozeß, der hier für das erste Beispielpaar vorgestellt werden soll. In den ersten beiden Beispielen wurden insbesondere die Themen: Formulartechnik und Eingabemasken, Tabellengestaltung, und teilweise auch Aspekte der Dialogsteuerung und der Benutzerunterstützung durch entsprechende Suchfunktionen etc. aufgegriffen. In einem ersten Schritt entwickelte ein Mitarbeiter unserer Forschungsgruppe ein Beispiel mit einer festgelegten Programmfunktionalität zur Durchführung bestimmter Verwaltungsaufgaben in einem Speditionsbetrieb. Diese nach „bestem Wissen und Gewissen" erarbeitete erste Version lieferte das erste „negative Beispiel". Ausgehend von der ersten Version wurde in Teamsitzungen schrittweise eine zweite, ergonomisch verbesserte Version entwickelt, die dann als „positives Beispiel" bewertet wurde. Dabei standen einige fiktive Arbeitsaufgaben, die mit dem Beispiel gelöst werden sollen, im Mittelpunkt der Umgestaltung. Für diese konkreten Aufgaben wurde versucht, die ergonomischen Kriterien umzusetzen.

Damit die Generalisierung des in den Beispielen angewandten Gestaltungswissens besser gelingt, sind Kommentare eingefügt worden. Die Kommentare geben in Kurzfassung noch einmal wieder, aus welchem Grunde eine bestimmte Gestaltungslösung zu empfehlen ist bzw. was - das negative Beispiel betreffend - an einer Gestaltungslösung genau zu kritisieren ist (siehe Abb. 2). Eine ihrer wichtigsten Funktion ist es, die Entwickler bei der Abstraktion von der Beispielebene in die Ebene der Gestaltungsprinzipien zu unterstützen.

Die Beispiellösungen liegen in der Beispiel-Komponente als Abbildungen einer Oberfläche auf einer Bildschirmseite vor (siehe unterer Teil der Abb. 2) und sind in einem speziellen Modus benutzbar. Innerhalb einer spielerischen Situation kann der Benutzer der Beispielsammlung (also die Entwickler) unterschiedliche Aufgaben eines Dispatchers einer fiktiven Speditionsfirma interaktiv mit der Software bearbeiten. Die festgelegten, einfachen Arbeitsaufgaben sind so ausgearbeitet, daß bestimmte Funktionen des Beispielprogramms benutzt werden müssen bzw. in der „negativen Variante" vermißt werden.

Einige Beispielaufgaben:

1. Auf Beschluß der Bezirksversammlung wurde in Unterhachingen die „Allee der Kosmonauten" umbenannt. In dieser Straße wohnt ein Fahrer unseres Unternehmens. Die Straße heißt jetzt "Astronautenallee". Bitte aktualisieren sie die Straßenbezeichnung in der Datei.

2. Unser Fahrer Paul Klotzky hat letzte Woche geheiratet und den Namen seiner Gattin angenommen. Er heißt jetzt "Frühauf". Bitte ändern Sie seine Personalangaben.

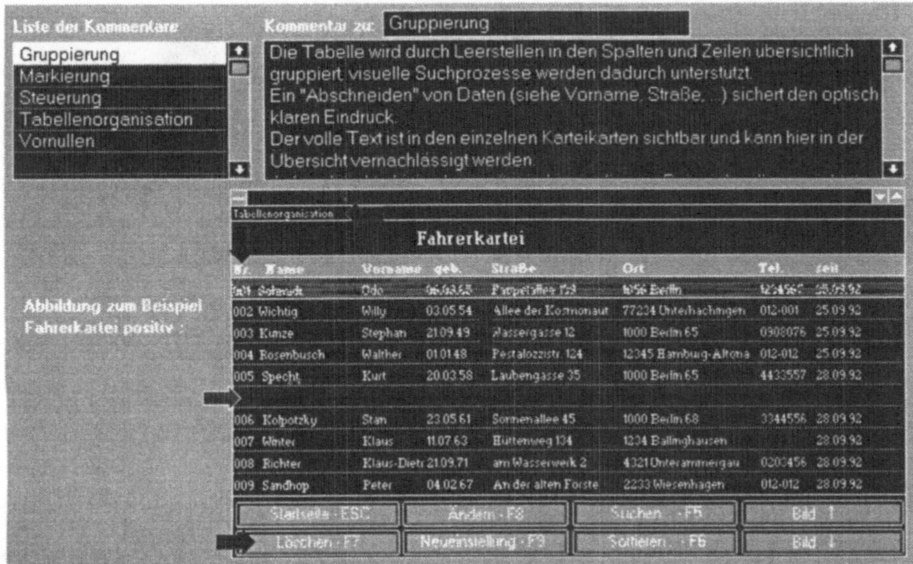

Abb. 2: Ausschnitt aus dem Beispielteil des Hypertextsystems. Die Kommentare im Feld
oben rechts beziehen sich auf einzelne Aspekte die mit den Pfeilen gekennzeichnet
sind.

Versetzt in die Lage des Benutzers, müssen die Entwickler mit der vorhandenen
Software arbeiten, so bemerken sie gestalterische Mängel der
Benutzungsschnittstelle, die sie ansonsten übersehen oder als trivial abtun würden.
Viele Mängel werden erst deutlich, wenn die Aufgabenlösung zwar prinzipiell
möglich - aber eben nicht angemessen erfolgen kann. In der Bearbeitung von
alltäglichen, einfachen Aufgaben soll der Anwendungs-Entwickler im
selbstgesteuerten Lernprozeß die Vorteile einer aufgaben- und benutzerorientierten
Funktionalität und Oberflächengestaltung bei der konkreten Anwendung erleben.

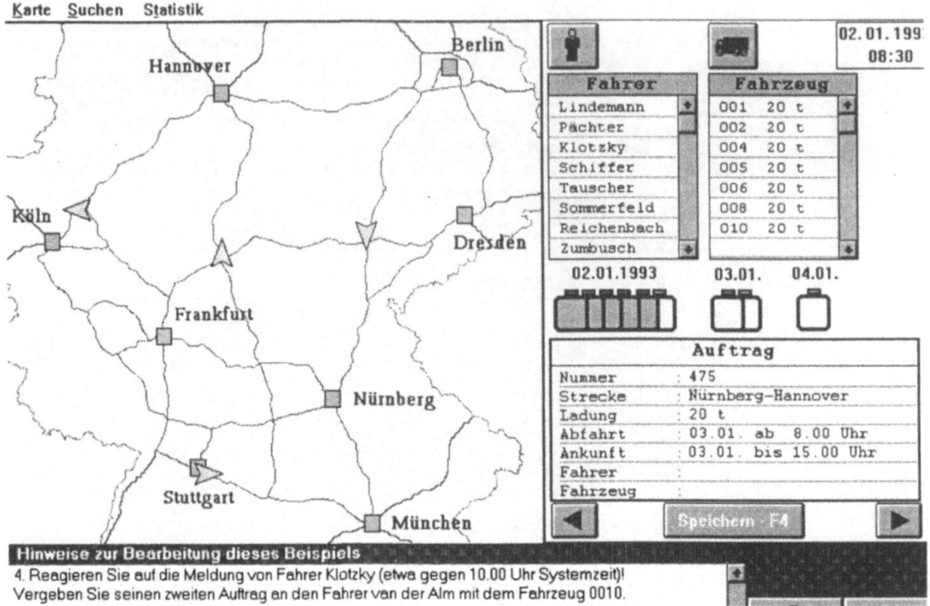

Abb. 3: Ausschnitt aus der lauffähigen Version des Beispiels „Dispatcher positiv". Im unteren Teil der Abbildung befindet sich eine Bearbeitungs-Aufgabe.

Resonanz und weitere Arbeitsschritte

Einen breiten Personenkreis konnten wir als Aussteller auf der CeBit'94 in Hannover erreichen. Die Resonanz war sehr positiv. Vor allem die Art der Wissensdarstellung mit einem Hypertextsystem weckte Interesse, das System zu testen bzw. zu erwerben. Das erhebliche Wissensdefizit war unseren Gesprächspartnern bekannt. Der Entwicklungsprozeß von *inra* schloß im Sinne eines iterativen Prototyping Evaluationsschritte ein. Studenten und Software-Entwickler testeten Teile vom *inra*, die Ergebnisse wurden in die aktuelle Version eingearbeitet. Nach dem vorläufigen Abschluß der Entwicklungsarbeiten ist eine stärker formative Evaluation vorgesehen. Das Unterstützungsmittel *inra* soll auf seine Gebrauchstauglichkeit am Arbeitsplatz getestet werden. Dabei soll die Qualität und Alltagstauglichkeit der einzelnen Teile und des *inra*-Systems insgesamt geprüft werden. Begleitbuch und Prototyp sollen Entwicklern einige Wochen an ihrem Arbeitsplatz zur Verfügung stehen. Durch die Variation von Anwendungsbedingungen und durch die Kombination von *inra* mit Methoden der Usability-Forschung erwarten wir Aussagen zur Wirkungsweise, zur Akzeptanz und zur Effizienz. Nach ersten Test-Erfahrungen mit der Software in der IQ-Media Berlin lassen sich zwei Benutzertyten erkennen: der „neugierige Praktiker" und der „zielgerichtete Analytiker". Beide verfolgen unterschiedliche Ziele und benutzen den *inra* auf verschiedene, charakteristische Weise (Adrian, 1994).

Literatur

[1] Adrian, V. (1994). Praxiseinsatz des Informations- und Beratungssystems *inra* in der IQ-Media GmbH. Vortrag auf dem WEDA-Abschlußworkshop am 7.10.94 in Stuttgart. (Unveröff.)

[2] Asymetrix (1991). ToolBook - Benutzerhandbuch. Bellevue: Asymetrix Corporation.

[3] Balzert, H. (1993). Der JANUS-Dialogexperte: Vom Fachkonzept zur Dialogstruktur. Vorabdruck zum Vortrag auf der Softwaretechnik-Tagung 1993 in Dortmund. (Unveröff.)

[4] Beimel, J., Hüttner, J. & Wandke, H. (1993). Kenntnisse von Programmierern auf dem Gebiet der Software-Ergonomie: Stand und Möglichkeiten zur Verbesserung. In A. Gebert & U. Winterfeld (Eds.) Arbeits- Betriebs- und Organisationspsychologie vor Ort. Bonn: Deutscher Psychologen Verlag.

[5] Cakir, A. (1991). Software-Ergonomie und Arbeitsorganisation - neue Regelungs-gegenstände im Arbeitsschutz in Europa 1992 - Was bringen die Europäischen Regelwerke für Bildschirmarbeitsplätze. Tagungsband, Berlin: Ergonomic Institut.

[6] Dzida, W. (1984). Das IFIP-Modell für Benutzerschnittstellen. Office-Management, Sonderheft.

[7] Hüttner, J. & Wandke, H. (1993) What do system designers know about software ergonomics and how to improve their knowledge? In H. Luczak, A. Cakir & G. Cakir (Eds.) Work With Display Units 92 (S. 304-308). Amsterdam u.a.: Elsevier

[8] Hüttner, J., Wandke, H. & Rätz, A. (1994). Benutzerfreundliche Software - Psychologisches Wissen für die ergonomische Schnittstellengestaltung. (in. Vorb.) Berlin: Bernd-Michael Paschke Verlag.

[9] Maaß,S., Ackermann,D., Dzida,W., Gorny,P., Oberquelle,H., Rödiger,K.-H., Rupietta,W. und Streitz, N. (1993): Software-Ergonomie-Ausbildung in Informatik-Studiengängen bundesdeutscher Universitäten. Empfehlungen des Fachausschusses 2.3 und des Fachbereichs 2 der GI. Informatik-Spektrum 16, 25-30.

[10] Mayew,D.J. (1992). Principles and guidelines in software user interface design. Englewood Cliffs : Prentice Hall.

[11] Moran, T.P. (1981): The command language grammar: a representation of the user interface of interactive computer systems. Int. J. Man-Machine Studies 15, S. 3 -50.

[12] Myers, B. & Rosson, M. (1992). Survey of User Interface Programming. In CHI-Proceedings, S. 195-202.

[13] Oberquelle, H. (1994) Software-Ergonomie lehren - Software-Ergonomie lernen. Ergonomie und Informatik 22, 24-26.

[14] Paul, H. (1993). Das Explorative Modell als konzeptioneller Ansatz zur Gestaltung interaktiver Systeme. In K.-H. Rödiger (Hrsg.) Software-Ergonomie 93 - Von der Benutzungsoberfläche zur Arbeitsgestaltung. (S. 77- 86) Stuttgart: Teubner.

[15] Paul, H. (1994). Exploratives Agieren - Ein Beitrag zur ergonomischen Gestaltung interaktiver Systeme. Ergonomie und Informatik, 22, 49-52.

[16] Reiterer, H., Oppermann, R. und Bleimann, U. (1992) Integration von software-ergonomischem Gestaltungswissen in ein User Interface Management System - User Interface Design Assistance (IDA). internes Projektpapier (unveröff.).

[17] Smith, L.S. & Mosier, J.N. (1986). Guidelines for designing user interface software. Bedford: The Mitre Corporation.

Hartmut Wandke & Jens Hüttner
Humboldt-Universität zu Berlin
Institut für Psychologie
Oranienburger Straße 18
10178 Berlin
e-mail: hwandke@psychologie.hu-berlin.de

Aufgabenbezogene Dialogstrukturen für Informationssysteme

Jürgen Ziegler, Christian Janssen

Zusammenfassung

Aufgabenangemessene Gestaltung von Dialogstrukturen ist ein wesentliches Ziel der Software-Ergonomie. Die Dialogstruktur soll eine effektive und effiziente Navigation des Benutzers im System ermöglichen. In diesem Beitrag wird dargestellt, daß existierende Ansätze der Aufgabenanalyse auf einzelne Perspektiven von Aufgaben beschränkt sind. Sie sind deshalb nicht für eine Beurteilung von Dialogstrukturen bei unterschiedlich ausgerichteten oder situativ stark variierenden Zielsetzungen des Benutzers geeignet, wie sie bei komplexen, ganzheitlichen Tätigkeiten typisch sind. Es wird ausgeführt, daß Aufgabenanalyse deshalb nicht auf die Untersuchung oder Modellierung einzelner Aufgaben oder Abläufe beschränkt bleiben kann, sondern zu einer abstrahierten Darstellung möglichst aller potentiell entstehenden Ziele des Benutzers kommen muß. Eine solche Darstellung wird als *Aufgabenraum* bezeichnet. Diesem werden eine Reihe von Grundstrukturen von Dialogen gegenübergestellt, die sich mit graphischen Benutzungsschnittstellen realisieren lassen. Die Eignung verschiedener Grundstrukturen von Dialogen für verschiedene Dimensionen des Aufgabenraums wird diskutiert. Die Folgerungen für die Systemgestaltung werden im Hinblick auf ein integriertes Gestaltungsvorgehen dargestellt.

1. Aufgabenangemessenheit und Systemgestaltung

Die Forderung nach einer aufgabenangemessenen Gestaltung von Informationssystemen stellt eines der Kernprinzipien der Software-Ergonomie dar und hat Eingang in alle wesentlichen Richtlinien und Normen gefunden [3, 5, 12, 23]. So fordert z. B. ISO [12] unter dem Aspekt der Aufgabenangemessenheit die effektive und effiziente Unterstützung des Benutzers bei der Aufgabenerfüllung.

Die konkrete Umsetzung dieser Forderung beim Entwurf von interaktiven Informationssystemen ist allerdings nach wie vor eine Problemstellung, für die bestenfalls in Einzelbereichen Lösungen vorhanden sind. Insbesondere zeigt sich, daß Aufgabenangemessenheit auf den unterschiedlichen Abstraktionsebenen eines Systems mit unterschiedlichen Gestaltungsaspekten in Verbindung zu bringen ist (vgl. [27]). Ein integriertes Vorgehen zur Realisierung aufgabenorientierter Systeme auf unterschiedlichen Ebenen liegt nicht vor.

Ein häufig verwendeter Ansatzpunkt besteht darin, Aufgabenstrukturen und -abläufe, die bei einer Tätigkeit vorgefunden werden, im System mehr oder weniger starr abzubilden. Die Leitvorstellung dabei ist die einer Hierarchie von Zielen oder Funktionen, die durch eine Zerlegung übergeordneter Ziele erhalten wird. Die vorhandenen Analyse- und Modellierungsmethoden unterscheiden sich insbesondere nach dem jeweiligen Analyseziel: Psychologisch ausgerichtete

Methoden wie GOMS [2] oder CCT [17] versuchen, mentale Repräsentationen von Aufgaben abzubilden. Software-technische Verfahren wie Strukturierte Analyse (s. z. B. [26]) sind auf die Erfassung benutzerunabhängiger „logischer" Anforderungen an die Systemfunktionalität ausgerichtet. Die mangelnde Verbindung dieser beiden Ansätze wird von verschiedenen Autoren festgestellt [4, 8].

Ansätze zu einer verbesserten Integration dieser Ansätze finden sich zum einen hinsichtlich der Einbeziehung von Benutzungsaspekten in das Entwicklungsvorgehen (s. z. B. [10, 24]), zum anderen im Bezug auf eine stärkere methodische Integration [16 , 22] für eine funktional orientierte Sichtweise. Für die Gestaltung objektorientierter Benutzungsschnittstellen wurden in den letzten Jahren Ansätze zur Ableitung von Dialogstrukturen aus softwaretechnischen Daten- bzw. Objektmodellen entwickelt [1, 7].

Ein gemeinsames Merkmal und zugleich Problem der genannten Methoden ist, daß sie jeweils *einen einzelnen Aspekt* der zu unterstützenden Aufgabe in den Vordergrund stellen, z. B. entweder das funktionale Ziel oder das Objekt der Aufgabe. Entsprechend führen die Ansätze zu funktional orientierten oder zu objektorientierten Grundstrukturen der Benutzungsschnittstelle [28].

Wie Hartson & Boehm-Davis [9] feststellen, liefern Aufgabenanalysemethoden immer nur einen beschränkten Ausschnitt des realen Aufgabenspektrums. Als Folge werden nur bestimmte Aufgabenstellungen in der realen Nutzungssituation wirklich effektiv unterstützt, andere Aufgaben können gar nicht oder nur mit Zusatzaufwand erfüllt werden. Dies ist besonders bei Tätigkeiten zu erwarten, die sich durch ein umfangreiches Aufgabenspektrum und hohe Variabilität und Flexibilität auszeichnen, Aufgaben also, die heute sowohl aufgrund neuer Organisationsansätze wie auch arbeitswissenschaftlicher Gesichtspunkte häufig gefordert werden.

In dem vorliegenden Beitrag soll ausgeführt werden, daß die Aufgabenanalyse für die Unterstützung solcher komplexer und flexibler Tätigkeiten nicht auf die Untersuchung oder Modellierung einzelner Aufgaben oder Abläufe beschränkt bleiben kann, sondern zu einer abstrahierten Darstellung möglichst aller potentiell entstehenden Ziele des Benutzers kommen muß. Eine solche Darstellung soll hier als *Aufgabenraum* bezeichnet werden. Diesem Aufgabenraum werden eine Reihe von Grundstrukturen von Dialogen gegenübergestellt, die sich mit graphischen Benutzungsschnittstellen realisieren lassen. Die Eignung verschiedener Grundstrukturen von Dialogen für verschiedene Dimensionen des Aufgabenraums wird diskutiert. Zunächst werden hierfür die verschiedenen Gestaltungsbereiche eines interaktiven Systems im Hinblick auf ein integriertes Gestaltungsvorgehen dargestellt.

2. Auf dem Weg zu einer integrierten Entwicklungsmethodik

In einer integrierten Entwicklungsmethodik wird zum einen ein Vorgehensmodell benötigt, das durch Benutzerbeteiligung eine ausreichende Aufgaben- und Benutzerorientierung gewährleistet, wobei in der Analyse und Anforderungsdefinition neben den software-technischen Schritten der Aufgaben- und Datenmodellierung insbesondere Aufgabengestaltungsschritte vorzusehen sind [1]. Ein Gesamtmethodenmodell ist in Abb. 1 dargestellt.[1] In der Analyse entsteht das essentielle Aufgabenmodell des Systems, das noch unabhängig von den verwendeten Technologien ist, also z. B. noch keine Aussage darüber enthält, welche Aufgaben vom Benutzer und welche vom Computersystem durchgeführt werden [18]. Parallel wird ein grobes Objektmodell (Objektabgrenzung und Objektstruktur) entwickelt.

Für die *Anwendungsspezifikation* (d.h. Anforderungsdefinition für die Anwendung) wird das essentielle Aufgabenmodell verfeinert, indem die essentiellen Aufgaben in Teilaufgaben aufgeteilt werden. Die Verfeinerung von Aufgaben und Teilaufgaben ist beendet, wenn alle Teilaufgaben eindeutig dem Benutzer oder dem Computersystem zugeordnet werden können oder Dialogaufgaben sind, deren Verfeinerung erst während des Entwurfs der Dialogabläufe stattfindet. Das bisherige Objektmodell wird verfeinert, indem die Attribute vervollständigt werden. Aufgrund der erfolgten Mensch-Rechner-Funktionsteilung werden die Funktionen der Objekte ermittelt.

Als verbindendes Glied zwischen Analyse und Detail-Spezifikation ist ein konzeptioneller Entwurf erforderlich, der zum Ziel hat, ein zu entwerfendes Informationssystem aus der Sicht der Benutzung zu beschreiben. Das hierbei entstehende konzeptionelle Benutzungsschnittstellenmodell (*Benutzungsmodell*) besteht aus Objekten, Funktionen und Objektstrukturen, mit denen die funktionalen Elemente der Anwendung an der Benutzungsschnittstelle abgebildet werden. Diese benutzungsorientierte „Architektur" der Benutzungsschnittstelle repräsentiert bereits Entwurfsentscheidungen hinsichtlich der Systemstruktur, ist aber noch frei von Annahmen über Realisierungsmittel, also hinsichtlich Darstellungsarten und Interaktionstechniken. Sie bildet einen sinnvollen Ausgangspunkt für systematisches

---

[1] Diese Arbeit basiert auf Ergebnissen des Projektes "TASK - Technik der aufgaben- und benutzerangemessenen Software-Konstruktion", das vom Projektträger "Arbeit und Technik" des BMFT gefördert wurde (Förderkennzeichen 01 HK 849 A2).

Prototyping, das begleitend zum Entwurf als Explorations- und Validierungsmittel
eingesetzt werden sollte (vgl. [6].

Hinsichtlich einer einfachen Realisierung von Benutzungsschnittstellen sind - sozusagen am unteren Ende eines Gesamtvorgehens - geeignete Spezifikations- und
Generierungsmethoden erforderlich, die ausführbare Beschreibungen für existierende Benutzungsschnittstellenwerkzeuge erzeugen. Hierbei können die Einzel-
aspekte der Erzeugung statischer *Sichten* (d.h. Bildschirmdarstellungen von konzeptionellen Objekten) und der Definition der Dialogdynamik unterschieden
werden, wofür bereits Methoden beschrieben wurden. (Siehe [25] zur Maskengenerierung, [13] zur Dialogspezifikation, [14, 15] zur Beschreibung des Gesamt-
zusammenhangs der Realisierungsmethode).

In den folgenden Abschnitten werden zunächst die für den konzeptionellen Benut-
zungsschnittstellenentwurf relevanten Aufgabendimensionen aufgezeigt. Danach
werden Klassen von Dialogstrukturen in Benutzungsschnittstellen, speziell graphischen Benutzungsschnittstellen definiert und mögliche Einsatzfälle diskutiert.

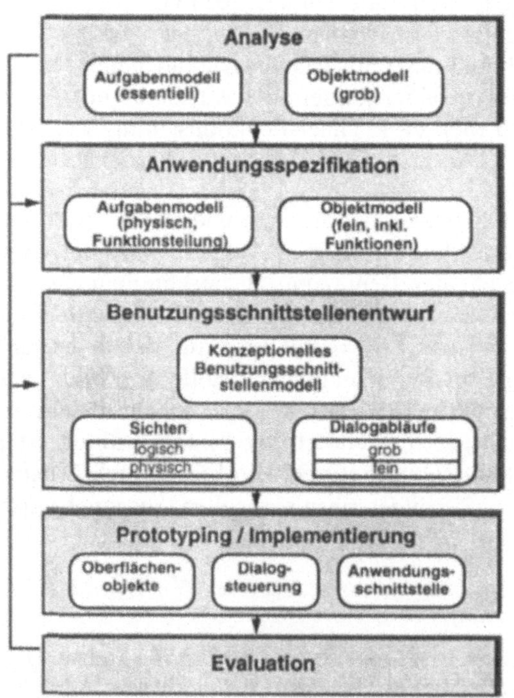

Abb. 1: Methodenmodell für die Entwicklung interaktiver Systeme

3. Das Konzept des Aufgabenraums als Basis für die Definition von Dialogstrukturen

Aufgabenanalysemethoden wie GOMS liefern an den funktionalen Zielen des Benutzers orientierte Beschreibungen. Diese Ziele sind zwar hierarchisch zerlegbar, in sich aber in der Regel nicht weiter strukturiert. Der Ansatz der Task-Action Grammar [19] bietet demgegenüber die Möglichkeit, Aufgabenparameter für eine weitergehende Abstraktion von den Einzelaufgaben zu nutzen und kann Aufgaben durch eine Kombination von Parametern beschreiben.

In den integrierten softwaretechnischen Vorgehensweisen werden mehrere Perspektiven auf das System nebeneinander modelliert, wobei die unterschiedlichen Modelle durch Regeln oder Werkzeuge konsistent gehalten werden müssen. Typisch ist eine Daten-, Funktions- und Dynamiksicht (Verhalten) auf das zu entwickelnde System, die allerdings getrennt voneinander vorliegen. Die Zusammenführung dieser unterschiedlichen Sichten an der Benutzungsschnittstelle stellt ein ungelöstes Problem dar. Dies gilt im wesentlichen auch für objektorientierte Analyseverfahren (vgl. [21]), wobei diese allerdings eine stärkere Integration des Objekt- und Funktionsaspekts liefern.

Es soll deshalb nun näher untersucht werden, durch welche unterschiedlichen Grundkomponenten Aufgaben definiert werden können. Diese bilden jeweils den Hauptausgangspunkt für die Zielsetzung des Benutzers. Bereits angesprochen wurde die mit einer Aufgabe verbundene *Funktion* im Sinne eines funktionalen Ziels. Eine typische Zielsetzung des Benutzers wäre z. B.: 'Ich möchte einen Brief schreiben'.

Eine andere Sichtweise ist der Ausgangspunkt vom Gegenstand bzw. *Objekt* der Aufgabe: 'Ich möchte die Akte der Firma XY sehen, um evtl. weitere Kontakte aufzunehmen'. Hier bildet ein Objekt die Invariante der Aufgabe, während das funktionale Ziel nicht oder nur unscharf definiert ist. Dies ist die typische Form bei objektorientierten Benutzungsschnittstellen, bei der Interaktionen mit der Identifikation eines Objekts beginnen.

Bei vielen Aufgaben bildet nicht die Identität, sondern der *Zustand* von Objekten den Ausgangspunkt für die Bearbeitung. Eine typische Fragestellung wäre: 'Ich will alle Rechnungen im Zustand ‹überfällig› bearbeiten' oder 'Liste alle Kunden im Raum Stuttgart auf'. Diese Konzepte werden seit langem bei Datenbankanwendungen verwendet, allerdings ist zu beachten, daß die interessierenden Zustände sehr variabel sein können, so daß ein flexibilisierter Zugriff in vielen Fällen erforderlich ist. Zustandsorientierte Zugriffe sind hilfreich, um den Ablauf von Aufgaben zu unterstützen, da alle Objekte eines gleichen Zustands gleichartige Folgebearbeitungen erlauben können.

Weitere aufgabenbestimmende Merkmale können die *Zeit* ('Ich will alle Vorgänge sehen, die heute zu bearbeiten sind') oder der *Ort* sein. Ortsbezogene Aufgaben treten insbesondere bei räumlich verteilten Systemen auf und spielen in der herkömmlichen DV keine Rolle. Bei einem integrierten elektronischen Haussystem etwa können aber durchaus Fragen auftreten wie 'Ich will den Zustand aller Geräte in der Küche überprüfen'.

Schließlich ist zu beachten, daß es beliebige, komplexe Zusammenfassungen von Objekten zu *Vorgängen* gibt, die häufig über längere Zeiträume hinweg bearbeitet werden und vorgangsorientierte Zielsetzungen des Benutzers auslösen können ('alle Objekte, die mit dem Lebensversicherungsantrag von X zusammenhängen'). Abbildungen solcher Konzepte finden sich z. B. in Workflow-Management-Systemen.

Führt man diese unterschiedlichen Ausgangspunkte für Aufgaben zusammen, so lassen sich Aufgaben als eine Kombination der Grunddimensionen

<Funktion, Objekt, Zustand, Zeit, Ort>

auffassen. Diese Dimensionen spannen folglich einen *Aufgabenraum* auf, der die einzelnen Aufgabeninstanzen beinhaltet. In jeder konkreten Aufgabeninstanz sind die Dimensionen mit spezifischen Ausprägungen belegt, wobei Dimensionen auch unbelegt bleiben können. Eine Aufgabe ‹Dokument Y löschen› z. B. wird durch eine Kombination aus einer Funktions- und einer Objektkomponente gebildet. Der Benutzer wird nun situationsabhängig die einzelnen Komponenten mit einer unterschiedlichen Gewichtung belegen, so wird im Zusammenhang mit dem Löschen alter Dokumentversionen die Funktion das größere Gewicht besitzen. Ein Funktionsmodus 'Löschen' wäre hier eine effektive Unterstützung, wobei natürlich sorgfältig gegenüber anderen Kriterien wie Konsistenz oder Erlernbarkeit abgewogen werden muß.

Die zielbestimmenden Aufgabendimensionen können also vom Benutzer situativ nicht nur mit verschiedenen Werten sondern auch mit unterschiedlichen Gewichten belegt werden. Entsprechend der Dimension, die jeweils das stärkste Gewicht hat, sind verschiedene Dialogstrukturen der Benutzungsschnittstelle geeignet, um das Aufgabenziel am effektivsten erreichen zu können. Solche Dialogstrukturen werden im nächsten Abschnitt beschrieben. Neben der variierenden Gewichtung der Dimensionen müssen für den Entwurf noch weitere, aus dem Arbeitskontext abzuleitende Merkmale berücksichtigt werden. Hierzu zählt insbesondere die erwartete Häufigkeit einer bestimmten Zielausprägung. Für selten auftretende Ausprägungen wird man häufig spezifische Unterstützungsfunktionen zugunsten einer geringeren Systemkomplexität weglassen.

Das Konzept des Aufgabenraumes erlaubt es, potentielle Zielsetzungen des Benutzers in systematischer Weise zu synthetisieren und nicht nur existierende Konzepte oder Abläufe zu analysieren und ggf. zu variieren. Somit kann eine beliebige Vielfalt an Szenarien generiert werden, für die jeweils die Eignung des zu entwerfenden Systems überprüft werden kann.

4. Dialogstrukturtypen bei graphischen Benutzungsschnittstellen

Im folgenden wird gezeigt, wie aus dem Aufgabenmodell in systematischer Weise angemessene Dialogstrukturen abgeleitet werden können. Das Benutzungsmodell (vgl. Abschnitt 2) zeigt durch die Definition von Zugriffspfaden die wesentlichen Navigationsmöglichkeiten des Benutzers im System im Sinne der Erreichbarkeit auf, ohne die Dynamik der Dialogabläufe im einzelnen festzulegen. Die möglichen Navigationspfade zu einer zu bearbeitenden Sicht, z. B. einem Formular zur Bearbeitung der Attribute eines Objektes, können hinsichtlich der unterschiedlichen Arten von „Einstiegspunkten" differenziert werden. Dabei können bei graphischen Benutzungsschnittstellen analog zu den im vorigen genannten Aufgabendimensionen folgende wesentliche Fälle auftreten:

- *Objektorientierte Navigation* beginnt mit der Auswahl einer bestimmten Objektklasse und identifiziert dann in einem oder mehreren Eingrenzungsschritten das interessierende Einzelobjekt oder die Objektmenge. Im Einstiegsfenster werden dabei die von der Aufgabenstellung her primär relevanten Objekttypen als Sammelobjekte angeboten. Erst nach der Identifikation der Einzelobjekte oder Objektmenge wird die beabsichtigte Bearbeitungsfunktion im Kontext einer Objektsicht ausgelöst. Weitere Dialogpfade ergeben sich, sofern aufgabenabhängig benötigt, entlang der Relationen im Objektmodell.

- Ein *zustandsorientierter Zugriff* ist dadurch gekennzeichnet, daß eine Sicht alle Objekte eines Typs repräsentiert, die bestimmte Merkmale erfüllen bzw. sich in einem bestimmten Bearbeitungszustand befinden. Diese Möglichkeit wird bei gegenwärtigen graphischen Benutzungsschnittstellen noch selten eingesetzt; sie bietet aber die Möglichkeit für die Realisierung aufgabengerechter Abläufe.

- Bei *funktionsorientierter Navigation* wird im ersten Schritt eine Bearbeitungsfunktion bestimmt, ohne die Objektmenge oder das Objekt vorher näher zu spezifizieren. Hier bildet jeweils eine bestimmte Aufgabe den Einstiegspunkt für einen Dialog.

- Eine *vorgangsorientierte Navigation* verwendet als Startpunkt ein Strukturierungsobjekt, das alle Objekte beinhaltet, die für einen bestimmten Geschäftsprozeß verwendet werden. Dabei können alle drei bereits aufgeführten Zugriffsarten innerhalb dieses Vorgangsobjektes für den weiteren Bearbeitungsweg eingesetzt werden. Wird ein Vorgang selbst wieder als Objekt gesehen, ergibt sich hier eine Spezialform der objektorientierten Navigation.

Für graphische Benutzungsschnittstellen ist die *objektorientierte Dialogstrukturierung* als Grundmuster typisch, die dem Prinzip der Direkten Manipulation [20]) entspricht. Diese Form eignet sich besonders für benutzergesteuerte Aufgabenstellungen, z. B. das gezielte Bearbeiten einer bestehenden Adresse eines Versicherten (Abb. 2), und gewährleistet eine hohe Flexibilität, was die Aufgabenabläufe betrifft. Der einzelne Datensatz eines Versicherten wird dabei über einen Selektionsdialog (z. B. „Versicherten Öffnen") aufgerufen (Abb. 3) und erst sekundär die Bearbeitungsfunktionen aufgerufen.

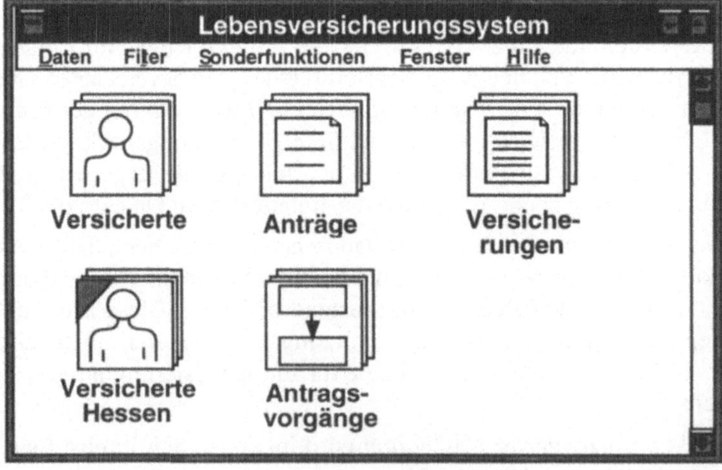

Abb. 2: Beispiel für ein Einstiegsfenster

Für eine zustands- oder merkmalsabhängige Strukturierung des Objektraumes eignen sich sogenannte *Filterobjekte*, die - fest eingestellt oder benutzerdefiniert - eine Gesamtobjektmenge nach einem bestimmten Kriterium einschränken. Ein Beispiel in Abb. 2 ist das Objekt „Versicherte Hessen", das z. B. zuständigkeitsabhängig definiert sein kann, und die Suche in der Gesamtmenge der Versicherten (über einen Suchdialog, siehe Abb. 3) erleichtert.

Im Gegensatz zu der freien Navigation im Objektraum sind für manche Aufgabenstellungen stärker geführte, *funktionale Dialoge* wünschenswert, etwa wenn formalisierte Abläufe vorliegen (vgl. Greutmann [7]). Ein Beispiel bei Finanzdienstleistern ist die Durchführung des Jahresabschlusses. Dieser sollte direkt als Funktionsaufruf (z. B. im Menü des Hauptfensters) ausgeführt werden und ggf. mit Parametern versorgt werden können.

Vorgangsobjekte können dazu dienen, bei komplexeren Aufgaben die Überwachung und Prüfung von Konsistenzbedingungen gegenüber der rein objektorientierten Navigation zu vereinfachen. Beispielsweise werden bei der Beantragung einer Lebensversicherung zunächst die Daten für den Versicherten und die Antragsdaten erfaßt und anschließend eine Gesundheitsprüfung für die zu

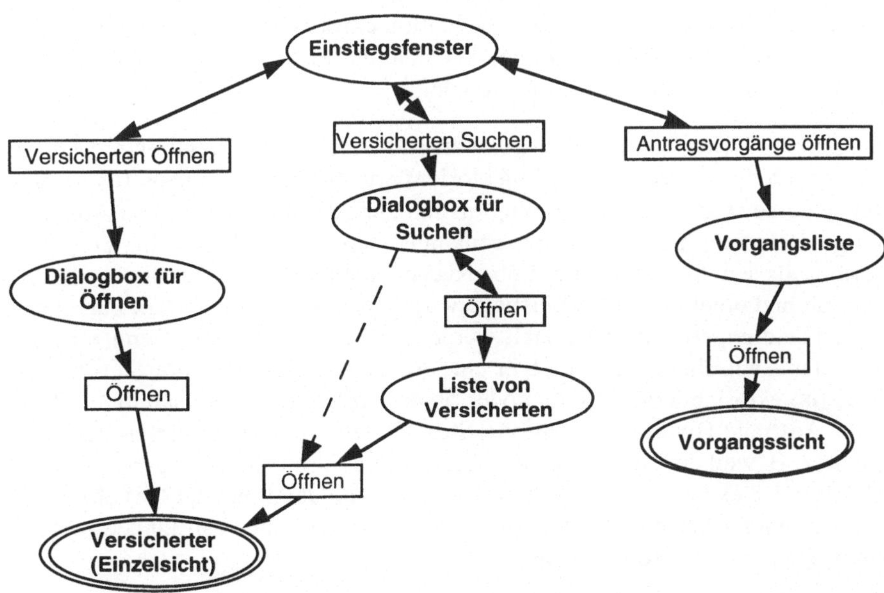

Abb. 3: Grundmuster für objektorientierte Dialogabläufe unter Einschluß von Vorgangsobjekten
(Dialognetz nach [13])

versichernde Person angefordert, bevor schließlich der Vertrag ausgestellt wird. Die Überwachung offener Vorgänge kann in einem entsprechenden Vorgangsobjekt erfolgen, das über eine Liste zugänglich ist. Beispielsweise kann für „Antragsvorgänge" in Abb. 2 eine solche Vorgangsliste und aus der Liste heraus die einzelnen Vorgänge geöffnet werden (Abb. 3). Die Teilobjekte für Antragsvorgänge (z. B. „Versicherter", „Vertrag") sind dann über die Vorgangssicht aufrufbar (der Unterdialog kann aus Platzgründen nicht gezeigt werden).

5. Einsatzerfahrungen und Schlußfolgerungen

Das hier dargestellte Konzept eignet sich für den Entwurf interaktiver Informationssysteme, mit denen vorwiegend Verwaltungsaufgaben bearbeitet werden, beispielsweise im Versicherungs- oder Produktionsplanungsbereich. Bei solchen betrieblichen Informationssystemen sind heute im Gegensatz etwa zu PC-Standardsoftware häufig noch funktionsorientierte Dialogstrukturen mit entsprechenden Einschränkungen der Flexibilität anzutreffen. Beim Übergang auf graphische Benutzungsschnittstellen werden nun auch zunehmend objektorientierte Strukturen eingesetzt, obwohl dieser Übergang für viele Entwickler noch nicht selbstverständlich ist. Die objektorientierte Strukturierung von Dialogen in graphischen Benutzungsschnittstellen hat sich als leicht erlernbare und flexible Grundstruktur in Anwendungsprojekten bewährt. Die Evaluation eines Prototypen für eine Anwendung im Versicherungsbereich zeigte, daß sich die Benutzer im System sofort zurechtfanden, und Testaufgaben lösen konnten.

Analog zu der Kritik von Hutchins, Hollan und Norman [11] an Direkter Manipulation jedoch wird über die rein objektorientierte Navigation hinaus für manche Aufgaben eine stärkere Unterstützung bis hin zur Automatisierung von Aufgaben benötigt. Beispiele hierfür sind der genannte Jahresabschluß und die Vorgangssteuerung für den Abschluß einer Lebensversicherung. Werden parallel funktionsorientierte und objektorientiert Einstiege vorgesehen, kann in manchen Fällen eine größere Handlungsflexibilität erzielt werden, etwa wenn sich der Benutzer eines PPS-Systems entscheiden kann, ob er eine Auftragsbestätigung sofort ausdrucken möchte (objektorientierte Struktur) oder ob er den Sammeldruck sämtlicher bearbeiteter Aufträge (funktionsorientierte Struktur) zu einem bestimmten Zeitpunkt vorzieht, z. B. weil der Drucker gerade mit anderen Formularen belegt ist. Das Beispiel zeigt, daß es bei der Dialogstrukturierung - und damit bei der Gestaltung des Aufgabenraums - keinen "one-best-way" gibt, sondern unter Umständen mehrere Dialogwege parallel anzubieten sind.

Insgesamt trägt der vorliegende Beitrag zum Schließen der Lücke zwischen Software-Ergonomie und Software-Technik bei, indem der Entwurfsraum für die Dialoggestaltung durch die Charakterisierung der wesentlichen Dialogtypen in Abhängigkeit von den Aufgabendimensionen strukturiert wird. Weiterführende anstehende Arbeiten erstrecken sich auf die vollständigere Erprobung des Entwurfskonzeptes in konkreten Projekten. Ferner ist die Übertragung des Ansatzes auf Bereiche außerhalb der betrieblichen Informationssysteme geplant, etwa auf den Bereich von Steuerungssystemen im häuslichen Bereich.

Literatur

[1] Beck, A.; Janssen, Ch. (1993): Vorgehen und Methoden für aufgaben- und benutzerangemessene Gestaltung von graphischen Benutzungsschnittstellen. In: Coy, W. u.a.: Menschengerechte Software als Wettbewerbsfaktor. Forschungsansätze und Anwenderergebnisse aus dem Programm "Arbeit und Technik", Stuttgart: Teubner, 200-221.

[2] Card, S.K.; Moran, T.P. & Newell, A. (1983): The Psychology of Human-Computer Interaction. Hillsdale, NJ: Lawrence Erlbaum.

[3] DIN (1988): DIN 66234, Teil 8: Bildschirmarbeitsplätze - Grundsätze ergonomischer Dialoggestaltung. Berlin: Beuth Verlag.

[4] Eberleh, E., Arend, U., Kasten, C., Strothotte, T., Ziegler, J. (1991): Integration software-ergonomischer Forschungsergebnisse in die betriebliche Software-Entwicklung. Diskussionsunterlagen. In: Ackermann, D., Ulich, E. (Hrsg.) Software-Ergonomie ´91. Stuttgart: Teubner, 141-151.

[5] EG (1990): Richtlinie des Rates vom 29. Mai 1990 über die Mindestvorschriften bezüglich der Sicherheit und des Gesundheitsschutzes bei der Arbeit an Bildschirmgeräten (90/270/EWG). Amtsblatt der Europäischen Gemeinschaften, Nr. L 156/14.

[6] Floyd, C. (1984). A systematic look at prototyping. In: Budde, R.; Kulenkamp, K.; Mathiassen, L.; Züllighoven, H. (Eds.). Approaches to Prototyping. Berlin, Heidelberg: Springer.

[7] Greutmann, Th. (1993): Datenmodellierung und aufgabengerechte Dialoge: ein Synchronisierungsproblem. In: Rödiger, K.-H. (Hrsg.): Software-Ergonomie ´93. Stuttgart: Teubner, 99-110..

[8] Hartson, H.R., Hix, D. (1989): Toward empirically derived methodologies and tools for human-computer interface development. International Journal of Man-Machine Studies 31, 477-494.

[9] Hartson, Rex H.; Boehm-Davis, Deborah (1993): User-Interface development processes and methodologies. Behaviour &Information Technology 12 (2), 98-114.

[10] Hoyos, C. Graf; Holz a.d.Heide, B; Ortlieb, S. (1993): Eine iterative Software-Entwicklungsstrategie mit gezielter Benutzerbeteiligung und systematischer Evaluation der Benutzungsfreundlichkeit. In: Frese, M., Kasten, Chr., Skarpelis, C., Zang-Scheucher, B. (Hrsg.). Software für die Arbeit von morgen. Berlin: Springer, 497-525.

[11] Hutchins, E.L., Hollan, J.D. und Norman, D.A. (1986): Direct Manipulation Interfaces. Norman, D.A. und Draper, S. (Hrsg.) User Centered System Design - New Perspectives on Human-Computer Interaction, Hillsdale, London: Lawrence Erlbaum, 87-124.

[12] ISO (1990): ISO Standard 9241: Ergonomic Requirements for Office Work with Visual Display Terminals. Part 10: Dialogue Principles. Draft International Standard. International Standards Organisation.

[13] Janssen, C. (1993): Dialognetze zur Beschreibung von Dialogabläufen in graphisch-interakti-
      ven Systemen. In: Rödiger, K.-H. (Hrsg.): Software-Ergonomie ´93. Stuttgart: Teubner, 67-76.

[14] Janssen, C., Weisbecker, A., Ziegler, J. (1993a): Generating User Interfaces from Data Models
      and Dialogue Net Specifications. INTERCHI´93 Proceedings. New York: ACM, 418-423.

[15] Janssen, Ch.; Weisbecker, A.; Ziegler, J. (1993b): Generierung graphischer Benutzungs-
      schnittstellen aus Datenmodellen und Dialognetz-Spezifikationen. In: Züllighoven, H.; Alt-
      mann, W.; Doberkat, E.-E. (Hrsg.) Requirements Engineering '93: Prototyping, Stuttgart:
      Teubner, 335-347.

[16] Johnson, P.; Drake, K.; Wilson, S. (1990): A Framework for Integrating UIMS and User Task
      Models in the Design of User Interfaces. in: Duce, D.A.; Gomes, M.R.; Hopgood, F.R.A.; Lee,
      J.R. (ed.) User Interface Management and Design, Berlin, Heidelberg, New York: Springer,
      203-216.

[17] Kieras, D. & Polson, P. (1985): An approach to the formal analysis of user complexity. Int. J.
      Man-Machine Studies, 22, 365-394.

[18] McMenamin, S.M.; Palmer, J.F. (1988): Strukturierte Systemanalyse. München, Wien: Carl
      Hanser Verlag.

[19] Payne, S.J. ; Green, T.R.G. (1986): Task-Action grammars: A model for the mental represen-
      tation of task languages. Human-Computer Interaction, 2, 93-133.

[20] Shneiderman, B. (1983): Direct Manipulation - A Step Beyond Programming Languages.
      IEEE Computer 16 (8), 57-69.

[21] Stein, W. (1994): Objektorientierte Analysemethoden - Vergleich, Bewertung, Auswahl.
      Mannheim u.a.: BI-Wissenschaftsverlag.

[22] Sutcliffe, A. G.; McDermott, M. (1991): Integrating methods of human-computer interface
      design with structured systems develepment. in: Int. J. Man - Machine Studies, Vol. 34, 631-
      655.

[23] VDI (1988): Software-Ergonomie in der Bürokommunikation. Verein Deutscher Ingenieure,
      Richtlinie 5005. Berlin: Beuth Verlag.

[24] Wassermann, Antony I.; Pircher, Peter A.; Shewmake, David T.; Kersten, Martin L. (1986):
      Developing Interactive Information System with the User Software Engineering Methodology.
      IEEE Transactions on Software Engineering12 (2) Feb. 1986, 326-345.

[25] Weisbecker, A. (1993): Integration von software-ergonomischem Wissen in die Systement-
      wicklung. In: Rödiger, K.-H. (Hrsg.): Software-Ergonomie ´93. Stuttgart: Teubner, 299-310.

[26] Yourdon, E. (1989): Modern Structured Analysis. Englewood Cliffs: Prentice Hall.

[27] Ziegler, J. (1988): Software-Ergonomie in der Bürokommunikation. VDI-Berichte 716, Düs-
      seldorf; VDI-Verlag, S. 141-159.

[28] Ziegler, J. (1993): Entwurf graphischer Benutzungsschnittstellen. In: Ziegler, J.; Ilg, R.
      (Hrsg.): Benutzergerechte Software-Gestaltung - Standards, Methoden und Werkzeuge. Mün-
      chen, Wien: Oldenbourg.

Jürgen Ziegler, Christian Janssen

Fraunhofer-Institut für Arbeitswirtschaft und Organisation (IAO)
Nobelstrasse 12c, D-70569 Stuttgart
Tel.: +49 711 970-2334, +49 711 970-2330, Fax: +49 711 970 2300
Email: Juergen.Ziegler@iao.fhg.de, Christian.Janssen@iao.fhg.de

Software-Ergonomie in
Banken und Versicherungen

Udo Bleimann, FH Darmstadt
Angelika Herzig, Schweizerische Kreditanstalt, Zürich

Zusammenfassung

Banken und Versicherungen entwickeln die intern benötigte Software in der Regel im eigenen Unternehmen. Auf diese Weise verfügen viele dieser Dienstleistungsunternehmen über EDV-Abteilungen, die von der Größe her mit mittelgroßen bis großen Software-Firmen verglichen werden können.

Im Workshop sollen nun Fragestellungen erörtert werden, die die Software-Ergonomie aus dem speziellen Blickwinkel der Banken/Versicherungen betrachten. Hierzu werden ca. zehn Kurzdarstellungen von verschiedenen Banken und Versicherungen in Form von Thesen gegeben, die Inhalte anschließend nach Bereichen strukturiert und dann schließlich gemeinschaftlich mit allen Teilnehmern diskutiert.

Der Workshop versucht ein wenig die Lücke zu schließen, die sich aus der realen Situation der Software-Entwickler bei Banken und Versicherungen einerseits und aus den wissenschaftlichen Erkenntnissen zur Software-Ergonomie andererseits zweifellos ergibt. Häufig bereiten die Fragen, die für den Wissenschaftler trivial, schon lange gelöst erscheinen, in der Praxis den Entwicklern große Probleme. So sind z.B. Style Guides - sei es firmenspezifisch oder einem Industriestandard angepaßt - bei den Firmen wohl gar nicht vorhanden bzw. gerade in Entwicklung, während sie aus theoretischer Sicht bereits überholt sind. Auch über Anwendungen, die sich an zeichenorientierten Terminals orientieren (3270 o.ä.) und deren Integration in die neue graphische Anwendungswelt, macht sich die Wissenschaft keine Gedanken mehr. In der realen Unternehmenssituation liegt hier aber ein Schwerpunkt der Arbeit.

Weitere Fragen, die im Workshop angesprochen werden sollen, sind etwa:
- Entwicklung von Graphical User Interfaces (Erfahrungen, Projekte)
- User Interface Standards: Lücken, Interpretationsspielraum, Geltungsbereich

- Applikationsentwicklung mit BenutzerInnen: Was klappt, was geht schief?
- Usability-Testing / Usability Labor
- Werkzeuge für Prototyping und Produktentwicklung
- organisatorische und ökonomische Aspekte (Stellenwert im Unternehmen etc.).

Zusammenfassend werden mit dem Workshop folgende Ziele verfolgt:

1. Es soll den TeilnehmerInnen die Gelegenheit geboten werden, sich über die spezifischen software-ergonomischen Belange der EDV-Entwicklung bei Banken und Versicherungen austauschen zu können.

2. Es soll aufgezeigt werden, mit welchen (ungelösten) Problemen sich Software-ErgonomInnen und Userinterface-DesignerInnen bei Banken und Versicherungen auseinandersetzen. (Diese Probleme könnten z.B. als Fragen an die Wissenschaft weitergegeben werden.)

3. Auch über den Workshop hinaus soll (sofern das Bedürfnis vorhanden ist) ein Forum zum Austausch von Informationen zwischen den einzelnen Banken/Versicherungen geschaffen werden.

Prof. Dr. Udo Bleimann

Fachhochschule Darmstadt

Fachbereich Informatik

Schöfferstr. 8 b

64295 Darmstadt

Tel: 06151 / 16-8418

Fax: 06151 / 16-8935

Umsetzung der EU-Richtlinie 90/270/EWG
Herausforderung für die Software-Ergonomie

Michael Burmester, Reiner Wieland-Eckelmann

Zusammenfassung

Seit 1. Januar 1993 gilt die EU-Richtlinie 90/270/EWE "... über Mindestanforderungen bezüglich der Sicherheit und des Gesundheitsschutzes bei der Arbeit an Bildschirmgeräten" in Europa und muß von den einzelnen Mitgliedsstaaten der europäischen Union umgesetzt werden. Die EU-Richtlinie schreibt ein hohes Maß an Sicherheit und Gesundheitsschutz bei der Arbeit an Bildschirmen fest. Arbeitgeber sind danach u.a. verpflichtet, Arbeitsplatzanalysen durchzuführen, hohe körperliche und psychische Belastungen auszuschließen und Endanwender bei der Gestaltung von Bildschirmarbeitsplätzen einzubeziehen.

Im einzelnen wird die Umsetzung des Arbeits- und Gesundheitsschutzes in Bezug auf die verwendete Hardware (Bildschirm, Tastatur, Arbeitstisch oder Arbeitsfläche, Arbeitsstuhl), die Umgebung (Platzbedarf, Beleuchtung, Reflexe und Blendung, Lärm, Wärme, Strahlung, Feuchtigkeit) und die Mensch-Maschine-Schnittstelle gefordert. Während es für die Hardware und die Umgebung bereits weitgehend definierte Anforderungen, Kriterien und Verfahren gibt, so ist die Situation bei der Umsetzung des Arbeits- und Gesundheitsschutzes in Bezug auf die Mensch-Maschine-Schnittstelle weniger klar umrissen. Ungeachtet dessen werden in der EU-Richtlinie weitreichende Forderungen bezüglich der Mensch-Maschine-Schnittstelle aufgestellt: Es wird die Anpassung der Software an die "auszuführende Tätigkeit", Benutzerfreundlichkeit und Adaptivität, benutzergerechte Informationsdarbietung und Rückmeldung über laufende Prozesse im System sowie die Anwendung der "Grundsätze der Ergonomie" auf die Software gefordert.

Der geplante Workshop soll die Thematik der Umsetzung der EU-Richtlinie aus der Perspektive der Software-Ergonomie beleuchten.

U.a. sollen folgende Themen während des Workshops diskutiert werden:
-   SANUS-Ansatz der Umsetzung der EU-Richtlinie in Deutschland (Verfahren, Hilfsmittel, Instrumente)
-   Zusammenhang der Gestaltung der Mensch-Maschine-Schnittstelle mit Belastung und Beanspruchung an Bildschirmarbeitsplätzen
-   ISO- und DIN-Normen als Basis zur Umsetzung der EU-Richtlinie
-   Arbeitsorganisation und Aufgabenangemessenheit
-   Vermittlung EU-Richtlinien-relevanter Inhalte der Software-Ergonomie in der Praxis

- Umsetzung der EU-Richtlinie schon während der Software-Entwicklung
- Prüfung der Konformität zur EU-Richtlinie.

Im Rahmen des vom Bundesminister für Forschung und Technologie geförderten Verbundvorhabens SANUS (Sicherheit und Gesundheitsschutz bei der Arbeit an Bildschirmen auf der Basis internationaler Normen Und Standards) sollen Hilfsmittel, Leitlinien und Methoden zur Umsetzung der EU-Richtlinie entwickelt und erprobt werden. Diese Verfahren werden in mehreren Pilotprojekten mit unterschiedlichen betrieblichen Profilen eingesetzt und so angepaßt, daß die Verfahren in der Praxis auch ohne Expertenwissen anwendbar sind. Darüber hinaus werden Strategien entwickelt, die die Aufklärung der einzelnen Mitarbeiter an Bildschirmarbeitsplätzen und die Weiterbildung von Projekt- und Ausbildungsveranwortlichen in Unternehmen sicherstellen.

Am Verbundprojekt SANUS beteiligen sich vier Forschungsinstitute: das Institut für Allgemeine Psychologie und Methodik der TU Dresden, die Fachgruppe Arbeitswissenschaft der TU Ilmenau, die Projektgruppe MenBIT der Bergischen Universität Wuppertal und das Institut für Arbeitswissenschaft und Technologiemanagement der Universität Stuttgart als Verbundkoordinator. Diese Institute arbeiten in enger Kooperation zusammen mit den Softwarehäusern ISA (Stuttgart) und ELK (Gladbach), den Beratungshäusern und Qualifizierungsexperten ATB (Chemnitz), ibek (Karlsruhe) und GSM (Stuttgart).

Der Workshop wird von zwei Vertretern des SANUS-Verbundprojektes organisiert: Dipl.Psych. Michael Burmester von der Universität Stuttgart (IAT) und Prof. Dr. Reiner Wieland-Eckelmann von der Bergischen Universität Gesamthochschule Wuppertal (Projektgruppe MenBIT). Bis zur Software-Ergonomie-Tagung werden Teilnehmer für den Workshop gewonnen, die im Gebiet der Software-Ergonomie tätig sind und zur Thematik der Umsetzung der EU-Richtlinie aus der Perspektive der Software-Ergonomie einen Diskussionsbeitrag leisten können und diesen auch als ein Positionspapier zur Verfügung stellen.

Dipl.-Psych. Michael Burmester

IAT Institut für Arbeitswissenschaft und Technologiemanagement

Universität Stuttgart, Nobelstr. 12, 70569 Stuttgart

Tel: 0711 / 970-01

Software-ergonomische Qualitätssicherung

Wolfgang Dzida
GMD

Zusammenfassung

Qualitätssicherung ist gemäß ISO 8402: "All those planned and systematic actions necessary to provide adequate confidence that a product . . . will satisfy given requirements for quality". Die Qualitätssicherung von Software ist somit keine rein software-technische Angelegenheit.

Die Sicherung der software-ergonomischen Qualität wird von Qualitäts-Managern inzwischen als Voraussetzung angesehen, um auf dem Softwaremarkt wettbewerbsfähig zu sein. Anwender und Benutzer fordern Konformität mit internationalen ergonomischen Normen, da sie verpflichtet sind, die Einhaltung der Prinzipien software-ergonomischer Gestaltung von Bildschirmarbeitsplätzen zu beachten. Hersteller müssen sich vertraglich verpflichten, daß die ausgelieferte Software diesen Prinzipien genügt, obwohl viele Hersteller noch kein ausgereiftes Konzept für die ergonomische Qualitätssicherung haben.

Aber auch Anwender wissen in der Regel zu wenig über die Methodik dieser Qualitätssicherung, die nicht nur am Produkt, sondern auch am Arbeitsplatz der Benutzer anzuwenden ist. Meist ist unklar, wie Hersteller und Anwender bei der Anwendung von Maßnahmen der Qualitätssicherung kooperieren müssen, wann und wie die betroffenen Benutzer zu beteiligen sind und in welchem Umfang diese Maßnahmen eigentlich notwendig sind.

Auf dem Workshop zur software-ergonomischen Qualitätssicherung sind Kurzreferate vorgesehen, die einige methodische, software-technische und organisatorische Grundlagen aufarbeiten helfen. Diese Grundlagen reichen von der Arbeitsanalyse und -gestaltung über die Konformität der Entwurfsentscheidungen mit Benutzerforderungen und Normen bis zur Bewertung der Gebrauchstauglichkeit eines am Arbeitsplatz installierten Produkts.

Software-ergonomische Qualitätssicherung wird auch mit Blick auf die einschlägigen Normen ISO 9001 und ISO 9000-3 diskutiert. Diese Normen enthalten viele

Forderungen, die bei der software-ergonomischen Qualitätssicherung berücksichtigt werden können. Allerdings sind noch viele Lücken vorhanden, weil ergonomisches Gestalten und Evaluieren von Software über die Produktqualität hinausreichen.

Obwohl software-ergonomische Qualitätssicherung bisher kaum professionell angegangen wurde, hat sich unversehens ein neues Berufsfeld ergeben, für das Fachleute gebraucht werden. Die Grundlage für die Tätigkeit dieser Experten sollte jedoch nicht nur in den Normen, der Bildschirmrichtlinie und den nationalen Verordnungen gesehen werden. Rechtsnormen mögen als Durchsetzungsinstrumente taugen - bei der Sicherung ergonomischer Qualität ist jedoch vornehmlich Sachverstand erforderlich. Der Workshop bietet Gelegenheit, im Spannungsfeld zwischen fachlichen Grundlagen und Durchsetzbarkeit zu diskutieren.

Dr. Wolfgang Dzida

GMD mbH

SET.SKS

Schloß Birlinghoven

53754 Sankt Augustin

Tel: 02241 / 14-2275

e-mail: Wolfgang.Dzida@gmd.de

Software-ergonomische Probleme und Lösungen in CAD/CAM-Anwendungen

Dr. J. Springer
RWTH Aachen

Zusammenfassung

Der in der Vergangenheit zu beobachtende Trend hin zu immer mehr Funktionalitat in CAD/CAM-Systemen wird zunehmend durch Entwicklungen hin zur Aufgaben- (Prozeß-) und Benutzerorientierung abgelöst. Integrierte Konstruktionssysteme, die Zusammenführung verschiedener Datenbestände wie Geometrie und administrative Informationen, Features als hochwertige Informationsbestände etc. führen dazu, daß die Benutzung einerseits komplexer wird, andererseits besser die praktischen Aufgaben von Konstrukteuren, Arbeitsplanern und anderen in ihrer Breite durch die Systeme abgedeckt werden. Die Software-Ergonomie gibt hier Hinweise für eine unter Benutzungsaspekten optimierte Gestaltung von CAD/CAM-Systemen. In dem geplanten Workshop sollen gemeinsam zwischen Anwendern und Herstellern von CAD/CAM-Systemen Benutzerprobleme und bereits realisierte und geplante Benutzungskonzepte, aber auch Visionen" für eine Verbesserung der Benutzung von CAD/CAM-Systemen entwickelt werden. Eine weitere Entwicklung in diesem Zusammenhang ist, in welcher Form inner- wie auch überbetriebliche kooperative Prozesse zwischen Konstrukteuren, aber auch zwischen Konstrukteur und Arbeitsplaner, Konstrukteur und Einkauf etc. unterstützt werden können. Einer Integration von Kommunikationssystemen in bestehende CAD/CAM-Anwendungen kommt hier eine zentrale Bedeutung zu. Zu untersuchende Fragestellungen im Zusammenhang mit Benutzungsaspekten sind u.a.

- wie kann eine betriebliche Anpassung von Software möglichst durch die Benutzer selbst vorgenommen werden und welche Hilfsmittel werden hier von CAD/CAM-Systemen zur Verfügung gestellt?

- Welche "Styleguides" müssen bei CAD/CAM-Anwendungen berücksichtigt werden?

- Wie können zunächst heterogene Anwendungen (CAD, CAM, PPS, EDM - (Engineering Data Management) - etc.) integriert und eine konsistente und durchgängige Benutzung sichergestellt werden?

- Wie können EDM-Systeme eingeführt und mit CAD verknüpft werden und welche Anforderungen aus Benutzersicht bestehen dabei? In moderierten Sitzungen werden Benutzungsprobleme bei CAD/CAM-Systemen aufgearbeitet und Strategien für eine Verbesserung der Benutzungsqualität entwickelt. Je nach Teilnehmeranzahl werden mehrere Arbeitsgruppen zu einzelnen Problemschwerpunkten gebildet.

Dr. Johannes Springer

IAW Lehrstuhl und Institut für Arbeitwissenschaft

der RWTH Aachen

Bergdriesch 27, 52062 Aachen

Tel: 0241 / 80 7953

e-mail: jspringe@iaw-1.iaw.rwth-aachen.de

Nicht-visuelle graphische Benutzungsoberflächen für Blinde

Gerhard Weber
Universität Stuttgart

Zusammenfassung

Der nicht-visuelle Zugang zu text-basierten Benutzungsoberfläche ist seit Ende der 70er Jahre kontinuierlich für blinde Benutzer verbessert worden. Seit kurzem sind auch Zugangssysteme auf der Basis von synthetischer Sprache und Brailleausgabegeräten für graphische Benutzungsoberflächen entwickelt worden. Hierbei wird auf die Umsetzung graphischer Informationen verzichtet und eine text-basierte Darstellung dynamisch aus exisitierenden Anwendungsprogrammen abgeleitet. Graphische Darstellungen werden dabei im allgemeinen kaum zugänglich.

Zur Umsetzung von Graphiken in eine nicht-visuelle Darstellung werden taktile Darstellungen und Modelle verwendet. Z.B. wird spezielles Butangas-haltiges Papier benutzt (sog. Schwellpapier), um reliefartige Strukturen durch einen dem Fotokopieren ähnlichen Vorgang von Papiervorlagen zu erzeugen. Diese Technik eignet sich jedoch nicht für den interaktiven Zugang zu Graphiken.

Durch Einsatz nicht-visueller Interaktionsmethoden kann ein interaktives Erkunden von Graphiken ermöglicht werden. Dabei können sowohl taktile Medien als auch akustische Medien eingesetzt werden. Im Workshop werden Beiträge mit verschiedenen Lösungsansätzen diskutiert. Dabei kann von folgenden Quellen ausgegangen werden:
- existierende Graphiken in Papierform (z.B. in Büchern),
- existierende Graphiken in einer Bildbeschreibungssprache,
- Graphiken in graphischen Benutzungsoberflächen und
- Graphiken, die fallweise aus Geoinformationssystemen erzeugt werden (Karten).

Der Dialog über Graphiken ist bei einer taktilen Darstellung durch Zeigeinstrumente zu unterstützen, um die flächenhafte Ausdehnung einer fühlbaren Darstellung auch interaktiv ausnützen zu können. Sprachbasierte Dialogformen können kaum mittels

Zeigehandlungen unterstützt werden. Jedoch erlaubt die Integration von taktilen mit akustischen Darstellungen multimodale Interaktionsformen, die zur Eingabe auch mit Zeigeinstrumenten arbeiten.

Verschiedene Anwendungen für den Einsatz von Graphiken werden in diesem Workshop vorgestellt und diskutiert.

Der Bedarf an weiteren Anwendungen mit einem verbesserten Design nicht-visueller Interaktionsmerthoden wächst, da mit der zunehmenden Akzeptanz von graphischen Benutzungsoberflächen durch blinde Benutzer die Erwartungen steigen. Weder die Grenzen beim Zugang zu Graphiken aus wahrnehmungsphysiologischer Sicht noch die Grenzen der automatischen bzw. halbautomatischen Umwandlung von Graphiken sind klar bestimmt, so daß die Teilnehmer aufgefordert sind ihre Vorstellungen einzubringen.

Dr. Gerhard Weber

Institut für Informatik

Universität Stuttgart

Breitwiesenstr. 20-22

70565 Stuttgart

Tel: 0711 / 7816-335

Fax: 0711 / 7816-340

e-mail: weber@ifi.informatik.uni-stuttgart.de

Entwicklungsunterstützung für ergonomische Benutzungsschnittstellen

Michael Herczeg, Alcatel SEL, Stuttgart
Kaisa Väänänen, ZGDV, Darmstadt
Jürgen Ziegler, FgH-IAO, Stuttgart

Zusammenfassung

Seit langem werden Werkzeuge zur Spezifikation und Entwicklung von Benutzungsschnittstellen aus technischer Sicht eingesetzt. Diese Hilfsmittel unterstützen verschiedene Phasen des Entwicklungsprozeßes oder auch verschiedene software-technische Komponenten einer Benutzungsschnittstelle. In der letzten Zeit wird zunehmend versucht, durch den Einsatz von Werkzeugen auch die ergonomische Qualität von Benutzungsschnittstellen zu verbessern oder abzusichern.

Zur Realisierung dieses Ziels werden ganz unterschiedliche Ansätze verfolgt, die z.B. von der Verwendung von Bausteinkonzepten über die automatisierte Generierung von Benutzungsschnittstellen bis hin zu Beratungs- oder Beurteilungskomponenten reichen. Dazu werden Methoden und Vorgehensweisen durch Werkzeuge unterstützt, die sich speziell auf die Gestaltung der Benutzungsschnittstelle richten. Welcher dieser unterschiedlichen Ansätze am stärksten zu einer benutzergerechten Gestaltung beiträgt, ist gegenwärtig eine offene Frage.

Der Workshop soll die Möglichkeit bieten, die Stärken und Schwächen der verschiedenen Ansätze zu diskutieren, Anforderungen an zukünftige Systeme zu präzisieren und den Forschungsbedarf in diesem Feld zu beschreiben. Dabei sollen auch die aus neuen Interaktionformen, wie Multimedia oder Hypermedia, resultierenden Anforderungen beleuchtet werden. Als Fragestellungen werden u.a. die folgenden Themen angesprochen:

- Style-Guides vs. ergonomische Softwarebausteine für Benutzungs–schnittstellen.
- Schränken Werkzeuge die Möglichkeiten zur ergonomischen Gestaltung ein oder können sie diese absichern?
- Generieren vs. Konstruieren von Benutzungsschnittstellen.

- Inwieweit helfen "intelligente" Unterstützungssysteme dem Entwickler?
- Für welche Phasen oder Rollen im Entwicklungsprozeß ist welche Unterstützung wesentlich?

Der Workshop geht von einigen kurzen Positionsdarstellungen von Vertretern unterschiedlicher Richtungen aus. Anschließend sollen in der Diskussion unter den Workshop-Teilnehmern die Möglichkeiten zur Unterstützung benutzerorientierter Entwicklungsprozesse bewertet und Anforderungen an zukünftige Techniken und Werkzeuge bestimmt werden.

Jürgen Ziegler
Fraunhofer-Institut IAO
Senefelderstrasse 26
70176 Stuttgart

Tel: 0711 / 970 2334
Fax: 0711 / 970 2300
e-mail: Juügen.Ziegler@iao.fhg.de

Software-ergonomische Aspekte von CSCW-Systemen

Jürgen Friedrich
Universität Bremen

Zusammenfassung

Die klassische Software-Ergonomie beschäftigt sich mit der menschengerechten Gestaltung von Einzelplatzsystemen. Angesichts der Forderung, diese durch Computer zu unterstützen (CSCW, Groupware) stellt sich die Frage, ob und wie die Software-Ergonomie den Anforderungen dieser neuen Klasse "kooperativer Systeme" gerecht werden kann. Müssen neue (gruppenorientierte) Kriterien für die Analyse und Gestaltung von CSCW-Systemen entwickelt werden? Reichen die bisher beteiligten Wissenschaften zur Fundierung von Groupware-Systemen aus oder müssen neue Disziplinen (z. B. Sozialpsychologie, Theorie verteilter Systeme) bei der Herausbildung einer "Groupware-Ergonomie" berücksichtigt werden? Der Workshop soll dazu beitragen, die Beantwortung derartiger Fragen durch die Einbeziehung von Fachleuten aus Wissenschaft und betrieblicher Praxis voranzuteiben.

Jürgen Friedrich

Universität Bremen Fachbereich 3

Bibliothekstr. 1

28359 Bremen

Tel: 0421 / 218-3395, -2488

Fax: 0421 / 218-3308

e-mail: friedrich@informatik.uni-bremen.d

Arbeitsunterstützung durch Software
Perspektiven im Forschungsprogramm "Arbeit und Technik"

Christoph Kasten, Constantin Skarpelis
DLR Projektträger "Arbeit, Umwelt und Gesundheit"
des BMFT

Zusammenfassung

Angesichts der hohen und noch ständig wachsenden Bedeutung der Software für die Arbeitswelt als Basis- und Schlüsseltechnologie für Anwendungen in nahezu allen Bereichen der Wirtschaft bilden die Gestaltungsfragen dieser Technik einen wichtigen Schwerpunkt im Programm "Arbeit und Technik" (AuT). Die Themen der meist interdiziplinären AuT-Projekte beziehen sich sowohl auf die Einführung von Software-Systemen in die betriebliche Praxis als auch auf die Gestaltung der Benutzungsschnittstelle, die Optimierung der Software-Funktionalität und die Verbesserung des Software-Entwicklungsprozesses selbst unter intensiver Beteiligung von Benutzern.

Die Förderung im Programm "Arbeit und Technik" zeichnet sich im Vergleich zu anderen Fördermaßnahmen durch einen ganzheitlichen Ansatz aus: Die Zusammenhänge und Wechselwirkungen zwischen Organisation, Qualifikation, Gesundheitsschutz und Technikgestaltung werden gezielt thematisiert.

Seit Beginn der Förderung 1986 wird diese regelmäßig bilanziert und in ihren Themenstellungen und Zielsetzungen aktualisiert.

Anläßlich der Software-Ergonomie '95 werden im Rahmen eines Workshops Forschungs- und Entwicklungsdefizite arbeitsbezogener Software-Gestaltung, der weitere Handlungsbedarf und mögliche neue Themenfelder und Ansätze der Förderung diskutiert. Dabei soll der potentielle Beitrag innovativer Softwaretechnik zur Unterstützung qualifizierter Arbeit im Blickpunkt stehen. Mit einleitenden Statements von Experten und Diskussionsbeiträgen der Workshop-Teilnehmer sollen folgende vier Themenfelder eines möglichen künftigen Engagements des Programms "Arbeit und Technik" erkundet werden:

Methoden und Werkzeuge zur ganzheitlichen Software-Gestaltung

Die Qualität von Softwaresystemen - auch besonders unter dem Aspekt der benutzungs- und aufgabengerechten Gestaltung - wird in hohem Maße durch die Güte des Erstellungsprozesses hinsichtlich der eingesetzten Vorgehensweisen, Methoden und Werkzeuge bestimmt. Das Programm "Arbeit und Technik" hat daher eine Reihe von Förderaktivitäten zur Unterstützung des Software-Entwicklungsprozesses aus der benutzer- wie auch aus der entwicklerseitigen Perspektive unternommen. Daneben sind im EU-Bereich und vom Bundesministerium für Bildung, Wissenschaft, Forschung und Technologie eine Reihe weiterer Fördermaßnahmen zur Qualitätssteigerung der Software-Entwicklung ergriffen worden, die sich auf neue Methoden und Werkzeuge der Softwaretechnik konzentrieren.

Erkennbare Defizite der Software-Entwicklung, soweit sie sich auf Anforderungen der benutzungsgerechten Gestaltung und der Arbeitsqualität der Software-EntwicklerInnen beziehen, bestehen weiterhin in erster Linie in der prozeßintegrierten ergonomischen Qualitätssicherung und Systemevaluation, den Methoden und Werkzeugen für Systeme zur Unterstützung kooperativen Arbeitens und anderer innovativer Softwaretechnologien, den Gestaltungsansätzen der Adaptierbarkeit von Software-Systemen und der Qualifikation der Software-EntwicklerInnen.

Kooperation, Kommunikation und Vernetzung

Moderne Produktions-, Arbeits- und Dienstleistungskonzepte stellen Arbeit in Gruppenzusammenhängen immer mehr in den Vordergrund. Dies gilt für inner- und überbetriebliche Kooperationsformen; sei es entlang einer Wertschöpfungskette, wie z.B. die Kooperation der Konstrukteure von Zuliefer- und Produktionsunternehmen, oder aber auch "quer zur Wertschöpfungskette" wie die Kooperation der Disponenten verschiedener Verkehrsträger.

Viele technische Probleme der Vernetzung sind inzwischen erkannt und zum Teil gelöst, Fragen des Datenschutzes und der Datensicherheit zumindest erkannt und erste Konzepte zur Softwareergonomie angedacht. Gruppenzusammenhänge und Arbeit in Gruppen unterliegen aber auch dynamischen Prozessen, deren Regeln daher emotional gestaltet sind. Die Probleme werden unter dem Stichwort der "C-Words" summiert: conversion, conflict, comitment und combat als Beispiele. Einen ersten Analysebeitrag hierzu liefert ein im Programm "Arbeit und Technik" gefördertes Vorhaben der Universität Bonn.

Zukünftige multimediale Technikanwendungen und CSCW- und Workflow-Systeme werden diese Regeln der Kooperation und Kommunikation berücksichtigen müssen und können erheblichen Forschungs- und Entwicklungsbedarf bedeuten. Weitere Erkenntnisdefizite liegen in der Definition und Umsetzung der neuen Anforderungen an die Softwaregestaltung und Hardwarekonfiguration, die sich aus den Wechselwirkungen zwischen Arbeitsaufgabe, Kooperation und Technikgestaltung ergeben.

Benutzungsgerechte Technik durch Softwareunterstützung

Organisatorische Entwicklungstendenzen zur Dezentralisierung und damit verbundene Reintegration von Tätigkeiten aus vor- und nachgelagerten Betriebsbereichen sowie Instandhaltungs- und Qualitätssicherungsfunktionen in den Produktionsprozeß verlangen neue Formen an Softwareunterstützung im Rahmen der Informations- und Kommunikationstechnik, die eine effiziente Anpassung an stärker variierende Benutzungsanforderungen ermöglichen. Dies gilt z.B. für die Ausfüllung erweiterter Handlungs- und Entscheidungsspielräume, die kooperative Planung und Entscheidungsfindung, die Diagnose und Instandhaltung von Maschinen und Anlagen.

Intelligente Softwarelösungen sind auch für die benutzungsgerechte Gestaltung der Maschinenarbeitsplätze erforderlich. Dies betrifft die Entwicklung programmgesteuerter Betriebsmittel ebenso wie die Modernisierung konventioneller Maschinen. Die hier integrierte Software kann z.B. die Prozeßtransparenz und Ein- und Ausgabemöglichkeiten verbessern, Wahlmöglichkeiten im Hinblick auf die Bearbeitungsstrategie und den Automatisierungsgrad anbieten, exploratives Handeln unterstützen, die Integration des Planens und Bearbeitens fördern, die sukzessive Übernahme komplexer Bearbeitungsaufgaben ermöglichen, die Anpassung an unterschiedliche Eingangsvoraussetzungen der Benutzer erleichtern und das Lernen im Arbeitsprozeß individuell unterstützen. Insbesondere der Arbeitsplatzwechsel in hochautonomen Organisationseinheiten erfordert individuell anpassungsfähige Maschinen- und Steuerungskonzepte und Planungs- und Steuerungssysteme, die den gemeinsamen Optimierungsprozeß durch eine Gruppe unterstützen.

Aktuelle Vorhaben im Programm "Arbeit und Technik", die richtungsweisend für eine benutzungsgerechte Technikgestaltung durch Softwareuntersützung sind, betreffen z.B. doppelfunktionale Maschinenkonzepte, Facharbeiter-Informations-Systeme und kooperationfördernde Fertigungsleitstände.

Zukünfige Entwicklungen von Softwaretechniken

Neue Softwaretechniken wie Multimedia, Virtual Reality, Fuzzi-Logik oder Expertensysteme halten nur langsam Einzug in die Betriebe. Die tatsächlichen Anwendungsmöglichkeiten sind bei weitem nicht vorstellbar, aber wesentliche Gründe für die Zurückhaltung liegen sicherlich darin, daß die Auswirkungen dieser Techniken bislang kaum absehbar sind und daß Konzepte zur arbeitsorganisatorischen Einbettung fehlen.

Erste Ansätze hierzu werden zur Zeit erarbeitet. So sollen zwei Vorhaben zum Einsatz von Multimedia im Vertrieb, in der Produktion bzw. im Krankenhaus Hinweise zur Softwaregestaltung und -anwendung liefern.

Im Rahmen eines Verbundvorhabens wird ein modulares Expertensystem entwickelt und erprobt, wobei qualifikatorische und organisatorische Gesichtspunkte einen breiten Raum einnehmen.

Christoph Kasten
DLR Projektträger "Arbeit und Technik"
Südstr. 125
53175 Bonn

Tel: 0228 / 3821- 134
Fax: 0228 / 3821- 107

Berichte des German Chapter of the ACM

Band 27: Remmele/Sommer, Arbeitsplätze morgen
Tagung II/1986 vom 11. bis 14. 3. 1986 in Marburg. 431 Seiten, DM 78,–/ÖS 609,–/SFr. 78,–

Band 28: Balzert/Heyer/Lutze, Expertensysteme '87
Tagung I/1987 am 7./8. 4. 1987 in Nürnberg. 493 Seiten, DM 82,–/ÖS 640,–/SFr. 82,–

Band 29: Schönpflug/Wittstock, Software-Ergonomie '87
Tagung II/1987 vom 27. bis 29. 4. 1987 in Berlin. 512 Seiten, DM 82,–/ÖS 640,–/SFr. 82,–

**Band 30: Winkler, Proceedings of the International Workshop on Software Version
and Configuration Control**
January 27 – 29, 1988 Grassau. 478 Seiten, DM 78,–/ÖS 609,–/SFr. 78,–

Band 31: Dillmann/Swiderski, WIMPEL '88
Tagung I/1988 vom 28. bis 30. 6. 1988 in München. 479 Seiten, DM 78,–/ÖS 609,–/SFr. 78,–

Band 32: Maaß/Oberquelle, Software-Ergonomie '89
Fachtagung vom 29. bis 31. 3. 1989 in Hamburg. 509 Seiten, DM 88,–/ÖS 687,–/SFr. 88,–

Band 33: Ackermann/Ulich, Software-Ergonomie '91
Fachtagung vom 18. bis 20. 3. 1991 in Zürich. 383 Seiten, DM 84,–/ÖS 655,–/SFr. 84,–

Band 34: Friedrich/Rödiger, Computergestützte Gruppenarbeit (CSCW)
Fachtagung vom 30. 9. bis 2. 10. 1991 in Bremen. 314 Seiten, DM 69,–/ÖS 538,–/SFr. 69,–

Band 35: Hoffmann, Eiffel
Fachtagung am 25./26. 5. 1992 in Darmstadt. 112 Seiten, DM 42,–/ÖS 328,–/SFr. 42,–

Band 36: Schweiggert, Wirtschaftlichkeit von Software-Entwicklung und -Einsatz
Fachtagung am 21./22. 9. 1992 in Ulm. 272 Seiten, DM 62,–/ÖS 484,–/SFr. 62,–

Band 37: Ludewig/Schneider, Software Engineering im Unterricht der Hochschulen SEUH '92
Workshop am 27./28. 2. 1992 in Stuttgart. 132 Seiten, DM 46,–/ÖS 359,–/SFr. 46,–

Band 38: Raasch/Bassler, Software Engineering im Unterricht der Hochschulen SEUH '93
Workshop am 25./26. 2. 1993 in Hamburg. 190 Seiten, DM 49,–/ÖS 382,–/SFr. 49,–

Band 39: Rödiger, Software-Ergonomie '93
Fachtagung vom 15. bis 17. 3. 1993 in Bremen. 330 Seiten, DM 78,–/ÖS 609,–/SFr. 78,–

Band 40: Coy/Gorny/Kopp/Skarpelis, Menschengerechte Software als Wettbewerbsfaktor
Arbeitstagung am 27./28. 1. 1993 in Bonn. 647 Seiten, DM 138,–/ÖS 1076,–/SFr. 138,–

Band 41: Züllighoven/Altmann/Doberkat, Requirements Engineering '93: Prototyping
Fachtagung vom 25. bis 27. 4. 1993 in Bonn. 383 Seiten, DM 88,–/ÖS 687,–/SFr. 88,–

**Band 42: Hartmann/Herrmann/Rohde/Wulf, Menschengerechte Groupware –
Software-ergonomische Gestaltung und partizipative Umsetzung**
Workshop am 20./21. 9. 1993 in Bonn. 356 Seiten, DM 90,–/ÖS 702,–/SFr. 90,–

Band 43: Hußmann/Paech, Software Engineering im Unterricht der Hochschulen SEUH '94
Workshop am 24./25. 2. 1994 in München. 178 Seiten, DM 54,–/ÖS 421,–/SFr. 54,–

Band 44: Spillner/Breymann, Software Engineering im Unterricht der Hochschulen SEUH '95
Workshop am 23./24. 2. 1995 in Bremen. 141 Seiten, DM 48,–/ÖS 375,–/SFr. 48,–

Band 45: Böcker, Software-Ergonomie '95
Fachtagung vom 20. bis 23. 2. 1995 in Darmstadt. 424 Seiten, DM 98,–/ÖS 765,–/SFr. 98,–

Preisänderungen vorbehalten

 B. G. Teubner Stuttgart